Power over Ethernet Interoperability

About the Author

Sanjaya Maniktala is Director, Communications and Power, at Microsemi Corporation. He has held lead engineering and managerial positions in India, Singapore, and Germany as well as the United States at companies such as National Semiconductor (now Texas Instruments), Power Integrations, Freescale Semiconductor, Fairchild, Siemens AG, Artesyn (now Emerson Electric), and Broadcom. Mr. Maniktala is the author of *Switching Power Supply Design & Optimization* (McGraw-Hill, 2005) and holds several patents including the Floating Buck regulator topology.

Power over Ethernet Interoperability

Sanjaya Maniktala

McGraw Hill

New York Chicago San Francisco
Lisbon London Madrid Mexico City
Milan New Delhi San Juan
Seoul Singapore Sydney Toronto

The McGraw·Hill Companies

1 2 3 4 5 6 7 8 9 0 LPI/LPI 1 9 8 7 6 5 4 3

ISBN 978-0-07-179825-9
MHID 0-07-179825-0

Sponsoring Editor
Michael Penn

Editing Supervisor
Stephen M. Smith

Production Supervisor
Richard C. Ruzycka

Acquisitions Coordinator
Bridget L. Thoreson

Project Manager
Yashmita Hota,
Cenveo® Publisher Services

Copy Editor
Erica Orloff

Proofreader
Megha Saini,
Cenveo® Publisher Services

Indexer
Robert Swanson

Art Director, Cover
Jeff Weeks

Composition
Cenveo® Publisher Services

This book is printed on acid-free paper.

Contents

Preface

December 2007, San Jose, California: It seems a long time ago. I walked into a big networking company to head their small Power over Ethernet (PoE) applications team. Surprisingly, I hardly knew anything about PoE prior to that day, having been a switching-power conversion engineer almost all my life. But it seemed a great opportunity to widen my horizons. As you can see, one notable outcome of that seemingly illogical career choice five years ago is the book you hold in your hands today. I hope this small body of work goes on to prove worthy of your expectations and also of all the effort that went into it. Because, behind the scenes, there is a rather interesting story to relate—about its backdrop, intertwined with a small slice of modern PoE history, punctuated by a rather restive search for our roots and our true heroes, one that takes us back almost two centuries.

PoE seemed exciting from the outside, certainly enough for me to take the plunge. Intuitively, it represented the union of two huge, hitherto parallel worlds of modern development—power and networking. It seemed obvious that the scion too would one day mature into a fine young adult. Soon after taking up the mantle, and the gauntlet, I somehow managed to resuscitate my skeletal systems team. Despite that near-heroic effort, even on a good day, we remained at best a freshly greased jalopy, lurching dangerously from side to side under the uneven weight of a large heirloom we discovered in our backseat—one which they insisted was a qualified principal hardware engineer reporting to me, and I of course begged to differ. On occasion, I had silently prayed that one day this artifact would materialize in our PoE lab, and even be spotted moving slowly but surely toward that dusty ol' Le Croy oscilloscope! But that was never meant to be, I was only dreaming. My job was not destined to get any easier. Finally in 2010, defying all odds (even a sneaky curator along the way), we somehow managed to get our rickety vehicle, not only to the finish line, but *first*—in the entire industry. We had received the first-ever American safety certification given to a PoE chip, from Underwriters Laboratories (UL) under their UL 2367 category for

30-watt PoE applications. Yes, it was my tiny team, all pumped up with steroids, which did that.

What exactly did it bring to us anyway? Bragging rights for sure. But beyond that, with a UL-certified PoE chip under our belts, we could enable our customers to drop all their port fuses, a savings of at least 5 cents on every port on almost all their Ethernet switches and hubs. That was huge from a commercial viewpoint. Not to be caught resting on our laurels, barely a few months later, in 2011, we acquired another industry-first safety certification—the very first PoE chip certified under the UL 2367 category, but now using four-pair construction, for enabling 60-W PoE. Keep in mind that, to date, a 60-W IEEE standard does not exist. This certification was initiated under a very specific customer request, to support a radical market-driven standard being pushed by a major original equipment manufacturer (OEM). That standard, alternatively called UPoE or UPOE, stands for Universal PoE.[1] It seems that just before this OEM reached out to us in distress, the very launch of UPOE from their side had been put on indefinite hold on account of UL refusing to certify product safety based on their existing PoE solution. I was asked to explain, to a rather sleepless customer late at night, why we stood a far better chance than the competition of getting them through the UL safety barrier, and also how much time that would likely take, considering my previous experiences related to safety testing and of achieving UL certification. The happy ending was straight out of Bollywood: the OEM did finally release its new UPOE products, but now completely based on our PoE chips, the ones we managed to get certified for up to 60 W barely a few weeks after that telephonic lullaby. It was delightfully obvious that at least for now, UL certification was no longer just a case of saving money on fuses—it had turned out that no one could even legally sell a 60-W PoE solution in the marketplace without *our* UL-certified chip, the only one out there at the time. And so, for a brief moment, our rundown vehicle had dared to pretend it was a Ferrari.

After this industry-leading UPOE chip certification,[2] my company's fortunes in this particular business unit turned north, though I doubt anyone around us really understood the technical reasons for their unexpected windfall, who delivered it to them, or under what circumstances. PoE was still considered a niche area (a hobby post, in effect). To make matters worse, PoE is largely analog electronics at work, whereas a good part of our modern world is fashionably digital. In a way, we were therefore low-tech, at least to most people at the helm. The PoE systems team was, I suspect, not really considered smart enough to even pass judgment on any one of the hundreds of

[1]http://www.excitingip.com/2320/universal-power-over-ethernet-upoe-technology-delivers-up-to-60w-of-power-per-port/.
[2]http://blogs.cisco.com/tag/upoe/.

"revolutionary" ideas emanating from just one person firmly ensconced in the hallowed office of the chief technology officer. To make matters worse, the systems team was now making incredibly feeble attempts at "incremental ideas" such as the industry's first commercially successful single-pair PoE (now appearing in the emerging automotive PoE market too[3]—see Fig 13.16). So, finally, inundated with tons of congratulatory messages that flowed in (*not*), shortly thereafter in early 2012 I walked out of what I consider an inconsequential meeting with someone in urgent need of a backbone, to join another company down the road, and to continue my PoE explorations without being forced to make professional compromises. It seemed a natural career move for me, because this new company had recently acquired an innovative Israeli company, which was one of the original pioneers of PoE, as detailed in Chap. 1 of this book. It would help me "close the loop," technically speaking. I could return to the roots of it all and enhance my perspective too. I could also practice both my power-management and PoE skills under one roof and under one business unit. And last but not least, it also allowed me to complete this book relatively unfettered, rather than having to watch over my shoulder for beady eyes in suits.

In Chap. 1 of this book, I have tried, first and foremost, to give credit where due, perhaps because I have felt cheated too at times. But I have also tried to quantify exactly what we, in PoE, owe our past lives. I am a firm believer that people must declare their inheritance before showcasing their wealth. We need to recognize and reward the truly deserving persons, even if they are long gone. Posthumous recognition does have tremendous value. Many have fought just for a lasting legacy all their lives, and died for it too. To that end, I have carefully retraced the history of PoE, and its underlying innovations, deep into the 19th century. My research points to the startling revelation that PoE started not at some high-stakes IEEE forum in the 21st century (you knew that of course), but almost 200 years ago. The real heroes of our times, largely unsung and unknown today, resided in that era. On closer examination, a lot of our modern achievements pale in comparison to theirs, especially when we learn that they had almost no available resources at their disposal, except sheer resourcefulness and dedication. Once we become aware of our rich legacy, we realize it is our lack of historical perspective which often makes us rather self-indulgent about our own achievements, and often puts us in the embarrassing position of crowning the wrong heroes of our times—perhaps the first ones who just happened to walk through the door, or the ones who spoke the loudest, or perhaps the ones who were just adept at using modern media or communications to stay in our faces all the time, or worse, the postman (the marketing or sales guy).

---

[3]http://www.broadcom.com/products/Physical-Layer/BroadR-Reach-PHYs.

As an example, in Chap. 1 I point out that modern PoE is built on a singular principle of data and power sharing the same lines (or medium). This principle, often called *phantom power* today, is much like two persons sitting on the same seat of a train, oblivious of each other (with due apologies to J. K. Rowling). I go on to show how the well-known Wheatstone bridge can be mentally morphed to describe this fascinating phantom circuit principle, a fact recognized very early by a man called John Joseph Carty.[4] Carty, a former engineer at Bell Telephone Company, went a step further and replaced the familiar resistors of the traditional Wheatstone bridge with inductors. Voilà! In doing so, he had laid the foundations of phantom power feeding, without which PoE as we know it today would not have existed. This sequence of events and the birth of the phantom circuit is recounted in a scanned image of *The New York Times* dated July 9, 1911, available at *The New York Times* archives (it has no direct mention of PoE, though, which is understandable back in 1911).[5]

All of Carty's patents can still be found, though as scanned images only.[6,7] We conclude that 1886 is the exact year the phantom power principle of modern PoE emerged. The National Academy of Sciences started giving out the John J. Carty Award for the Advancement of Science, the first recipient being Carty himself, just after 1932, the year Carty died.[8–10]

Incidentally, the vaunted twisted-pair cable, which we use freely today in Ethernet, came from Alexander Graham Bell (hark: the telephone guy was the cable guy too!)—in 1881.[11] It is almost embarrassing to sense that we may have been patting ourselves on our backs all along for their achievements, or patting an unknown person with 100 dubiously acquired and questionable patents. Carty and Bell would surely turn in their graves.

We will notice that through time immemorial, genuine contributors to society always do it for love, not for money. Money is incidental in their minds. And that is how some of them brought us to the point at which modern telecommunications really started to take off. The ancient patents on which our modern networking world is based

---

[4]http://en.wikipedia.org/wiki/John_J._Carty.

[5]http://query.nytimes.com/mem/archive-free/pdf?res=FA0816F73F5517738DDDA00894DF405B818DF1D3.

[6]www.uspto.gov.

[7]http://www.google.com/patents/US348512andhttp://www.google.com/patents/US353350 (though there is a spelling mistake by the Google optical character recognition software, so his last name is erroneously spelt Caety instead of Carty).

[8]http://www.nasonline.org/programs/awards/john-j-carty-award.html.

[9]The fascinating Franklin Institute repository: http://www.fi.edu/learn/case-files/carty/file.html.

[10]http://www.fi.edu/learn/case-files/carty/full/clipping.oct2.p1.jpg.

[11]Look for patent number US000244426 at www.uspto.gov. Or go from here: http://www.uspat.com/bell/.

are probably not too many, their value is not proportional to their numerical count, but those key patents are indeed rock-solid as it turns out today. They have not only stood the test of time, they have changed our times completely—almost 200 years later and still counting. At the end of the day, any innovation or idea, however it was acquired or pitched to the general public, must pass a litmus test: *can it hold up to (impartial) technical scrutiny*, not just within the organization or close-knit community it was supposedly created and peer-reviewed in, but by the larger scientific and engineering community? And not just today, but a hundred years from now? That is the test which ultimately distinguishes the real Alexander Graham Bells and J. J. Cartys of yesteryears from some unprincipled modern wannabes. We remember the sad case of Milli Vanilli in the music world.[12] We also had the sad case of Janet Cooke in journalism.[13] More recently, we had the extremely sad case of Lance Armstrong in sports.[14] Extremely sad for us that is, not for them—because we let them do it to us for that long. There are also perhaps eight impostors in modern science listed on the Web.[15] That is ignominy, not fame. So we ask: are there any such waiting-to-happen scandals in our brave new networking world? That could be terribly embarrassing to all, especially if it turns out that we gave them a stage to strut around on, padded them with generous financial incentives, and then created an ecosystem around them based on completely inadequate checks and balances, one which even protected them ferociously.

I did a fairly comprehensive Web survey while writing this book and came across a bunch of recently filed PoE patents at the United States Patent Office Web site.[16] I learned that a staggering number of patents listed there in PoE came from just one person. I counted over 350 U.S. patent applications pending, and another 138 U.S. patents already granted by the office. A rate of one patent per week or so it seems, judging by the dates. Is he the new Edison? The new Carty? The new Bell? How about the father of energy-efficient Ethernet? Time will tell of course. But I did start to wonder if innovation had become a numbers game now. Are 100 patents better than one? Couldn't that one patent happen to be Alexander Graham Bell's famous patent, number 174,465, issued on March 7, 1876—the one that brought the telephone into our houses?[17] I wondered if these 100 modern patents will change the world, or just the lifestyle of their inventor (and perhaps the inventor's mentors and carefully chosen co-inventors too). With these closing thoughts, let us now turn our

[12]See http://en.wikipedia.org/wiki/Milli_Vanilli.
[13]See http://en.wikipedia.org/wiki/Janet_Cooke
[14]See http://en.wikipedia.org/wiki/Lance_Armstrong
[15]http://healthland.time.com/2012/01/13/great-science-frauds/
[16]www.uspto.gov
[17]See http://www2.iath.virginia.edu/albell/bpat.1.html.

attention to more recent PoE patents, to see how they are perhaps shaping our world and how they might contribute to the future growth of technology. Here is a tiny sampling. All are available at www.uspto.gov or Google Patents at www.google.com/patents. I do caution: please read the original filings carefully for yourself; judge for yourself eventually. Assume I am making off-the-cuff and ignorant remarks. Because the truth is PoE is still evolving, and so are we.

1. U.S. patent number 5,065,133 on November 12, 1991. "Method and apparatus converting digital signals to analog signals and simultaneous transmission of AC power and signals over wire conductors."[18] This patent is perhaps the earliest modern reference to injecting (AC) power on to the center taps of data transformers, similar to what we do in PoE today.
2. U.S. patent number 5,994,998, on November 30, 1999. "Power transfer apparatus for concurrently transmitting data and power over data wires."[19] This injects DC power on the center-taps of data transformers, as in modern PoE.
3. U.S. patent number 6,115,468, on September 5, 2000. "Power feed for Ethernet telephones via Ethernet link."[20] This injects DC on the center-taps of Ethernet data transformers, as in PoE today. An extension of the previous ideas.
4. U.S. patent number 8,026,635 B2, on September 27, 2011. "Power over Ethernet power sourcing equipment architecture for variable maximum power delivery."[21] This says that if a PoE chip inside the power sourcing equipment (PSE) has integrated pass-FETs and thus gets too hot, the PSE chip can be designed to include a control for driving an external FET which can be switched in parallel to the main (integrated) FET. The port current will thus get split, a fraction of it going through the main (internal) FET, so it will not get too hot—problem solved. We will learn in Chap. 6 of this book that there are limits on port current in standards-compliant PoE. By switching in an external paralleled FET, we actually lose information about the net port current, because only the current in the internal FET is being monitored by the PSE. This idea could work if the PSE can somehow accurately sense the current in the external FET too, but that aspect is not addressed.
5. U.S. patent number US 7,956,616, on June 7, 2011. "System and method for measuring a cable resistance in a power over

[18]See http://www.google.com/patents/US5065133.
[19]See http://www.google.com/patents/US5994998.
[20]See http://www.google.com/patents/US6115468.
[21]See http://books.google.com/patents/US8026635.

Ethernet application."[22] This invention uses the available time between end of classification and power-on to measure the cable resistance (see Chaps. 4 and 5 of this book). It also thinks there is an available time slot between detection and classification, though, after detection, if you raise the port voltage, any normal PD will assume the PSE is doing classification. Such minor details aside, as the PSE raises the port voltage V_{PORT} above about 22 V (see same chapters), a short-circuit module (SCM) inside the Powered Device (PD) suddenly conducts and applies a 22-V zener across the port, causing a certain port current to flow. The inventor bypasses the problem of infinite currents resulting from short cables (under an applied voltage source as explicitly stated) and proceeds to eliminate the zener drop (diode offset) by using $R_{cable} = (V_2 - V_1)/(I_2 - I_1)$, essentially subtracting two potentially infinite numbers to produce a finite number always. But we also need a differential current sensing concept spread across the time dimension now, so that infinite numbers don't need to be processed or stored at all by the chip. It is an inarguably awesome display of Ohm's law being used to clobber conventional number theory.[23] We can almost feel the infectious eagerness with which the inventor's mentors must have pursued this idea relentlessly through their brilliant patent review committee and highly paid law firm. And even I admit, in such cases, the typical five-figure incentive stock option award per patent seems rather incongruous. Taking the idea to the bench, a typical PSE cannot distinguish between a short-circuit inside a PD versus a short-circuit on the cable. So it will jump in to protect from both, by applying current limiting, eventually turning the port off, as is also evident from Chap. 8 of this book. The PSE will try again and again to power up with this new-fangled PD placed on the other side. Interoperability is expected to be the first casualty. This may work with a proprietary PSE, but it is not backward compatible.

6. U.S. patent number US 8,217,527 B2, on July 10, 2012. "Midspan powering in a Power over Ethernet system."[24] The idea of this is that you can inject power into the data pairs by inserting a center-tapped transformer en route to the PD, rather than injecting PoE at the starting point (the switch) itself. Midspan manufacturers may have used this idea for years.[25] Though this inventor omits citing the other patent for whatever reason, he or she says that this innovation involves

---

[22]See http://books.google.com/patents/US7956616.
[23]http://www.suitcaseofdreams.net/infinity_paradox.htm.
[24]See http://books.google.com/patents/US8217527.
[25]See U.S. patent number 7299368 at http://books.google.com/patents/US7299368.

"insertion of inductance at the Midspan to overcome killer patterns that can cause baseline wander." Baseline wander is described in Fig. 9.2 of this book. It is not clear what exact inductance the inventor is proposing at the Midspan level, except to suggest it is *above* 350 µH, same as in any Endspan. In Chap. 9 of this book we will learn that all practical data transformers (up to 100 Mbps) have an inductance of over 350 µH, because that happens to be the *minimum* specified value. In Fig. 2.9 of this book I have given out another such method of PoE injection on the data pairs using a Midspan. Hopefully no one will ask you to pay for that. It is common sense. This inventor proposes the usual known inductance value of data transformers, but since it appears that somewhere in the Ethernet standards it was not explicitly spelled out that a data transformer could be found inside a Midspan, or alternatively, someone did not explicitly write that a Midspan could use a transformer (to inject PoE), the inventor has claimed rights to it. Though, one option, perhaps, was to simply bring it to the notice of the IEEE committee to plug, as most others did. But perhaps this inventor was not present in those IEEE meetings. That would explain it. So, the final remaining question is: Is this an innovation which will change the course of technology?

7. U.S. patent number US 8,082,453 B2, on December 10, 2011. "Power sharing between Midspan and Endspan for higher power PoE."[26] Though filed as a separate idea, and much earlier in time, it actually seems like an extension of the previous idea, described in number 6 above. Clearly, these two patents (numbers 6 and 7) have been culled from one. This part-patent is remarkable. The inventor implies that instead of just using a simple Midspan to inject power, we can send power from the switch over the spare pairs to the Midspan. The Midspan will pass through the power coming over the spare pairs, but will inject power on to the data pairs as described in the previous patent (number 6). In effect we now have power on all four pairs going into the PD, which is strictly not standards-compliant (is that what the patent implies?). Note that "N-pair" delivery systems are discussed in detail in Chap. 13 of this book. One statement in the patent is noteworthy (see the link in this entry):

> For example, one scenario that could occur when no power coordination between Endspan and Midspan PSE is used includes both Endspan and Midspan, each independently powering the

[26]See http://books.google.com/patents/US8082453.

PD, thereby providing the PD more than two times its required power. However, with Endspan having the ability to configure the first and second output powers, Endspan can have either of the two PSEs, but not both, power the PD, thereby reducing by a half the overpowering inefficiency.

In Chap. 3 of this book we will learn that the basic idea behind back off is simply to never allow the Midspan and Endspan to power up the cable simultaneously. But here, both the PSEs (Midspan and Endspan) have somehow powered up. So the new problem seems to be that the two PSEs are not just making available, but providing, twice the power needed by the PD. This is akin to food being forced into someone's mouth against his or her wishes, or grain being jammed into a harvesting thresher, never mind the actual load across the PD. This spooky overstuffing of the PD results in a catastrophic situation, which the inventor labels "overpowering inefficiency." I am the first to admit the inventor has coined a novel term here, which may in fact highlight certain process efficiencies and/or deficiencies of our times, and may lead to a far more energy-efficient Ethernet.

8. U.S. patent number US 8,217,529 B2, on July 10, 2012. "System and method for enabling power applications over a single communication pair."[27] This concerns power delivery over a single pair. The problem with one-pair PoE is, as explained in Chap. 13 of this book, that with one pair we no longer have two available center-tapped nodes for connecting the PoE forward and return wires. Some PoE engineers have therefore copied the phantom circuit approach used in audio microphones and in landline telephones as shown in Fig. 13.16 of this book. In this particular patent, a separate winding is placed around the data transformer to counterbalance the flux produced by the PoE current, thus ensuring we do not need to oversize the data transformers. However, how do you derive power to drive this PD flux-cancellation circuitry (and successfully power up), without having already powered up? Usually, we cannot afford to bombard magnetic cores with 350 mA rather than the 8 mA they are probably designed for. And that is what this patent set out to accomplish. Did it do that? Does it work? When will it get built and tested?

Our analysis leads to the conclusion that though there are some great ideas and some *potentially* great ideas out there, there is also a great need for fresh ideas and thinkers to synergize with the remarkable

[27]See http://books.google.com/patents/US8217529.

legacy we inherited. There is still lingering ambiguity in the IEEE 802.3at standard. There is thus a strong need for a book like this one. I hope you will use it to innovate in a way that makes future generations look back at us proudly, not askance. Patents are consulted forever. They form our combined legacy, and are also a fogless mirror of our times. Therefore, the underlying process behind creation and innovation is very much worth defending every single day. I do feel the number-of-patents situation, in particular, stoked by generous incentive schemes which reward quantity not quality, needs to be scrutinized. Or we will end up with (more) inventions which aren't and inventors who didn't.

This book does happen to be the very first book on the subject. I apologize in advance for any unintentional mistakes or misinterpretations. There are no references other than the IEEE standard to consult really. The dust has still not fully settled. Please double-check and validate everything for yourself. I hope this book will help you in that process and that it will also be fun and enlightening.

Sanjaya Maniktala

Acknowledgments

This book wouldn't have been possible without outstanding support from some key people listed as follows.

1. Bryan Young, who led the PoE team at Underwriters Laboratories in San Jose, California, and Pete Johnson of Sifos favorably reviewed an initial book proposal.
2. Daniel Feldman has had a lot to do with pushing the frontiers of PoE in various forums, with infectious passion and humility.
3. Others who helped make this book possible and kept me in good spirits include Fred Schindler, Paul Pickle, Amir Asvadi, Tamir Reshef, Nadav Barnea, Pavlick Rimboim, Iris Shuker, and the one-and-only Fernando Serpa. Many others, too; I apologize if I forgot to list all of them here.
4. I am very appreciative of the outstanding help from the team involved in the production of this book, especially Yashmita Hota from Cenveo Publisher Services, Noida, India. They made it very easy for me.
5. My writing career really started around the year 2003, when I approached Steve Chapman, sponsoring editor at McGraw-Hill. Almost a decade later, he still remembered me and made this book possible too. He is by now a legend in the publishing field for power electronics.
6. Steve assigned sponsoring editor Joy Evangeline Bramble to this book initially, and though she has gone on to other things, I still remember her outstanding support and infectious enthusiasm.
7. I must thank my indefatigable wife Disha and daughter Aartika for once again letting me drift away for months while writing this magnum opus. It is absolutely certain by now that none of my books would have been remotely possible without their unstinting consideration. Or the warmth of my two furry little canine friends, Munchi and Cookie. Not to forget my

long-departed friends, Chippy and Monty, who really got me to this point years ago.

8. Good engineers are like Sheffield knives. They get honed by repeated contact with dull and impenetrable objects. It gets their creative juices flowing and they feel reinvigorated and recharged, whereas the dull object only gets duller. With that positive perspective in mind, I am grateful to some highly placed people who inadvertently helped make this book so much better and purposeful eventually. I also acknowledge that I learned a lot about PoE in their labs while on the job. I was privileged to lead a small team of gung-ho engineers, who, barring one intensely mollycoddled curio (the inspiration behind Chap. 12), were all ethical, competent, honest, and enthusiastic. Yes, we were outsiders, but together we drove deep into the frontiers of an emerging technology for years, without ever pulling over to the curb to collect quick cash rewards, or file tons of dubious patents to harness incredibly generous stock options, or simply make a cheap grab for credit or power (such opportunistic activities most engineers usually leave to their indolent marketing guys, past and present, of course). It is therefore to my former engineering colleagues that a very special thank you goes out. They helped a lot in making this book what it is, and I know they have also considerably enriched a very nascent field from deep within the shadows, where they unfortunately still remain despite my best efforts, directed to the very top, to fix the system and make them feel hopeful enough to resume actively filing good and honest ideas and patents so they too can be deservedly better known. I certainly learned a lot about PoE rubbing shoulders with them every day for almost five years. Way to go guys!

Power over Ethernet Interoperability

CHAPTER 1

The Evolution of Power over Ethernet

PART 1 AN OVERVIEW OF ETHERNET

Introduction

We seem to be in one of those rare moments in our history when absence has paradoxically emerged as the strongest proof of omnipresence. Since its rather humble beginnings back in 1973, Ethernet has metamorphosed so dramatically, it bears almost no resemblance today to its origins. As testimony to its ever-increasing popularity, 95 percent of all local area networks (LANs) in existence today are estimated to be *Ethernet*-based. But the irony is Ethernet may likely never have been around in its current form had it not been conceived way back then in its now-extinct form.

The basic purpose of Ethernet remains unchanged—to connect computers, printers, servers, and so on in LANs so they can exchange information and services among themselves. Although Ethernet continues to be a *packet-based* computer-networking technology, even its packets of data ("frames") are no longer exactly as originally envisaged. So much has changed in Ethernet that some very notable people have gone as far as to say Ethernet is (now just) a *business model*.

But first, to dispel a popular notion in a timely manner: Ethernet is *not* the Internet and vice versa (despite sounding similar). Internet actually came into our lives a bit *after* Ethernet did—specifically in 1989 when the first commercial Internet service provider (ISP), aptly named "The World" offered its services to the general public. Internet is, quite literally, an *inter-network*, a global *network of networks*. By now, most of the networks linked by the Internet do happen to be *Ethernet*-based, and nearly all Internet traffic starts or ends on an Ethernet connection. So it is obvious that a good part of the reason for the explosive growth of Ethernet is that it so naturally complemented the exponential growth of the Internet over the last few decades.

More recently, another very similar symbiotic growth pattern has quietly emerged from the shadows, perhaps a bit unnoticed. Riding on the remarkable growth story of Ethernet, figuratively and literally, Power over Ethernet (PoE) is now seeing a huge upswing of its own and seems to be now driving growth by itself. New families of PoE-capable network appliances and Ethernet equipment have suddenly started appearing. An ever-increasing percentage of "ports shipped" today are PoE-enabled, and PoE seems to be fast-becoming a default choice for Ethernet ports. It is therefore increasingly important for us to recognize PoE for what it is—a very fast-emerging technology. We should try and understand it much better before it gets ahead of us.

From the viewpoint of chip and systems designers, some rather unusual challenges are associated with implementing PoE. At a higher level, PoE represents the cusp of two major, hitherto parallel worlds of electronics development: *power* and *networking*. PoE is, in ways, the virtual confluence of the Applied Power Electronics Conference, (APEC) and Interop® (the annual networking expo at Vegas). It symbolizes the convergence of digital and analog hardware and software, power and data, and so on. It is exciting—and therein lie the challenges too.

If we look around, we may notice that things are not going so well on the *human* plane. Hardware personnel are still quite prone to ignoring software personnel; analog designers often shun digital designers, and so on—and vice versa of course. In the oft-repeated (though infamous) words of a famous analog legend of Silicon Valley, the late Bob Widlar: "Any idiot can count up to 1." No surprise that if we look around, we may notice several mutually suspicious "knowledge cliques" hovering around us. Unfortunately, *skill segregation* (specialization) is not commensurate with our modern world of increasing convergences. Personally, as systems and chip designers, we just cannot afford to allow our skill sets to become so increasingly finely tuned and subdivided anymore.

We need to illustrate this issue a little better, lest it sound hyperbolic to some. In 2006, one such engineer, let's just call him Bob for now (another Bob), recalled an ancient bench struggle he had faced decades ago. These are his words [italics added by the author of this book]:

> I designed this thing. It took me forever, *a good portion of a year*. I finally had it, and I got it working with my test programs. It would work, but it only worked for 15 minutes and *then it would go "bah,"* and all the lights would go in the wrong direction and then it would die. Then I would reboot, and then it would run again. I had these extensive diagnostics, random patterns, and everything, to just test the hell out of the thing, different word lengths, test every edge case. Then it would run just fine, over and over again through every—and then it would just die. *I worked on it for a month and I couldn't figure it out*...I went down to the other end.... Tom looked at

> it, and he asked me a few questions, and then he said, 'Well, I know what's wrong.... *you don't have any bypass capacitors on it.* Now I know you just took a bunch of courses in digital electronics but *at some point everything is analog*. So what's happening, Bob, is when certain patterns get into your registers and they all go 1, and all the transistors turn on, they take too much current, too many electrons, and then the voltages start to droop because you've sucked in all those electrons and then all the digital devices start to malfunction. So what you need to do, Bob, is *sprinkle some bypass capacitors here and there* to store up some extra charge for those cases when you have lots of 1's in your registers.'.... I went quickly back to the lab and got my soldering iron, and I soldered on—I probably put a bypass capacitor on every third socket to this huge board. And I plugged the sucker in and *she worked for the next 13 years*. Anyway, I learned a lesson there—the *analog digital*—which came in handy later because *Ethernet is a combination of analog and digital itself.*

Now add to the anecdote two not-so-obvious facts:

1. PoE had not even come into the picture at the time of my bench-wrestling story, and the design "gotchas" were already appearing fast on the horizon.
2. The engineer in question was none other than *Robert ("Bob") Metcalfe*, considered today to be the "inventor of Ethernet."

So, we can imagine, that if Metcalfe was confused, more so *without PoE on the scene* yet, what we may experience *today*. That thought is humbling. Therefore, one of the key objectives of this book is to try and narrow down the differences between power and networking and analog and digital, and so on. Because if we don't, we may take almost forever to get to a *successful*, commercial product down the road (*cable* in this case).

One last question before we move ahead: *Who invented Power over Ethernet (PoE)?* Was it Bob? Or John? Was it Harry? Why not Jane? Actually, *none of these*. The correct answer lies buried somewhere in the *19th* century—*J. J. Carty* was his name. And that is a true eye-opener, especially to some gung-ho "modern-day" engineers. In fact, it's actually over a hundred years ago that the basic principles underlying PoE started to take shape. So, in that sense, Ethernet actually came *after* PoE. It's funny how the world goes round and round. If not, it would probably all go "bah," in the words of Metcalfe.

And so, for just a moment longer, let's return to the future—that is, back to *Ethernet*.

A Brief History of Ethernet

Ether Network, Ether Net, EtherNet, Xerox Wire, X-wire... That's what modern Ethernet *almost* got called. And had it, very likely it could have ended up in our collective memories (and Wikipedia) today as the "the now-defunct proprietary networking technology from Xerox Corporation." In fact, even the term "Ethernet" was originally a registered trademark of Xerox Corp. Fortunately for all of us, and perhaps even for Xerox, Xerox got talked out of all its rights on this subject by Robert Metcalfe himself and agreed to work with Digital Equipment Corporation (DEC) and Intel to spread its version of networking technology far and wide. Shortly thereafter in 1979 Metcalfe founded 3-Com Corp (last acquired by Hewlett Packard in 2010) but continued to work with the consortium, informally called DIX (for DEC, Intel, and Xerox). Together, they published the first formal (industry) Ethernet standard, DIX V1.0 in 1980. Meanwhile, the Institute of Electrical and Electronic Engineers (IEEE), in an effort to standardize LAN, had their very first meeting on this subject early the same year (1980). Because of the timing, some say that the well-known modern Ethernet standard 802.3 got it basic name—802 coming from February '80 or 2/80. Others say 802 just happened to be the next-available number in the normal sequence of IEEE standards. Either way, DIX approached IEEE to help standardize (their version of) Ethernet. But things were just as they are on any typical day at IEEE even today. In walked pushy General Motors with a rival LAN proposal called the Token Bus. Not to be outdone, IBM walked in with their Token Ring. As a result, IEEE bowed and decided to standardize *all three* proposed LAN standards. And that is how the IEEE "dot committees" got created: 802.3 was Ethernet, 802.4 was Token Bus, and 802.5 was Token Ring. These were all later blessed for international acceptance by the International Standards Organization (ISO), and became, respectively, ISO 8802-3, 8802-4, and 8802-5. Yet despite the initial boost, the latter two eventually died from "natural causes," though Token Ring does seem to have hung on rather stubbornly—for close to 15 years as per Metcalfe's estimate. Some think it is still around somewhere. But no one contests the fact that, in contrast, Ethernet grew extremely rapidly.

So, why did Token Ring in particular, lose out to Ethernet? One reason was that IBM was charging prospective vendors heavily in terms of royalties for producing Token Ring cards and medium attachment units (MAUs), or simply, the cable-driver electronics (akin to *transceiver* or PHY in modern lingo) placed between the controller card and the cable. This made Token Ring equipment too expensive overall. A Token Ring card itself could cost 5 and 6 times as much as an Ethernet card. Add to that more expensive cabling, and Token Ring literally priced itself out of the market.

Besides cost, a big advantage of Ethernet, going forward, was its inherent *flexibility*. On December 21, 1976, in an internal memo at Xerox, Metcalfe explained a key advantage of Ethernet in the following words (italics inserted here belong to the author of this book):

> The OIS [Office Information Systems] protocol is based on *distributed* many-to-many communication as required by *incrementally grown* and

increasingly interconnected office systems, rather than *hierarchical* mainframe-centered data processing systems.

The way Metcalfe originally visualized Ethernet was a *single long coaxial* cable connecting many computers (workstations) and printers. In modern terms, this is a "bus topology." A new device (computer, printer, and so on) could be simply added on, as the need arose, using a "vampire tap." This contained a needle (let's call it a Dracula tooth to be visually clear and consistent) that would get clamped down and penetrate the coaxial cable to make contact with its inner conductor, while the outer shield of the cable would connect to the outer shield of the newly added segment. So the network could be built up steadily over time, rather than needing a big central infrastructure right off the bat (preguessing future needs). This proposed type of LAN architecture was envisaged to grow along with the size of the organization.

The bus topology is somewhat akin to a giant plumbing system with a main water pipe...(but with data, not water) flowing down, with several feeder connections on it along the way (see the hybrid architecture example in Fig. 1.1).

It is indeed ironic that in later years, the basic framework of Ethernet (in terms of its *packet-based architecture* and supporting techniques) proved flexible enough to allow moving *away* from the original bus topology concept to a "star topology." In this new architecture, every computer (or networking device) gets connected via a dedicated cable plugged into a central switch or hub. In terms of hardware expandability, the star topology is not as flexible as the bus topology, but it can provide much higher speeds, besides other advantages. As PoE designers, we should also realize that the bus topology could never have supported PoE as it is today. The star topology is a prerequisite for power over data cables, one cable for each end-device. So clearly, things seem to have gone in the right direction, both for data and power. That's survival of the fittest.

Another reason for the continuing rise of Ethernet was that Ethernet got suddenly empowered along the way by something called a "switch." This concept seems to have originally come out of a startup called Kalpana (Hindi for imagination or vision), cofounded by Vinod Bhardwaj, an entrepreneur of Indian origin. Kalpana was acquired by Cisco in 1994, ten years after Cisco itself was born. The switch eliminated a very basic problem of *collisions*, which was slowing down networking in general. The switch eventually helped achieve much higher data rates. It soon became unstoppable because it catered to the rapidly growing need for speed.

What are collisions? Quite similar to what you would expect would happen if hundreds of cars were let loose in both directions on a *single lane*. More formally expressed, collisions can be explained as follows: (Data) collisions occur when, for example, all the computers

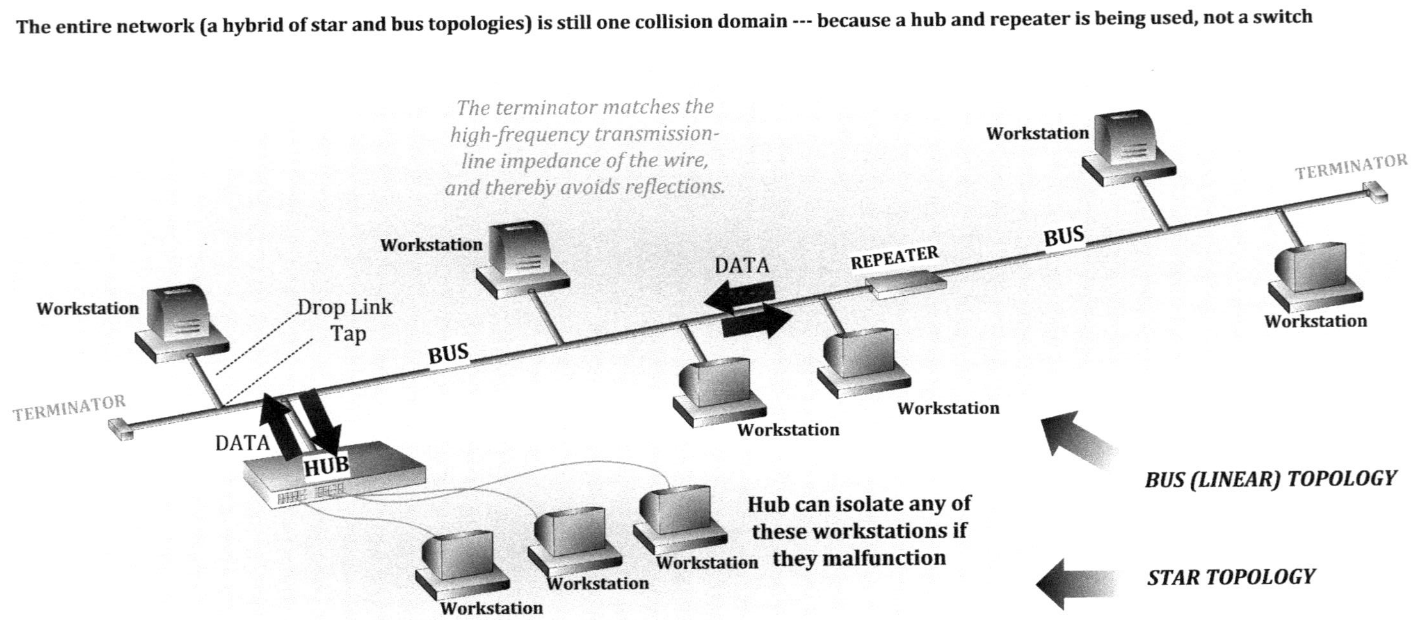

FIGURE 1.1 Older Ethernet LAN architecture showing bus/star topologies and hub/repeater.

on a shared line start "talking" (transmitting data packets) simultaneously. What results is almost noise (garbage). It is also very similar to a whole bunch of people brought into a small room, each person trying to talk over everyone else's head to someone else in the room. Pretty soon, with all the din and shouting, no one really understands a thing anymore. We have all been there and perhaps done that too. To avoid this unpleasant and *inefficient* situation, the connected computers need to detect that a collision is occurring, then *back off* and try a little later again. In doing so, they must *not* try *simultaneously* again, or a clash will recur, slowing down all communication for an even longer time. On the other hand, if they wait too long before trying once again, the communication slows down again. But if they try too fast again, they run the risk of overlapping (talking simultaneously), causing more delays. And so on. That is where Metcalfe originally came into the picture. His claim to fame is U.S. patent number 4,063,220, titled "Multipoint Data Communication System with Collision Detection." This is basically a (statistical) *algorithm to back off* and try again in an optimum manner. In contrast, and in a more *deterministic* manner, IBM's LAN architecture had a software "token" that was moved around in a circle of connected computers, and so whoever had possession of the token at a given moment, got the right to transmit. It was a little like the game of "passing the parcel" at a typical children's birthday party. Token Ring did seem to be superior to Ethernet initially, especially to engineers who felt uncomfortable with the lack of clearly defined timings characteristic of Metcalfe's software-based algorithm.

Note that in 1985, IEEE finally published a portion of the ongoing standard pertaining to Ethernet: IEEE 802.3 titled "Carrier Sense Multiple Access with Collision Detection ('CSMA/CD') Access Method and Physical Layer Specifications." We can see the title does not even mention the word "Ethernet." But Metcalfe's original term did catch on in a big way. And so the IEEE 802.3 standard was, and still is, referred to as the Ethernet standard.

CSMA/CD can be explained in informal language as follows:

1. CS: Carrier Sense (*Hey, do I hear someone talking?*)
2. MA: Multiple Access (*Careful, we can all hear what each one of us is saying!*)
3. CD: Collision Detection (*Hey, we're both talking now—stop!*)

The underlying logic of CSMA/CD is

1. If the medium is idle, transmit *anytime*.
2. If the medium is busy, wait and then transmit *right after*.
3. If a collision occurs, send 4 bytes of "jam" signal to inform everyone on the bus, then back off for a random period, and after that, go back to Step 1 above.

The last point is pretty much the social technique we use rather intuitively, in a normal, polite group or conversation, say at a dinner party in the evening, as opposed to trying to talk, or rather shout, at each other in a crowded bar (perhaps with 100 dB of rock music playing in the background).

As mentioned, with the entry of the switch, the basic problem of collisions was eliminated altogether. It was a huge boost for Ethernet, both in terms of speed and market popularity, and eventually this allowed Ethernet to move up to much higher speeds: 1 Gbps (gigabits per second), 10 Gbps (over copper), and beyond, as of today. Though in the first step, Ethernet just went from 10 Mbps (megabits per second) to 100 Mbps. That actually happened without even requiring a switch, just by using a "hub," described further below. Yet, even that was enough to start making the 16 Mbps of the IBM Token Ring architecture obsolete. The Token Ring concept, however, did seem to get a fresh lease on life in Hewlett-Packard's (HP's) 100-Mbps LAN architecture called 100VG-AnyLAN in 1995, but that was virtually extinct by 1998 too.

The advantage of "switching" in the area of Ethernet can be explained in modern terms as follows: Today, every device on a network has a unique self-assigned identifier, a hardware address, called its Media Access Controller (MAC) address. The media in this particular case is the copper wire, over which signals are sent. When a computer sends data on the network, it sends it in packets (frames). Each packet carries information describing the source and (intended) destination. Network switches (or just switches) are devices smart enough to read the MAC addresses and direct the packets from the source to the intended recipient device. In other words, switches do *not* broadcast data to all and sundry. Unlike a hub, switches do not talk loudly as over a public announcement (PA) loudspeaker, preventing others from talking when they want to. But with a little thought, it should also be clear that for switching to be successful, the shared bus topology of the original Ethernet needs to be replaced by the *star topology*, in which each computer (or node) gets its own Ethernet cable, and all Ethernet cables get plugged into the switch (into its available ports, or jacks). With such a topology, conversation can be *physically* directed from source to destination in a planned manner over dedicated cables. Note that hubs also connect to computers in star topology; the difference being that hubs are not smart enough to avoid collisions completely by inspecting and directing packets back and forth in a planned manner, as switches do. So (IEEE-compliant) hubs will just use CSMA/CD. On the other hand a switch is not even aware of CSMA/CD. It doesn't need to be.

Long before switches and hubs appeared on the scene, there were "repeaters." These devices just amplified the signal for sending across longer distances over copper, and extended the geographical reach of the LAN. No intelligence was built into repeaters. A little later, since it was getting difficult to troubleshoot and isolate problems that were occurring on the shared wire, hubs were introduced, with the basic intention of creating a certain amount of *segmentation* (or segregation), along with security, within the LAN. Now, instead of several computers all hanging off one shared line, we could have several hubs connected to this line. Computers would then get connected to the hub in a star topology as shown in Fig. 1.1. Now if a computer malfunctioned, or perhaps just "talked too much," its hub could detect the condition, and shut off all communication to it (or to any other errant network device). This would isolate that device from the shared bus and prevent the bus from going down. Keep in mind, however, that with all hubs still using CSMA/CD, the entire LAN was still one big "collision domain."

At some stage bridges were introduced. The purpose was to create separate, *smaller* collision domains within the same LAN. Note that the entire LAN is still one, big "broadcast domain," but it now consists of not one, but several collision domains interconnected by bridges. This segmentation was found to be very helpful in a situation in which a very large number of computers were sharing the same line. Because, if they all tried to talk, even with a good collision detection/avoidance algorithm in place, they would eventually slow down the entire line considerably. So it was thought much better to have, say, two separate shared lines (or buses), with a bridge to pass data back and forth *when required*. Like two, bustling cities on either side of a river, connected by a bridge. People in either city lead separate lives, except when people from one city want to, or need to, go over to the other side. That's when they actually cross the bridge. Consider the contrasting case, with residents of both cities cramped into one city only, the traffic situation and congestion will only get much worse.

So looking back, the repeater was the dumbest of all. In between there were the bridge and the hub. The switch emerged as the smartest. Then, with the advent of the Internet, a device called a router appeared. It is even smarter or more powerful than a switch. As far as simple traffic routing *within* the LAN is concerned, a router operates exactly as a switch, learning the location of the computers on the LAN and routing traffic precisely to those computers. But the actual routing, as carried out by router, unlike as in a switch, is *not* based on MAC addresses, but on *Internet Protocol* (IP) addresses. Because ultimately, routers don't just allow different devices on a given local area network to communicate with each other, as switches do, but also allow *different networks* to communicate with each other over the Internet. To communicate between different networks, routers must have the ability to talk to other routers too (using IP addresses). In effect, a router becomes an interface between its LAN and the Internet.

When a router initially attempts to connect to the Internet, it requests an (external) IP address—one address for the entire LAN, much like a single postal address on a street for a huge building complex, though there may be several individual apartments or single homes within that complex (that will eventually need letters delivered to and collected from). The request for this single external address is made by the router to a Dynamic Host Configuration Protocol (DHCP) server somewhere in the ISP's network. The router also distributes internal (local) IP addresses to all the devices (clients) on its LAN, to identify them (similar to assigning house/apartment numbers for residents). To accomplish communication between all these client devices and the Internet, the router uses network address translation (NAT). NAT involves modifying the source and destination IP addresses *within the packets*, so as to direct traffic appropriately between LAN clients and servers/devices on the Internet. NAT is in essence, a way to map all the devices within a network to a single external IP address. Why is it useful and/or necessary? In layperson's terms, if, for example, we want to connect just one computer in our home to the Internet, we do not even need Ethernet (no switches, hubs, routers, and so on)—just a direct connection and the use of Transmission Control Protocol/Internet Protocol. (TCP/IP) But what if we want not one, but three computers in our home to connect to the Internet (and talk among themselves too if possible)? And suppose we have just one incoming cable or digital subscriber line (DSL) connection. We do not want to pay our ISP for installing three separate lines/connections. Maybe they cannot do so either. So, as far as they are concerned, with the help of a router, all three computers can be made to appear as a single IP address (and we will of course get billed for just one connection/computer). Internally however, inside our home, the router distributes IP addresses to the three (or more) home computers. In effect, the router creates a small, switched LAN within our home, but also handles the back-and-forth exchanges from the home computers to the Internet via the IP's server, and all that in a manner that is transparent to the IP server. The IP server "thinks" it is dealing with just one computer inside our home (one IP address). In a sense, this is socially acceptable, mutually agreed-upon *deceit*. Summarizing, NAT becomes necessary when the number of IP addresses assigned by the ISP is less than the total number of computers on the LAN that need Internet access.

This natural morphing/evolving of Ethernet, combined with the growth of the Internet, contributed to the impressive rise of Ethernet. But another key reason for its success over rival architectures was that it became low-cost down the road. The star topology was the main enabler of that. Communication became possible (though over shorter distances of up to 100 m), using cheap, dedicated, "twisted-pair" copper wiring, as compared to the far more expensive coaxial cables required by rival LAN technologies (and by Ethernet itself originally). Note that *two* twisted pairs per connection were used for point-to-point communication: one for receive and one for send. So transmission and reception could now occur simultaneously too.

In other words, there were absolutely no collisions anymore: CSMA/CD could be forgotten forever, with just the flick of a switch (literally)! But note that still, data transfers were *uni*directional: Each pair worked in only one direction. Later in an attempt to achieve 1000 Mbps (1 Gbps), the electronics were made smart enough to ensure bidirectional communication over a single twisted pair by the use of a hybrid circuit as we will soon learn.

When it first started, Ethernet was just 2.94 Mbps, but that was because 2.94 Mbps happened to be the available system clock on Metcalfe's computer. Soon 10Base-2 and 10Base-5 appeared. Both these implementations ran at 10 Mbps, the first over about 200 m (hence the 2) of thin coaxial cable, the latter over 500 m (hence the 5) of *thicker* coaxial cable. But both required *coaxial wire*, and both are obsolete now. The follow-up IEEE 802.3 Ethernet standard was called 10Base-T, where "T" stands for twisted pair. This is also 10 Mbps, but, as mentioned, works over two pairs of twisted pair copper wiring (of American Wire Gauge number 26, or AWG 26). The deal breaker for Token Ring, however, was not 10Base-T, but 100Base-TX, which is 100 Mbps Ethernet over (two pairs of) twisted-pair wiring. This is called Fast Ethernet and is still popular today.

Modern Three-Layer Hierarchical Network Architecture

Having understood networking topologies, routers and switches, and so on, we take a quick look at Fig. 1.2, which represents a typical, modern three-layer Ethernet network architecture, mainly popularized by Cisco. It also includes a connection to the Internet. The basic purposes of the three layers can be summarized as follows.

Core Layer

This is the high-speed "backbone" of the "inter-network." The core layer is critical for maintaining interconnectivity between distribution-layer devices. The core must be available readily (and immediately), and also have some built-in redundancy to avoid a single failure bringing the whole network down. The core connects to Internet resources. It aggregates the traffic from all the (lower) distribution-layer devices. Core-layer switches/routers must therefore be capable of forwarding large amounts of data very quickly. And for that reason, core switches are more hardware-based than software-based. This helps reduce latencies that can arise from large number-crunching within software programs.

In small-business establishments, such as those called SMB (small and medium business), or equivalently SME (small and medium enterprise), the core and distribution layers may be one: as a single, "collapsed-core," layer.

Note that, sometimes, people call the core layer the "edge layer," and that can get confusing as indicated below.

Distribution Layer

The distribution layer aggregates the data received from the (lower) access-layer switches before it gets transmitted to the core layer for routing to the final destination(s). This layer also controls the flow of all network traffic in general, choosing the best (optimum) routes to send data between users on the LAN, also applying any relevant policies. For example, in a university we may want to separate the traffic according to faculty, students, and guests. Note that the switches used in this layer are typically high-performance devices too that have high availability and redundancy to ensure reliability.

Access Layer

The access layer interfaces with end devices, such as PCs, printers, and IP phones, to provide access to the rest of the network. This layer can include routers, switches, bridges, hubs, and wireless-access points (WAPs). The main purpose of this layer is to provide a means of connecting devices to the network, and also controlling which devices are allowed to communicate on the network at any given moment (their "access" privileges for example). Note that since end devices reside in this layer, PoE capability is most likely to be provided in switches, hubs, routers, and WAPs operating on this layer. This is also sometimes called the "access edge," or just the "edge," and it gets confusing because the core layer is also sometimes called the edge layer.

PoE is today provided even in switches meant for the upper (non-access) layers. For example, PoE capability may be present in the core-layer switches too, for powering customer-premises equipment (CPE). This would include any terminal and associated equipment, located at the subscriber's premises, connected to a carrier (or ISP's) telecommunication channel at the point of demarcation (i.e., where the line connects to the home/business wiring, and responsibility for maintenance gets handed over from provider to the customer/subscriber).

What Exactly Is "Ethernet"?

With all this evolution, what really was, or is, "Ethernet"?

In 2006, Metcalfe said: "Ethernet is [now] a *business model.*" Metcalfe probably rightfully meant that Ethernet is now almost a *brand-name* of sorts, and bears almost no resemblance to what it originally was.

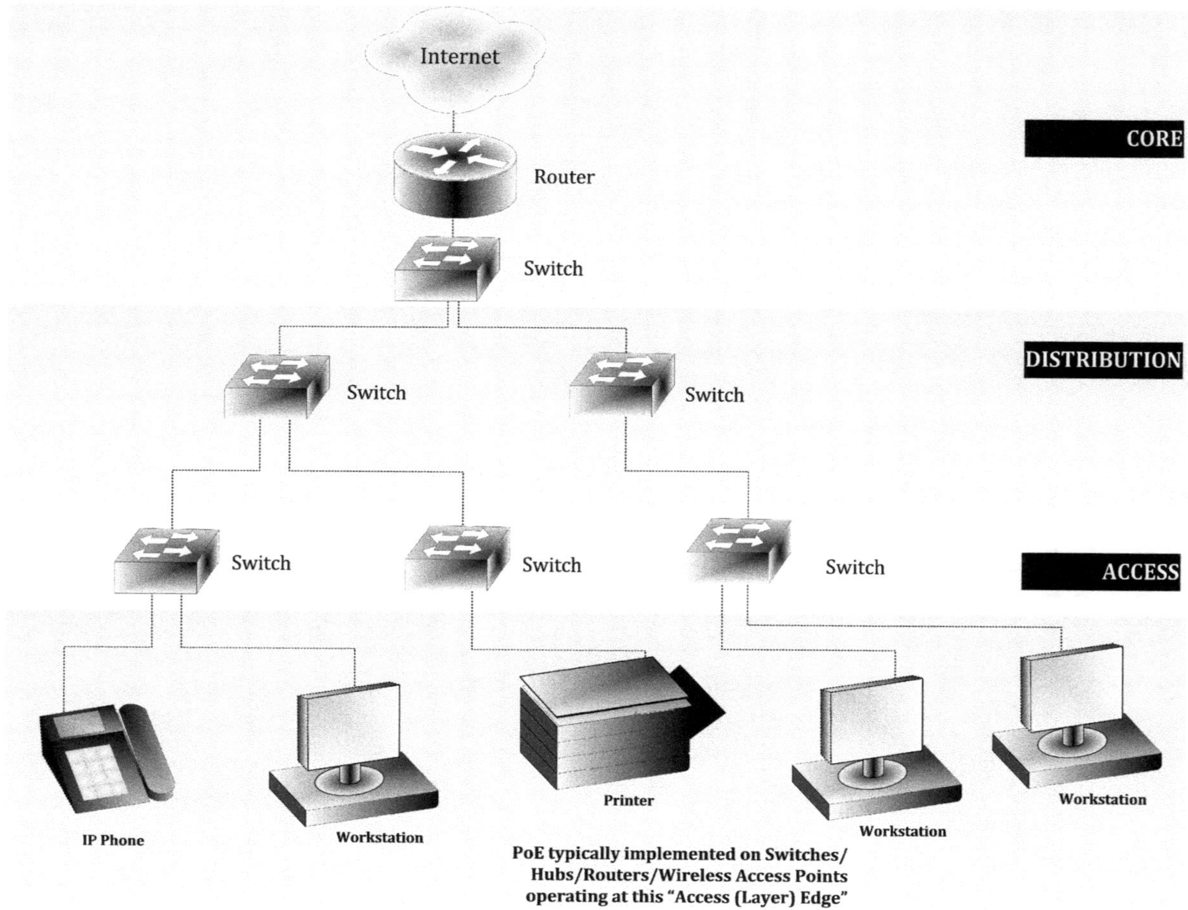

Figure 1.2 Hierarchical three-layer network architecture (and where PoE fits in).

But what did Metcalfe have to say about it in 1973? On May 22 of that same year, he wrote a world-changing E-mail within Xerox, Palo Alto, California—an E-mail which set the Ethernet ball rolling.

Did we just say "E-mail"? That's obviously not accurate. Because Metcalfe was *going* to enable E-mail *soon*, but till then "E-mail" (at least as we know it today) did not exist. So we just dodged a trick question.

It is more accurate to say something like this: "On May 22, 1973, Metcalfe hunched over an IBM Selectric typewriter using a spinning Orator ball, and talked about his vision of the future." In reality, his (almost) first sentence was *"I propose we stop calling this thing 'The Aloha network.'"*

The Aloha network (ALOHAnet) had been developed between the years 1968 to 1971 at the University of Hawaii. It was a radio-frequency link to connect the university facilities across different islands. Necessity is obviously the mother of invention. Metcalfe's system was an improvement over that, since it (eventually) detected and avoided collisions (his patent). But to make it clear to others that the system could support any computer, not just Alto (the Xerox workstation), Metcalfe chose to create a deliberately vague name based on the word "ether." In ancient times, people were not comfortable with the concept of a vacuum (complete nothingness). So "luminiferous-ether" was imagined to be the medium through which electromagnetic waves could propagate through space. In a similar fashion, Metcalfe envisaged a generic "physical medium" carrying bits of data to all stations (nodes in modern terminology). He explains that in the 1973 memo: "While we may end up using coaxial cable trees to carry our broadcast transmissions, it seems wise to talk in terms of an ether, rather than 'the cable,' for as long as possible. This will keep things general. And who knows what other media will prove better than cable for a broadcast network; maybe radio or telephone circuits, *or power wiring*, or frequency-multiplexed CATV, or microwave environments , or even combinations thereof." This book's author inserted the italics in the above statement.

Note very carefully that Metcalfe had already envisioned power and data sharing the same lines. But he was not the first as we will see.

Data over power cables, or power over data cables—what is the big difference?

Power-line carrier communication (PLC) has been around in a basic form since 1920s. A (modulated) wave of very low frequency was injected into high-tension power lines using coupling capacitors. It provided very basic, one-way communication/control. It was used for activating remote relays, public lighting, and so on. In the 1970s, Tokyo Electric Power Co. reported successful bidirectional operation to read and control power meters remotely. "Baby alarms" have been available as consumer products since 1940. The author too had built several pairs of "baby (monitoring) phones" in the mid-1980s. These were small, short-distance, power-line carrier walkie-talkies, based on FM (frequency modulation using voltage-controlled oscillators) to transmit voice over home mains-AC wiring, followed by phase-locked loops inside the receivers, which were typically based on low-cost LM565/LM567 chips for decoding the modulation. Like walkie-talkies, both stations could not transmit at the same time (that would result in noise), and their best use was for monitoring purposes (one-way). In mid-1980s, research commenced into the use of existing electrical grids to support data transmissions using modulation of base frequencies up to 500 kHz. This was, however, still one-way communication. In 1997, the first tests for long-distance, bidirectional data transmissions over high-tension lines took place in Europe. Closer to our homes and times, today we have the most widely deployed power-line networking standard from HomePlug Powerline Alliance. We also have Broadband over power line (BPL), and so on. New devices from Netgear and others try to turn every AC outlet in our homes into a potential Ethernet jack. These devices comply with the IEEE draft P1901 standard and typically work up to 500 Mbps. Colloquially, this is often called Ethernet over power lines.

Thinking of other media as Metcalfe had imagined, Ethernet has now evolved into Ethernet over optical fiber too. For example, we now have the 1000 Mbps standard called 1000Base-F, where F stands for fiber. Of course we also have 1000Base-T and 1000Base-TX over twisted pair (copper). The general nomenclature being used (summarized as best as possible under the changing and evolving landscape) is presented in Fig. 1.3. In Fig. 1.4, we present an overview of communication standards, including non-Ethernet standards such as digital subscriber line (DSL), since, along with Ethernet, they remain a popular choice for data communication *over copper*.

A summary of the key Ethernet-over-twisted-pair (Base-T) standards that we will run into when designing PoE products is listed as follows (clearly PoE can't be used over fiber!):

1. 10Base-T: 10 Mbps (megabits per second) over 100 m of standard Ethernet cable consisting of four twisted pairs. Note that only *two* twisted pairs are used for data (data pairs). Two are just unused (spare pairs). Further, the data is conveyed unidirectionally on each of the two data pairs. That means one pair is dedicated to sending signals in one direction, while the other pair communicates in the opposite direction. It therefore is one-way on each pair and two-way on two pairs.
2. 100Base-TX: 100 Mbps (megabits per second) over 100 m of cable consisting of four twisted pairs. Once again, only *two* unidirectional twisted pairs are used for data. This is sometimes called Fast Ethernet and is the most prevalent Ethernet standard today.

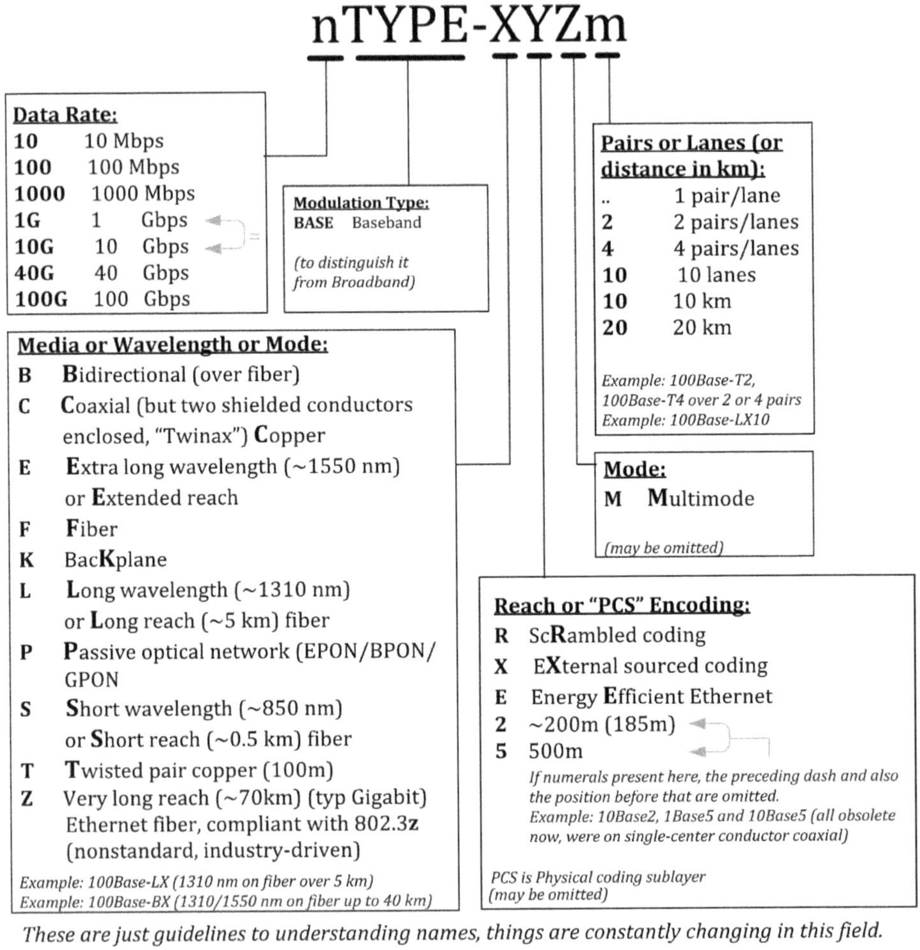

FIGURE 1.3 Ethernet nomenclature.

3. 1000Base-T: 1000 Mbps (megabits per second) over 100 m of cable consisting of four twisted pairs. Here all *four* twisted pairs are used for data. Further, each pair is bidirectional, which means two-way signaling on each pair, four pairs in parallel for higher speed. This is sometimes called 1 GBase-T or Gigabit Ethernet. The key advantage is that it can use the same cabling infrastructure as commonly used for 100Base-TX.
4. 1000Base-TX: 1000 Mbps (megabits per second) over 100 m of cable consisting of four twisted pairs. Theoretically, this was intended to save the cost of the electronics (the PHYs and so on), because though all *four* twisted pairs were to be used for

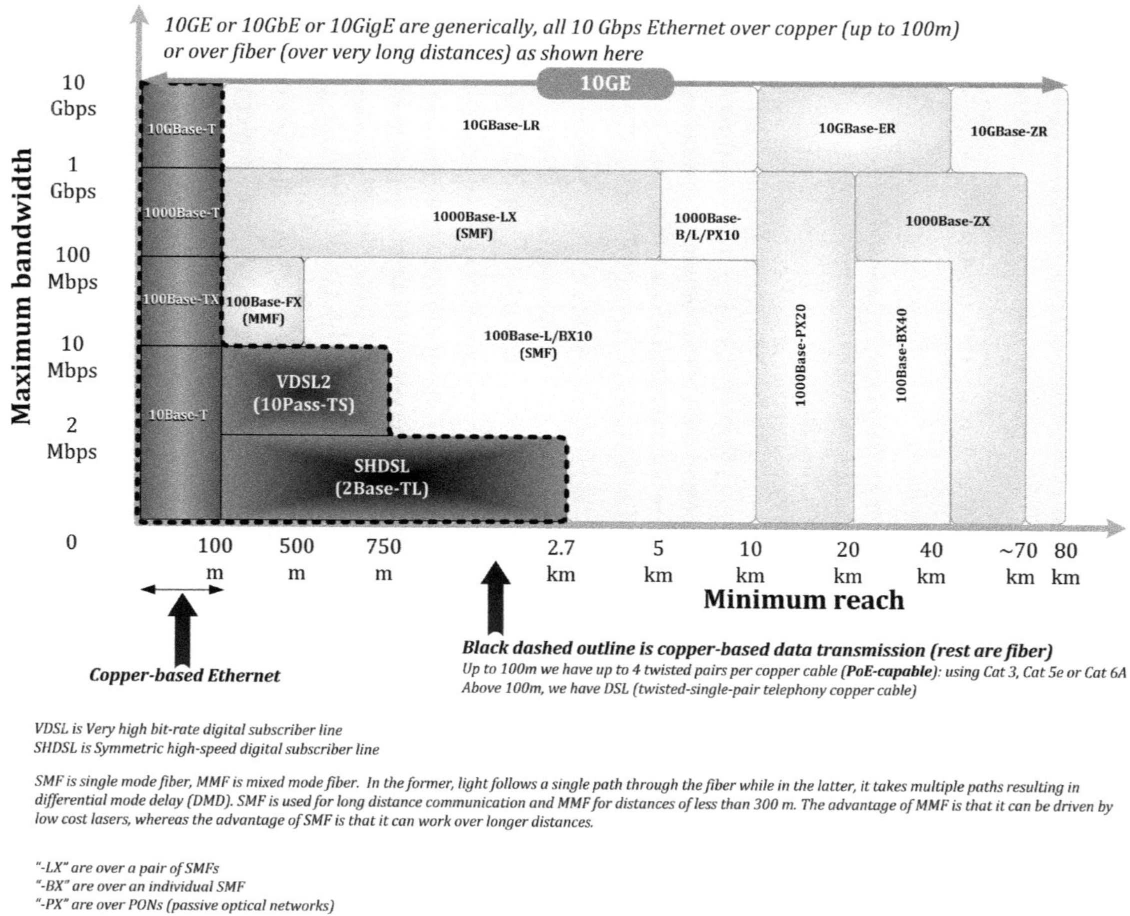

FIGURE 1.4 Overview of Ethernet and related data transmission technologies.

data as in 1000Base-T, *each pair was to remain unidirectional,* as in 10Base-T and 100Base-TX. However, to compensate for the lowering of electronics capability, the cable data capability had to be correspondingly raised. In other words, 1000Base-TX requires more expensive cabling than 1000Base-T or 100Base-TX, which is just not easily available. Therefore, for all practical purposes, 1000Base-TX is now considered a commercial failure and effectively obsolete.

NOTE *The "T" at the end, as in 10Base-T, comes from twisted pair.*

NOTE *"Base" as in 10Base-T stands for* ***Baseband****. A Baseband network is one that provides a single channel for communications across the* physical medium *(the common Ethernet cable in the case of copper transmissions) so only one device can transmit at a given time. Devices on a Baseband network are permitted to use all the available bandwidth for transmission (no sharing of bandwidth is necessary). The opposite of Baseband is Broadband. Broadband implements multiple channels, typically using frequency- or time-division multiplexing techniques. A typical example of a Broadband network is cable or satellite TV. Here bandwidth is shared.*

A list of key acronyms is provided in Table 1.1. One key acronym we run into all the time in PoE is PHY. In Ethernet, on one end of the cable we can have a digital-line driver, on the other end a digital receiver. In general, we could have digital transceivers (a combination of transmitter and receiver) at both ends. Generally speaking, these cable drivers/transceivers are referred to as PHYs, which literally stands for *phy*sical-layer drivers/transceivers. The physical layer in the case of Base-T Ethernet is simply the twisted-pair copper cable. In Base-F applications, the physical medium is the fiber-optic cable. In either case, the driver/transceiver is called the PHY.

What Is Interoperability?

Continuing Metcalfe's 2006 interview, he went on to say: "What the word Ethernet actually means today is six things... (1) It begins with a *de jure* standard made by a legitimate standards body, in this case the IEEE 802. (2) The implementations of that standard, painfully arrived at over years, are *owned by private companies*... (3) *Fierce competition* among the purveyors of the standard with their various implementations... (4) *Evolution* of the standard based on how things look after it

IEEE	Institute of Electrical and Electronics Engineers	EFMC	Ethernet in the First Mile over Copper
Cu	Copper	EFMF	Ethernet in the First Mile over Fiber
CO	Central Office	PoE	Power over Ethernet
LRE	Long-Range Ethernet (Cisco)	PoE+/PoEP	Power over Ethernet Plus
FTTH	Fiber to the Home	UPOE	Universal Power over Ethernet (Cisco)
FTTB/C	Fiber to the Building/Curb	PSE	Power-Sourcing Equipment
MDI	Medium-Dependent Interface	PD	Powered Device
PHY	Physical-Layer Device ("PHYceiver")		
PI	Physical-Interface (e.g., Cu)		
MII	Medium-Independent Interface		

TABLE 1.1 Key Acronyms to Keep in Mind

gets shipped, that is in the marketplace.... (5) Maximization of backward compatibility.... (6) an ethic in the competitive marketplace, *where it is not allowed to be incompatible.*

The last sentence points us to what we call interoperability today. The IEEE glossary defines this term as: *the ability of two or more systems or components to exchange information and to use the information that has been exchanged.* Wikipedia says it is the *ability of diverse systems and organizations to work together* (to "interoperate"). [Italics added by the author of this book].

For us, this basically means that equipment from Manufacturer A should "play well" with corresponding equipment from Manufacturer B, and also with Manufacturer C, and so on, because all this various equipment supposedly complies with the same governing standard. So provided the standard itself was carefully debated and formulated to start with, especially in terms of what is really crucial or important to overall performance, and hopefully is unambiguous to help reduce the possibility of mismatch (where it matters), then no interoperability issues should arise *in principle.* But the truth is there are lingering ambiguities in all standards. Also there are some subtle interpretation issues we need to consider very carefully. In this book we will attempt to show not only how to design good and reliable PoE equipment but also ensure they work and play well together. Hence the title of this book too.

Note *To put things in perspective, Metcalfe is also well-known for his prediction that the Internet would suffer a catastrophic collapse in 1996. He also promised to eat his words if it did not, and indeed he tried to when, in 1997, he took a printed copy of his column that had predicted the collapse, put it in a blender with some liquid and then consumed the pulpy mass. Metcalfe is also known for his harsh criticism of open-source software. In particular he had predicted that Linux would be finished after Microsoft released Windows 2000. He had said it was "utopian balderdash," and likened it to communism. He also predicted the end of wireless networking in the mid-1990s: "after the wireless mobile bubble bursts this year, we will get back to stringing fibers... bathrooms are still predominantly plumbed. For more or less the same reason, computers will stay wired." This is all available on Wikipedia.*

PART 2 THE HISTORICAL EVOLUTION OF PoE

Introduction

In Power over Ethernet (PoE), power and data are sent together down a standard Ethernet cable. The first formal PoE standard, IEEE 802.3af, was ratified in 2003, applicable to devices requiring up to 13 W. IEEE 802.3at followed in 2009, bringing into its fold higher-power devices, up to 25.5 W. The IEEE 802.3at standard actually contains two clear application categories. The first 13 W is as measured at the end of a 100-m cable, called Type-1, or "low-power." This was the same as in IEEE 802.3af. But it also introduced a new category for 25.5 W at the end of 100 m and called it Type-2 or "medium power." So, the "AT standard," as it is often colloquially called, is supposed to be just an "enhancement" of the previous AF standard, but it actually encompasses the previous standard and, in effect, *supersedes* it.

As we look back at the development of Ethernet in Part 1 of the chapter, and the advent of PoE, we can't but help feel all these events seem very recent, the underlying technology very modern. But as mentioned, the basic idea of sending information and power simultaneously over copper didn't even start with Metcalfe's 1973 memo, *it is actually almost two centuries old.*

It turns out that a surprising amount of ideas, tricks, and techniques that are in use today, not only in PoE, but in the general area of networking, can be traced back to a small group of incredibly resourceful engineers, scientists, innovators, and entrepreneurs, working against immense odds in what we perhaps consider a rather obscure moment in history. It is to this motley group that we owe many of our much-vaunted successes of today, and perhaps more to come. In contrast, the much-touted achievements of modern-day pioneers, many claiming a huge impact on mankind and society, pales into insignificance and borders self-promotion if not ignorance.

History can be not only entertaining and enlightening as a conversation topic over coffee, but very useful too. For example, not too long ago, two digital subscriber line (DSL) world-speed records were set in quick succession. DSL is a digital-transmission technology over existing telephone copper wiring, but it is different from Ethernet since raw data is not sent down the line; instead, *modulated* data is sent on a high-frequency carrier-sine wave (quite like a radio).

These DSL breakthroughs occurred just when the world seemed poised to conduct a perfunctory "let's get-it-over-with" funeral ceremony for DSL. The soothsayers were already starting to say: FTTH (optical *fiber to the home*), with speeds up to 100 megabits per second (Mbps), is the future, whereas DSL is a relic of the past. But all that changed suddenly over just a few months in 2010. In April, a DSL speed record of 300 Mbps was set (over 400 m of standard telephone wire) by the legendary Bell Labs (which is now part of Alcatel Lucent). That compared very well to the maximum prevailing DSL speeds of just around 10 Mbps typically (maximum 40 Mbps). Then in October of that year, equally unexpectedly, Nokia Siemens Networks announced a staggering 825 Mbps (over 400 m of telephone wire), bringing DSL close to the threshold of gigabit (1000 Mbps) over copper. In the process, something else also happened: copper had just become the "cockroach of telecom"—do what you like, you just can't make it go away.

Perceptive observers noticed something else in the twin DSL breakthrough announcements—a common underlying feature. Both companies had declared they had used something called *phantom DSL*. What exactly is that? Bell (Alcatel Lucent) elucidated further by admitting that they had exploited a 100-year-old networking trick. At first it seemed a little unusual to see such forthright candor and self-abnegation in our modern times. But it turns out they had every reason to be both candid and proud because that particular networking trick had also originated from an ex-Bell employee: named John Joseph ("J. J.") Carty, way back in 1886. In fact, we will soon learn that PoE is also based on the same phantom circuit principle. It is interesting to realize that this is really does make the very basis of PoE an offshoot of J. J. Carty's mind from way back in the 19th century.

We start to discover that nothing is as completely modern as we were hitherto inclined to believe. Also, both networking and PoE share a common heritage. Knowing that fact, we can hardly argue that a deeper knowledge of "past tricks" won't serve us well going into the future. That is why we too have chosen to take the historical path toward explaining PoE in this book.

Blasts from the Past

To many of us today, the 19th century swirls with names we've never heard of, and perhaps don't care to either. Emile Baudot, Claude Chappe, Cyrus West Field, William Thomson, John Joseph McCarty,

Oliver Heaviside, and so on, to name a few. Wait a minute: Doesn't the term "baud rate" sound very similar to the first name? Indeed, Baud rate did come from Baudot's work. Similarly, "modern" transmission line equations came from William Thomson and Oliver Heaviside over the period 1855 to 1885. William Thomson and Cyrus West Field were the pioneers behind the early transatlantic cables. Modern transmission-line equations were a direct result of their efforts to understand long-distance propagation of (telegraph) signals across these new "submarine" (underwater) cables. Incidentally, Thomson is also responsible for our "modern" temperature scale because of his discovery of absolute zero in 1848. And much more, in fact. It is therefore indeed surprising that most of us don't even have an inkling who Thomson was! But perhaps this rings a bell: William Thomson was subsequently knighted and became Sir William Thomson. A little later he took on the title *Lord Kelvin*. And *that* we may have heard of!

Not to forget Thomas Edison (1847–1931), with 1093 U.S. patents under his belt, considered the fourth-highest inventor in history. Especially in Edison's case, it was never a case of quantity over quality, or claiming innumerable "inventions" just to rake in the money from "incentive" corporate restricted stock units (RSUs). Incentive to cheat? Edison has to his credit the incandescent electric bulb, the phonograph, a motion picture camera, the first public, power-generation company, the electrical stock ticker, a quadruplex telegraph, and so on. We should, however, not forget that the key to Edison's fortunes was actually *telegraphy*. He learned the basics of electricity during years of working as a telegraph operator, and later he applied that knowledge to the telephone too. For example, the famed carbon transmitter (telephone mouthpiece) found in telephones, even until a few decades ago in many parts of the world, came from Edison. This author too remembers taking apart a standard Delhi City landline phone in mid-1970s to study its carbon microphone, complete with tightly packed carbon granules and all. That was Edison immortalized.

Engineers in that bygone era achieved a lot with almost nothing in their hands. Certainly they had no Internet to scour for information, much less to communicate with—no Wikipedia, Google, Facetime, Skype, Twitter, E-mails, IMs, nothing at all… horrifying as it may seem. They probably had to undertake long journeys by horse-drawn carriages or small boats just to arrive at the door of some eccentric, perhaps even suspicious, visionary or financier, hoping to generate fleeting interest in working together toward some vaguely defined mutual advantage. But these were still only relatively minor communication issues compared to the fact that both their hands were, technically speaking, tied firmly behind their backs. Think about it: What were their available resources at the time? In the 19th century, electricity had barely been harnessed, much less fully understood. Ohm's law arrived in 1827, Kirchhoff's circuit laws in 1845. There were no vacuum tubes lying on rough-hewn work-tables, certainly no semiconductors, let alone 40-nm (nanometer) monolithic integrated circuits (ICs). There was barely an incandescent lamp in sight: Even the carbon-filament lamp (from Edison) came in 1879, the vacuum tube much later in 1906. Plastics had yet to be invented: The first plastic from a synthetic polymer was Bakelite, in 1907. Centralized electricity generation and distribution had just gotten off the ground—in 1881. What could modern-day greats like Henry Samueli (founder of an organization that proudly claims it is "connecting everything") have achieved under these circumstances? The truth is: probably zilch. No wonder we too instinctively start to think what could these poor 19th-century guys have done other than bow their heads and pray?

To our complete astonishment, on August 16, 1858, 18 years before even the invention of the most basic telephone, the first transcontinental message was being sent, not by horses or ships, but *telegraphically*, using electricity coursing through 2500 miles of copper lying deep under the Atlantic ocean. That momentous event, akin to landing a man on the moon in its time, set the stage for scenes of unprecedented jubilation and rejoicing across America and Europe. Alas, all for just a fleeting moment in time because this brand-new, very-first, transatlantic submarine cable failed in barely a month, but for reasons that can hardly be considered related to any fundamental design infirmity, technical oversight, or even ignorance. At least not ignorance on *both* sides of the cable. Many historians have concluded that the untimely demise of the cable was the handiwork of one person: Doctor Edward Wildman Whitehouse. A medical doctor by profession, he was the assigned engineer at one end of the long cable, with a theory of electrical propagation that can be best summarized in a few words (his) as follows: "the further that electricity has to travel, the larger the kick it needs to send it on its way." Banking on this little tidbit of knowledge, reportedly impervious to others around him, Whitehouse started zapping the cable using induction coils, with voltages of up to 2000 V—about four times larger than the cable was meant to carry. Thousands of miles away, stationed on the other side of the cable, *not* linked by a 3G network, Skype, telephone, or even a *telegraph* (the latter is what they were trying to barely get working at this point), Lord Kelvin reportedly had a chance to get through to Whitehouse, literally and figuratively—right untill the moment Whitehouse seems to have concluded with an impressive demonstration of a phenomenon we call "dielectric breakdown" today. Admittedly, there is no smoking-gun evidence in the form of a viral YouTube video, but Whitehouse is largely believed to have been the one to firmly kick the month-old transatlantic cable into the annals of history (or whatever constituted history way back then). However, to be fair to him, the person with the ultimate responsibility for the debacle was arguably Cyrus West Field, the famed entrepreneur, who recruited Whitehouse in the first place. But as often happens today, Whitehouse was the (only) one who got fired. Cyrus got himself a second and third chance to succeed. And he did, rather spectacularly.

After a few years' delay on account of the Civil War, completely undaunted and undeterred by the previous cable failure, and despite having been thoroughly ostracized by neighbors and generally labeled a charlatan across the globe, Cyrus West Field, working with Thomson again (not Whitehouse this time), succeeded in laying not one, but *two* brand-new transatlantic cables, in 1866—in a procedure that is mind-boggling to read about even today. These new cables served their purpose for around a decade thereafter, and in doing so, they spurred a revolution in lifestyle that had hitherto never been seen before. That was truly a societal change. Wikipedia has a page dedicated to a 1998 book called *The Victorian Internet: The Remarkable Story of the Telegraph and the Nineteenth Century's On-Line Pioneers* by Tom Standage. The book reveals some of the astonishing similarities in the rise of the 19th-century telegraph and the rise of the Internet in the late-20th century. The central idea of the book, Wikipedia points out, is that of these two technologies, it is the telegraph that is the more significant, because the ability to communicate globally in real-time was a *qualitative* shift at the time, while the change brought on by modern Internet is merely a *quantitative* shift. Roll over Samueli.

Whether we agree with that viewpoint or not, a historical perspective invariably creates a very interesting entry point into the heart of what we instinctively consider to be modern technologies.

Don't SWER No More

A very, very long time ago, telegraph systems were based on just *one* copper wire laid down over several miles. Metal poles buried in the ground on both sides completed the return path of the current (through moist subsoil, water, sea, or even ocean). This is called for *single-wire earth return* (SWER). This single-conductor principle was used extensively in power distribution systems even later, and is still considered an effective and economical choice for rural electrification in remote and backward locations. The same single-conductor principle is also often used today for modern light-rail systems, remote water pumps, and so on.

Unfortunately, completing a return path through (earth) ground creates a current loop with a huge arbitrary, almost *undefined*, and possibly varying enclosed area. This makes the entire system susceptible to picking up extraneous disturbance and noise (of the electromagnetic variety)—it is a big antenna, courtesy of Ampere's circuital law combining forces with Lenz's/Faraday's law of induction. In modern parlance, we always need to ensure our systems have adequate *electromagnetic immunity*.

The terms *immunity* and *susceptibility* are used equivalently to describe the ability of equipment to function acceptably in a typical electromagnetic environment. But why did the vague boundaries of SWER *not* pose much of a problem with the telegraph? Because telegraphy is essentially *digital* in nature! Yes, digital was there long before analog. The dots and dashes can be thought of as a string of ones and zeros. We know all too well today that digital systems are inherently more noise-resistant than analog systems—same as in the 19th century. Therefore, telegraph systems worked quite well within the rather vague physical boundaries of SWER. Unfortunately, the inadequacies of this single-ended architecture were thoroughly exposed when analog (voice) signals were attempted to be transmitted over the existing telegraph-wiring infrastructure following the invention of the telephone in 1876. Plagued with strange noises, the solution emerged in quite quickly too, in the form of the return copper wire, proposed in 1881 by the very same J. J. Carty mentioned previously. And that same year, Alexander Graham Bell, the man behind the telephone (or the talking telegraph as it was initially called), filed a patent for the *twisted* return wire—which is basically what we call *unshielded twisted pair* (UTP) today. From Part 1 of this chapter we remember that is what drove Ethernet into all-time popularity. In one word: *cost*. UTP still happens to be the most cost-effective, most common type of Ethernet (and telephone) cable in use today.

Some may argue that Ethernet is *digital* too, so why can't we still use SWER? There are several reasons for that:

1. In our modern world of ever-decreasing voltage sources, we now have digital thresholds that are very close together compared to the higher-telegraph voltages, so noise immunity is not so good either.
2. At the high data-transfer speeds we are talking about, we can no longer afford too many errors caused by noise.
3. By using the ground for high-frequency data, we would create a huge amount of electromagnetic radiation that would impact neighboring (sensitive) equipment. This is discussed in Chap. 2.

Historically, compared to SWER, the return-wire concept (metallic circuit), when proposed, seemed to imply *double the cost* of copper, so the idea was obviously met with some high initial (human) resistance. But it was also quickly apparent that a copper return wire was simply unavoidable for ensuring acceptable performance in telephony. However, there was another major breakthrough toward the end of the 19th century, in the form of Pupin (loading) coils that helped greatly. These are coils of very large inductance inserted every few thousand feet (typically 88 mH every 6000 feet) over the entire length of a long telephone cable. The discrete inductors couple electrically with the existing distributed cable capacitance, creating LC-type transmission-line effects, similar to what we rely on in modern high-frequency data transmissions, but now effective at very low (audio) frequencies. This "pupinization" of telephone wiring, as it was called, allowed voice frequencies to travel much greater distances—an alternative to blindly increasing the thickness of copper just to lower the DC resistance for achieving comparable propagation distances. Pupinization is said to have saved up to 75 percent of the projected copper costs associated with telephone cabling.

All this is just a fascinating example of the prolific ideas swirling and accruing rapidly in the 19th century. Riding on such clever breakthroughs, by the close of the century, there was an almost complete conversion from grounded circuits (SWER) to metallic circuits (those with copper returns). And with that, the *twisted pair* rose to supremacy. As indicated previously, not only does Ethernet uses it today—so does DSL.

The Twisted Pair and the Principle of Immunity

The basic principle behind the twisted pair and its various implementations is shown in Fig. 1.5. At the very top of the figure is an analog signal being transmitted from the microphone of a traditional telephone to the loudspeaker on the other side.

NOTE *We are ignoring another clever technique here for the time being, by which we combine, and then later separate, the loudspeaker signal from the microphone (transmitter) signal, over a single twisted pair—this involves an innovation called the "hybrid transformer." It is discussed in more detail in Chap. 13.*

In the cases that follow in Fig. 1.5, the signals are digital, but the underlying principle is the same. We see the noise spikes (small triangles) riding *equally* on both constituent wires of the twisted pair (same amplitude and same direction/polarity). The rationale behind that is that since the wires are twisted uniformly, they get exposed *equally* to disturbances—without any preference to either wire. Otherwise, we could well ask why the noise pickup is *different* on one of the two wires if nothing distinguishes one from the other. In other words, just by plain symmetry, the two wires of the twisted pair must have identical noise pickups. In modern terminology, the disturbance/pickup is called *common-mode*. We can, however, ask in a common-mode case: Is the noise voltage identical on both wires *relative* to what exact potential? The answer to that is the noise spikes are identical with respect to (earth) ground. And that is what we mean by

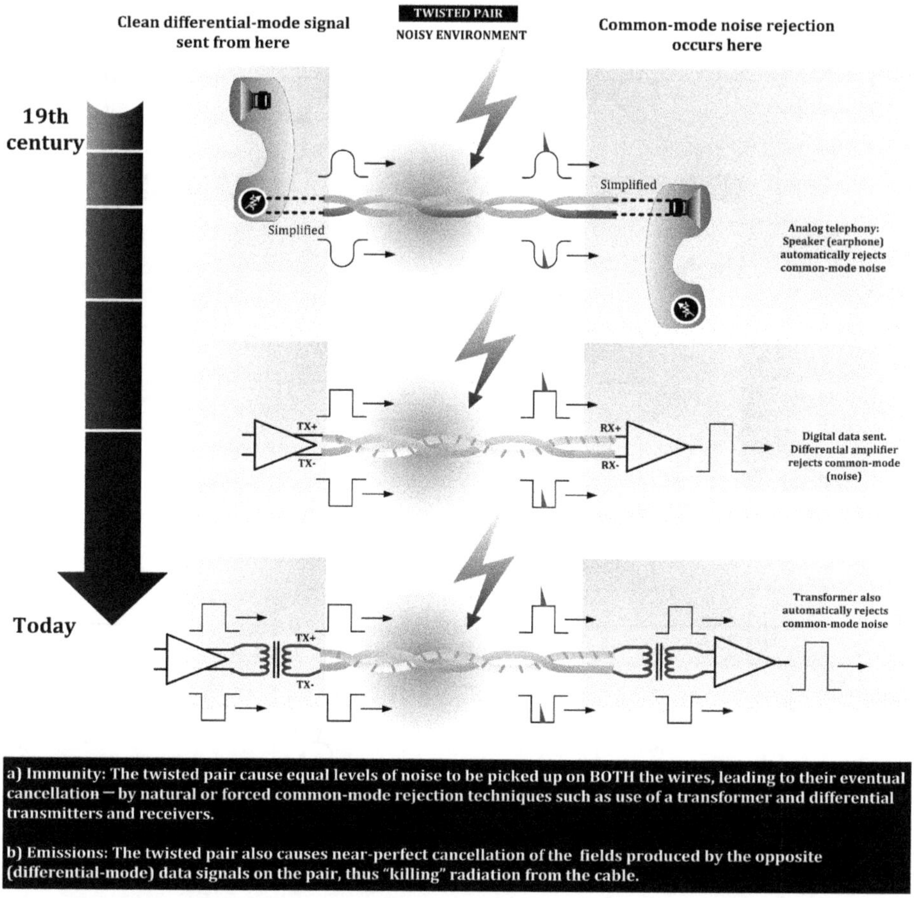

FIGURE 1.5 How the twisted pair, along with a differential amplifier and a transformer, helps reduce electromagnetic noise pickup.

common-mode. We can alternatively say that there is no voltage difference between the two wires because of noise spikes, and so the noise pickup is *not differential-mode*.

Differential-mode implies the opposite of common-mode: At any instant, common-mode consists of two *equal* signals with the *same* polarity, whereas differential-mode consists of two *equal* signals of *opposite* polarity, and in both cases the referred-to voltages are with reference to earth-ground potential. In Fig. 1.6 we see more clearly what exactly are differential-mode (DM) and common-mode (CM) currents. We have marked the noise source generically as N inside a circle. We also see how in a general case of mixed-mode (MM) currents, we can split the currents into their DM and CM components. Keep in mind the sign logic we are using in the numerical

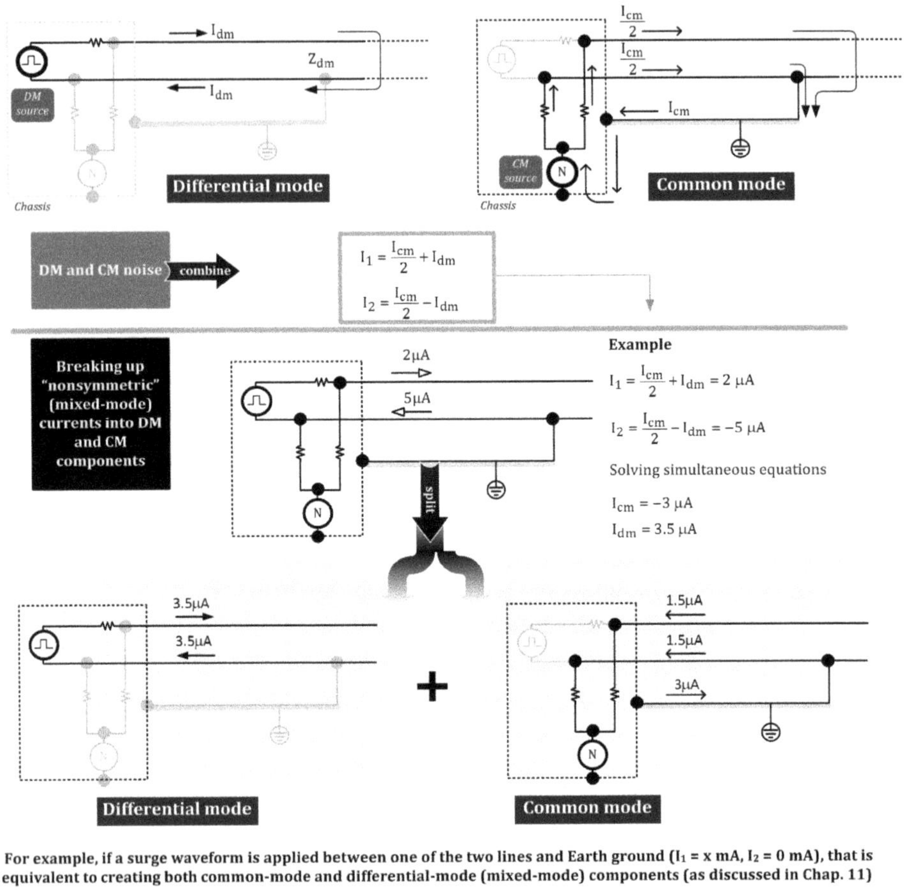

FIGURE 1.6 DM and CM currents (top), and splitting mixed-mode (MM) currents into constituent DM and CM components.

example: Any current from left to right is positive, and from right to left is negative.

Noise pickup is common-mode (with a negligible differential-mode component), only provided the wires are twisted tightly together. If not, we can certainly get unintentional asymmetry, which will lead to a small unintentional differential-mode noise component (as indicated by the numerical example in Fig. 1.6). But why is that so scary anyway? The problem with that scenario is the actual signal is (by design) that transmitted down the wire in a *differential* fashion (explained further below). If the noise has a differential component, it will end up interfering with the actual (useful) signal. In that case, we could well ask how any circuit would "know" what constitutes signal and what constitutes noise? In a good setup, signal and noise are distinguishable (and separable) only because one is purely differential-mode (the signal), while the other is purely common-mode (the noise). In other words, noise and signal are made to reside in separate and distinguishable domains using special techniques. Then, with appropriate circuitry, we accept one of them (signal) and reject the other (noise).

Common-Mode Rejection by Coils/Transformers and Other Techniques

Besides the most basic requirement of a twisted pair, what special techniques were we discussing previously ? Let's list some of them here.

1. We start by revealing the simplest technique to separate signal from noise, applicable to both analog (telephony) and digital (Ethernet). Visualize the following situation: If both ends of a magnetic coil, such as in the loudspeaker of a telephone (the first schematic in Fig. 1.5), or one of the windings of say, a data transformer (the last schematic in Fig. 1.5), are raised or lowered in unison by *exactly the same amount* (and that by definition is common-mode), no corresponding current will be produced in the coil. Why? Because current only flows if there is a voltage *difference* (delta) present, and in this case we have no voltage difference *across* the ends of the coil (the voltages at its two ends are changing in unison by equal amounts and with the same polarity). In other words, only a *differential*-mode signal applied across the ends of a coil/winding will produce a delta V with a resultant current flow. Common-mode will do nothing here. That is common-mode rejection, by definition.

   Alternatively, if the noise picked up is purely common-mode, as is true for the twisted pair in Fig. 1.5, the loudspeaker will not emit any sound corresponding to the noise. Only the voice signal, which is applied differentially across the twisted pair by the microphone on the other side of the cable, will be heard through the loudspeaker.

   A very similar situation arises in a transformer. If no current flows through a winding on one side, no voltage or current can appear on its other side—because transformer action requires that a (time-varying) current flow through one winding, creating an induced voltage across the isolation barrier on the other winding. In other words, both coils and transformers have inherent common-mode rejection properties. This property is commonly used in Ethernet today, as it was in analog telephony over a hundred years ago. In telephony, voice-frequency transformers were typically placed at certain key positions, such as in telephone exchanges.

They were called repeating coils at the time because their main application was to inductively transfer (or repeat) the signal from one telephone circuit (branch) into another. But the actual signal came through clearly to the other side *minus* (a good deal of) noise. So they were also used for common-mode noise rejection. Some people express this property of a transformer in a slightly different manner by saying "repeating coils (isolation transformers) break up ground (earth) loops," (and that leads to cleaner signal transmissions, with no funny, buzzing sounds in telephony, and so on).

2. Ground loops are nothing but a path for common-mode noise/signals to flow. So breaking up a ground loop, however we do it, is tantamount to enforcing common-mode rejection. One way to do that is by using data transformers as explained previously. But it is also obvious we should *avoid making any direct galvanic connection* to earth (ground) on the line (cable) side. We will learn that PoE stages, which are always located on the line side, are for that reason *never* connected directly to earth (ground). Blocking capacitors of very small capacitance, called Y-capacitors, are the only link from PoE (line) side to earth (ground) (some amount of capacitance to ground is deemed necessary for overall EMI-suppression purposes). On the other side of the data transformer (the driver side), there is however almost invariably a direct physical connection to earth (ground) (via the AC wiring) for safety reasons (to protect the user from electric shocks). We will discuss the safety and isolation aspects in greater detail in Chap. 10.

3. Let's briefly summarize here: SWER systems *depend* on ground loops to work. We tried to eliminate that ground loop by the use of the twisted pair (metallic circuit). The data transformer helped further in that mission because it enforced a break in any inadvertent ground loop. In addition, we learned that we should avoid physical connections to earth (ground) and instead use Y-capacitors to connect anything on the line (cable) side to earth, (ground). In other words, we need to isolate the line and line-side circuitry from earth (ground).

   On top of this, capacitor injection is another technique that can be used to inject a signal into the twisted pair (instead of using drive transformers). In that position, the capacitors will block DC and thereby help break up any ground loops that may form through the line driver side. Unfortunately, a capacitor is a high-frequency bypass, because the impedance of a capacitor is $1/(2\,\pi f \times C)$ and if f and/or C is large, the impedance is very low. In other words, capacitors do block DC, and break up DC-ground loops, but they also permit high-frequency or AC-ground loops to continue to exist. In contrast, a drive transformer is much better at common-mode rejection and breaking up of ground loops. So a drive transformer is the preferred choice in Ethernet. Capacitor coupling is not as effective.

Note that, theoretically, the twisted pair can be directly driven as shown in the middle of Fig. 1.5. But it is now obvious that is not a good idea for breaking up ground loops, so that is certainly one reason it is never used. But we should note that there is another subtle reason to avoid direct drive too: Under a fault condition, a direct-drive line driver (transceiver) can be easily damaged. Capacitor coupling or transformer coupling, on the other hand, are relatively fail-safe since they both end up blocking any DC, which will likely result in the case of most common types of fault conditions.

4. All these techniques—twisted pairs, data transformers/coils, and so on—are part of our growing war-chest of tricks for separating noise and signal, thereby ensuring "signal integrity" over long distances. What other techniques can we use? As indicated in the schematics of Fig. 1.5, the most obvious way of rejecting common-mode noise is to use *differential* stages, both for transmitting the signal at one end (differential driver), and for receiving it at the other (differential amplifier).

However, just so we do not lose track of the bigger picture here, we need to emphasize once again that in all the schematics of Fig. 1.5, actually depend on noise being picked up identically (common-mode) on both wires. That is the key advantage of a twisted pair. So, the twisted pair is a basic requirement. Combined with the repertoire of related techniques, such as described previously, we then continue to restrict noise to the common-mode domain and the (useful) signal to the differential-mode domain. Eventually, that is what makes noise and signal distinguishable, ultimately filterable and separable.

Immunity and Emissions

We have been mentioning "immunity" in previous sections without having spelled it out very clearly so far. We also referred to "a typical electromagnetic environment." What do these terms mean and relate to?

Electromagnetic immunity/susceptibility (EMS) is one side of the total coin called *electromagnetic compatibility* (EMC). See the left side of Fig. 1.7. On the other side of the EMC coin lies electromagnetic emission (EME). For example, an intentional/unintentional electromagnetic emitter (Device A) sends out electromagnetic interference (EMI)

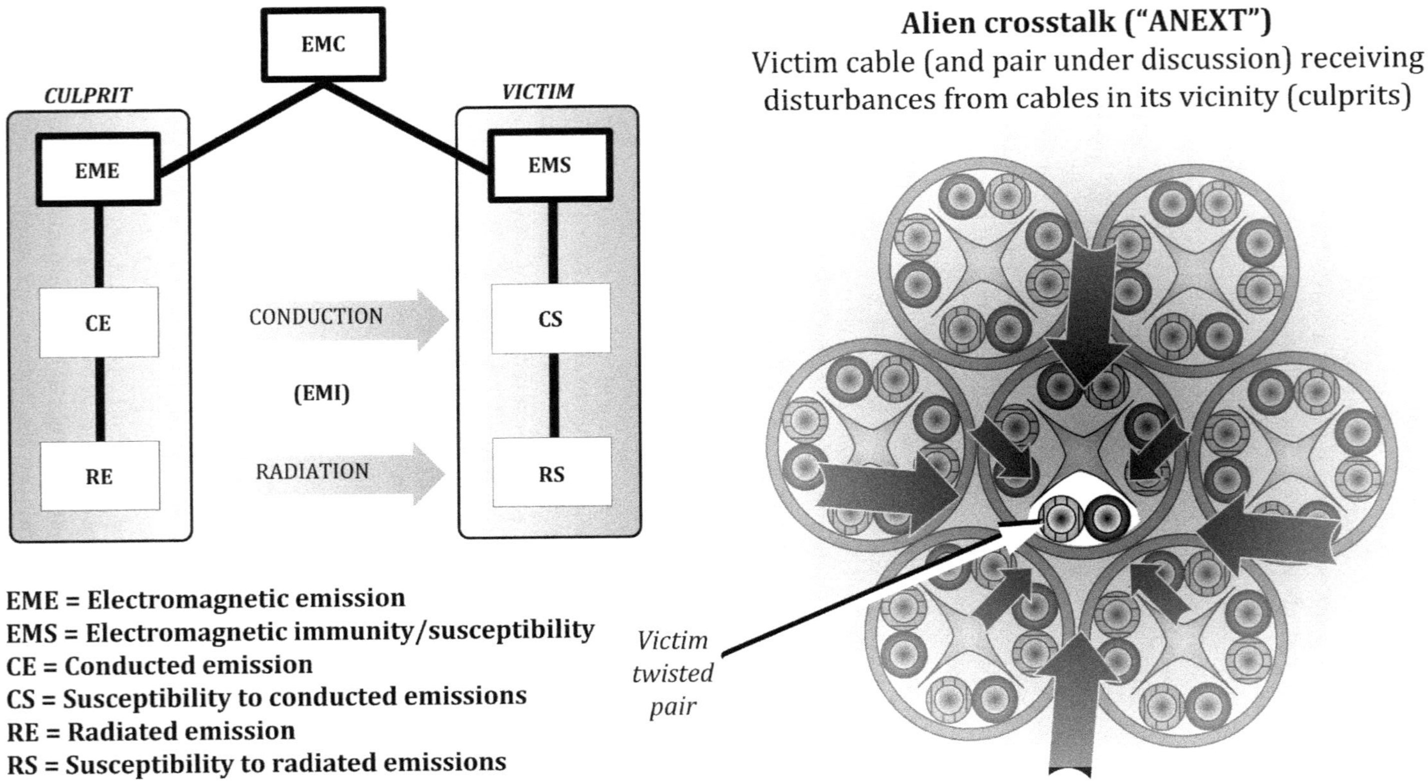

FIGURE 1.7 Concept of EMC and cross talk in cables.

all around it. Another device (Device B) in its immediate vicinity should not only continue to work well when faced with this impinging EMI (provided, of course, the levels of that are not excessive), but must itself not emit significant amounts of EMI, so as to allow other devices in its vicinity, such as Device A, to function acceptably too. These are just basic good-neighbor principles at work—within the EM environment.

To clearly define and regulate both aspects of EMC, there are well-known European Norms (EN) in Europe and Federal Communication Commission (FCC) standards in United States, but detailed discussions about regulatory EMC regulations are out of our scope here because that hardly concerns PoE, which can be considered largely *DC-based* (and we know DC does not radiate). The important thing to remember for our humble purpose here is that, in general, a good RF antenna is not only a good *receiver* of EMI, but a good *transmitter* too. Similarly, a bad receiver of EMI is a bad emitter of EMI too. So, for example, we know that a long, single-conductor wire is a good antenna, both for transmission and reception. However, perhaps rather nonintuitively to us at first, a long twisted-pair cable is a relatively bad antenna, or at least not as good an antenna as we may have expected, based on its good length. But at least one thing remains true and consistent (though it is almost coincidental in this case): The twisted pair cable, as used in an Ethernet environment, happens to be a bad antenna for both reception and transmission (of EMI). We will explain the reasons for all this below.

When used in Ethernet applications, the twisted pair not only rejects incoming EMI (in effect, it provides system immunity), but does not radiate too much itself (so it doesn't test the immunity of neighboring devices too severely either). The reason for the low emissions comes from the fact that the magnetic fields produced by each wire of the twisted pair, when driven with purely *differential* signaling, are in opposite directions with equal magnitudes at any given instant. So they mutually cancel each other out—there is no *net* (resultant) field, at least not in theory. But yes, if the differential-mode signal has an inadvertent *common-mode* component to it, not due to noise this time, but from design-related issues, (such as inherent imperfections in the differential nonideal driver), the cable will end up radiating somewhat. Similarly, if the noise pickup is purely common-mode, it does get rejected very well as discussed previously, and that gives us immunity. But if the noise has a small differential-mode component (e.g., caused by poor twisting in a certain area of the cable, such as at a very sharp bend), the system *will* see some noise getting mixed in with the signal, and that will, in effect, lower, the overall system immunity. In other words, *signal integrity* will be compromised.

On the right side of Fig. 1.7, we see at the cross section of four twisted pairs of a typical Ethernet cable (viewed from the top). With four twisted pairs in every cable, and several cables in a bundle (as the cables go out from their hub/switch to the workstations), there can be significant "pickup" via radiation between *adjacent* twisted pairs. This is called cross talk. In effect, it degrades signal integrity, affects data transmission capability, and eventually reduces its reach (distance). So for all the reasons described above, the twisted pair will help significantly reduce cross talk too, since interference from adjacent pairs, in the same cable or from adjoining cables, is basically noise pickup from the viewpoint of any given twisted pair under study.

Note *When the disturbance is from pairs in surrounding cables, the word for that is alien cross talk (ANEXT or AXT). We also have near-end cross talk (NEXT), and far-end cross talk (FEXT), which refer to the cross talk from the other three pairs of the same cable as the victim. NEXT occurs when a receiver overhears a signal being sent by a transmitter positioned at the same end of the cable as the receiver, whereas FEXT occurs when the overhead transmitter is located at the opposite end of the cable, away from the receiver.*

Note *In Fig. 1.7, the victim cable has been shown with exactly six disturber cables surrounding it. We realize though that in a typical Ethernet cable bundle there may be many more cables surrounding any given cable. However, as we can see, by sheer geometry, six cables will completely surround a given cable with no intervening gaps, so they effectively shield the victim from the effects of outer cables. Therefore, for studying ANEXT, the standard setup is as displayed—with exactly six cables surrounding the cable containing the victim pair. This is often called the 6-around-1 configuration.*

We thus see that the magical unshielded twisted-pair cable (UTP) helps achieve *both* immunity and low EMI (with the help of all the supporting techniques as explained previously). It turns out that UTP is not only low-cost, but high-performance too. It's like a free lunch, in effect. And that's why it contributed significantly to the explosion of Ethernet, compared to rival LAN proposals. Note that there is no need for any separate external shielding either. A shield may only complicate matters by providing an alternative and ambiguous return path for the CM and DM currents. So it is no surprise that both the coaxial cable and the shielded twisted-pair cable (STP) are almost dead (for this purpose). In contrast, UTP abounds all around us.

Twist Rate and Wire Diameter

An extremely important characteristic of a twisted pair is related to the basic question: How twisted is it? Cable categories have been defined in the standards, and these eventually relate directly to a certain twist rate, or twists per unit distance (distance is measured in inches or feet, for example). We do not need to go into too much networking detail here, but it is good to keep in mind that the greater the twisting, the better the performance of the cable in general.

For PoE, the twist rate is not of any *direct* concern, except perhaps in some esoteric system-design matters as discussed later. Because, in PoE, we are essentially concerned only with the DC resistance of the cables, not its reactive parasitic elements, like inductance and capacitance per unit length. However, in what may be considered a fortuitous coincidence, "good" cables from the viewpoint of *data,* are usually good for PoE too. Not because of the higher twist rate (rather, *despite it,* as discussed below), but because "good" cables are typically made of *thicker* wire. Greater wire thicknesses imply not only lower-DC resistance, but lower-AC resistance too. That helps both data and PoE. A thicker wire increases the useful signal received at the end of a 100–m cable, because it reduces Insertion Loss (reduces attenuation). For almost the same reason, in the case of PoE, a thicker wire allows for more power to be delivered at the end of the cable, because we have lower I^2R losses in the cable. In other words, in most cases, power and data capabilities of cables seem to go hand in hand: They end up dovetailing, much to our design satisfaction and ease.

We may notice, while playing around in the lab, that some of the twisted pairs of a typical Ethernet cable are easier to unravel. There is a good reason for that. If adjacent pairs have an exactly identical twist rate (pitch), we could end up with a situation in which wires of different pairs fall coincidentally almost adjacent to each other for the *entire* cable run, affecting differential signaling negatively, and increasing cross talk. To prevent this, Ethernet cable manufacturers use different twist rates for different pairs in a good cable, though all this is not usually declared or apparent to the user.

Unfortunately, a high-twist rate leads to a *longer* (unraveled) length of copper wire. This not only increases the AC and DC resistance somewhat, but also increases the propagation delay, which is the time taken by the signal to travel across the cable. Luckily, propagation delay by itself is usually of less concern than the *differences* in the propagation delays of adjacent pairs of a cable, which is called delay skew. By using differing twist rates on different pairs of a cable to reduce cross talk, we end up with larger delay skews. And that can become of serious concern, especially in high-definition video applications and/or very high-speed data transmissions. But twisting in general, despite this relatively minor disadvantage, has overwhelming advantages.

Categories of Ethernet Cable

In principle, we can implement Ethernet technology not only over unshielded twisted pair (UTP), but also shielded twisted pair (STP), coaxial wire, or even optical fiber. However, in this book we are going to focus only on the ubiquitous UTP, since for most applications, UTP happens to be the most cost-effective, popular, and prevalent choice. In this section we will list the key cable categories that make them either suitable or unsuitable for Ethernet applications. We have also summarized key applications versus cable categories in Table 1.2.

The Telecommunications Industry Association (TIA) categorizes cables depending on their data transmission capabilities (over 100 m of cable). The TIA is largely North American. In Europe, the corresponding standard is from the International Standards Organization (ISO). In Europe, the cable categories/components are called by different names, as we can see in Table 1.2. But they are U.S.-equivalent categories. Keep in mind that in Europe, telephone/Ethernet/AC color coding can all be quite different from the United States too.

The older U.S. Ethernet cable standard was TIA-568A, the more modern one is TIA-568B. These standards are the origin of the prefix "CAT" or "Category" that we will find on a typical (North American) Ethernet cable, expressing its rating and capability. The most common cable category in use until a few years ago was Category 3 (CAT3), which is considered good for 10Base-T Ethernet, or basic telephony applications. For 100Base-TX, the most common cable around is Category 5e (or CAT5e, in which "e" stands for *enhanced*).

What are the associated wire diameters? We note that TIA-cable standards are inherently pre-PoE, or datacentric. The good news is that since data and power capabilities do seem to dovetail, we can deduce the worst-case wire thickness (AWG) required for PoE power calculations. It eventually leads to

1. CAT3 (Class B) (16 MHz/16 Mbps): Typically AWG26 (worst-case) to AWG24. Used primarily for 10Base-T. The first PoE standard, IEEE 802.3af-2003, was written with this category in mind.
2. CAT5 (100 MHz): Typically AWG26 to AWG22. Rare or obsolete. Ignore.
3. CAT5e (Class D) (guaranteed 100MHz; typically up to 350 MHz): Typically AWG24 (worst-case) to AWG22. "e" stands for enhanced, which implies a higher twist rate and lower cross talk than CAT5. Used primarily for 100Base-TX, but can usually also support 1000 Mbps over 100 m by using all four pairs. The second (most recent) PoE standard, IEEE 802.3at-2009, was written with this category in mind.
4. CAT6 (Class E) (250 MHz): Typically AWG23 (worst-case) to AWG22. It is rarely used, since it was intended for 1000Base-TX, which is dead as discussed earlier. It also falls short of supporting 10 G (10,000 Mbps applications) over the full 100 m as required.

TIA and ISO EQUIVALENTS								
Frequency Bandwidth (MHz)	TIA Components	TIA Cabling	ISO Components	ISO Cabling				
16	Cat 3	Cat 3	Cat 3	Class B				
100	Cat 5	Cat 5	N.A.	N.A.				
100 (Typ 350)	Cat 5e	Cat 5e	Cat 5e	Class D				
250	Cat 6	Cat 6	Cat 6	Class E				
500	Cat 6A	Cat 6A	Cat 6A	Class E_A				
600	N.A.	N.A.	Cat 7	Class F				
1000	N.A.	N.A.	Cat 7_A	Class F_A				
APPLICATION CHART								
	Direction	**Cat 3**	**Cat 5**	**Cat 5e/ Class D**	**Cat 6 / Class E**	**Cat 6A / Class E_A**	**Class F**	**Class F_A**
Telephony (separate analog signals on each pair)	↔ ↔ ↔ ↔	•	•	•	•	•	•	•
10Base-T	→ ← — —	•	•	•	•	•	•	•

Table 1.2 Summary of Cable Categories and Applications

	Direction	Cat 3	Cat 5	Cat 5e/ Class D	Cat 6 / Class E	Cat 6A / Class EA	Class F	Class FA
100Base-T4	→ ← ↔ ↔	•	•	•	•	•	•	•
100Base-TX	→ ← — —		•	•	•	•	•	•
1000Base-T	↔ ↔ ↔ ↔			•	•	•	•	•
1000Base-TX	→ ← → ←				•	•	•	•
10GBase-T	→ ← → ←					•	•	•
Broadband CATV (with Ethernet on same cable)	→ — — —						•	•

TABLE 1.2 Summary of Cable Categories and Applications (*Continued*)

5. CAT6A (Class E_A) (500 MHz): Typically AWG23 (worst-case) to AWG22. This is a future specification, intended for 10 G applications. It is becoming increasingly popular in an attempt to "future-proof" new installations.

Why is CAT5/CAT5e so much better than CAT3 anyway? The minimum wire gauge is better for one. We can also intuitively understand that another key reason is the twist rate. Typically, CAT3 has three twists per foot, whereas CAT5/5e has about 2 to 3 twists per inch (10 to 12 times more than CAT3). In CAT5e has lower cross talk than CAT5. One way to reduce cross talk significantly is to use dissimilar twist rates in the pairs of a given cable.

PoE Cable Categories

From a PoE perspective we need to remember this:

1. IEEE 802.3af-2003 assumes a worst-case of AWG26 (CAT3).
2. IEEE 802.3at-2009 assumes a worst-case of AWG24 for higher-power applications (CAT5e).

(Keep in mind that AWG24 is thicker than AWG26.)

In terms of resistances:

1. IEEE 802.3af assumes that a 100 m CAT3 cable that has a worst-case (DC) loop resistance of 20 Ω. This is the cable resistance assumed for low-power applications (13 W at the end of the cable).
2. IEEE 802.3at assumes that a 100 m CAT5e cable that has a worst-case loop resistance of 12.5 Ω (for Type 2 medium power applications). This is the cable resistance assumed for medium-power (Type 2) applications (25.5 W at the end of the cable).

We will do some calculations later, based on the resistivity of copper. At this point the above information is enough, but we may also want to keep in mind that these resistance numbers are actually for 90 m of Ethernet cable plus a total length of 10 m of patch cables at either end. It also includes estimated contact resistances of connectors on both sides. Temperature variations are also included in these resistance numbers.

Bandwidth and Information Capacity of Cables

We may have noticed from Table 1.2 that CAT5e can support 1000Base-T (1 Gbps), even though it is only rated 100 MHz. We realize that we are using all four pairs of the 100 MHz cable for doing 1 Gbps, but we still can't seem to explain this rather big jump to 1 Gbps. There seems to be no obvious math here. And does that mean 100 MHz is really *not* equivalent to 100 Mbps as often assumed? Yes, *there is really no obvious relationship between bandwidth and maximum data rates.*

Historically, especially when used for RF purposes, the usable bandwidth (maximum frequency range) of a cable was supposedly related to the relative attenuation of different sine-wave frequencies as they passed through the cable. For example, we have for years used coaxial cable (RG-6) for cable TV (CATV), in which many stations are carried simultaneously up to very high frequencies (~1000 MHz). Also, the length of the cable really does not seem to profoundly affect its frequency characteristics; the length actually seems related more to the attenuation of the entire signal over very long cable lengths, and the sensitivity/design of the RF preamplifier/receiver to extract a "clean" signal from the noise. Yes, we do know today that the diameter of the cable is a key factor in determining its frequency characteristics (cutoff frequency).

What Metcalfe proposed in 1973 was a very different application of coaxial cable. First, it was now being used *not* for analog sine waves but for *digital* signals, with sharp "edges" containing a lot of high-frequency harmonic content. Second, it was being *shared* for data. So the final point-to-point data rates would be affected by the number of computers hanging off the bus. Clearly the concept of bandwidth and "information capacity" was evolving and developing.

Let us fast-forward to modern times where we have a star topology (no shared bus), and we are using twisted pairs, not coaxial, because that is what is relevant to us today. One of the most important and basic parameters that defines the final performance of telecommunications cabling is its *channel bandwidth*. This is the key differentiator between what we call CAT3 and what we call CAT5e, for example. The channel bandwidth is the frequency range over which the signal-to-noise ratio (SNR) is a positive quantity when expressed in decibels (dB); which basically just means the signal level is greater than the noise level. SNR is basically the same as the (power sum) attenuation-to-cross talk ratio (called PSACR or just ACR). For example, for a CAT5/5e channel, the objective is to have a PSACR greater than zero (a positive number in decibels) over a frequency range up to 100 MHz. That is, by definition, bandwidth. Note that in all cases, we are assuming 100 m cable length in Ethernet applications.

Coming to the information capacity of cables, people often equate 10 MHz bandwidth to 10 Mbps, 16 MHz to 16 Mbps, and so on. This, in fact may be true, but only *coincidentally* so. For example, a cable of 100-MHz bandwidth is *not* limited to 100 Mbps. It can usually go to much higher bit rates. We know that using all four pairs of CAT5e, rates up to 1 Gbps can be achieved. Many factors come into the picture in determining maximum bit rate. The upper megabits per second (data rate) achievable is very hardware-dependent for one. In addition, modern Ethernet PHYs (transceivers) use many novel techniques to extend data rates. These are out of the scope of this book, but if the reader is interested, he or she can refer to "Manchester coding" on the Internet, and branch out from there. Underlying all this, there is in fact a fundamental relationship between the bandwidth of a channel expressed in megahertz (MHz) and the maximum information capacity (or data rate) expressed in megabits per second.

A good analogy is the traffic flow on a major highway. Bandwidth is similar to the number of lanes of traffic on a highway. The data rate is very similar to the traffic flow (the number of vehicle crossing over per hour). So one obvious way to increase the traffic flow (data rates) is to widen the highway (increase bandwidth). But another way is to, say, improve the road surface, eliminate bottlenecks, use better signage, and so on (lower the cross talk, use special encoding schemes, and so on). It is therefore possible to pack more bits of information per Hertz of available bandwidth; but that requires a higher SNR.

NOTE *The mathematical relationship between bandwidth and information capacity was discovered in the 1940s by Claude Shannon, an engineer with Bell Telephone Laboratories. This is called the Shannon limit or the Shannon-Hartley theorem. It determines the maximum information rate for a noisy channel as a function of the available bandwidth and the SNR. DSL is also credited to Shannon. As per Wikipedia, "the theory behind DSL, like many other forms of communication, can be traced back to Claude Shannon's seminal 1948 paper: A Mathematical Theory of Communication… He is also credited with founding both digital computer and digital circuit design theory in 1937, when, as a 21-year-old master's student at MIT, he wrote a thesis demonstrating that electrical application of Boolean algebra could construct and resolve any logical, numerical relationship. It has been claimed that this was the most important master's thesis of all time." For such contributions, Shannon is often called "the father of information theory."*

Effect of Temperature on Cable Performance

Ideally, we want the network to be unaffected by our decision whether to run power down the cable (use PoE) or not. We want power and data to be separate and as transparent from each other as possible. Otherwise, for one, troubleshooting can become very challenging. Many techniques and tricks are employed to make the separation of power and data over Ethernet cables a reality, and we will discuss some of these later in this chapter. But there is an obvious manner in which they can interfere, and we will discuss that here.

Signal strength is a critical factor in overall network performance. A lower Insertion Loss is the functional equivalent of a strong signal at the receiver end. We prefer thicker conductors because that lowers Insertion Loss and thus helps improve the SNR, thereby increasing immunity to external and internal noise sources. We also realize that cables with a lower Insertion Loss will be able to support longer distances. What we may not immediately recognize is that good cables also support a higher-operating temperature range. Cables are often installed in ceiling spaces, air plenums, and riser shafts, where the ambient temperature is much higher than in a typical air-conditioned environment. A study performed by the Lawrence Berkeley National Laboratory at the University of California revealed that temperatures in plenum spaces of medical buildings could reach as high as 49°C on a hot day in the middle of summer. We can expect that in tropical countries and/or in warehouses and factory environments, even higher cable temperatures will be encountered. Add to that possible self-heating if we are also sending PoE down the cables.

Keep in mind that Ethernet cables are typically rated *only* up to 60°C. In the long term, high temperatures can adversely affect the life expectancy of the cabling. In the short term, performance can be severely affected because the resistivity of copper increases significantly with temperature.

Let us do the math here. The resistance of copper goes up 4 percent every 10°C. For example, if a certain cable has a resistance of 10 Ω at 20°C, then at 30°C the resistance is $10 \times 1.04 = 10.4$ Ω. What is the resistance at, say 60°C? Note that some wrongly say that since $60 - 20 = 40$, the resistance has gone up by $4 + 4 + 4 + 4 = 16\%$, which gives $10\ \Omega \times 1.16 = 11.7\ \Omega$. That is not quite correct! The actual increase needs to be calculated based on the cumulative factor: $1.04 \times 1.04 \times 1.04 \times 1.04 = 1.17$, which leads to an increase of 17 percent, which in turn leads to $10\ \Omega \times 1.17 = 11.9\ \Omega$. Agreed, it doesn't seem to be much different from the 11.7 Ω calculated by the previous (incorrect) method, but in general, the first method is inaccurate and can produce noticeable error.

Knowing that the resistance of copper goes up 17 percent from 20°C to 60°C (a rise of 40°C), and since DC losses depend on I^2R and are clearly proportional to R, we expect cable losses related to PoE to also go up 17 percent for the same temperature rise (for a given maximum current, I).

From the viewpoint of data/signal transmissions, the Insertion Loss also goes up proportionately. But note that Insertion Loss is usually expressed in decibels (dB). So raising the temperature by 10°C, leads to an increase in Insertion Loss by the amount $20 \times \log(1.04) = 0.34$ dB. Similarly, going all the way from 20°C to 60°C, the Insertion Loss increases by $20 \times \log(1.17) = 1.36$ dB. In decibels we can just add up numbers. So we could have written the increase in Insertion Loss from 20°C to 60°C as $0.34\text{ dB} + 0.34\text{ dB} + 0.34\text{ dB} + 0.34\text{ dB} = 1.36\text{ dB}$. The "wrong math" would have given us $20 \times \log(1.16) = 1.29$ dB, noticeably different from the correct answer of 1.36 dB.

What do these numbers really imply? Consider a cable of 90-m length at 20°C. If we raise its temperature up to 40°C (a rise of 20°C), the resistance goes by a factor 1.04 × 1.04 = 1.082. That is just 8.2 percent higher. But the Insertion Loss also goes up by the same factor. So to have the same transmission performance at 40°C as a 90-m cable at 20°C, we need to reduce the length of the cable by the very same factor too: that is down to 90/1.082 = 83 m. So just a 20°C rise has impacted the data reach by 7 m. In other words, if the 90-m wire was just acceptable ("marginal") for a given application at 20°C, it will certainly have serious trouble in the form of data bit-errors as the cable heats up, unless we started off with a smaller cable (83 m in this case) than was just *adequate* at 20°C (90 m).

As mentioned, the increase in temperature of the cable may be caused by rising ambient temperatures, but also due to self-heating from PoE losses. Since this will also cause an increase in Insertion Loss, to truly keep data and power separate (transparent from each other), we need to account for PoE-induced temperature rise upfront: if necessary by using a *better-quality* (nonmarginally-compliant) cable.

Cable Temperature Rise Caused by PoE

We need to know the expected temperature rise caused by PoE self-heating so we can estimate more accurately the maximum temperature of the cable, and thus prevent deterioration in signal-transmission capabilities (increase in Insertion Loss).

We will keep this simple. Temperature rise and the maximum allowable PoE current was the subject of several committees, reports, and intense discussions, especially during the creation of the IEEE 802.3at standard. But the dust has settled, so it is enough to just quote the results that matter to us going forward.

Initially, during the creation of the older (IEEE 802.3af) standard, the logic was very simple. The TIA liaison reported that existing infrastructure was rated for an absolute maximum of 500 mA on any one conductor. That was the starting point. Keep in mind that at this stage, the assumption was CAT3 cabling with AWG26. Now, as we will soon learn, although a normal PoE connection uses both conductors of a twisted pair in parallel, the committee decided *not* to allow twice the current per pair (2 × 500 mA = 1 A). The reason is a) active current balancing is not present, so we can't say for sure how the PoE current will actually distribute on the two wires of the center-tapped pair, b) in addition, we may also have a defective connector, with continuity on only one conductor. And so if 1-A wire were to be allowed on a twisted pair, we could, under faulty connector conditions, get 1 A flowing through only one conductor. That would be unsafe. So the absolute maximum current was fixed at 500 mA *per twisted pair*. To comply with this absolute maximum, a fairly fast-acting current limit with a typical ± 50 mA tolerance (± 10% of 500 mA) needs to be set. Its nominal (center) value must be at 450 mA. Because then we get a practical current limit lying anywhere in the range 450 ± 50 mA, or 400 to 500 mA. In other words, with tolerances considered, the lowest level of the current limit could be worst-case 400 mA. Now coming to normal operation, we typically also want to include an overload region just above the normal continuous current rating. This will allow a typical device running off PoE power to draw momentary surges of power if necessary (as per the normal operating profile of most devices), without the port being shut down by the activation of the current limit. So if we plan on a 50-mA overload region (below the lowest value of current limit), we get the normal continuous current rating of the cable as 400 mA – 50 mA = 350 mA. And that's how 350 mA was fixed as the maximum continuous PoE (DC) current in IEEE 802.3af. It corresponds to 13 W at the end of 100-m of CAT3 cable as we will soon see.

When the AT standard (IEEE 802.3at) was being drafted, the cable category under discussion was CAT5e (for medium-power/Type-2 applications). TIA guidance recommended a maximum temperature increase of 10°C because of PoE self-heating in a typical cable bundle, up to an absolute maximum cable temperature of 60°C, which is the maximum temperature rating of most Ethernet cables. But that implies that the maximum ambient temperature is restricted to 60 - 10 = 50°C (for Type-2 applications). We then have the desired headroom of 10°C for PoE self-heating, without exceeding the rating of the cable. With several tests on cable bundles, the committee found that 600 mA is a good value, since it gives about a 7.2°C rise. Yes, there is some *additional* built-in headroom here, since the temperature rise is less than 10° C, but that is certainly nice to have and can only help in extending the life of the cabling. And that's how, in a nutshell, 600 mA was fixed as the maximum continuous PoE (DC) current in IEEE 802.3at for Type-2 (medium-power) applications. It corresponds to 25.5 W at the end of 100 m of CAT5e cable, as we will soon see.

NOTE *The temperature rise of 7.2°C is actually for the case of power applied through only two pairs of the four available pairs of an Ethernet cable. Specifically, we have 600 mA flowing in a forward direction through one pair, and the same current returning through the other pair. Two pairs are always unused in normal IEEE-compliant PoE, whether Type-1 or Type-2. But as an experiment, if all four pairs are energized with 600 mA (1.2 A going forward through two pairs, and 1.2 A returning through the other two pairs), the observed temperature rise is 10°C. We see that this temperature rise is also acceptable as per broader TIA guidelines. And that is the reasoning driving some new industry standards for four-pair PoE. For example we have recently seen, Universal Power over Ethernet,*

(UPOE) from Cisco. This corresponds to twice the output wattage, that is 25.5 × 2 = 51 W at the end of 100 m (of CAT5e cable). But keep in mind that still, IEEE PoE standards apply only to 2-pair PoE. four-pair PoE is not covered by the standard, nor is it ruled out. For example, Section 33.1.4.1 of the AT standard deliberately kept the door open for that future possibility.

Some caution needs to be applied in interpreting the listed results and recommendations. First, we are not allowed to increase the max current (above 600 mA) if the ambient is somehow known to be lower than 50° C. This, in effect, means we are not just concerned about the actual average temperature of the cable bundle, but its temperature gradient or *rise* (δT) above ambient too. A higher temperature rise will create hot spots inside the cable bundle, possibly degrading the life of the cabling infrastructure. For that reason, a cable temperature rise greater than 10°C above ambient is not allowed under any circumstances. Second, nor is there some simple formula to allow us to lower the max current judiciously, allowing us to raise the ambient above 50°C, though still staying less than 60°C (by the use of some derating curve). There is simply no derating curve presented in the IEEE standard. The rules were created to keep things simple as far as possible, and also ensure life expectancy of the cabling infrastructure.

The bottom line is we are not allowed to go above 50°C ambient for both Type-1 and Type-2 applications, nor above 350 mA and 600 mA for Type-1 and Type-2, respectively.

With that background and understanding of how the twisted cable was adopted and used in modern Ethernet, we get back to another key reinvention from the past, the center-tapped transformer. It is an implementation of the "phantom" circuit principle we had mentioned previously in connection with the 2010 DSL breakthroughs.

The Center-Tapped (Hybrid) Transformer and the Phantom Circuit

In the top schematic of Fig. 1.5, we see the signal from the microphone being transmitted down a twisted pair to a loudspeaker inside a remotely located telephone. That ensures Person A can talk to Person B. But what about *reverse* communication? We need Person B to talk to Person A too, and *simultaneously*. We could use a second twisted pair for that. See the uppermost schematic of Fig. 1.8. That works, but it is neither smart nor cheap. Can we save one twisted pair? The historical answer to that was the hybrid transformer. It is presented in a very simplified form in the second schematic in Fig. 1.8 (shown in more detail in Chap. 13). We are not going to do any math here, but we should notice the separation of the microphone

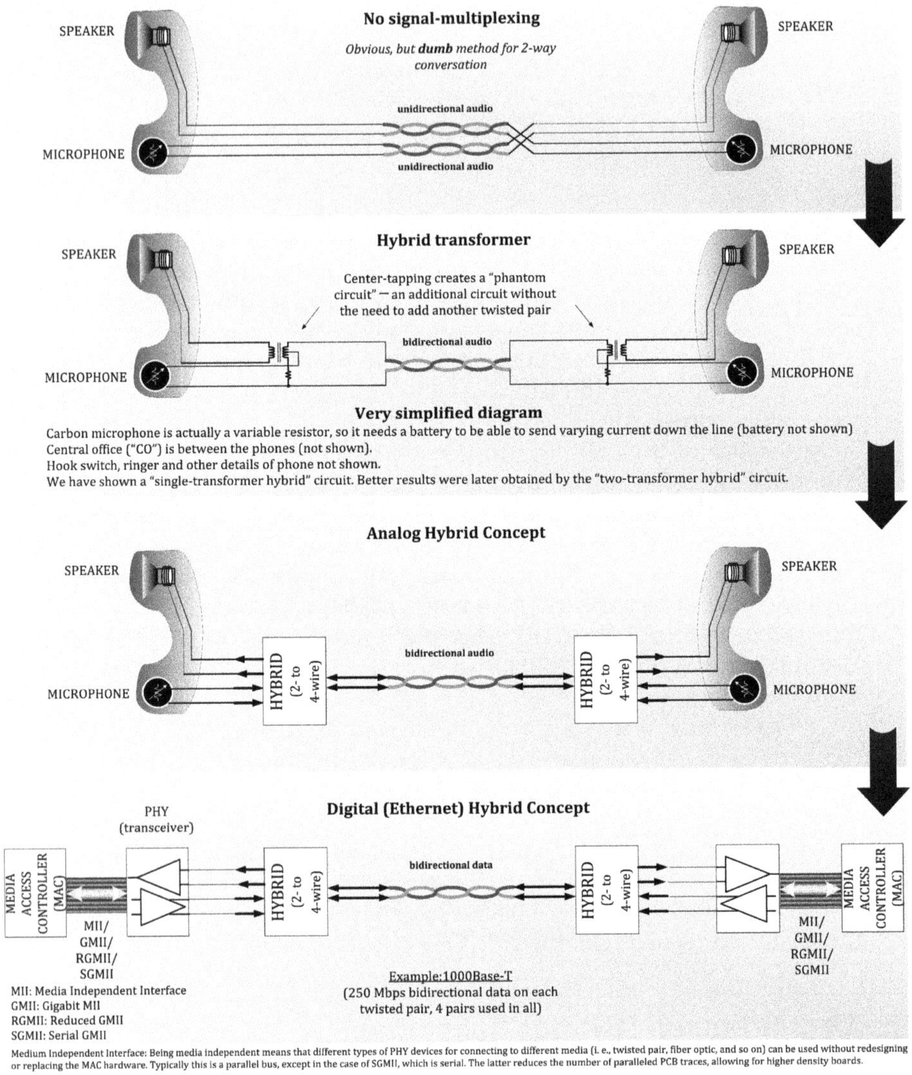

FIGURE 1.8 Development of the hybrid (2- to 4-wire) concept from telephony to Ethernet.

from the speaker by center-tapping. This is a clue to the overall concept used here. Subsequently, with the advent of electronics, the hybrid transformer disappeared and was replaced by active circuitry, though the circuit block was still aptly called a "hybrid" circuit. It is basically a 2- to 4-wire (or in the reverse direction a 4- to 2-wire) multiplexer of sorts. The same concept is used in 1000Base-T, in which all four pairs of the Ethernet cable are used for data transmission, and each pair is *bidirectional* as shown. Of course, we cannot connect the output of a differential transmitter directly to the differential receiver located next to it, for that will lead to the digital equivalent of "audio feedback"—the familiar howling we often hear during stage shows when the microphone happens to catch (and amplify) its own sound coming from the speakers. Clearly, we need to insert a separator/ multiplexer, as shown. As indicated, for historical reasons, this multiplexer is also called a "hybrid" in Ethernet terminology.

Returning to the second schematic from the top in Fig. 1.8, we see that we have multiplexed two audio signals on the same twisted pair by using a center-tapped transformer. This could be one of the earliest such circuits discovered and used. In effect we are creating an additional circuit (we can call it a phantom or ghost circuit) that rides on top of the existing twisted-pair circuit. It is somewhat like two people sharing the same seat on a bus, unaware of each other. In the process, we are saving a seat (a twisted pair in this case).

To really understand the principle behind the phantom circuit, we need to open our old high-school physics book to a page we have likely forgotten long ago: the *Wheatstone Bridge*. Historically, that's exactly how the principle of phantom circuits was first discovered and analyzed. In Fig. 1.9, we have selected a special case of the Wheatstone Bridge with all its bridge resistances *exactly equal*. By simple voltage-divider principles, we realize that the voltage at the common node between R1 and R2 is going to be $V/2$, where V is the battery voltage. Similarly, the voltage at the common node between R3 and R4 is also $V/2$. And the voltage *across* the load resistor (which is the resistor shown inside the gray box of the other cross-branch) is therefore $V/2 - V/2 = 0$. In other words, *no* current will flow through this load resistor.

Then we do a little "morphing" to show that, in fact, both the load resistor and the battery are in *equivalent* cross-branches of the bridge, even though they may have been sketched in a seemingly different way (with one of them appearing to be inside the bridge, the other outside). In reality, their positions are, fully interchangeable—using *identical* bridge resistors, they are, in effect, *identical positions*. In other words, we could replace the load resistor with another battery too (or a general voltage source), and the two voltage sources will never drive currents *into* each other. We could mix and match and have, for example, a DC source in one cross branch and an AC source in the other. One could even be a voltage source, the other a current source, and so on. We will discover that the two sources in the cross-branches of this "equalized" Wheatstone Bridge *never interact with each other*. In effect they are mutually independent. Each is a "phantom" to the other. So neither sees the other. Of course the resistors "know better." Each source will drive its corresponding current contributions through the four resistors of the bridge. To calculate the total currents in the resistors, we need to calculate the current contributions corresponding to one source being present, *assuming the other source is not even connected*. Then we add their individual contributions to get the net current in

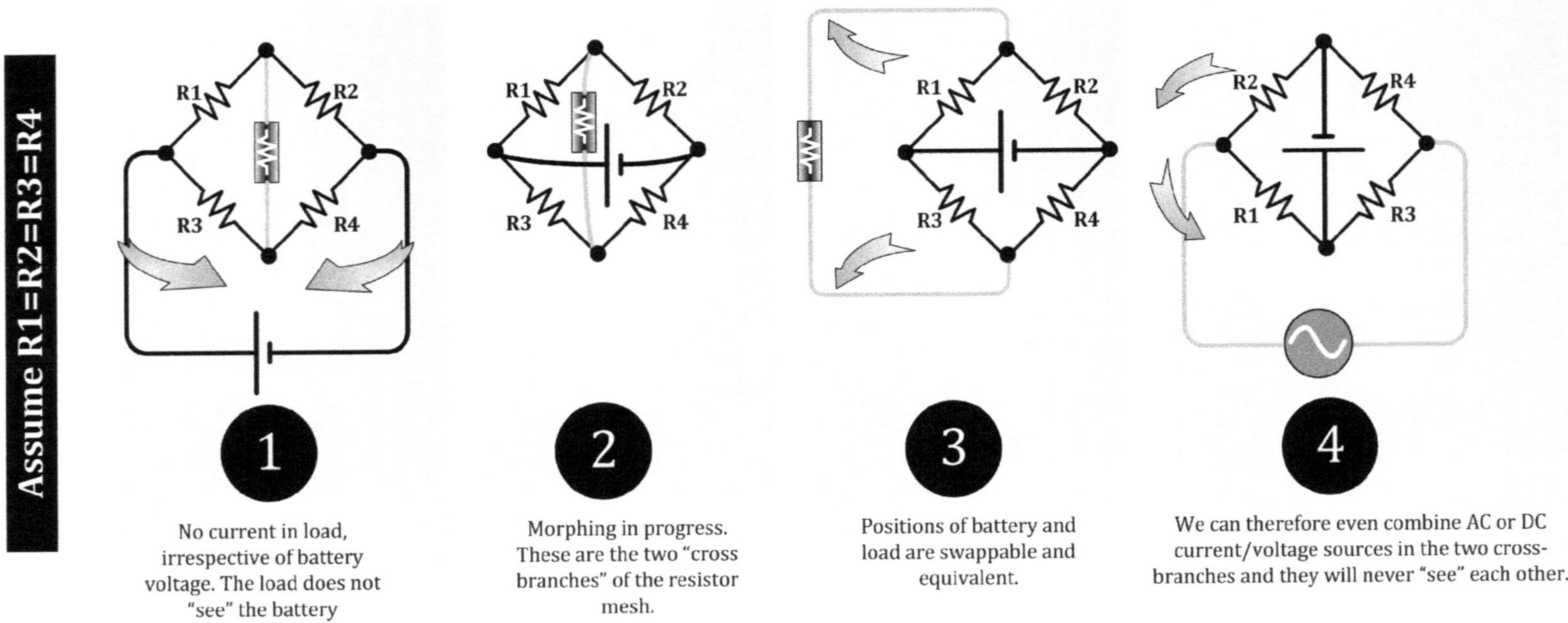

FIGURE 1.9 The two cross-branches of a Wheatstone Bridge with equal resistances have equivalent positions and are independent of each other.

each resistor. This process is shown with a numerical example in Fig. 1.10. We thus see that the two voltage sources (persons) can share the same bridge resistors (seats), but remain independent (unaware) of each other. Right out of the pages of a Harry Potter novel!

In Fig. 1.11 we take this bridge morphing further in a few simple steps. In the first schematic (marked 1), we create two independent circuits: one involving an AC signal (in this case a telephone signal actually, symbolically indicated by the circle with a "T" inside), the other with a battery and load resistor in series with each other. The two circuits are independent, as mentioned previously. We now also clearly see that the current flow produced by either source does not go through the other source. In the next schematic (marked 2, we show that we could do the same thing, not by using resistors but by using identical *inductors*)

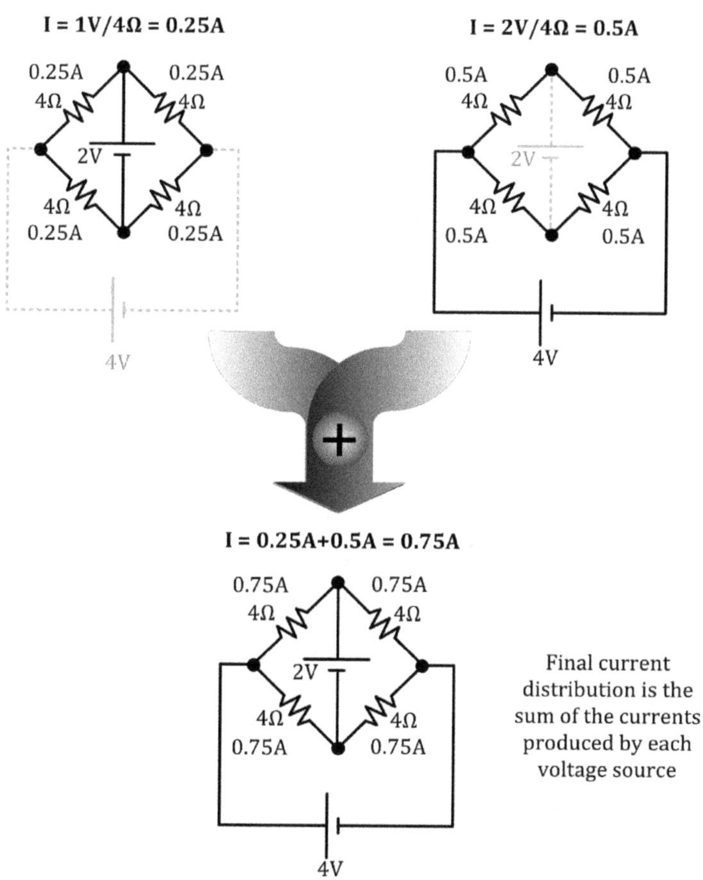

FIGURE 1.10 Numerical example of how two voltage sources in the cross-branches produce currents independent of each other, which can then be added up to get the net current in each resistor.

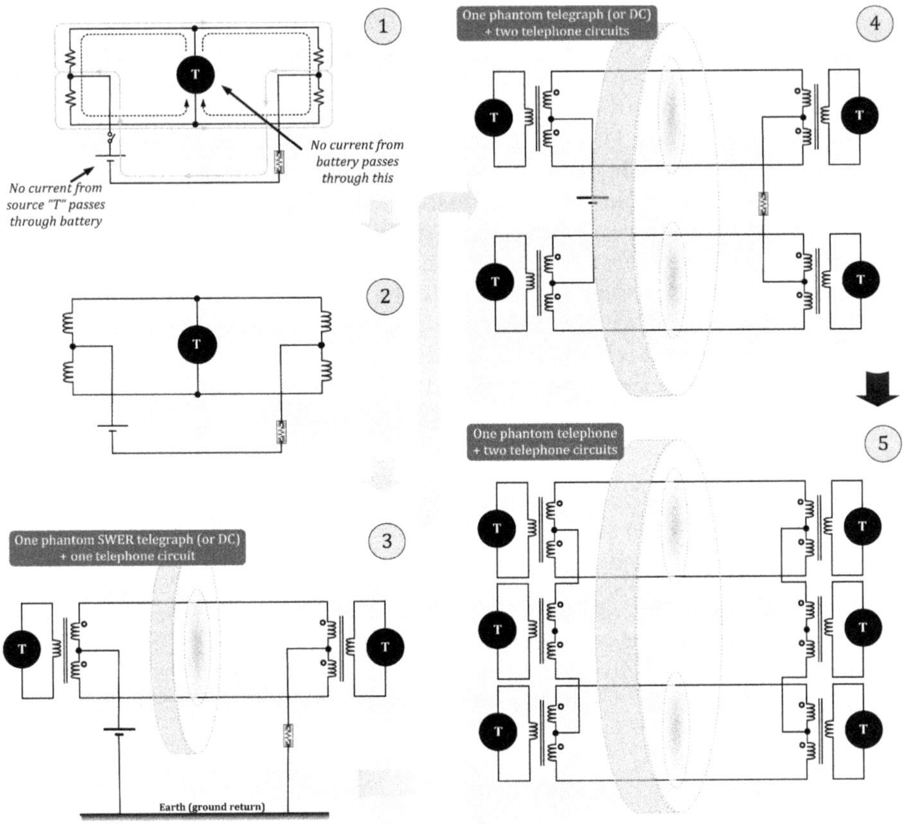

FIGURE 1.11 Morphing the Wheatstone Bridge to produce several exemplary phantom circuits.

(though in this case we are assuming there is some real-world winding resistance present for ensuring that sufficient impedance is presented to the DC source; otherwise it will get shorted out). In the next schematic (marked 3), we replace the inductors by transformers. The long connecting wires are now considered part of a (single-pair) cable. In this case, if the cable is long enough, and assuming the resistances of its two wires are equal, we do *not* need the transformer windings to have any resistance at all. This particular schematic can serve as a telegraph (or switched-DC) circuit based on SWER architecture, in combination with a normal telephone circuit.

To remove the SWER architecture, in the next schematic (marked 4), we introduce *two* identical Wheatsone Bridges, corresponding to the case of *two* twisted pairs. We see that the two twisted pairs carry not only two telephone circuits, but a phantom telegraph/DC circuit with a proper return wire (not requiring a ground return). That is, in fact, almost exactly what we do in PoE today. (See top of Fig. 1.12.)

CENTER-TAPPING A DRIVE TRANSFORMER

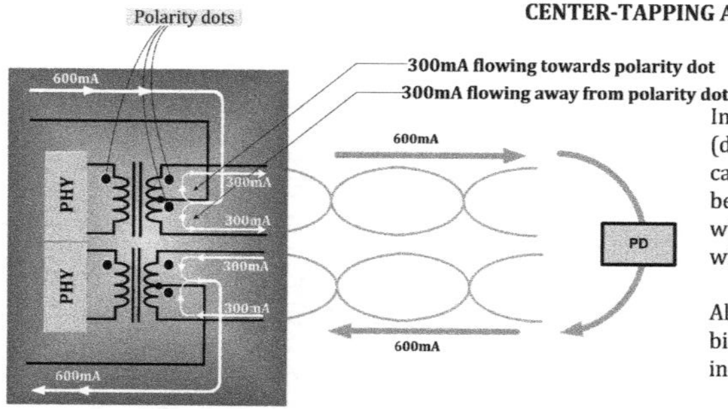

In the ideal case, current (outgoing or incoming) splits up in halves of equal magnitude and opposite direction (direction with respect to polarity dots). So the corresponding flux contributions are also equal and opposite, canceling out and *leaving no net DC flux in the transformer core on account of PoE.* So the core does not need to be increased in size to handle any extra energy. It remains as small as without PoE. However the copper windings will got hotter on account of PoE, so they will need to be thicker. To accommodate the thicker windings, the size of the core *may need to be increased slightly compared to non-PoE case.*

Also, in reality, *current imbalance* can occur, and if that is excessive, it can cause saturation of transformer (DC bias created by net current in one direction). This can affect data transmissions. So the core *does* need to be increased in size somewhat to account for *imbalance* too.

CENTER-TAPPING AUTOTRANSFORMERS

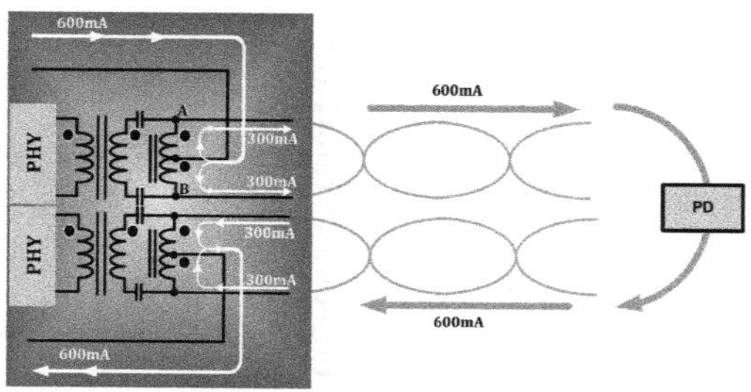

In this case, an "autotransformer" (center-tapped inductor) is used to inject the PoE current. In the *ideal case,* current splits up in halves of equal magnitude and opposite direction. So the corresponding flux contributions are also equal and opposite, canceling out and *leaving no net DC flux in the autotransformer core on account of PoE.* So this autotransformer core is also as small as the drive transformer, though to accommodate the windings for PoE, its size may need to be a little larger than the drive transformer. In reality, we also have to worry about PoE *current imbalance in the autotransformer.* However, *the drive transformer sees almost no PoE current* because **(a)** in the ideal case, the voltage at "A" equals the voltage at "B." If there is a small imbalance, it can drive a small DC bias current (difference current) through the drive transformer (but not the full PoE current). So the copper of the drive transformer need not be thicker, but to handle the small DC bias, it may be slightly larger **(b)** however, by putting blocking caps as shown, no DC bias current at all can flow through the drive transformer, and the latter can be just as small as without PoE present . *This helps the AC performance of the drive transformer.* **However, recently, some 1000Base-T magnetics vendors are omitting the blocking caps entirely (bypassing them).**

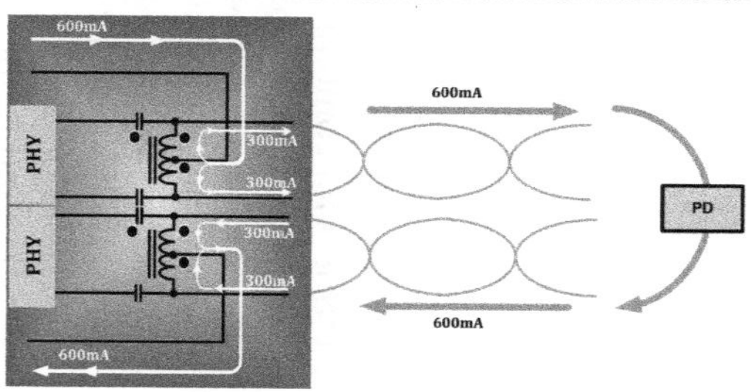

In this case too, an "autotransformer" (center-tapped inductor) can be used to inject the PoE current. But a drive transformer can be omitted, though at the expense of good noise and common-mode rejection properties of transformers. Capacitor coupling is used in **Backplane Ethernet** (802.3ap draft standard).

FIGURE 1.12 Flux cancellation principle for ensuring small-sized magnetics in phantom-powered PoE center-tapped drive transformers and autotransformers.

We will explain Fig. 1.12 in more detail shortly. For now, coming back to Fig. 1.11, note that in the last schematic (marked 5), we have replaced the DC source by a new (phantom) telephone circuit. This is in fact the historical way in which *three* telephone circuits were created out of just *two* existing twisted pairs—in effect, providing a free (phantom) telephone circuit without the added cost and complexity of actually laying out a new twisted pair across several miles.

We have discovered the sheer resourcefulness and ingenuity of engineers working in that obscure period from the late-19th and early-20th centuries. Keep in mind that the phantom DSL breakthroughs of 2010 we talked about earlier are based on the phantom circuit principle, and in fact, perhaps both the DSL breakthroughs are very similar to the last circuit discussed (with some proprietary enhancements to reduce "cross talk" and so on). The underlying idea of using transformers instead of resistors in the Wheatstone Bridge came from J. J. Carty in 1886. That was the 100-year-old networking trick Alcatel Lucent talked about.

Methods of Injecting PoE via Phantom Power

Schematic 4 in Fig. 1.11 is the basic way of inserting PoE by phantom power. The key difference, or rather addition, is a pass-FET in series with the battery on the left side and another pass-FET in series with the load on the other. The purpose of these pass-FETs and related circuitry is basically to impart some intelligence to the power-delivery scheme, and that is discussed in great detail in the following chapters.

In Fig. 1.12, in the upper portion, we have drawn this popular way of injecting (and extracting) PoE (DC power) on an Ethernet cable via the center-taps of the data transformers. We have learned that this is the way telegraph and telephone circuits have been historically combined for over a century. Nothing new here. We should have known it would work from way back in *1886*. Metcalfe also talked about the possibility of combining data and power lines in his 1973 memo but did not mention phantom power specifically.

The question is how did this evolve in PoE more recently?

One of the earliest references to center-tapping of drive transformers for combining power and data (digital) is U.S. patent number 5,065,133 filed in 1989. This one is assigned to The Siemon Company, and the inventor is Gary Howard. Its basic intent is simply to increase the reach of digital signals over unshielded twisted pairs by creating an "enhanced analog signal," by suitably mixing the digital signals with AC power. With some tuned-impedance matching, this creates transmission-line effects and extends the range of the digital signals, which would otherwise get severely attenuated, if not distorted, in UTP cable. The method of combining data and AC power as per the patent, is basically schematic number 5 in Fig. 1.11, except that the inventor has used the phantom circuit (in the middle), not for a third telephone line (voice), but for transmitting AC power. In other words, the center circular source shown marked "T" should now be marked "AC" instead. The AC frequency could be derived from household mains wiring, but it is better to use higher frequencies for reducing the size of filters, and so on. The idea of extracting the AC power and using it for remote powering is also mentioned in this patent. This was indeed clever and ground-breaking.

The above patent was later cited by a key U.S. patent number 5,148,144, filed in 1991. This one was assigned to Echelon Systems Corp., Palo Alto, and the inventors are Philip H. Sutterlin et al. This seems to be one of the first showing center-tapped data transformers for DC remote powering. It also contains a very good discussion on the advantages of phantom powering, and for the first time perhaps, it talks of how center-tapping avoids "core saturation." Keep in mind that when J. J. Carty introduced his idea in 1886, they were using audio-frequency transformers that were big and bulky to start with. These also had many turns on them, and so there was plenty of DC resistance to limit the currents. These transformers were also wound on iron-cores, and we know today that iron-cores can support very high flux densities without core saturation. So the whole idea of reducing transformer core size and also avoiding core saturation, was of little concern back then. No saturation was likely ever observed. So it seems plausible that the 19th-century inventors were themselves unaware of the biggest advantage that center-tapping brought to the table: avoiding core saturation. But with ever-decreasing sizes of components today, it is something we are very cognizant of today. Especially when using the tiny ferrite/powdered iron/Kool Mu cores found in typical data transformers today.

We now realize that center-tapping has a big advantage in avoiding core saturation when injecting power—AC, DC, or PoE. If not for center-tapping, we would need to significantly increase the size of the drive-transformer core for the sake of adding PoE capability. That would be hardly desirable. For one, the AC characteristics of such a bulky drive transformer will get severely compromised as pointed out in the patent number 5,148,144 too. Besides, a typical switch/hub will have 4, 8, 24, 48, or 96 ports. Each port has two drive transformers at least, and so in the interest of keeping overall size of equipment manageable, we need to keep the magnetics very tiny, despite introducing PoE. And center-tapping is the way to do that.

What determines the size of a transformer? Every core has a physical limit as to the amount of energy it can store. Larger cores can store more energy. Flux ($\Phi = B \times \text{Area}$) corresponds to stored energy. It is proportional to ampere-turns. So from ampere-turns, we can estimate core size. A complete treatment of magnetics can be found in this author's *Switching Power Supplies A–Z* book.

For calculating ampere-turns at any given moment, we need to (algebraically) sum up the product of the current in every winding placed across a core and the number of turns of that winding (ΣNI). The sign of the current is determined with respect to the polarity dots of the winding, as shown in Fig. 1.12. In center-tapping, *in an ideal case*, current splits up exactly equally in the two halves, and these current components have opposite polarities (away from the dot, toward the dot). So they cancel out completely in terms of the flux in the core. For all practical purposes, the core does not "see" the PoE current, unless there is an *imbalance*, and then it would just see the "difference current" (difference of the magnitudes of the currents in the two halves). We also realize, the transformer core would need to be a little larger to handle real-world imbalances as discussed in more detail in Chap. 9. However the copper windings *do* see the full PoE current, and there is no "cancellation" at work here, because heating depends on I^2R, and the squaring of current masks its sign anyway. So the copper windings *will* need to be somewhat thicker for a "PoE-capable" drive transformer. To accommodate these thicker windings, on rare occasions, the core size may need to be increased just a little, to provide a larger "window." But in general, center-tapping for the purpose of introducing phantom power causes almost no increase in the size of the magnetics. This is what was so explicitly pointed out, apparently for the first time, in U.S. patent number 5,148,144. To quote from that patent (italics inserted by this author):

> ....prior art systems are not without their disadvantages. One major disadvantage of this type of prior art system lies in the fact that *the transformer must be sized to handle the DC current without saturating*. In general, a transformer which can accommodate DC currents without saturating *has much poorer AC characteristics than one in which does not have to handle any DC current*. These degraded AC characteristics are manifested by poor communications signal quality and by a limited bandwidth.... To overcome the difficulties associated with providing power and communications along the same cable, some practitioners have chosen to provide separate conductors for power and message delivery.... A further problem of conventional power distribution approaches is that they tend to make inefficient use of cable.... .Therefore, what is needed is a means of providing power and communications over the *same* cable network.... the present invention provides a wire-based communications network in which power and message information is delivered over the same cable network with improved AC characteristics. The enhanced communication capabilities of the present invention permit greater communication speeds and transmission over greater distances.

Do we need to always center-tap the drive transformers for injecting PoE? No, we can also use center-tapped *inductors* instead. These are also called "autotransformers." See the lower portion of Fig. 1.12

for a breakdown of their pros and cons. The first patent that seems to have talked about this alternative method for phantom powering via center-taps is called "Power transfer apparatus for concurrently transmitting data and power over data lines," filed on May 29, 1997. It bears the U.S. patent number 5,994,998, naming David Fisher et al., from 3Com. There were several continuation patents of this initial patent, extending to U.S. patent number 6,710,704 filed in 2002 and more recently U.S. patent number 6,989,735 filed in 2004.

The first mention of the possibility of using phantom power for PoE at the IEEE 802.3af meetings seems to have been on March 10, 1999, by Nick Stapleton of 3Com. On July 6 of the same year, Amir Lehr from PowerDsine (now part of Microsemi) mentioned the possibility. In fact 3Com and PowerDsine were the key companies at the time, urging IEEE to standardize PoE. Later, Yair Darshan of Power Dsine (Microsemi) became one of the key technical persons involved in the development of the PoE standards, along with Fred Schindler of Cisco.

In this manner, PoE got built from the ground up—the "ground" in this case being J. J. Carty's idea from 1886. That idea had worked spectacularly for telephony, later for power over data in Ethernet, and finally data over data in phantom DSL too. It is the old "networking-trick" in its various incarnations, but the same basic principle.

NOTE *The key concern in phantom circuits is that they truly remain "phantom" to each other. In other words, in our case, data should be completely unaffected by power, and vice versa. Unfortunately, the latter is almost a fait accompli,* the former is typically *not: Power can easily affect data. We saw in this chapter when we learned that an increase in temperature of the cable caused by to PoE self-heating will cause an increase in the Insertion Loss, which can affect the reach/quality of data transmissions. We also made initial assumptions about how well-matched/equal the resistances of the Wheatsone Bridge were. Because if they are not, we will get current through the cross-branches, and so, in effect, the two cross-branches will interfere with each other. That means the two subcircuits are no longer very good "phantoms" to each other. We can thus understand that any asymmetry in, say, the center-taps, or even in the wire resistances of the twisted pairs of the cable, can lead to PoE severely affecting data transmissions. These nonidealities will be discussed later in more detail.*

PoE Chip Vendors: The Emerging Landscape of PoE

This happens to be the first book on the subject. We can ask how did the general technical community, more specifically, PoE engineers, survive and learn so far? The answer is with the help of some very useful technical information on PoE and the related IEEE standards available on several major chip vendors' Web sites. We end this chapter by listing such vendors. Most of them have been a huge part of the technical community at large and the growth of PoE as a field. This book, too, has relied heavily on their technical information in an effort to disseminate and "put it all in one place." The acronyms PSE and PD (see Table 1.1) are further explained in the next chapter.

1. Microsemi (PowerDsine): The pioneers of PoE currently have PSE chips with both internal and external pass-FETs. They also have PD chips, both with only the front-end pass-FET, and also with integrated PWM (DC-DC converter) controller stages.
2. Texas Instruments (TI, along with recently acquired National Semiconductor): They currently have PSE chips with both internal and external pass-FETs. They also have PD chips, both with only the front-end pass-FET, and also with integrated PWM (DC-DC converter) controller stages.
3. Linear Technology: They currently have PSE chips with both internal and external pass-FETs. They also have PD chips, both with only the front-end pass-FET, and also with integrated PWM (DC-DC converter) controller stages.
4. Silicon Labs: They currently have PSE chips with both internal and external pass-FETs. They also have a highly integrated PD chip with on-board bridge rectifiers, front-end section (with pass FET), and a complete DC-DC switcher (including the switching FET).
5. ST Microelectronics: They currently only have PD chips, both with only the front-end pass-FET, and also with integrated PWM (DC-DC converter) controller stages.
6. Broadcom Corp.: Integrated-FET PSE-chip vendor. No further details. Extremely secretive. "Protects" datasheets and App Notes in an electronic documents safe ("docsafe") under heavy surveillance. Known to have unsuccessfully tried to convict departing employees for "espionage"—those who "suspiciously" downloaded files from docsafe (see Tien Shiah case on Google).
7. Akros Silicon: The most highly integrated PD chips with on-board bridge rectifiers, front-end and integrated PWM (DC-DC converter) controller stages. Also includes on-board isolation barrier and secondary-side buck switchers.

And that completes our discussion on the evolution of PoE. In the next chapter, we move on to more specific implementation details.

CHAPTER 2

Overview of PoE Implementations

Power Sourcing Equipment and Powered Devices

As explained in Chap. 1, Ethernet has evolved into a star topology, in which a number of end-point devices are connected to a switch or hub. Looking at things from the perspective of data (just Ethernet), on one side of a given cable we have the switch/hub, and on the other side we have the data terminal equipment (DTE). Looking at this arrangement from the perspective of power (Power over Ethernet), we have Power Sourcing Equipment (PSE) on one side of the cable, and on the other side a Powered Device (PD). See Fig. 2.1.

In general, the DTE could be a workstation/computer for example, or perhaps a printer. It could also be a more modern device like an Internet Protocol (IP) phone or camera. The key question to ask is: Is it a *powered DTE* or not? This is equivalent to asking: Does the DTE contain a PD or not? If affirmative, we can go into finer subdivisions such as: Is it a Type 1 PD or Type 2 PD? And so on. Or: What is its "class" (power category)? We will explain all these subdivisions shortly.

In PoE, all the power that flows down the cable comes from a DC source. In this book, we may refer to this **DC source as "48V"** *(inside quotation marks)* for historical reasons. The terminology is actually symbolic. In reality, the actual voltage range may not even include 48 volts, as we will soon learn.

Note that a typical workstation (computer) is not a *powered* DTE. In other words, it will not "ask" for any power from the cable. In general, that could be because the available power from the cable is inadequate to meet the power requirements of the device to start with, so the device was just not designed to be operated off PoE. But it is also possible that the device is "pre-PoE," which means that when it was designed, PoE wasn't even around (at least not as we know

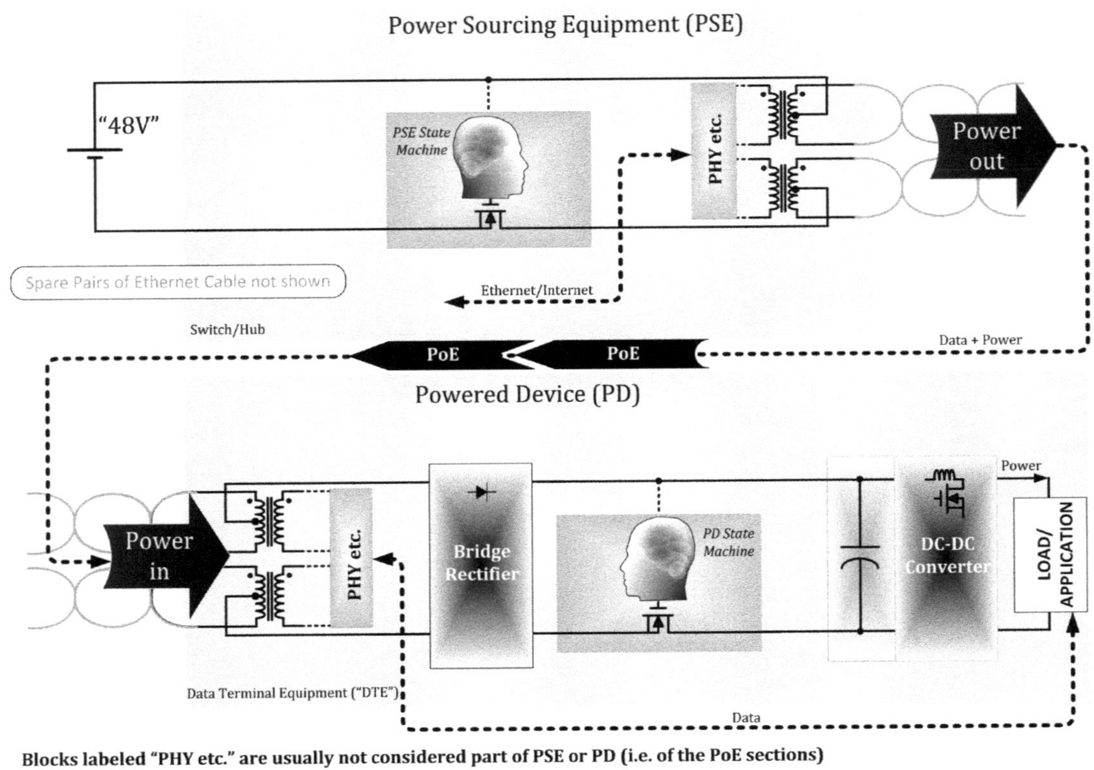

FIGURE 2.1 Typical (simplified) PSE and PD.

it today). In either case, the possibility is high that the device input circuitry (from its Ethernet jack) may get damaged if we were to place a high voltage (for example "48V") straight on the cable ("line"). That is one reason why the IEEE standard so carefully documents how the PSE is supposed to energize, and later quickly de-energize, the Ethernet cable as required. Other related concerns, not necessarily subsidiary or of lesser importance, include for example, not sending too much current down the line, even under single-fault conditions, to avoid overheating and potential fire hazard. Other concerns, like long-term life and reliability, isolation, user safety, and so on, will be covered later.

In general, to provide the necessary control function in a PSE, the DC voltage source ("48V") is always in series with a pass-FET. The Gate of this pass-FET is then carefully controlled by a "smart-block," which complies with the IEEE PoE standard—its state machine diagram specifically. This is discussed further in Chap. 8.

From a higher level, we can look upon the PSE as very simply: a "48V" DC source in series with a smart-FET—in essence, a "switched battery."

A couple of clarifications on terminology and some nit picking too:

1. Does the PSE include the "48V" DC source? A battery/source is a part of any power-sourcing block, but the term "PSE" usually refers to *all the rest* (i.e., *without the power supply*).
2. On the other side of the cable, the PD typically contains a DC-DC converter. But the IEEE standard does not really discuss the converter's specific construction or its specifications, other than some caveat concerning its input capacitor value, or its start-up delay, as explored later. So PD or PD interface, most often refers to everything *except* the DC-DC converter ("switcher"). The PD interface (with the pass-FET inside, but not the switcher) is often called the PD "front-end" too, or the "hot-swap controller."

NOTE *Keep in mind that, in effect, the PSE is just a "door" that opens and closes. Its pass-FET conducts to let the incoming voltage/current through onto the cable, or stops conducting to de-energize the cable. The PSE cannot, and does not* actively condition or regulate *the "48V" rail, that being the function of the DC voltage source itself ("48V"), not the PSE. So, for example, if there is too much noise and ripple on the Ethernet line coming from the PoE sections, we should first check the switching power-supply design, not the PSE, as some engineers mistakenly do.*

PoE supports modern *powered* DTEs. Examples of such devices are IP phones, IP cameras, and wireless access points (WAPs). In such situations, we could also colloquially say that the DTE is PoE-capable, or PoE-enabled. It will certainly contain a PD, as defined by the IEEE PoE standard, to *intelligently* extract power from the cable.

Why did we say "intelligently"? When the IEEE committee sought to standardize PoE in the form of an open standard, there were already several devices out in the field based on *nonstandard* implementations of PoE (primarily from Cisco). There were also many "pre-PoE" devices, some of which could get damaged by high DC voltage on the line. We will discuss these legacy devices later, but the important thing to note here is that any proposed PoE open standard would need to recognize and identify such devices, and respond appropriately to them. To help in the mutual identification (discovery) of PoE-capable devices on both ends of the cable, the remote PoE-enabled device, called a Powered Device (PD), also needs to have a pass-FET—controlled by another "smart block" as indicated in Fig. 2.1. From a higher level, a PD can be considered a *switched load* (just as a PSE is a switched battery). The logic within the smart blocks, residing at the PSE and PD ends, is the essence of modern PoE, as embodied in the IEEE 802.3af and 802.3at standards.

A note on terminology: As per the general IEEE Ethernet standard 802.3, the Ethernet registered jack/socket (RJ-45, as it is also called, as discussed later) is, in effect, the medium dependent interface (MDI). The term medium or MDI is intentionally broad, to cover the various manifestations of "ether" that Robert Metcalfe visualized in 1973 (see Chap. 1). In our particular case, the "medium" is just the twisted-pair (copper) cable. Further, an *MDI with PoE present* on it is called a power interface (PI). Of course, this has to be copper; we can't send power down a fiber-optic cable.

It is interesting that the IEEE 802.3af/at standard does not mention or use the words "PoE" or "Power over Ethernet" in any of its sections (except for an unlinked keyword in the AF standard). The title of the PoE standard is also just "DTE Power via MDI." One related question is would *this* book sell better had it been titled *DTE Power via MDI Interoperability Guide?* Perhaps not. We will therefore try to avoid some of the complicated names and acronyms that the IEEE standard uses and assume that most readers will understand what is being referred to by common sense. Our ultimate purpose is to try and *simplify* the standard, thus making it easier for laypersons to understand. Obfuscation is being shunned.

Lastly, keep in mind that in Fig. 2.1 we have a special case in which both the PSE and PD inject and accept power, respectively, from the *center-taps of the data transformers*. That is not the *only* way to implement phantom power as discussed shortly. Also, besides phantom power, there are other ways of sending power over the cable as we will soon see.

The Input Voltage Source and Corresponding Power Levels

In PoE, the first requirement is a power source at one end of the Ethernet cable to inject power into it. Historically, this DC source used to be a battery of nominal value "48V," but now it is almost invariably the output of a ("silver-box") AC-DC switching power supply. To supplement the AC-DC power supply, there may be an uninterruptible power supply (UPS) present somewhere. Its purpose is to maintain the DC source in the event of a power outage.

Why do we say "48V"? Because safety agencies have decided that 60 V is the maximum voltage safe for a user to inadvertently touch without fear of electrocution. So to keep a little headroom (a safety margin of 5%), the IEEE 802.3af/at standard fixed the *maximum* acceptable voltage at 57 V for all PoE applications.

Typical lead-acid batteries are "12 V" nominal, but they can vary between 10.5 V (dead) and 12.7 V (fully charged). Five batteries in series is out of the question since at the upper end that could exceed 60 V (check: 5×12.7 V = 63.5 V). But four serially connected 12-V batteries can be used. That corresponds to a nominal of 12 V × 4 V = 48 V. And that is where the "48V" originated. What about its minimum value? When fully discharged the voltage of four series batteries is 10.5 V × 4 = 42 V. To keep a little safety margin, to avoid dead batteries, the IEEE 802.3af standard fixed the minimum PoE range at 44 V. So the full DC voltage range fixed by IEEE 802.3af is 44 to 57 V.

We learned in Chap. 1 that 350 mA was the maximum continuous current deemed safe for CAT3 cabling. That is the cabling assumed for low-power applications (as per 802.3af). Since the voltage could fall to 44 V, the minimum guaranteed power into the cable is clearly 44 V × 0.35 A = 15.4 W. This is a "low-power" application. From Chap. 1, we also know that the IEEE standard fixed the total PoE *loop resistance* at 20 Ω for 100 m of CAT3 cable. So cable losses are $0.35^2 \times 20 = 2.45$ W. In other words, the minimum guaranteed power at the other end of the cable that is available to a device is 15.4 – 2.45 = 12.95 W. This was rounded up by the IEEE 802.3at standard to 13 W and was called a Type 1 application. We can calculate that the minimum voltage at the remote (PD) end of a 100 m cable will be 12.95 W/0.35 A = 37 V. At zero (or very light) load, the voltage at the PD end can rise to the 57 V IEEE limit of the "48V" source (negligible drop across the cable). So the voltage range at the PD end for low-power (Type 1) applications is 37 to 57 V.

Summarizing, in a Type 1 application, the minimum guaranteed power of the CAT3 cable is 15.4 W at the PSE end and 13 W at the PD end. The DC voltage range is 44 to 57 V (PSE) and 37 to 57 V (PD). The current is fixed at a maximum continuous of 0.35 A.

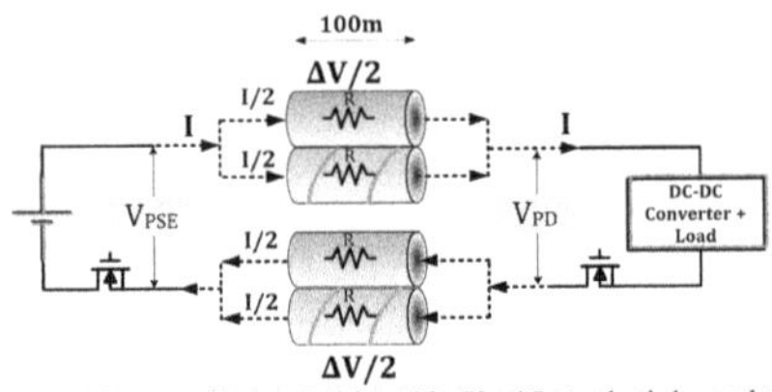

a) The AF standard used "V_{Port}" for the port voltage at either end of the cable. The AT standard changed that to V_{Port_PSE} and V_{Port_PD}. We are just calling these V_{PSE} and V_{PD} respectively, for simplicity.
b) These voltages do *not* include the FET drops on either side: they are measured at the "PI" (power interface).
c) The max power is calculated on the basis of the *lowest* PSE voltage. The power is kept fixed as V_{PSE} is raised. In other words, *the max continuous current is derated with voltage*. For example, for Type 1, the max continuous current at 44V is 0.35A, but at 57V, the max current is reduced to (44/57) × 0.35 = 0.27A. Check that this gives the same PSE-side wattage: 57V × 0.27A = 15.4W. Similarly, for Type 2, at 57V, the max current is 50/57 × 0.6 = 0.526A. Check: 57V ×0.526A = 30W.

TYPE 1 (low power)

R = 20 Ω (100m CAT3)
(AWG 26; d = 16 mils, 0.4 mm)
I = 0.35A (@ V_{PSE} = 44V)

R_{LOOP} = 2×(20/2) = 20 Ω †

Voltage drop across entire cable (loop):
V_{PSE} - V_{PD} ≡ ΔV= 0.35A x 20Ω = 7V

V_{PSE} can range from 44V to 57V

Min PD-side voltage = 44V - 7V = 37V ††

V_{PD} can range from 37V to 57V

<u>Power out from PSE</u>
44V x 0.35A =15.4W
<u>Power into PD</u>
37V x 0.35A = 12.95W †††

Pcable = 15.4 - 12.95 = 2.45W

Check:
Pcable = 0.35^2 ×20 = 2.45W

Loss in cable

TYPE 2 (medium power)

R = 12.5 Ω (100m CAT5e)
(AWG 24; d = 20 mils, 0.5 mm)
I = 0.6A (@ V_{PSE} = 50V)

R_{LOOP} = 2×(12.5/2) = 12.5 Ω †

Voltage drop across entire cable (loop):
V_{PSE} - V_{PD} ≡ ΔV= 0.6A x 12.5Ω = 7.5V

V_{PSE} can range from 50V to 57V

Min PD-side voltage = 50V - 7.5V = 42.5V

V_{PD} can range from 42.5V to 57V

<u>Power out from PSE</u>
50V x 0.6A =30W
<u>Power into PD</u>
42.5V x 0.6 = 25.5W

Pcable = 30 – 25.5 = 4.5W

Check:
Pcable = 0.6^2 ×12.5= 4.5W

Loss in cable

† *Total loop resistance is two resistors of resistance R in parallel with each other, in series with two resistors of resistance R in parallel (R/2 +R/2 = R)*

†† *In the AF standard this was erroneously given as 36V. It was corrected in the AT standard to 37V (for Type 1).*

††† *12.95W is correct as per the AF standard, but it was rounded up to 13W in the AT standard*

FIGURE 2.2 Voltage drop and power calculations in Type 1 and Type 2 applications.

See a full breakup and analysis for Type 1 in Fig. 2.2.

Later, when the IEEE 802.3at standard was being written, it was clear that lead-acid batteries were almost obsolete in most applications. So there was an opportunity to increase the minimum voltage. The current was fixed at 600 mA so as to keep the temperature rise of the cable to less than 10°C as discussed in Chap. 1. So, the minimum voltage was set a little higher than 44 V—at 50 V. Now the minimum guaranteed power was a round figure of 50 V × 0.6 A = 30 W. The minimum loop resistance of CAT5e cabling was fixed at 12.5 Ω in the 802.3at standard. So cable losses were $0.6^2 \times 12.5 = 4.5$ W. In other words, the minimum guaranteed power at the other end of the cable, available to a device, is 30 − 4.5 = 25.5 W. This was called a Type 2 application by the IEEE 802.3at standard.

Summarizing, in a Type 2 application, the minimum guaranteed power of the CAT5e cable is 30 W at the PSE end, and 25.5 W at the

PD end. The DC voltage range is 50 to 57 V (PSE) and 42.5 to 57 V (PD). The current is fixed at a maximum continuous of 0.6 A.

See a full breakup and analysis for Type 2 in Fig. 2.2.

As mentioned, for historical reasons, in several places in this book, the input voltage source may be stated as "48V." But that is purely symbolic. It should now be understood that the actual range of the "48V" rail is 44 to 57 V or 50 to 57 V (at the PSE end), depending on whether we are talking about low-power (Type 1) or medium-power (Type 2). At the PD end the corresponding range is 37 to 57 V or 42.5 to 57 V, for Type 1 or Type 2 respectively.

Clarification

We have realized we need a pass-FET in series with the DC source to carefully control the current in the cable (to turn the port "on" or "off," colloquially stated). That FET has a certain *nonzero* forward voltage drop when it is conducting, which could be typically 1 V (when passing ~ 0.6 A). The FET resistance (R_{DS}, or Drain-to-Source resistance) and its corresponding forward drop can also increase by typically 40 to 50 percent when the FET gets hot (after conducting for a short while). In addition, the FET is often combined with a sense resistor (typically 0.5 Ω), because the IEEE standard requires very accurate current monitoring (high resolution). In all, there may be a drop of 1 to 2.5 V between the DC source and the actual entry point of voltage on the cable (at the PI). It is impossible for any open standard to account for all these variations, which would also vary greatly from vendor to vendor. So the IEEE standard does not specify the voltage of the DC source. In other words, the specified range of 44 to 57 V is the voltage for Type 1 applications measured *after the PSE pass-FET and any sense resistor*. In fact even *after the Ethernet jack*, to eliminate variations in contact resistances and so on. It is just the *"port voltage"* to keep things simple. The same holds for the 50 to 57 V specified for Type 2 applications. This is all for the PSE end. At the PD end, the voltage range is once again where the copper ends, just *before* the Ethernet jack. This should all become much clearer looking at Fig. 2.2.

Corollary

If we are trying to support a Type 2 application, for example, we should not pick an AC-DC power supply with a set nominal output of "50V." We have to account for all the variations in the forward drops up front (including PCB drops), and also power supply tolerances, so as to guarantee that the voltage at the PI (port) is always above 50 V, otherwise IEEE compliance (and interoperability) may be jeopardized for Type 2 applications. For this reason, most vendors choose AC-DC power supplies with declared nominal outputs set somewhere between 52.5 and 54 V.

Features

Major switch vendors today want to estimate the losses in the PSE pass-FET and sense resistor (if present) up front. This helps allocate system power better. This topic falls under the larger topic of power management. The underlying motivation for it is that a typical built-in AC-DC power supply cannot support all its ports with full power simultaneously. So a switch vendor may request a new feature, one that is not *mandated* by the IEEE PoE standard: They may request the PSE controller monitor not only the port voltage (as required for IEEE compliance) but also the input ("48V") voltage rail. Then, using the difference between the two, it can compute the actual real-time voltage drop across the pass-FET and sense resistor. This helps power more accurately allocate and maximizes the number of ports that can be powered by a given AC-DC power supply.

Current Derating (Constant Power)

From Fig. 2.2 we also learn that the IEEE standard *fixes the PSE output wattage at lowest voltage*, then it keeps the wattage fixed as the voltage is raised. The reason for that is inside the PD is a DC-DC converter providing a regulated rail to whatever circuitry is being powered in the DTE. The input of any DC-DC converter is very close to a constant power input. If its input voltage rises, its input current falls, so as to keep its output power constant. Yes, *we are neglecting the fact that the series cable losses do not behave in this manner* (cable losses depend on I^2R, and that does not change with voltage). The IEEE standard assumes a constant wattage at the PSE-end, and calls for a *proportional decrease in the max continuous current of the PSE*. For example, in Type 1 applications, the max continuous current varies between 0.35 A and 0.27 A. Check: 57 V × 0.27 A = 15.4 W. Similarly, for Type 2 applications, the current varies from 0.6 to 0.53 A. The maximum port current is clearly not fixed as sometimes mistakenly assumed.

NOTE *Some PSE-PD schemes try to take advantage of the fact that at high voltages it is possible to send more power down the cable if we do not derate the current as per the IEEE standard. But they remain proprietary, and may not be fully IEEE-compliant.*

Center-Tapping (Alternative-A) Possibilities

As mentioned, in Fig. 2.1 we had a special case in which both the PSE and PD inject and accept power, respectively, from the *center-taps* of the data transformers. At the end of Chap. 1, we learned that center-tapping does not lead to bigger magnetics on account of adding PoE, as we may have intuitively imagined at first. The most popular method of adding PoE capability is via the center-taps of the drive transformers, as per Fig. 2.1. This is "phantom power" or "phantom feeding" (see Chap. 1) and is also referred to as Alternative-A (or Alt-A) in the IEEE PoE standards (i.e., power multiplexed with data).

Coming back to center-tapping, the IEEE standard does *not* preclude the possibility of using a *tapped inductor* (autotransformer) instead of an isolated transformer. Yes, we could also use discrete inductors instead of transformers or autotransformers, but the inductors *will* need to be very big to handle PoE in that case, since flux cancellation is not a possibility in any inductor (it has a single winding). All this is captured in the overall summary of center-tapping in Fig. 2.3. Also, when using the autotransformer method, we should, in principle, use blocking capacitors to prevent any small DC-current imbalance from flowing back into the drive transformers and changing their characteristics. However, many magnetics vendors are removing the blocking caps, especially for high-speed (gigabit) applications. They argue that only a small current flows in the drive transformer on account of imbalances, so if the drive transformer can handle that small DC bias, it is acceptable. However, the user needs to check the data performance of such an arrangement thoroughly.

PoE on Data or Spare Pairs?

PoE was originally intended for 10Base-T and 100Base-TX (10/100 Mbps) Ethernet applications, in which, as we learned in Chap. 1, (only) two pairs of the cable are used for data. These are called the data pairs; one pair is for transmitting, one for receiving. So two pairs of the Ethernet cable were unused, these are called its "unused pairs" or "spare pairs." That is strictly from the viewpoint of *data*. When the first IEEE PoE standard (the AF standard) was under debate, the only thing clear from the very start was that the scope of the PoE standard was restricted to *power over only two pairs*—one pair for forward-current flow (PSE to PD), one for return. But the question was: Which two pairs of the four should be used for PoE? The most obvious solution was actually *not* phantom feeding but simply using the *spare pairs* for PoE. That seemed easy, since it wouldn't need a transformer either; the entire question of saturating drive transformers and flux cancellation techniques, as in Fig. 2.2, would be moot. However, it was perhaps felt that by using up all available resources (in this case, all four pairs of the Ethernet cable for some purpose), the arrangement was *not future-proof*. There would be little room to grow later; it was almost a dead end.

Oddly, there may be a need someday to send power over a cable with only *one* twisted pair present. So, it seems a good idea that we should learn to couple power and data (on the same pair/s); otherwise, we will likely run into some limitations and problems down the road.

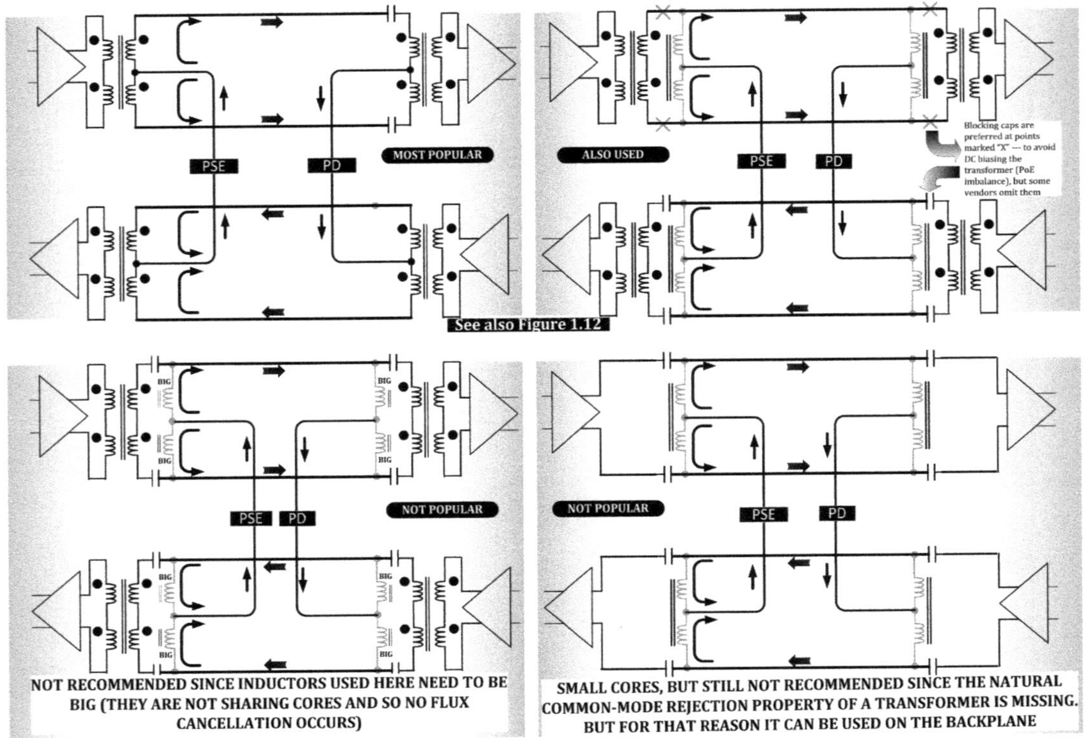

FIGURE 2.3 Center-tapping possibilities (Alt-A).

As discussed in Chap. 1, without perhaps being aware of it, the "100-year-old networking trick" (phantom circuits using transformers) got reincarnated by the PoE committee/task force. This was the phantom power principle, of course. As indicated in IEEE 802.3af/at, power over the data pairs is designated Alt-A as mentioned previously. The standard does allow the other (obvious) method for injecting PoE too, if so desired. Power over the spare pairs is called Alternative-B, or just Alt-B. See Fig. 2.4.

Note that when 1000Base-T came along, there were no spare pairs left, since data is now sent over all four pairs of the cable (for higher speeds), and so all four pairs of the cable have data transformers on either end of the cable. Does that mean there is no such thing as Alt-B in 1000Base-T applications? As per the IEEE 802.3at standard, *there still is*. The PoE terminology of 10/100 Mbps is maintained in 1000 Mbps applications, just to avoid confusion (but almost at the expense of causing *more* confusion!). So the PoE "AT" standard says that in 1000 Mbps applications, injecting PoE on what *would have been* unused pairs, had we been in a 10/100 Mbps application, is still called Alt-B. No direct PoE connection to the pairs is possible now, it *must* be via center-taps (phantom power on all pairs), even though it is called Alt-B. As a corollary, Alt-A is no longer exclusively power over data. In this case, Alt-B is *also* power over data.

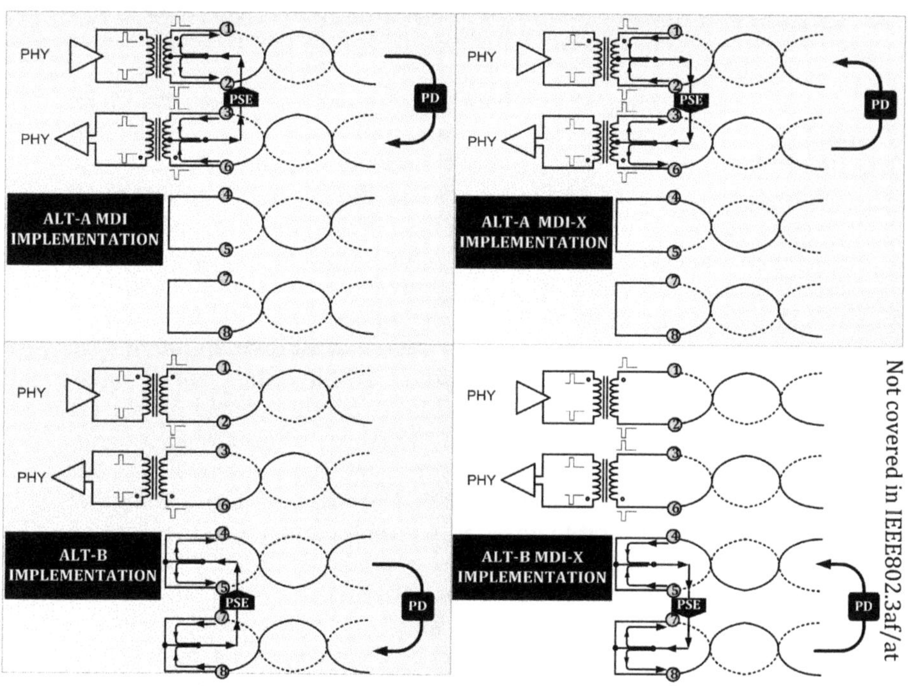

FIGURE 2.4 Alternative-A and Alternative-B.

Returning to 10/100 Mbps applications, there is additional granularity within Alt-A to note. In Alt-A, there are *two* possibilities, depending on the *polarity* of the PoE voltage as applied to the two data pairs. The overall Ethernet standard (802.3) has fixed which data pair of the cable is for transmitting and which one is for receiving ("transmit" and "receive" are both from the viewpoint of the switch/hub). But for PoE it is *not* fixed: We could place the positive polarity DC on the transmit pair, or on the receive pair. The choice is ours. This creates two PoE possibilities: Alt-A MDI and Alt-A MDI-X, as shown in Fig. 2.4. However, for Alt-B, the IEEE standard only specifies that positive polarity be connected to the pair 4 and 5, with negative polarity on pair 7 and 8. The opposite polarity case on the spare pairs is basically not an IEEE-compliant configuration. It may however work *in most cases*, because the input of a typical PD has bridge rectifiers on both the data and the spare pairs, so it will extract power from whichever pair and with any polarity. But strictly speaking, an IEEE-compliant PD need *not* have a bridge rectifier on the spare pairs—a simple diode will do. And in that case, if we reverse the polarity of Alt-B at the PSE end, the PD will not receive power.

Pin Numbering, Colors, and Registered Jacks

Previously, we had already started referring to the pin numbers and also their functions for data transfer and for PoE. Let us complete that discussion.

The familiar socket (or "jack") in Ethernet is called the RJ-45, and it is actually an evolved form of a previous family of registered (modular) jacks (or RJs), all slightly smaller on the sides than the Ethernet jack. In Fig. 2.5, we see one of the earliest/simplest of all was the RJ-11, or more accurately a 6p2c connector, which stands for six positions, two contacts (two copper wires). Typically, these two contacts only had red and green wires and could serve one analog telephone because there was just one twisted pair. The pin numbering was equally simple: Pins 1 and 2 were, as shown, in the middle of the connector. After that, we could also have the 6p4c or RJ-14. There is also the 6p6c or the RJ-25. Note how the twisted pairs were created in each case: They were all axial, or symmetrical (mirror-reflected) around the geometric center axis. The innermost was one twisted pair; then the two wires added to the sides formed another twisted pair. Another two wires added on the outside of that formed another twisted pair. But later, two more contacts were needed to support one more twisted pair for Ethernet in particular. For that reason, the existing RJs were expanded *on the sides*, making way for one contact on each side. Then came the question of how to *pair* (twist) the wires. The simplest way was to make a new twisted pair out of the two new wires, but for high-speed data purposes, the two

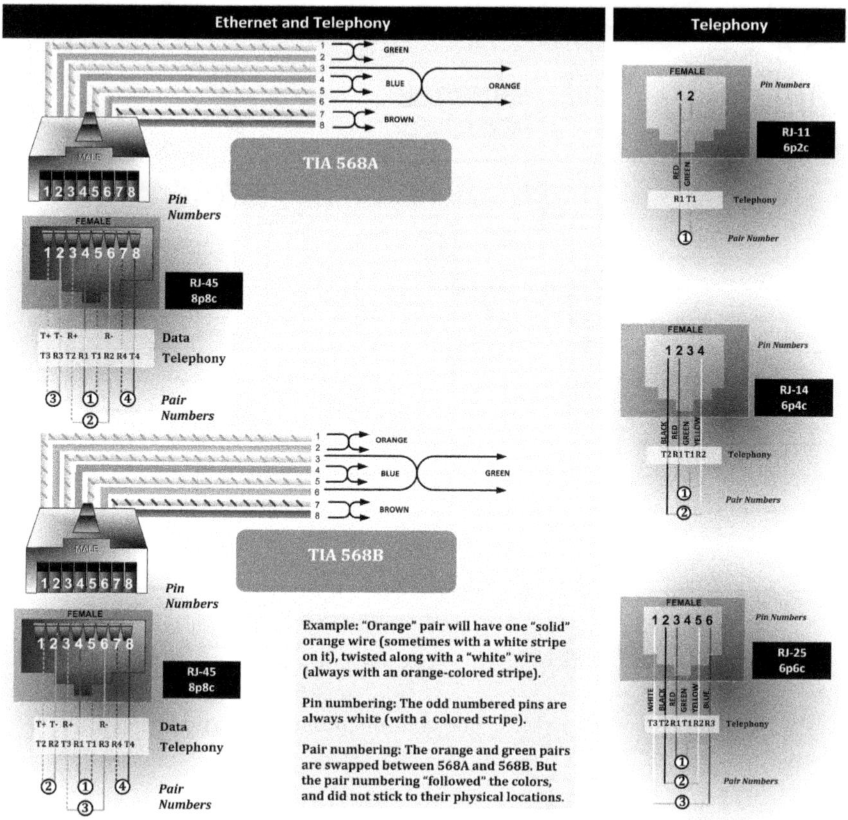

FIGURE 2.5 Pin numbering and color-coding conventions from telephony to Ethernet.

new wires would be too far apart physically (at the roots), and there would be a fair amount of distance before they could be brought together and twisted. That could not be very good for signal integrity. So the axial method gave way to the special winding technique shown on the left side of Fig. 2.5. Now for the RJ-45 (8p8c) we have the two outermost wires on *each* side forming separate twisted pairs (1-2 and 7-8). After that, the remaining wires follow the old axial/symmetrical method. Once again, we have the innermost wires being twisted together (4 and 5). That leaves 3 and 6 for the last twisted pair.

Note that the first Ethernet cabling standard was TIA-568A, which was, in effect, superseded by TIA-568B in 2001. The latter was largely unchanged, and though it became TIA-568C, for all practical purposes we can consider TIA-568B as the prevalent cabling standard. A key change in going from 568A to the new 568B standard was that the colors orange and green were swapped.

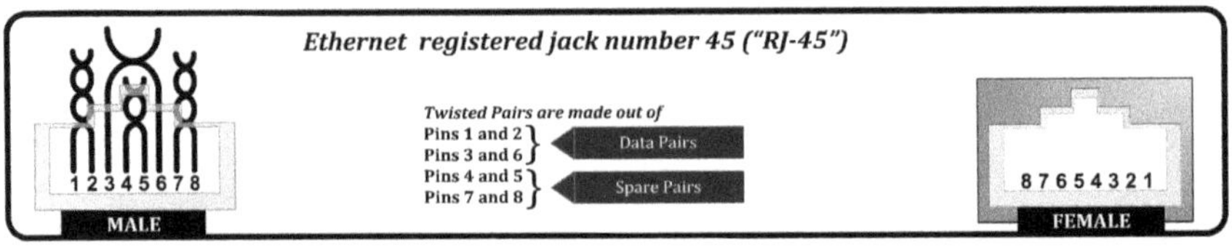

Any empty field in table implies "NOT USED"

	ETHERNET COLORS		VOICE		ETHERNET		PoE		
	TIA 568A/ 8p8c/RJ-45	TIA 568B/ 8p8c/RJ-45	Telephone/ 6p6c (RJ-25)		10/100 Mbps	1000 Mbps	Alt-A MDI	Alt-A MDI-X	Alt-B (MDI)
PIN 1 (DATA)	White (green stripe)	White (orange stripe)	Unavailable		Transmit +	+BI_DA	+48V	48VRTN	
PIN 2 (DATA)	Green	Orange	White	Tip3	Transmit -	-BI_DA	+48V	48VRTN	
PIN 3 (DATA)	White (orange stripe)	White (green stripe)	Black	Tip2	Receive +	+BI_DB	48VRTN	+48V	
PIN 4 (SPARE)	Blue	Blue	Red	Ring1		+BI_DC			+48V
PIN 5 (SPARE)	White (blue stripe)	White (blue stripe)	Green	Tip1		-BI_DC			+48V
PIN 6 (DATA)	Orange	Green	Yellow	Ring2	Receive -	-BI_DB	48VRTN	+48V	
PIN 7 (SPARE)	White (brown stripe)	White (brown stripe)	Blue	Ring3		+BI_DD			48VRTN
PIN 8 (SPARE)	Brown	Brown	Unavailable			-BI_DD			48VRTN

Pin numbering above is as per 8p8c connector (RJ-45). Colors are also only for RJ-45 (Ethernet plug).
All odd-numbered pins have white with stripes. Even numbers are usually solid colored, but may have white stripes on them.
568A was the older wiring standard, 568B is more recent.
In 568A and 568B, orange and green colors are swapped (the only difference).

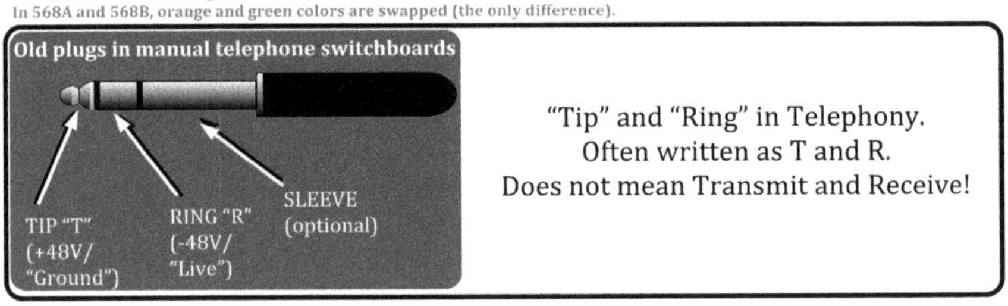

Figure 2.6 Summary of pin functions with PoE added on.

In an RJ-45 (for Ethernet), the data pairs are always 1 and 2 and 3 and 6. The spare pairs are always 4 and 5 and 7 and 8. So, the manner in which Alt-A and Alt-B were added on to these is tabulated in Fig. 2.6 for easy reference. Note that we have also shown prior art (telephony) for comparison. Note that the words "tip" and "ring," often abbreviated to T and R, are sometimes confused with transmit and receive.

Telephone Cable to Ethernet Cable

We will summarize some of the key points to keep in mind about cabling and jacks before we move on. In Fig. 2.5, on the right side, we have the registered jacks (RJs) used in telephony since the 1970s. The simplest version could involve just one twisted pair per cable/RJ, and that is RJ-11, or a 6p2c telephone jack: referring to its 6 positions (available slots) and its 2 contacts (number of wires present). Another twisted pair could be added to the cable, and that would make the same jack into a 6p4c jack, called an RJ-14. One more twisted pair would make this a 6p6c jack, or an RJ-25. At this point, the physical place in this particular RJ would have run out, since it has a maximum of six positions available and we have used up all of them in the form of six wires/contacts. We can add another twisted pair by simply "widening" the same jack, not changing anything else. And that actually takes us to the standard RJ-45 (8p8c) used in Ethernet. That is also the reason we may discover, while accidentally trying to in a telephone into the back of our computer, that *a male RJ-11 plugs right into the standard Ethernet jack (RJ-45)*. This property can actually be used to implement/test proprietary "single-pair" data communications on standard telephone-cabling infrastructure.

To specify how the pairs are created for Ethernet applications (in an RJ-45), we have to refer to the relevant Ethernet cabling standards. The first standard that described Ethernet cabling was TIA-568A in 1991, updated in 1995. As mentioned earlier, this was eventually superseded by TIA-568B in 2001. A key change in both these standards, compared to the historical practice of axial pairing, is shown in the pairing diagrams on the left side of Fig. 2.5. Two of the innermost pairs follow the old method of axial pairing, but after that, the two extreme wires on *each side* are paired together. The current standard is actually TIA-568C, with slight changes from the 568B standard (568C introduced CAT6A, and discusses fiber optics).

Note that the formal pin-numbering and pair-numbering scheme changes completely as we go from one jack/configuration to another. The pair-numbering scheme changes even within the RJ-45, just by going from 568A to 568B; the pair (not pin) numbers 2 and 3 are swapped, for example. The color coding is also different in going from 568A to 568B; the orange and green pairs are swapped. What remains common to both is that the orange pair in 568A is still called pair 2 in 568B, and the orange pair of 568A is also pair 2 in 568B. The pin numbers are different. Similarly, green remains pair 3 in either standard. This may be all somewhat confusing at first. But it is useful to know this while in the lab.

Here are some of the key bullets:

- Orange is called pair 2, green is always pair 3. Their pin numbers, however, differ in 568A and 568B.
- Blue is pair 3; brown is pair 4. Their pin numbers are the same in both 568A and 568B.
- In Ethernet, any pair of color X consists of one solid wire of color X, combined with one white wire with stripes of color X on it. Sometimes, the supposedly solid-colored wire may have a white stripe on it too.
- Odd-numbered pins, that is, 1, 3, 5, and 7 are white (with appropriately colored stripes).

- Even-numbered pins, that is 2, 4, 6, and 8 are solid colors (may have a white stripe).
- In 10/100 Mbps, pins 1 and 2 are for transmit, pins 3 and 6 are for receive (from the viewpoint of the switch/hub). These are designated data pairs of the cable. *These are the pins used for phantom-powering in PoE as explained previously.*
- The twisted pair on pins 4 and 5 is *not* used for data in 10/100. Similarly, the twisted pair on pins 7 and 8 is also *not* used for data. Therefore, these are often called spare pairs.
- Ethernet cables can be used for telephony too. In that case, the pins are traditionally designated as T and R. These do not refer to transmit or receive anymore: they stand for tip and ring. These refer to the sections of the standard TRS (tip, ring and sleeve) plug used in manual switchboards: (see inset in Fig. 2.6). The idea was that while plugging, the first metal piece to make contact should not be "live" but "ground (earth)" for safety reasons. And the ground was traditionally the upper rail (higher potential) those days, unlike nowadays, where we commonly use the lower rail as our circuit ground.

NOTE *There has been a flip-flop on what should be the Earth-ground convention in telecom systems (also automotive). Most don't realize it, but in the earliest telegraph and pipeline systems, the ground was the lower potential rail (just as it is today: negative-ground convention). But, it was seen over time that any long metal object (for example, a copper wire, a pipeline with stray currents, or an electrified tram/train rail much later), with a higher potential on it with respect to earth (ground), degraded because of electrolysis. So, the polarity was changed, and the long metal object (pipeline or single conductor) was made negative with respect to earth (lower potential). In other words, the convention changed to positive-ground for years thereafter (the positive/upper rail of the battery was then connected by a water pipe deep into earth).*

In modern times, almost all circuits (on printed circuit boards and chip substrates) are negative-ground once again. So that is the preferred convention again. Most switches and hubs too have a metal enclosure (chassis) that is connected to earth-ground, which is also connected to the lower supply rail on the printed circuit board on the PHY (host) side. *In other words, we now typically use a negative-ground convention on* the PHY side. *But in reality, the grounding convention on the PHY side* does not matter *at all,* since it is isolated from the lines by the drive transformers, *unlike older telegraph/telephone systems. There is simply no possibility of electrolysis here. But, does it matter to the PoE sections? No, and for the same reason actually. The entire PoE circuitry, including the lines, is isolated from the earth-ground anyway (there is an isolation boundary in the data transformer). It is floating, so it does not matter whether we call the upper or the lower PoE rail "ground." It is the PHY (host) side that gets connected to the enclosure/chassis, which is decisively earthed through the main wiring or other means. The bottom line is that in Ethernet we can stick to the normal negative-ground convention of today. It really does not matter anymore as it did historically with no transformer coupling.*

NOTE *The standard TRS switchboard plug, used in the early 20th century, was later used for analog; our standard ¼-inch headphone plug is exactly the same as the early TRS switchboard plug. Its mono version (without the sleeve) was also commonly used in telephony. Typical 3.5- and 2.5-mm small audio plugs, such as those used in mp3 players, are just miniaturized versions of the early switchboard plug.*

NOTE *A 6pXc male (corresponding to RJ-11, RJ-14, or RJ-25) can be plugged into a standard female RJ-45 jack.*

Finally, to avoid confusion in Ethernet and PoE, the word "jack" is commonly used to refer to the female, whereas "plug" is used for the male.

Midspan or Endspan?

Once PoE is incorporated into a switch/hub, the switch/hub becomes PoE-capable, and as per IEEE terminology, it is then an endpoint PSE (or colloquially, an Endspan), because of its location at one *end* of the cable.

One question that arose during the creation of the IEEE PoE standard was would the standard *force* people to throw out all their old switches, hubs, and so on, and buy brand-new Endspans? That could prove very expensive. Wasn't there some way to simply *upgrade* existing installations if desired? The answer to that was the PoE Midspan (or injector). As its name indicates, it is a unit inserted somewhere between the switch/hub and DTE (PD). It contains a PSE by which it can inject PoE onto the cable. We therefore have two PSE possibilities in general: the *Endspan* and the *Midspan*, depending on their physical location and function.

This terminology was based purely on *physical location*. How about inside? How does a typical Endspan actually work? With a little thought we will realize that it *usually* makes more sense to use Alt-A inside an Endspan, though we could opt for Alt-B instead. How does a typical Midspan work? Here, Alt-A seems to be illogical in *most* cases, and so Alt-B seems a better choice. See Fig. 2.7. In other words, in a Midspan, there seems to be no reason to break up the path of data or introduce new transformers/autotransformers (for center-tapping)

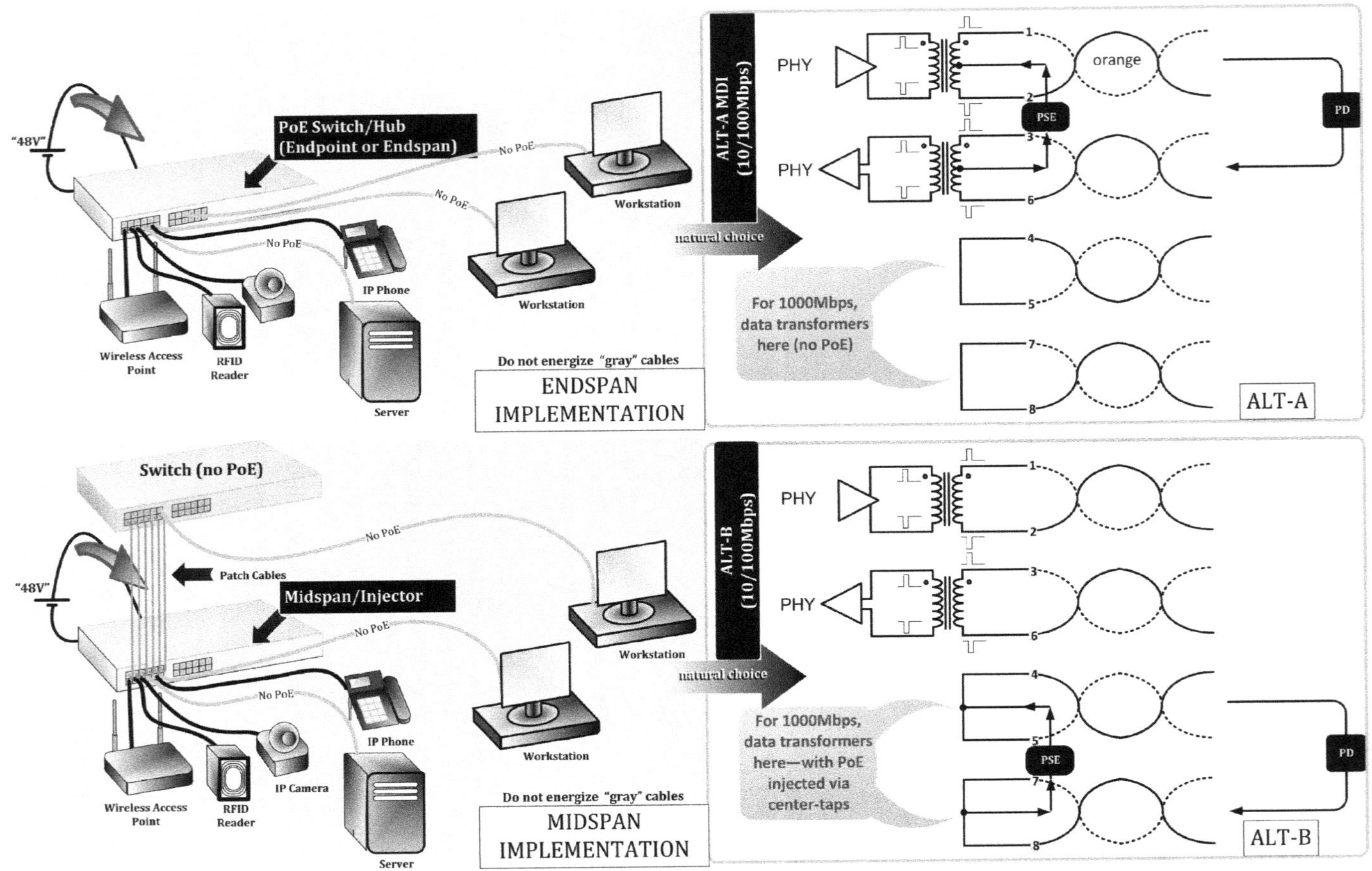

FIGURE 2.7 Endspans and Midspans with their natural choices (Alt-A and Alt-B, respectively).

to inject PoE. It seems to make more sense to allow the data to pass through the unit uninterrupted and use the spare pairs to add power with a direct connection as shown in the figure.

So, from Fig. 2.7 we see that Alt-A (power on data pairs) is the natural choice for Endspans *in most cases.* Similarly, Alt-B (power on spare pairs) is the natural choice for Midspans *in most cases.* But these natural choices can be changed on occasion.

In Fig. 2.8 we have shown more complete schematics of Endspans. Here are some points to note.

1. Terminations have been shown in this figure for the first time, consisting of several resistors and capacitors connected to the transformers. These are for EMI suppression purposes and matching of impedances, as discussed later in this chapter.
2. Earthing and isolation boundaries are also shown clearly. There are areas marked "safe to touch" because they are connected to earth-ground. Everything with a direct galvanic connection to the line (Ethernet cable) is relatively unsafe to

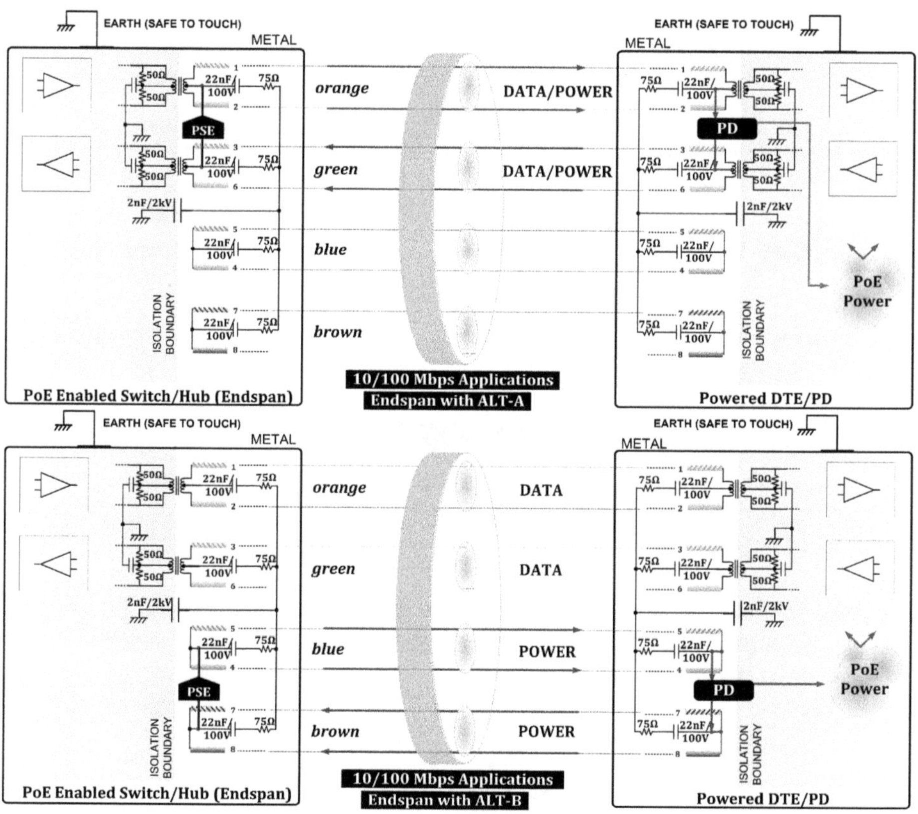

FIGURE 2.8 Complete Endspan schematics (Alt-A and Alt-B) showing terminations and Isolation boundaries.

touch (including the PoE circuitry). This will also be discussed in more detail later.

3. Note that a new symbol for a PSE has been introduced for convenience. As we have learned, a PSE is basically a switched battery, with a certain polarity. So the polarity of the newly-introduced PSE symbol mimics the way polarity of electrolytic capacitors is usually indicated (but with a pointy edge).

With the advent of 1000Base-T applications, all four pairs of the cable are used for data, so there are no spare pairs in reality. But as mentioned, names linger on, and IEEE 802.3at continues to refer to Alt-B. In this case, a Midspan unit *will* need to contain transformers/autotransformers to be able to inject power via center-taps.

In Fig. 2.9 we have shown more complete schematics of Midspans. Here are some points to note.

1. There is a lot of discussion on whether regular (isolated) drive transformers should be used in a Midspan, and if so, what is the best way to add terminations on either side. This aspect is best left to the signal-integrity and EMI engineers to decide. But in Fig. 2.9, we have shown a very simple implementation that most engineers seem to agree requires *no terminations*—by

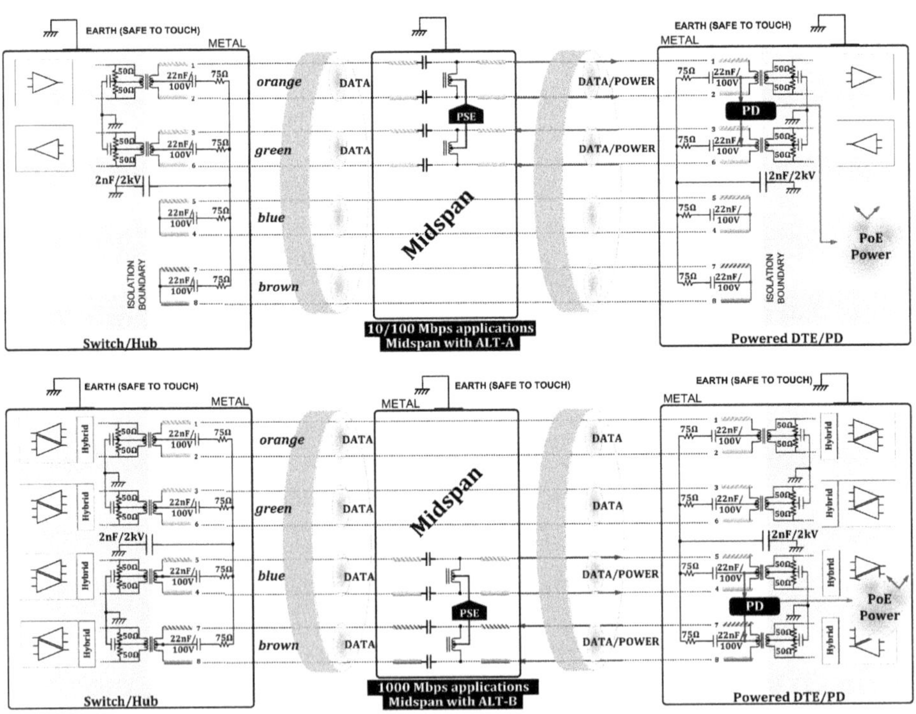

FIGURE 2.9 Complete Midspan schematics (10/100 and 1000 Mbps).

just using capacitor coupling with the center-tapped auto-transformer method of injecting PoE, as explained previously.

2. For 1000Base-T, *bidirectional* transceivers are used on each pair. To keep the send and receive blocks separate, a "hybrid" circuit is required. Incidentally, this is the electronic equivalent of the hybrid transformer used in analog telephony, as discussed previously in Chap. 1.

3. In terms of terminology, keep in mind that usually we tend to consider the terms MDI and PI equivalent when PoE is present (an MDI with PoE is a PI). But in a Midspan, the outgoing interface between the cable and the Midspan unit (the RJ-45 with PoE available on it) is considered a PI, not an MDI, since the latter also implies a PHY alongside. But there is no PHY in a typical Midspan.

Transmission Lines

Some basics should be briefly covered. Any two, long parallel conductors have an associated distributed inductance and capacitance, which imparts to them a characteristic impedance of $\sqrt{(L/C)}$. That is why it is often said that the impedance of a typical coaxial cable used for radio-frequency purposes (RG-6) is 75 Ω. The unshielded twisted pair inside the Ethernet cable has an impedance typically 100 ± 15 Ω above 1 MHz.

This leads to the theory of transmission lines. When an AC signal is sent down this transmission line, it can travel great distances. If the transmission line is ideal, that is, if it has zero AC and DC resistances, there are absolutely no losses in the cable, because pure reactive elements (L and C) can store energy but cannot dissipate any. So in an ideal case, the AC signal can travel infinite distances without any attenuation. In reality, it decays progressively with distance, because of resistance. Historically, transmission-line effects were discovered and used to tremendous advantage in the 19th century, by Lord Kelvin, Pupin, and others (see Chap. 1). Modern transmission-line equations were also originally called "telegrapher's equations."

One basic question is: Why are we restricted to 100-m cable length in Ethernet? The main reason is: DC resistance and consequent signal attenuation. And that is the reason why TIA-568B specifies a maximum DC resistance of *9.38 Ω per 100 m* (for CAT5e). In other words, the low-voltage differential signal applied on one side off the cable steadily decreases in amplitude as it propagates, and eventually it will fall to a level where it will be masked (overwhelmed) by noise. Basically it all becomes a question of signal-to-noise ratio (SNR). We can extend the reach of the signal further by just lowering the noise/EMI pickup. As explained in Chap. 1, that is a natural quality of a twisted pair. Note that besides DC resistance, AC resistance of the cable also comes into play, and AC resistance is a function of frequency. Higher frequencies will generally suffer more attenuation. So the square digital signal (with high-frequency harmonic content) will gradually distort as it propagates. In general, for a given cable length, we can send higher and higher data rates depending on how "good" the twisted-pair cable is. That is the difference between CAT3 and CAT5e for example.

No transmission line is infinite. So, it is important to terminate the transmission line correctly, with a matching impedance/resistance. Otherwise we will get reflections and standing waves in the cable, which will compromise signal integrity. We had shown typical RJ-45 (Ethernet) terminations in Figs. 2.8 and 2.9, but without any explanation as to how they worked. Yes, terminations are not PoE-dependent directly, but the presence of PoE can affect their performance and reliability. As PoE engineers, we should understand how terminations are designed/created and what their basic function is.

Terminations

We mentioned that the impedance of the twisted-pair transmission line is 100 Ω. In the topmost part of Fig. 2.10, on the left (PHY) side of the transformer, we have two 50 Ω in series between the two wires of the twisted pair. So we get a total of 100 Ω. We know that any impedance reflects across a transformer boundary according to the square of the turns ratio. The drive transformer in a typical case has a turns ratio of 1:1, so the 100 Ω on the PHY side appears on the line side as 100 Ω, and that is the appropriate termination for a 100 Ω line (it matches the impedance of the line as desired). Note that if there is a *common-mode* disturbance, the center node of the two 50-Ω resistors will not remain at a fixed potential, but will move up or down with respect to ground. That will produce a current flow through the capacitor in the middle of the two resistors. The current clearly depends on the common-mode noise/EMI component, and so by shunting that component to ground, in effect we have produced common-mode rejection. We saw in Chap. 1 that noise pickup on a (properly) twisted pair is common-mode, so by rejecting any common-mode component, we are in effect improving the SNR, and thereby improving the quality of the signal received at the end of the cable.

But, in an Ethernet cable, there are four twisted pairs, and that means several conductors in parallel, not just two. So, in fact, there are several possible transmission lines we can visualize. There is much discussion (and occasional disagreement) about how to analyze all this, but signal integrity issues are not part of our scope anyway. So, to put it very simply, just for understanding any possible impact from adding PoE, we say that two *adjacent twisted pairs* have a "pair-wise" characteristic impedance of around 150 Ω. That is why we have two 75-Ω resistors in series between any two pairs in Fig. 2.10.

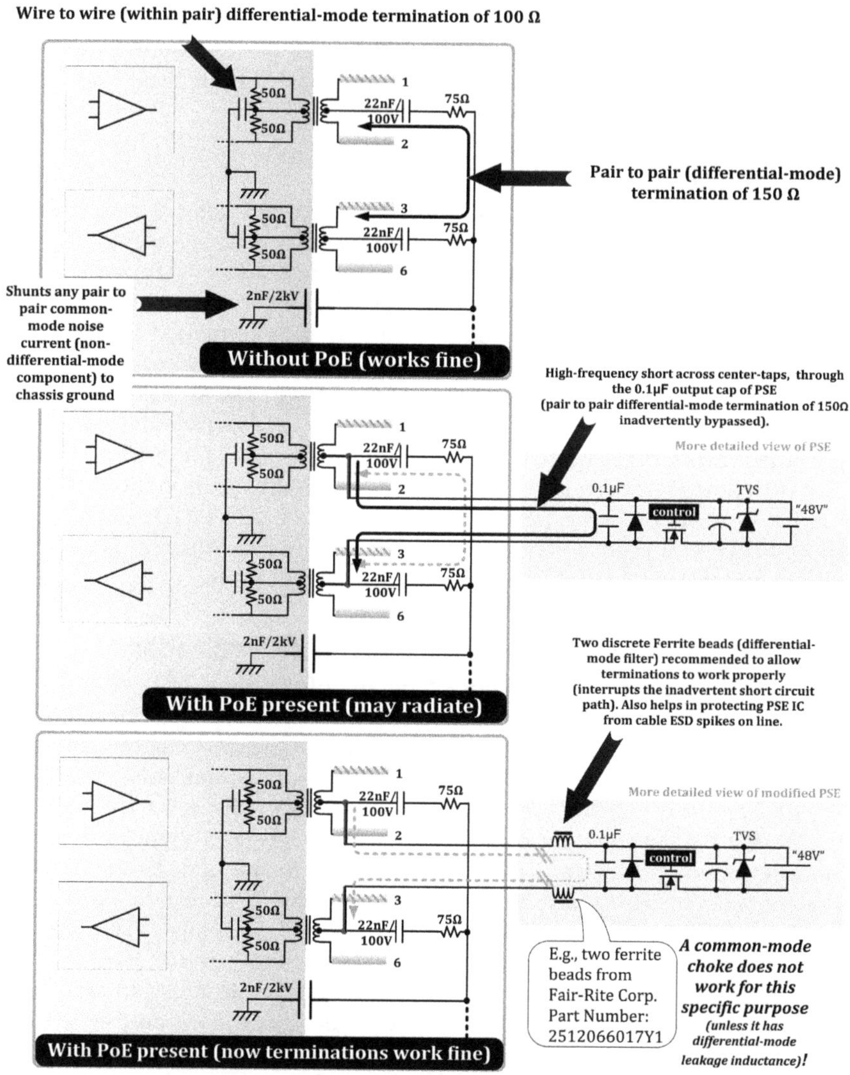

FIGURE 2.10 How inserting PoE function can affect the function of the terminations, and how to prevent that (on both sides of the cable).

Once again, if there is common-mode noise *between pairs*, the high-voltage cap in the middle (2 nF/2 kV typically) will conduct. The high-voltage cap thereby rejects pair-wise common-mode noise pickup. It is a high-voltage type for isolation reasons, not for EMI, and this aspect is discussed in Chap. 10.

In general, any capacitor connected from a part of any circuit to chassis/earth-ground for EMI suppression purposes is called a Y-cap. Pure differential-mode components on the line will not cause any of the Y-caps in Fig. 2.10 to conduct, but of course will pass small currents through the two 50-Ω or two 75-Ω resistors, thus offering the required transmission-line termination impedance to the "true" signals (which we know are differential-mode).

"PoE-Friendly" Terminations

Previously, we had mentioned "pre-PoE" devices could get damaged easily by high voltage on the line. The reason was those are what we now consider PoE-unfriendly terminations. These are shown in Fig. 2.10, and now more clearly in Fig. 2.11. The 75-Ω resistors are tiny low-wattage resistors that would get immediately damaged if high voltage was present at the center-taps. Today, almost all devices, whether they are themselves PoE-capable or not, are at least PoE-friendly, so their terminations will not get damaged even if the port gets inadvertently energized (by a malfunctioning PSE, for example). The difference between PoE-friendly and -unfriendly terminations is the 10- to 22-nF blocking capacitors as shown in Fig. 2.11. Almost all modern devices seem to have these now, whether the devices actually support PoE or not. The same is true on the PD side.

Adding PoE without Affecting Functionality of Terminations

In Fig. 2.10, we reveal more hidden details of a typical PSE. A key component on its output is the decoupling 0.1-μF (typical) ceramic capacitor. Any PSE needs to monitor port voltage and current for IEEE compliance. Since the lines can pick up substantial noise, this 0.1-μF cap is necessary for the analog-to-digital converter (ADC) inside the PSE to function properly. Unfortunately, as shown in the middle schematic of Fig. 2.10, this creates a high-frequency short at the center-taps. This short basically bypasses the 75-Ω resistors and reduces their expected functionality as explained previously. That is why a *differential-mode* (not common-mode) filter is needed here as shown in the lowermost schematic of the same figure. This can take the form of two appropriately sized ferrite beads on each output of the PSE. Many engineers seem to use a common-mode choke instead, almost by sheer habit, based perhaps on traditional signal-integrity practices. But in this position, a common-mode choke actually presents *no* impedance to what is really a *differential-mode* signal between the center-taps. Why? Because a common-mode choke presents an impedance to common-mode signals/noise only, not to differential-mode. Yes, a common-mode choke in this location may have some beneficial effect—it can prevent any common-mode noise from coming in from the *reverse* direction, that is from the "48V" power supply onto the lines. But that functionality should really be present *inside* the silver-box AC-DC power supply itself. A good telecom power supply is usually tested for EMI compliance (as per CISPR 22 or EN-550022) not only at its main-input side, but on its *outputs* too.

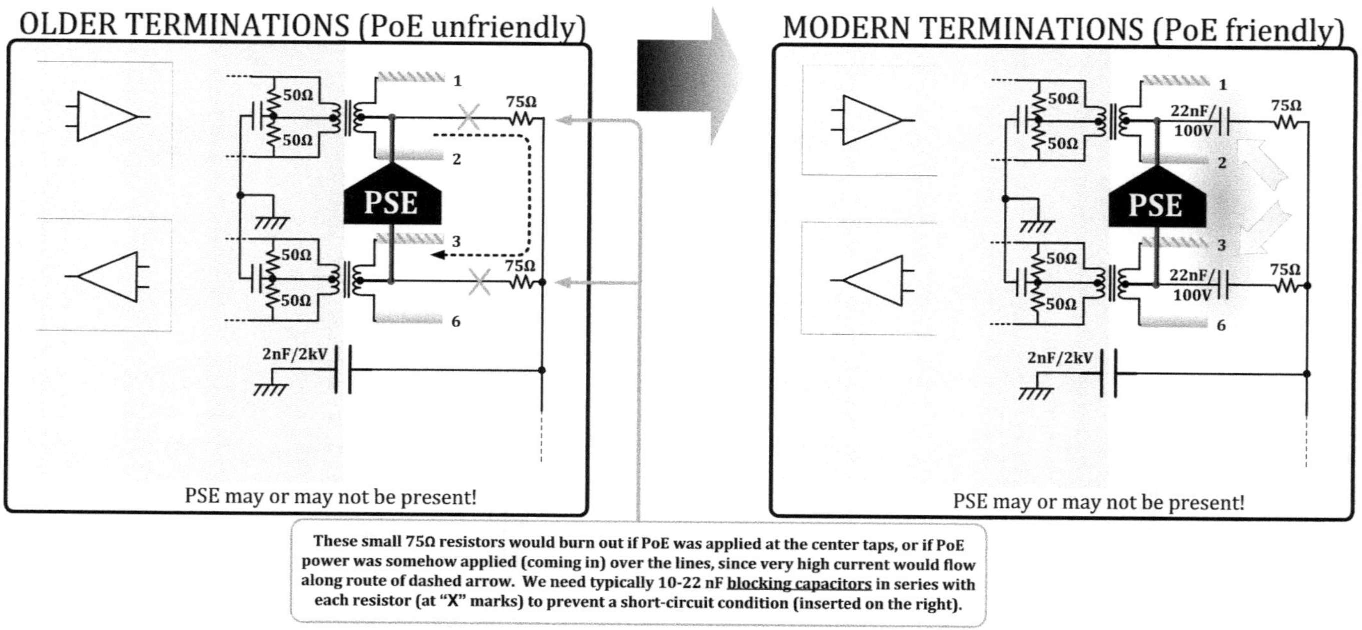

Figure 2.11 Pre-PoE devices had PoE-unfriendly terminations. Modern devices, whether PoE-capable or not, typically have PoE-friendly terminations to avoid accidental damage.

As mentioned, the PSE control section and FET don't *add* any noise to what is already coming in. They act as a "door" that opens or closes for the "48V." The PSE section does *not* actively regulate or condition the "48V" rail in any way. So common-mode filtering should reside in the AC-DC power supply, not in the PSE circuitry.

Types of Powered Devices

In Fig. 2.1, we had shown a very simplified PD. We now expand on that. At the top of Fig. 2.12, we have the most common implementation of a PD interface. We learned that to support Alt-A MDI and Alt-A MDI-X, a bridge rectifier is needed on the data pairs. However, to support Alt-B we actually only need a single diode on the spare pairs, since pins 4 and 5 are always of positive polarity with respect

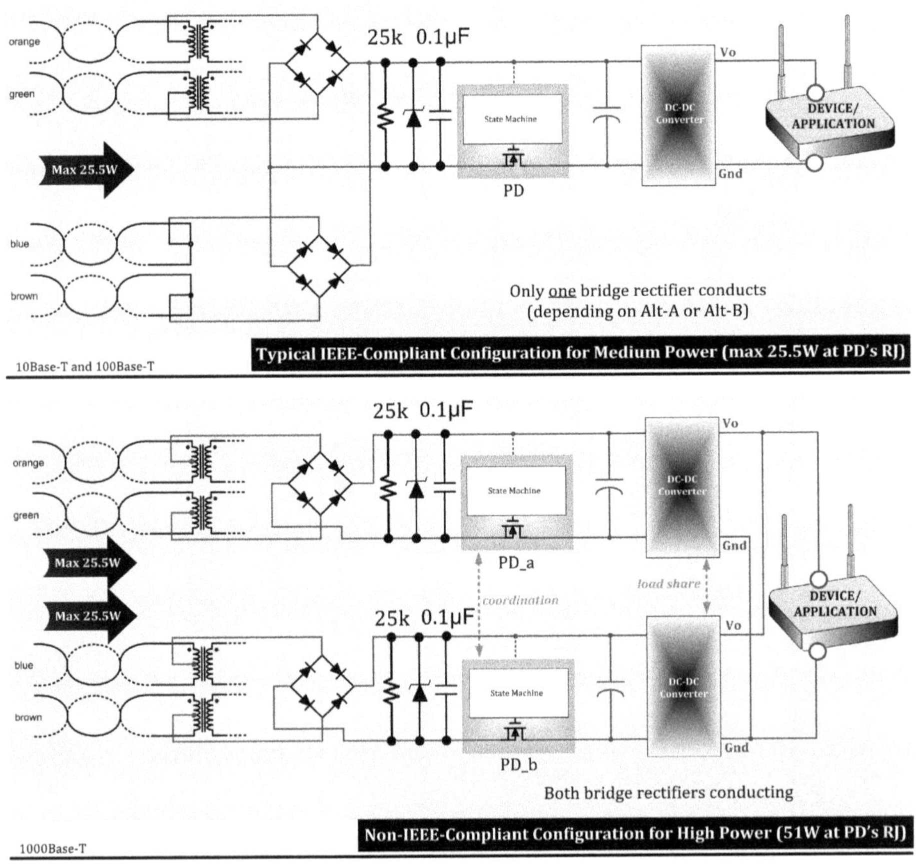

FIGURE 2.12 Typical PD-interface implementations.

to 7 and 8, as per IEEE requirements. Nevertheless, to keep diode drops and reverse leakages the same on both the data and spare pairs, and to thereby avoid strange incompatibility/interoperability issues, most commercial PDs have identical diode bridges on both sets of pairs, as shown in Fig. 2.12.

For low-power and medium-power applications, the diode bridges are OR-ed as shown in the upper schematic of the figure. That means the positive and negative terminals of the two bridge rectifiers are connected before the pass-FET. Depending on whether the PSE is sending power in Alt-A or Alt-B, only one bridge rectifier ends up conducting, and then on only two of its four diodes.

In high-power applications (industry-driven, not IEEE), the custom-PSE will send power down all four twisted pairs. In this case, *both* the diode bridges are intended to conduct at the same time. However, if one bridge conducts more readily, it can end up reverse-biasing the other bridge rectifier, cutting off the current through it entirely (or reducing it significantly, producing imbalance). This situation can easily happen because, though typical cabling standards specify 5 percent or less difference in the DC resistance of a given pair, between different pairs of a cable, the DCR difference can be almost twice that (closer to 8 percent). So for example, if the data pairs have lower DC resistance compared to the spare pairs, the diode bridge connected to these pairs will conduct more readily, reverse-biasing the bridge on the spare pairs. That would *not* make it a high-power application by definition. And even if somehow we do extract more power, it would be the result of excessive current through only two pairs of the cable instead of four. The acceptable temperature rise data that the IEEE committee depended upon while framing the AT standard is no longer valid in a case of such imbalance. So to properly share current between the data and spare pairs, the two diode bridges are *not* OR-ed in high-power applications, but go to separate pass-FETs (distinct though coordinated PD interfaces). From there they go to separate DC-DC converters. These converters are, in turn, often designed with *active* current-sharing circuitry, using load-share ICs, like the industry-standard UC3907 from Texas Instruments (TI). Also refer to the U.S. patent number 7,492,059 from Peker et al.

Note that, in general, outputs of multiple DC-DC converters (regulators) should never be tied directly together, since regulators are high-gain systems (for tight regulation), and therefore any slight differences in their output levels can easily cause severe output instability and oscillations.

The last thing we should notice in Fig. 2.12 are the 25-k resistors. These are signature resistors that serve to identify the PD to the PSE, as conforming to the IEEE 802.3af/at standard. The actual identification range and detection (discovery) process is discussed in the next chapter.

CHAPTER 3
Detection

Overview

IEEE 802.3af came into a world in which many devices were already out in the field working off power delivered via an Ethernet cable. Businesses had not been sitting around waiting for an open-standard to emerge and be gradually ratified. The idea of sending power over the same cables as data/voice was as appealing now as it had been over a hundred years ago. Why wait?

Initially, IP phones were meant to be supported, but later powered wireless-access points (based on 802.11) joined in too. And that was the only thing that was clear! A big problem was that all these "legacy" or "pre-standard" PDs, as they are called today, were based on proprietary implementations of PoE, such as Power over LAN (or PoL, from PowerDsine, now part of Microsemi) and Inline Power (ILP, from Cisco). To complicate matters, many devices also appeared that declared themselves to be draft-compliant—that is, compliant to some intermediate IEEE 802.3af version prior to its full ratification. It was quite reminiscent of the situation surrounding the rise of Ethernet at the expense of Token Ring and so on, just a few decades ago. Just a lot *more* confusing.

A major source of confusion was that even the proprietary PoE-type standards (like PoL and ILP) came into a world in which there were already many devices that *could not* accept any form of PoE—they could get damaged if high voltage was applied to the lines. For example, the vulnerability of "PoE-unfriendly" terminations was discussed in Chap. 2. To avoid such situations, and to apply power judiciously, each proprietary implementation had its own way of discovering or *detecting* whether there was a *compatible* PD on the other end of the cable or not.

> **Detection is the process by which the PSE asks:** "Who are you?" (In effect; are you "my type" or not?)

There are many obvious questions in such a matchmaking situation. For example, how *should* a pre-standard PSE, based on say some general PoE implementation which we call Type A here, respond to a PD based on another general PoE implementation, called Type B? It seems the best option is for a Type A PSE to *not* power up (or detect) a Type B (incompatible) PD, because that would at least avoid damaging the unknown/incompatible PD. And that did happen in fact. In this way, some higher form of coexistence had been achieved, but *interoperability* (the ability to play well together) was not. A Type A PSE could not operate a Type B PD, and so on. Therefore, at some point, PowerDsine (now Microsemi), in particular, started driving toward an open PoE standard, which evolved slowly into IEEE 802.3af.

It is possible that some major vendors did not favor interoperability, at least initially, so as to be able to promote their own brand/family of products at both ends of the cable. It became a veritable PoE battlefield, in the rather pointed phraseology extracted from an interesting "competitive positioning paper" that can be found on the web today (at ftp://ftp.ocs.ru/pub/gvozd-ftp/BS/BayStack%20Switches/BayStack%20460/poecisco_pss5.pdf). Dated February 18, 2003, from Nortel (now part of Avaya), titled "Cisco Catalyst Inline Power." Note that Nortel was already working closely with PowerDsine at the time on the issue of pushing through an open standard. Its claim in this paper was that Cisco Inline Power was proprietary and very different from the evolving open PoE standard. That perception probably led to the following statements (quoted straight from the above-mentioned Nortel paper):

> Customers who may believe that their Catalyst LAN environment is "IP Telephony-ready" soon find out that they are only ready for Cisco-based IP telephony (see Page 2 of above-mentioned positioning paper).
>
> ... Enterprises must be aware of Cisco's attempts to portray infrastructure upgrades as open and a preparation for IP Telephony when many of these upgrades only support "Cisco Telephony... (see Page 2 of above-mentioned positioning paper).
>
> Blindly following Cisco in these cases will cause enterprises to lose competitive advantage—to say nothing about the increased costs of the entire infrastructure when alternatives are rarely if ever considered... (see Page 4 of above-mentioned positioning paper).
>
> In a cruel twist of fate, even Cisco's 802.11 Access Point is not compatible with their own pre-standard PoE implementation (see Page 2 of above-mentioned positioning paper).

These only serve to highlight the confusion, and some bitterness too, with the existing situation at the time. But there was another "twist of fate" awaiting at the end of all this open-standard effort. Because the IEEE PoE standard came up with an entirely new way of "detection" to discover IEEE-compliant PDs (*only*) and power them up as per its own recommendations. From a higher level, looking down at this vast playground generically called "PoE," the IEEE standard was just another

PoE implementation, "Type C." One more had just gotten added to the existing confusion, and the confusion would continue until the other implementations died naturally, as is expected eventually whenever the high-adoption rate of an open standard takes over.

Note that IEEE 802.3af/at is not and was never intended to be a legal requirement. It is purely *voluntary*. The thinking was that its adoption is best left to *market-forces*, just the way Ethernet, well before PoE, had originally proliferated. Open standards would win eventually, based on the overall market desire for interoperability and lower costs. And that is, in fact what seems to have happened eventually. Today we see no switches/hubs based on PoE Implementations Type A or Type B. The winner is clearly Type C (the IEEE PoE standard).

After the AF standard was ratified, for almost a decade, IEEE-compliant PSEs were being asked by market forces to continue to support existing pre-standard devices out there in the field (like older IP phones). Granted it wasn't mandatory to do so, but it was still unavoidable. Fast-forward to a decade later (today), most of these pre-standard (legacy) PDs have reached the end of their useful lives and have almost all been replaced by IEEE-compliant versions. But prior to that, IEEE-compliant PSEs were being asked to offer a "legacy detection" option in addition to IEEE detection. Many such PSEs ICs are still around us today, though the most recently released PSE ICs are no longer supporting this feature of legacy detection.

Next, we will discuss how some of the pre-standard detection schemes worked.

Pre-Standard/Legacy Detection Schemes

These come in several types:

1. In Fig. 3.1 we have shown the key *non-IEEE* detection schemes. The top half shows the principle behind the well-known Cisco Inline Power ("ILP") detection. A low-frequency "ring signal" is injected on one pair and this gets looped back by a valid (ILP-based) PD. The ring signal returns to the PSE on the other pair, which then declares it to be a valid (ILP-based) PD and powers it on. This may need additional hardware, not usually part of most PSEs and PDs. For example, a ring generator is needed on the PSE side, and a low-pass filter, or normally closed relay contacts, on the PD side. When power is sent down the cable, it will be used to actively disconnect circuitry/relays on the PD side that are meant only for detection, so as to allow normal Ethernet traffic to commence.

2. In (non-ILP) "legacy" detection schemes used in most modern PSE ICs, a small probing current source called Idet, typically between 100 μA to 2 mA (max 5 mA), is used as shown in Fig. 3.2. This is not only current-limited but also typically

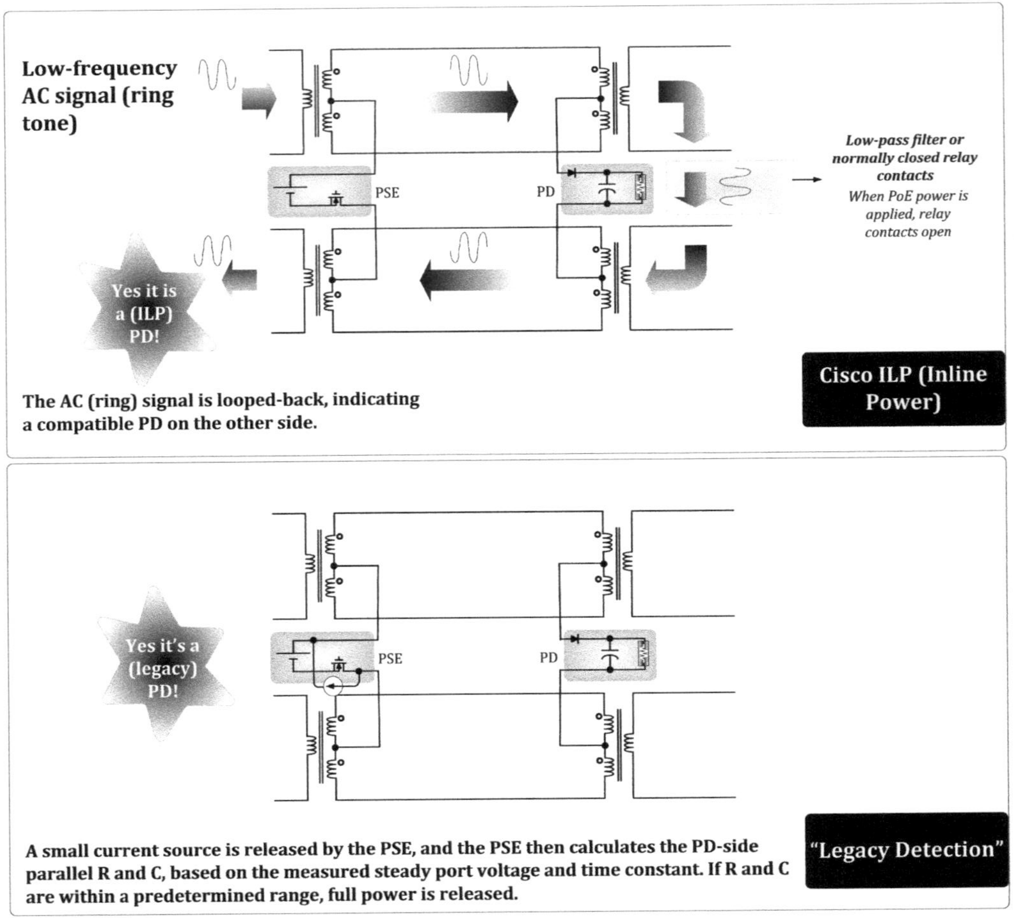

FIGURE 3.1 Pre-standard detection techniques.

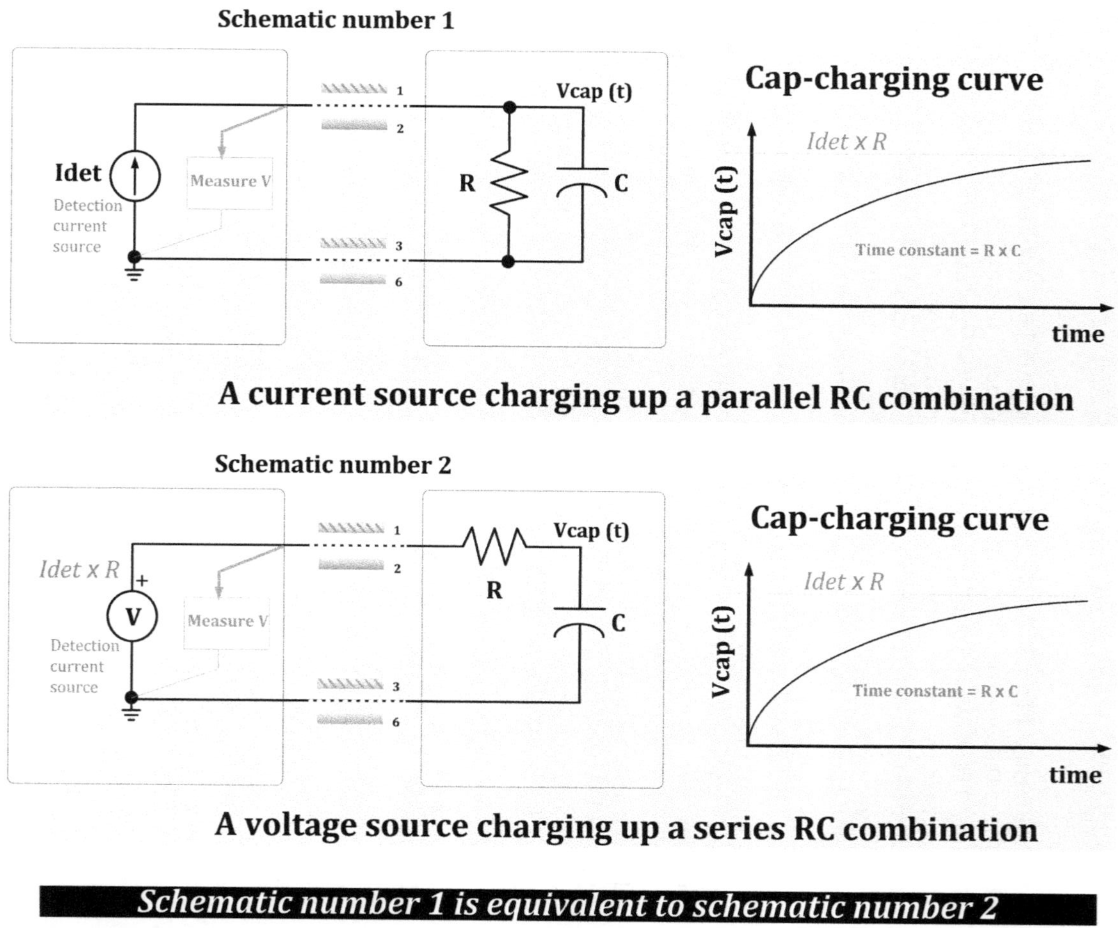

FIGURE 3.2 A typical legacy-detection (non-ILP) calculation scheme.

voltage-limited to less than 30 V (it is in effect a voltage-clamped current source). This arrangement is considered safe for older (PoE-unfriendly) terminations, if these ever happen to appear on the other side of the cable. The worst-case dissipation would then be (2 mA) × (30 V) = 60 mW—not enough to cause the small 125-mW termination resistors to burn out.

The legacy PD can be thought of as a parallel combination of some resistance in parallel with a bulk-decoupling capacitor. Note that any PSE always monitors the port (PI) voltage, and since the drop across the cables is negligible at these current levels, we can consider the PSE-side PI voltage reading as being almost the same as the PD-side cap-charging voltage. So, the PSE can monitor the *capacitor-charging curve*. First, based on the final settling voltage value $V = \text{Idet} \times R$, as read by the PSE, the PSE can evaluate R on the PD side. Second, by taking a few samples of the voltage during the charging/discharging process, the time constant $\tau = R \times C$ can be evaluated, from which we can plug in R (known) and thus calculate C. Finally, both the effective R and C that the PD presented to the PSE can be evaluated. These values are then compared to a predetermined range of acceptable values of R and C (and their combinations) for known legacy PDs, and the PD is powered up or rejected accordingly.

Cisco phones based on ILP could also be detected and powered up, not by using ringtones, but by probing their R-C signature (legacy detection) and "opening the door" for a valid combination of R and C.

Keep in mind that IEEE-compliance must not be jeopardized in the process. At least that is the key goal while creating a "good" legacy detection scheme. And it is usually borne out. But there are cases in which legacy detection has led to test failures on typical test equipment, such as from Sifos, which is almost a de facto requirement by now.

For example, suppose IEEE 802.3af/at says that a resistance of a certain value *must* be rejected (perhaps in conjunction with a certain capacitance). Then we should never allow that value, *even under legacy detection*. Because, if we accept that resistance value and power up the PD, the PSE will be considered IEEE-noncompliant. As a corollary, if there are any such legacy PDs, which even inadvertently encroach into the IEEE detection region, they must no longer be supported, even with legacy-detection mode enabled at the PSE side.

While testing a PSE for IEEE compliance, we are usually allowed to disable its legacy-detection option completely during the testing. That may be acceptable, but the best option is to *not* need to do that: that is, keep IEEE detection *and* legacy detection enabled simultaneously, yet achieve full IEEE compliance (via Sifos testing, for example).

However, there are PSE-controller ICs in the market today that extend their legacy detection into impedance areas which IEEE detection asks us to definitely reject (not even a "maybe"). For example, the IEEE standard asks us to definitely reject any port capacitance greater than 10 μF during detection. And it asks us to do this irrespective of the parallel resistance. So a PSE that allows the user to enable such a wide, almost indiscriminate "legacy detection" window runs the risk of failing IEEE compliance testing, unless the legacy detection option is disabled during IEEE testing. Though that is considered somewhat "acceptable," it is actually risky (for unsuspecting DTEs) to put such PSEs out in the open market with legacy detection enabled. A better option is to not only measure C but also R (appearing in parallel to C), and then consciously allowing only very explicit (and unique) *combinations* of R and C to be detected as legacy devices. This can only happen after thoroughly researching known legacy PDs in the market that need to be supported. It also needs firmware control to implement.

The IEEE committee, on its part, consciously formulated the IEEE-detection range so as to distinguish it consciously from known PDs and other common networking conditions. In other words, they tried to create a unique "signature" (like a thumbprint, see Fig. 3.3) for the IEEE-compliant PD, one that would distinguish it from virtually all other port conditions. This would allow most legacy devices to be supported by optional legacy detection if required, but also ensure that an IEEE-compliant PSE did not inadvertently apply power under any conditions, *except* to an IEEE-compliant PD. This detection process is discussed next.

IEEE Detection

In Fig. 3.3 we first provide a quick look ahead at the overall handshaking process between a PSE and PD prior to power-up, as defined in IEEE 802.3af. We note that the PSE is initially just looking to confirm whether the device on the other side is an IEEE-compliant PD or not. Then it examines it further to estimate its power class. We recall that

Detection is the process by which the PSE asks: ***"Who are you?"*** (and the PD then presents its detection signature; akin to providing a *thumbprint*).

To this we now tentatively add the next part of the handshaking process, which will be discussed in more detail in the next chapter.

Classification is the process by which the PSE asks: ***"What are you?"*** [and the PD then declares its power-rating category (class); akin to stating its size as shown in Fig. 3.3].

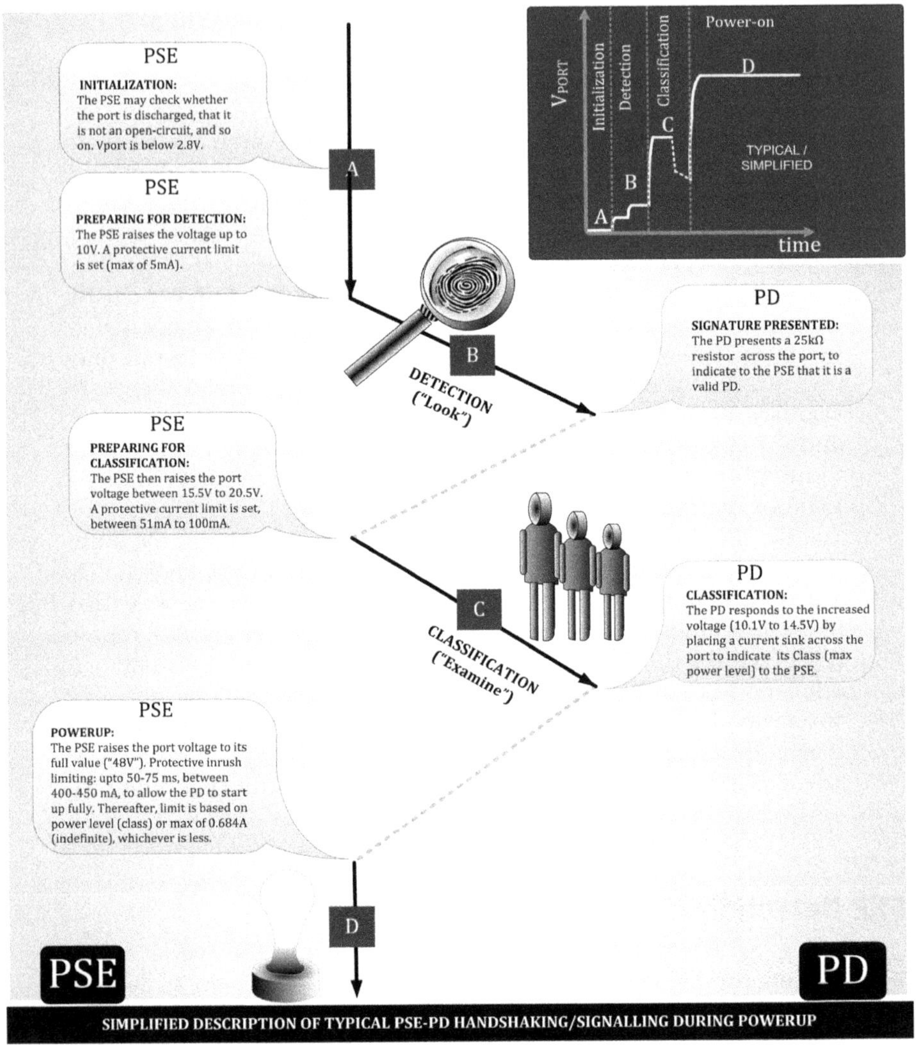

FIGURE 3.3 Simplified detection and classification sequence before port power-up.

In general, from the viewpoint of any PSE/switch, it may get to see several different impedances across its RJ-45. The question is what impedance can we define that will be unique enough to unambiguously identify a standards-compliant PD from all (or at least most) other common conditions? That unique impedance could serve as a signature (or thumbprint) of a standards-compliant PD. So we need to first quantify "all other impedances," also include the effects of tolerances, drifts, leakages, and so on, then ensure there is enough margin, and finally come up with a unique impedance to ascribe to the PD as its signature. This is how the process played out at IEEE. For example, if the cable coming from the switch is connected to an older network-interface card (NIC) inside a computer, with older PoE-unfriendly terminations as discussed in the previous chapter, then because of the absence of blocking caps, the PSE in the switch would "see" two 75-Ω resistors in series at the other ends, which is an impedance of 150 Ω. As explained, we do not want to power up fully with this NIC at the other end of the cable as we could easily damage it. But we could, in principle, "probe" it, with tiny current-limited voltage sources, or voltage-clamped current sources. A unifying moment was Robert Leonowich's presentation in July 2000 on signature margins. His graph is shown in Fig. 3.4. We see that it also included recognizing (and avoiding) impedance conditions in which a switch/hub/router gets connected by an Ethernet cable to another switch/hub/router, which in turn could be either "powered" (that means with a 48-V incoming supply on it) or "unpowered" (with no connection to the mains supply). In the former case, the polarities could also be reversed. Generically, this case is often called a *PSE-to-PSE (detection) case*. (See some of these in Fig. 3.4.) We will discuss this issue specifically in greater detail later as it is more a *systems-related* aspect. But for now, this sums up the general approach in which Leonowich created a hazard matrix, on whose basis the value "25 k" was identified as the nominal IEEE PD-signature resistor value (though the standard E96 series-resistor value, using 1 percent tolerance resistors, is actually 24.9 k, which is the value more commonly used on an actual board).

Observe from Fig. 3.4, the reverse polarity diode across a port. We now know this diode serves many purposes actually and helps system reliability. Some PSE vendors even suggest replacing this with a transient-voltage suppressor (TVS) diode, which also brings in zener-clamping behavior at port voltages typically above 60 to 63 V. We will discuss that in Chap. 11, but for now we keep in mind that the initial rationale for this diode was just to make the PSE look like a "short" under a reverse polarity applied on its output (such as the PSE-to-PSE "aided" case in the same figure). Thus each PSE would now see another PSE that it got connected to as a short (forward drop of the diode) and reject it as a prospective PD. Thus it would not inadvertently try to power it up.

How is PSE-side detection *implemented*? In general, the PSE puts out either tiny current or voltage sources on the line. Then based on the resulting voltage or current it reads, respectively, it can compute V/I to get the resistance at the other end (ignoring the relatively negligible cable resistance). If that calculated resistance turns out to be ~ 25 k (within a certain defined range), the pass-FET in the PSE will turn on fully, injecting power into the cable. Of course, prior to that, the PSE will significantly limit the applied voltages and maximum currents, so as to keep conditions extremely safe on the line during

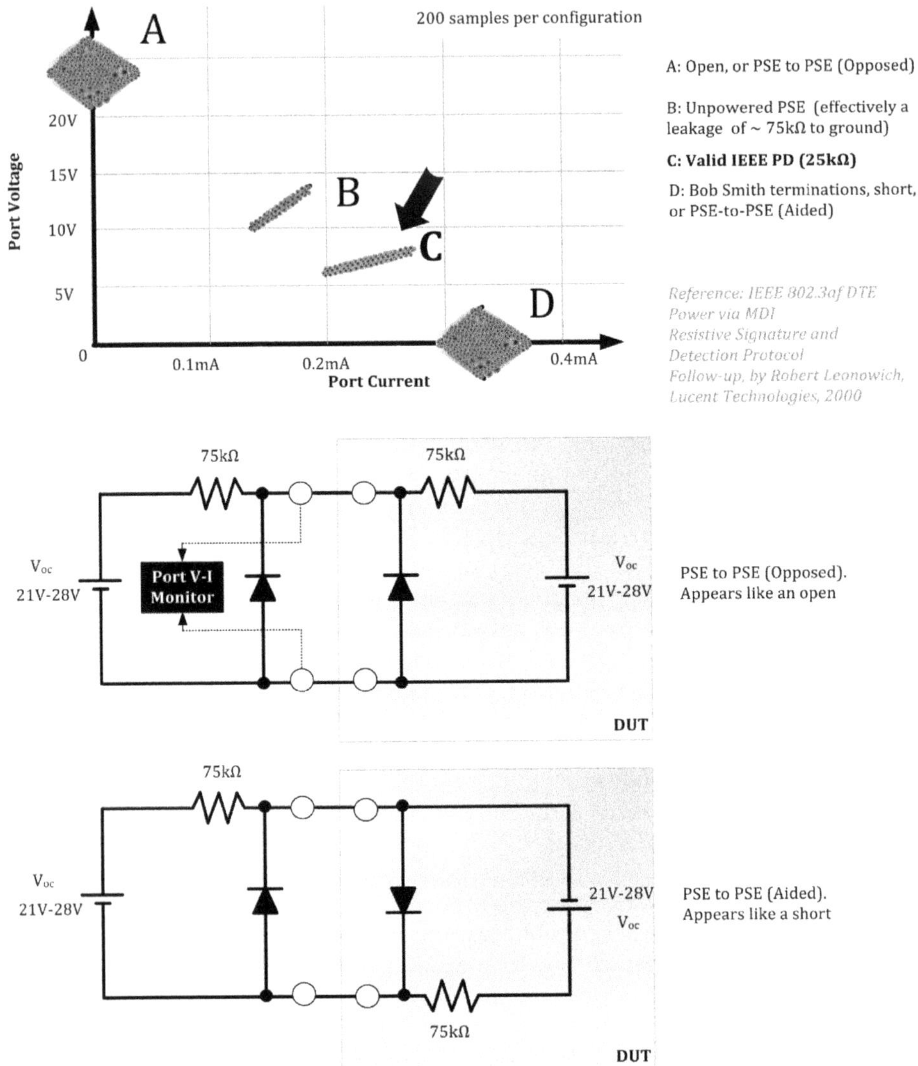

FIGURE 3.4 Hazard diagram and PSE-to-PSE equivalent circuits.

the discovery process, (because at this stage, it still does not know what lies at the other end).

On the other side of the cable, there is a PD which typically has *two* OR-ed input bridge rectifiers, only one of which actually conducts for 10/100 Mbps applications. *The 25-k signature is usually placed after the bridges,* so that a single resistor is all that is required for both power-over-data pairs (Alt-A) and power-over-spare pairs (Alt-B). See Fig. 3.5. Note that the IEEE standard *does* allow the signature

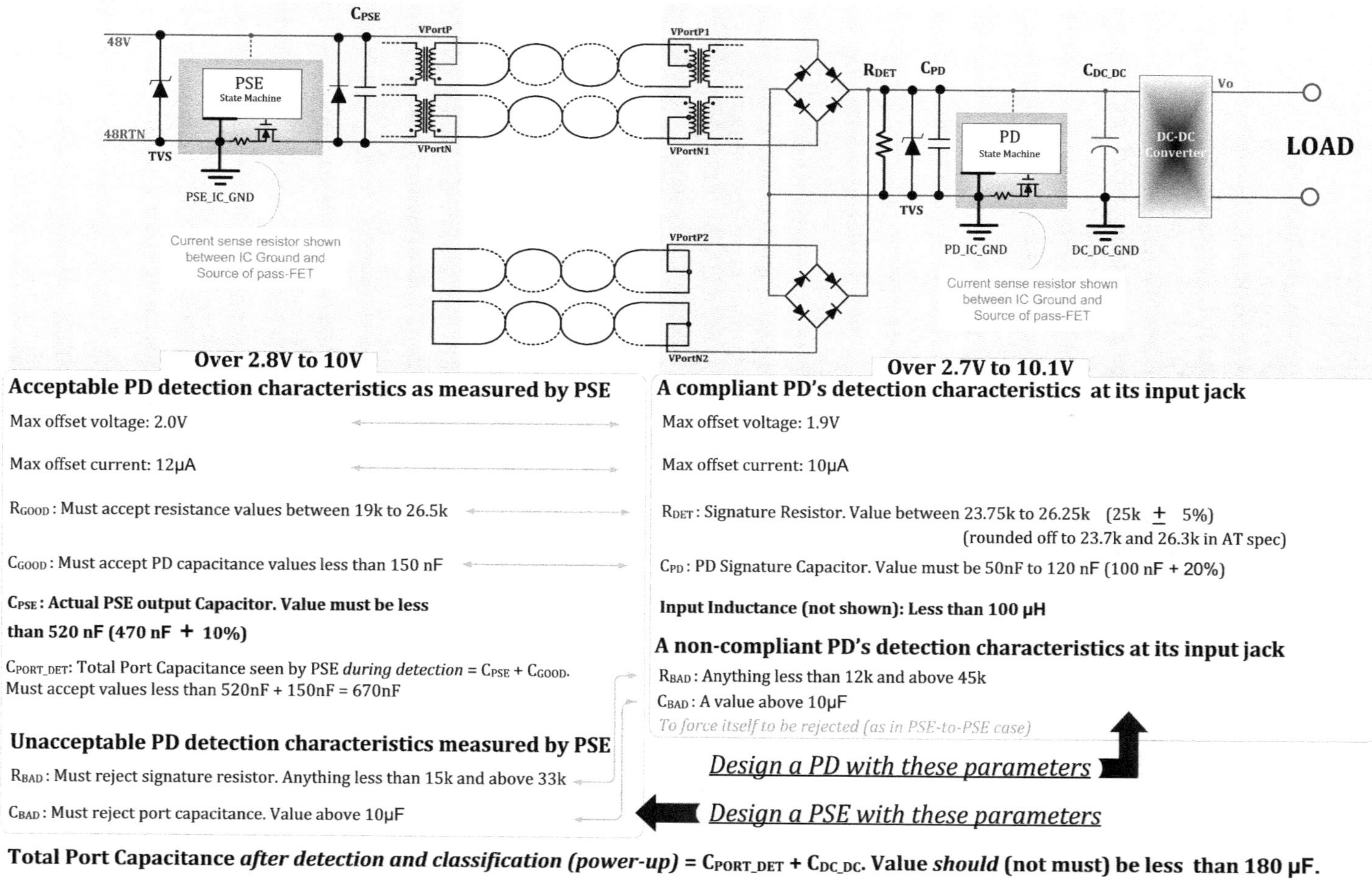

Figure 3.5 Detection characteristics from the viewpoints of the PD and the PSE.

resistors to be placed before the bridges too, but it is rarely ever done that way.

The DC-DC converter that needs to be powered up eventually by the incoming power will typically have fairly large capacitances at its input for effective decoupling and proper operation. But since we only want to present a 25-k signature resistor to the PSE during the detection process, the signature resistor is deliberately "shielded" (or isolated) from the DC-DC converter section by means of a nonconducting pass-FET in the PD. This is sometimes called the hot-swap FET for vague reasons. However, this FET does seem very similar to the one inside the PSE, but in reality, it is present for altogether different reasons. Note that this PD-side pass-FET section (without the DC-DC converter stage included) is usually just called a PD interface or PD front-end. It basically just contains the PD's pass-FET along with detection/classification circuitry. This FET is asked to conduct only when detection/classification is completed (successfully). And when it finally does conduct, the input cap of the PD's DC-DC converter suddenly gets "exposed" to the PSE. In other words, the PSE now sees a very large port capacitance in parallel to the signature resistor (the latter can now even be disconnected by the PD to save energy, because it serves no purpose anymore—but more on that important issue in Chap. 4). Note that at this power-up mode transition, a large-surge current flows into C_{DC_DC} to charge it up, and as we will soon learn, the IEEE standard specifies this in-rush current rather carefully too.

Besides asking for a signature resistor, the IEEE standard allows (and in fact asks for) small *decoupling caps* to be placed next to the PI (jack) on both ends of the cable. These are collectively often called "signature capacitors," because from the PSE's perspective, all these caps get lumped together and appear in parallel to the 25-k signature resistor of the PD during detection. These caps are usually ceramic/film caps and are considered completely necessary for proper functioning of either PI-side circuitry (on both PSE and PD sides) to avoid any noise from the lines affecting overall operation. The values of these signature capacitors are also carefully defined in the standard and summarized in Fig. 3.5.

In Fig. 3.5 we have introduced a key design and measurement concept, one that forms the reason for the differences in the PSE and PD ranges shown in the figure. We will explain this here.

For example, we design a PD in the following manner. We start off by ensuring the PD presents a signature resistance well within the allowed 25 k ± 5 percent (presented at its PI), and that includes tolerances of the signature resistor, leakages, temperature-related drifts, and so on. Typically, we just select a standard value of 24.9 k ± 1 percent resistor, but we *also try to ensure acceptably low leakages in critical spots, especially across the pass-FETs of the PSE and PD.* We also need to ensure that during detection, the PD appears as

25 k ± 5 percent at its PI over a certain specified low-voltage range of 2.7 to 10.1 V (conditions).

This is the basic manner in which we design a PD. But ultimately, it is the PSE that needs to confirm the PD's acceptability or not. How do we design the PSE? On the PSE side, the readings will be affected by the PSE's own measurement capabilities/tolerances (for example, the resolution of its analog-to-digital converter), plus other drifts, cable leakages, and so on. We therefore need to set the measured-pass range to be wider than the design range. In particular the IEEE standard allows the PSE to accept as a valid PD, any resistance within 25 k + 6 percent, –24 percent (i.e., 19 to 26.5 k). Note that this range is *not* centered around 25 k. In fact the measured reading is allowed to be significantly *lower* than 25 k (up to –24 percent). This low-side headroom actually represents an additional allowance for cable insulation deterioration over time. We can visualize that any cable leakage will appear as a parallel resistance across the 25-k signature, lowering its measured value at the PSE-side significantly. In general, the range of any *measurement* is set *wider* than the *design* range.

What about the conditions under which measurements are done (by the PSE)? The PD's 25 k ± 5 percent resistance needs to be guaranteed over the range 2.7 to 10.1 V. In other words, the PSE should not make the mistake of checking the resistance at an applied voltage of say, 10.2 V for example. That is not correct because the signature value of the PD is not guaranteed at 10.2 V. Yes, some PD ICs may guarantee R_{DET} as high as 11 V (or beyond), and that would certainly be good design practice, but the IEEE standard does not mandate it, so the PSE should not count on it either. On the other hand, a PSE measurement of R_{DET} at 10 V is OK (guaranteed). Keep in mind that the IEEE standard thus sets the PSE detection range as 2.8 to 10 V, whereas the PD signature resistance (minimum) design range is slightly wider: 2.7 to 10.1 V. Summing up, conditions for measurements need to be set tighter (narrower) than conditions for design, and the measurement range should be set a bit wider than the design range. That is the basic philosophy behind Fig. 3.5.

Why 2.8 V? This is basically the lowest-measurable voltage level as per the standard. Falling below this value amounts to a complete reset (in effect, 0 V). Therefore, this particular value shows up at different points in the IEEE standard.

In Fig. 3.6, we present the key information of Fig. 3.5 concerning detection in an easy-reference wall chart for greater convenience and better visualization. Note there are areas with a "question" mark (?). Colloquially, these are considered gray or "don't care" areas. Circuit-comparator thresholds are typically set in these areas to guarantee the required ranges as marked with ticks or cross signs. The latter are colloquially referred to as "must accept" and "must reject" areas, respectively.

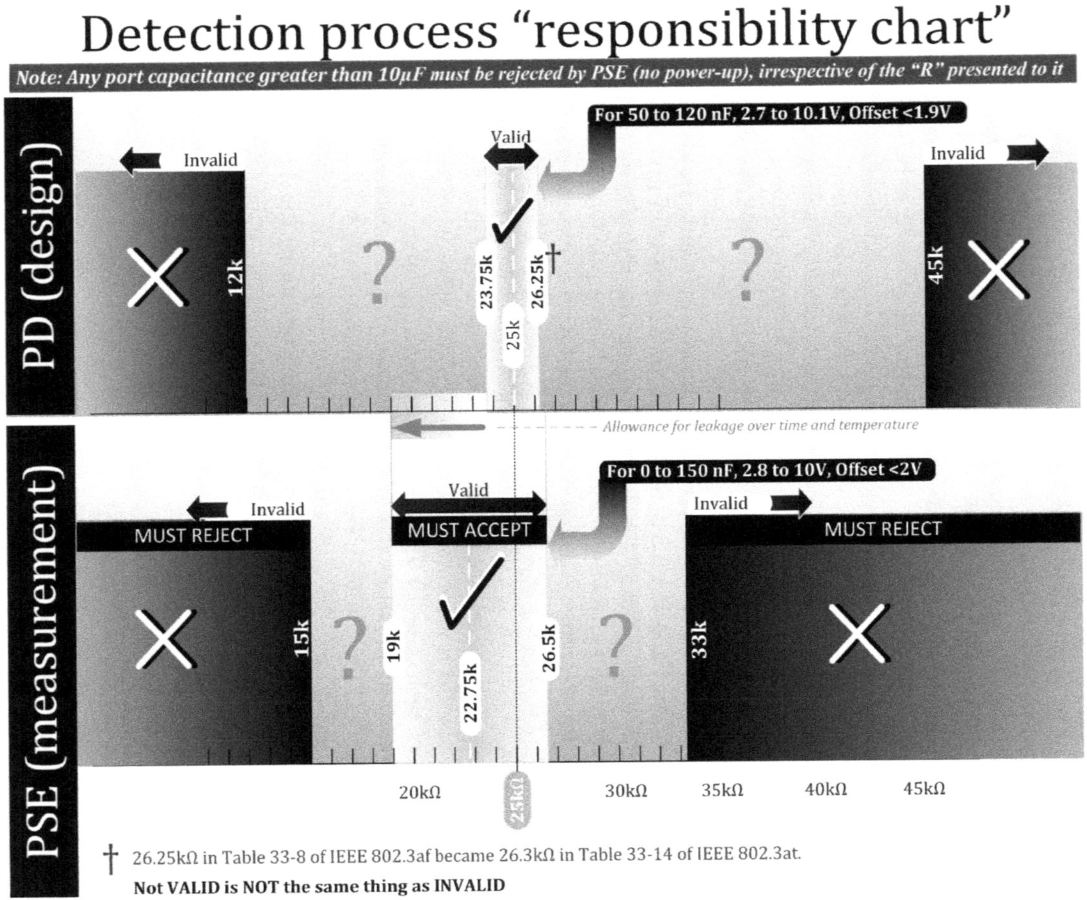

FIGURE 3.6 Graphical reference-chart for detection.

In Fig. 3.5, we should also pay close heed to the fact that the grounds on either side of the PD pass-FET in particular must be *separated*. We want to completely isolate the PD's circuitry, especially its DC-DC converter, from the front-end section. That is how we expect to see a simple 25 k signature resistor during detection. That is why the capacitors on each side of the PD's pass-FET must **not** be accidentally connected to the same ground symbol in a schematic, or the signature resistor will *not* be detected properly and the PD will *not* turn on. This casual schematic mistake has, in fact, often been falsely considered an interoperability issue by some PD-device manufacturers. But while PSE vendors struggled to make their devices more tolerant, often risking IEEE-compliance, all the PD vendor really needed to do was recheck the grounds on their schematics and fix the problem.

Most IC-fabrication processes prefer to use N-channel (low-side) pass-FETs for the PSE or PD, and then connect the source of the pass-FET directly to IC ground, or through a current-sense resistor as shown in Fig. 3.5. This Source-side IC ground forms the substrate of almost all known PoE chips.

We must also remember that any bridge rectifier produces *two* forward drops in series with the current. On the PSE side, this DC offset or signature-offset voltage tolerance, is called Vos, and is set to 2 V. On the PD side however, the maximum-allowed offset is 1.9 V, in keeping with the philosophy outlined previously.

To eliminate the error caused by the diode offset, and also to reject noise, the standard requires the PSE take *at least two measurements*, which are spaced at least 1 V apart (and also at least 2 ms apart), and to use the following equation to calculate the resistance:

$$R = \frac{\Delta V}{\Delta I} = \left|\frac{V_2 - V_1}{I_2 - I_1}\right|.$$

So we look for the *incremental* resistance (slope of the V-I curve).

NOTE *The standard also limits the maximum slew rate during detection steps to less than 100 mV/μs. This is easy to meet with current sources at least. From I = C×dV/dt, we can see that even with the maximum upper-current-limited source of 5 mA, we can meet the slew-rate limit with just 50 nF of port capacitance. In all practical implementations, there is at least 0.1 μF port capacitance from the PSE side and another 0.1 μF (minimum 50 nF as per the standard) at the PD side. A total of 0.2 μF in all known practical implementations. So the dV/dt aspect is not of concern in most detection schemes.*

NOTE *As mentioned, the standard includes a possible offset current too (10 μA or 12 μA at the PD or PSE ends, respectively). We understand that the DC-offset voltage is theoretically the voltage with no current* *flowing at all. Theoretically, we could also have some leakage current flowing with zero voltage applied. But there was some debate and disagreement about the entire concept of "Ios"at the IEEE discussions leading up to the AT standard. Some even asked to delete Ios altogether as they felt it was not real or relevant. Their viewpoint was that with no PSE voltage applied, the only way a current could flow was if the PD was somehow a source itself. But it couldn't be as it has no 48 V connected to it. But there could be situations, however, where a PSE was connected via a Midspan on the spare pairs and had acted as a source of leakage current. In any case, the Ios spec is still there in the AT standard, and there are test setups which apply this offset current in a compliance-test suite too. But normally this offset current poses no problem in any common detection scheme.*

NOTE *In creating the hazard matrix, Leonowich had originally just used a standard-DC voltage source with a limiting resistor of 75 k (the "probing resistance"). But at some stage, for obscure reasons, this number seems to have gotten changed in the standard to 45 k. However, the basic purpose remained unchanged: The high value was meant to avoid false PSE-to-PSE detection and resulting (spurious) power-up. This means that the IEEE PoE standard asks that if we connect PSE A to PSE B, PSE B must look like a resistor greater than 45 k to avoid being detected as a valid PD by PSE A. Similarly from the viewpoint of PSE B, PSE A must not appear less than 45 k either.*

But there is also the issue of what happens if legacy detection is enabled in PSE A and/or PSE B. The IEEE standard does not talk about legacy detection anyway, so what can happen in such a case is also undefined. Different interpretations of that can, however, cause false detections and interoperability issues between switches/hubs of different vendors in PSE-to-PSE configurations. If legacy detection is disabled or not supported (increasingly so today), this is of little concern.

Technically speaking, to avoid PSE-to-PSE false detection as a PD, a PSE can not only present itself as a high resistance (> 45 k) during its detection sequence (to avoid being mistaken for a PD), but it can also present itself as a *lower* resistance (< 12 k), and other PSEs should reject that value too. Unfortunately, there are PSE chips out there that apply a rather strong internal bleeder of around 10 k temporarily between successive failed detection attempts to initialize the port voltage, and there are also some PSEs out there from other vendors that do *not* reject 10 k with legacy detection–mode enabled. This has led to some well-known interoperability issues in the past.

There is perhaps no right answer here unfortunately—no clear culprit in sight. It is part of the confusion surrounding the introduction of any "standard" right in the middle of an already-ongoing PoE-era.

In general, it is fair to say that PSE-to-PSE detection issue remains one of the most complex and often barely resolved or even unresolved systems issues. It will be discussed in more detail in Chap. 11.

Practical Voltage and Current Limits during Detection

Besides ensuring detection, there are safety concerns too. What are the worst-case PSE-side voltages and currents allowed by IEEE on the line during the detection process? These are the key limits during detection.

1. Maximum (open-circuit) voltage $V_{oc} = 30$ V.
2. Short-circuit current $I_{sc} = 5$ mA

Open circuit is defined by the standard as any measured port resistance over 500 k.

Note that there is an odd/contradictory situation here. Can we use the maximum safe-current source of 5 mA to probe a *valid PD*? A valid PD would by definition present 25 k on the port. However, 5 mA through it gives a voltage drop of 5 m × 25 k = 125 V! But obviously that is not only unsafe (above 30 V), but outside the valid PD upper-detection limit of 10 V. So 5-mA current limiting is not a practical target. Working backward, if we are to keep a max voltage of 10 V across 25 k, the max current is 10/25 k = 400 μA. So typically, to account for tolerances and so on, no PSE will be designed with detection-probing sources of more than about 350 μA (nominal). That is the practical upper limit.

Some Practical Detection Techniques

In the inset in Fig. 3.3, we have shown the port voltage during the initial sequence. Region B is detection. The port voltage starts from zero (cable not energized) toward fully applied "48 V." In this simplest case of detection, just *two* steps of voltage or current are being applied, and they are applied on the "charging" curve (ascending staircase). As per the IEEE standard, two steps (or *samples*) are a minimum, since we need them to eliminate diode offset from the bridge, as discussed previously. The standard also asks we ensure that the voltage *levels we use for our calculations* are more than 1 V apart. Using $(V_2 - V_1)/(I_2 - I_1)$, we thus get R. This may sound very basic, and it is, but it is a viable detection method used by many PoE-controller ICs.

Note the emphasis on the voltage *levels* (for sampling) being more than 1 V apart, not necessarily the voltage *steps* above. For example, nothing stops us from applying, say four voltage steps only 0.75 V apart, as long as we calculate R based on *alternately numbered* steps: $(V_3 - V_1)/(I_3 - I_1)$ and $(V_4 - V_2)/(I_4 - I_2)$. The difference between V_3 and V_1, and between V_4 and V_2, is 1.5 V, and that is OK as per the IEEE standard. This was the basis of U.S. patent number 7,711,967 from Woo et al.

There are also PSE chips, like the LTC4266, that use an alternative four-point detection scheme, consisting of two descending steps based on current sources, followed by two descending steps based on voltage sources.

We need not even have *visible* steps really. We can, in principle, use a weak *linearly increasing current* source, and record the corresponding (and proportional) linearly increasing PI voltage live as we go along (1 V samples apart though). If the rise time of this ascending-PI voltage is large enough, port capacitances will play almost no part, and so we will get a proportional increase in voltage for a proportional increase in current. We can then calculate the detection resistor using $R = \Delta V/\Delta I$.

If we collect multiple R readings, we can ultimately average them out to get the final R and thus decide whether it is a valid PD or not. But in the process we can also ensure that several readings for R are within, say 1 to 2 percent of each other. We can implement some form of arbitration too. All these methods are a good way to further reject the effects of noise on the detection process.

There are some historical intellectual-property issues here too that we are ignoring, but in general, today, current sources are usually preferred over voltage sources. The reason is the signature resistor and PD-signature capacitance both usually lie *after* the bridge rectifier. The voltage across the signature section is not necessarily the voltage being monitored by the PSE. The PSE is measuring the port voltage on the *other* (anode) side of the bridge. So there is a possibility that the bridge can get *reverse-biased* with a higher voltage across the PD-signature section and a lower voltage across the port. However, by using current sources to probe the PD instead of voltage sources, we manage to *force* the probing current through the bridge, this being a basic property of a current source, and so the bridge does not get reverse-biased. With this method, the voltage read by the PSE will be very close to the actual voltage across the signature section of the PD (though with some diode offset, which is rejected by calculating the incremental resistance as explained previously).

We need not apply steadily-increasing-current sources in steps (a staircase going *up*). We could even apply them in descending order starting with the maximum value first (staircase going *down*). Because, provided we are using current sources, the diode bridge will not get reverse-biased, and this would also work.

Note the slightly rounded edges of the applied voltage steps in Fig. 3.3, caused by the small signature (port) capacitance. We realize that if the port capacitance is too big, as in the case of legacy devices, then the associated time constant will be very large and stable readings at each step will perhaps never be obtained quickly enough. Keep in mind that as per the IEEE standard we must complete the entire detection process (whether successful or not) within 0.5 seconds (T_{DET} timer). This upper boundary serves to limit the time constant, the number of voltage/current steps, and the maximum-port capacitance. For example, the time constant RC for R = 25 k, and with a not-unacceptable (gray-area) port capacitance of a little less than 10 μF (more than 10 μF is deemed unacceptable and *must* be rejected, irrespective of the value of R, as per the standard) is about 0.25 s. We also know that *three time constants* (in this case, a total of 0.75 s) are needed to get to within *5 percent* of the settling value in any RC charging/discharging curve. But clearly, 0.75 s is already in excess of the maximum detection time of 0.5 s. In other words, we cannot get any meaningful results with large capacitances, even if they are in the gray area (i.e., not definitely disallowed). On the other hand, with a maximum "must accept" theoretical port capacitance of 670 nF (as explained in Fig. 3.5), the time constant 25 k × 670 n = 17 ms. And this works well: each current source must be applied for at least 51 ms (**three time constants**). With four charging steps followed by a discharge interval, we will be at about 250 to 300 ms for a complete detection cycle, which is less than 500 ms and therefore acceptable.

The next improvement is possible by trying to be *doubly sure* there is a valid (IEEE-compliant) PD on the other side before powering up. Because the standard allows for erring on the side of caution. The answer to that extra robustness is double-detection. In particular, this is very useful in avoiding false detection and port power-up in the very difficult PSE-to-PSE case discussed earlier. Double-detection be implemented in an IEEE-compliant manner in two ways actually.

1. In one method, provided an existing detection sequence took less than 250 ms, we just repeat that twice (within 500 ms). We should thus get two (or four) very close values of R, unless noise is clouding matters (literally). Mentally we can consider the two successive detection sequences (occurring within 500 ms) as just *one detection sequence*: The reason is we do in fact have enough latitude within the IEEE standard to come up with *many unique detection schemes,* as long as we comply with the general rules as already discussed.

   As mentioned, we need not repeat the same method in fact. Some PoE-controller ICs (example from Linear Technology) complete the first detection phase using current sources (they call it "forced current detection") and then within the remainder of the 500 ms, they apply voltage sources (they call it "forced current detection").

2. There is a loophole in the IEEE standard we can use. The standard allows that we do *not* power up even after a successful detection. This will be clearer when we discuss the IEEE state

machine in Chap. 8. For example, we could even wait endlessly after a successful detection, suggesting that we just do not have the necessary power budget for powering up an additional PoE port. That is allowed. Alternatively, we can do the following: complete one full detection sequence within 500 ms, *not power up* (on grounds of insufficient available power), and then repeat the same detection sequence in the next 500 ms. After that, we could ensure the two results (for *R*) are very close to each other and both are also within the IEEE PSE range (19 to 26.5 k). In effect we are taking 1 s to do detection. But there is a loophole in the standard that allows that! Finally, if all these filter conditions are met, we can decide whether to power up the port or not.

NOTE *The AT standard does mention: "if the PSE returns to the IDLE state (that is for PI voltages below 2.8 V), it shall maintain the PI voltage at* V_{RESET} *(below 2.8 V) for a period of at least* T_{RESET_MIN} *(15 ms) before starting a new detection cycle." Though this is stated in the context of Type 2 devices between classification and power-up, it is a good idea to adopt this rule in all cases between successive detections too, and for Type 1 and Type 2, to avoid strange interoperability/test issues.*

We thus realize we can come up with many innovative detection techniques *within the bounds of acceptability* as laid out by the IEEE standard. The biggest source of trouble could be in *passing* compliance tests on automated-test systems, such as those from Sifos Inc. The reason for that is that any automatic tester is designed to cater to several vendors' designs, but can rarely be intelligent enough to anticipate and account for all possible variations up front. So a constructive discussion with the test-equipment manufacturer is encouraged at such junctures. They will normally be able to "tweak" their firmware to allow for a new detection scheme, provided we can convince them we are complying with the IEEE standard.

In Fig. 3.7, we finally clarify the charging/discharging time-constant constraint issue when using current sources during detection. The math behind charging and discharging of a parallel RC combination, using current sources, was already indicated in Fig. 3.2. That math is valid here too. Knowledge of the time constant is essential in setting aside enough time for each applied step, to allow the voltage to settle and be read correctly by the PSE. In Fig. 3.7 we summarize the simple math behind a simple PD detection scheme, including a diode-bridge voltage offset and detection timer. More complex detection schemes are all based on this underlying reasoning.

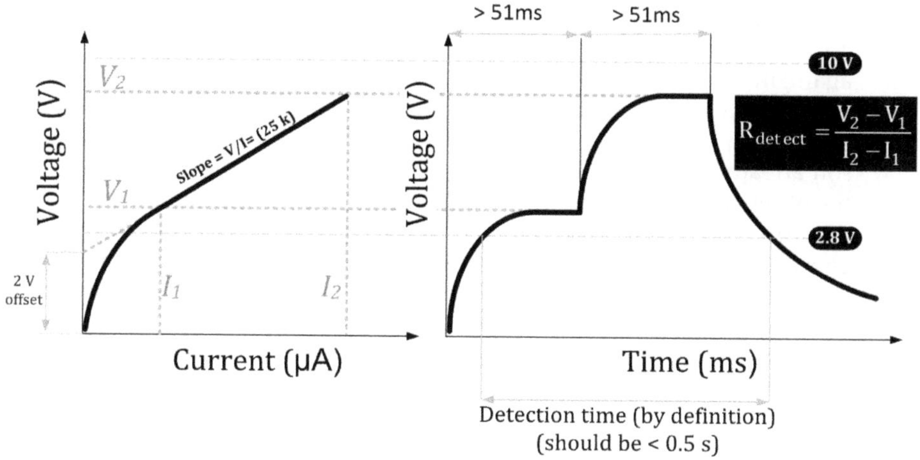

Example: Selected current sources are 200 μA and 300 μA

Passing through 25 k resistor,
V_1 is 200 μ x 25 k = 5 V
V_2 is 300 μ x 25 k = 7.5 V

Add to this 2 V max offset. Then applied PI voltages are 7 V & 9.5 V (less than 10 V as required).

If offset is zero, applied PI voltages are 5 V & 7.5 V (above 2.8 V as required).

Voltage difference between steps is 2.5 V (greater than 1V as required).

Worst case RC time constant is 17 ms based on R = 25 k and C = 670 nF. Need 3 time constants to get to within 5 % of final value. So each step should be at least 17 x 3 = 51 ms.

FIGURE 3.7 An example of a basic IEEE-compliant detection scheme.

Predetection/Open-Circuit Detection/Initialization

In Fig. 3.3, Region A is something we have avoided discussing so far. We now explain this a bit more.

For added safety, we want to make sure that there is something like a signature resistor on the other side of the cable, before even attempting to measure it accurately. In particular, we want to guard against short circuits or open circuits. The former could be anything less than 200 Ω (representing the two 75-Ω series-termination resistors of a NIC). The latter would be by IEEE definition anything above 500 k. The simplest way to check for the former is to put an exploratory current source out on the cable and make sure the voltage exceeds a certain low threshold after a certain settling time. This is the inspiration behind U.S. patent number 8,097,982 from Louis Joseph Maggiolino. Many vendors, such as Silicon Labs, have working nonproprietary variations of that in their chips already. To check for an open circuit, we apply a weak current source once again; in this case if the voltage exceeds say 15 or 20 V, we know that we have an open circuit.

During Initialization, we can also check if the port is for any reason already powered. For example, another PSE may be present on the other end of the cable instead of a valid PD, and we could likely discern its presence and avoid powering up ourselves by simply checking for preexisting voltage on the cable. Any such voltage could be the result of the other PSE constantly attempting detection on the cable, or a PSE which may even have applied full power over the cable. Or there could just be some residual charge remaining after quickly unplugging and plugging back a cable. Or there could be some other unknown fault. The PSE chip could try to initialize the port by applying a bleeder resistor of large value (> 45 k) across the port, waiting a while, then checking that the port voltage is indeed almost zero (< 2.8 V).

None of this is actually demanded by the PSE state-machine diagram in either the AF or AT standards, but it makes logical sense and is good from a systems perspective. Further, it helps test-equipment vendors initialize their equipment. So they know when exactly to start checking for formal IEEE detection (to verify $T_{DET} < 0.5$ s for example). Test equipment will often first look to see a threshold of 15 to 20 V being crossed, indicating a successful open-circuit detection. Then the next time the voltage exceeds 2.8 V, that would be the start of IEEE detection.

Detection Back Off

One question remains: After an "unsuccessful" detection attempt (resistor was not 25 k, i.e., an invalid signature), how quickly can another detection attempt be carried out? The answer to that depends on whether the PSE is delivering power over the data pairs (Alt-A) or over the spare pairs (Alt-B). For Alt-A, there is no back off requirement. But since most PSE chips will carry out port initialization/pre-detection/open-circuit detection all over again, the next detection will naturally take some time. But a typical PSE design will try to complete a second detection *within about a second* after the end of the preceding failed detection. Section 33.2.4.1 of the AT standard has this suggestion:

> If a PSE performing detection using Alternative A detects an invalid signature, it should complete a second detection in less than T_{DBO_MIN} *after the beginning of the first detection attempt*. This allows an Alternative A PSE to complete a successful detection cycle prior to an Alternative B

PSE present on the same link section that may have caused the invalid signature.

What is the concern here? The worry is that the cable from an Endspan PSE may have been connected to a Midspan PSE en route to the PD. So now two PSEs are on the same link, trying to provide power to the same PD. The standard wants to give priority to the Endspan, and therefore, concluding that the failed detection could be the result of the voltages/currents injected by the other PSE across the same 25 k signature resistor, it asks that the Alt-B PSE "back off" to allow the other PSE to go first. So it sets a "back off time" of minimum 2 s (T_{DBO_MIN} = 2 s) for an Alt-B PSE. Note that the T_{DBO} timer starts at the very moment the detection fails, not from the start of the failed detection attempt as somewhat implied by the extract from Section 33.2.4.1.

Since the Alt-A PSE on the same link has been given the go-ahead for at least 2 s after its failed detection attempt, it needs to use that period to complete a successful detection and power up the port before the interference from the Alt-B PSE resumes. When the Alt-A PSE successfully powers up the port, it will automatically reverse-bias the bridge connected to the spare pairs, and so the Alt-B PSE will never see a valid detection signature (until the Alt-A PSE is powering the port).

But will that prevent the Alt-B PSE from trying again and again to detect? Not really, because the standard says that an Alt-B *may* (that means *optionally*) omit the minimum back off time of 2 s *provided it sees an open circuit* (> 500 k). Chances are high that if its bridge on the PD side gets revere-biased, it will very likely see an open circuit and so it will try and try again to detect. Therefore, it is probably a good idea to keep the > 2 s back off timer for an Alt-B PSE always, whether it sees an open circuit or not. This will prevent unnecessarily quick repeated detection attempts.

Detection Signature Resistor Disengagement

At this point, it is a good idea to take a step back at a very basic issue: *the presence of the 25-k detection resistor.* In almost all cases discussed so far, and also in related literature, this resistor is shown as a stand-alone component just past the bridge on the PD-chip-side. It is always present. So when the voltage ramps up into the classification range (~18 V), it continues to draw current, of about 18/25 k = 0.7 mA. Not a big deal, and though it adds some current to the constant class-current-sink, its effect is less than 1 mA. When the port is fully powered up, it continues to draw about 50 V/25 k = 2 mA. In terms of wattage this is 50 V × 2 mA = 100 mW. Not entirely negligible. Keep in mind that we need to draw about 0.5 W to keep the PD-PSE connection "alive" in the first place. Nevertheless, it did lead to some people wanting to shut off this "wastage of power," and they proposed the 25-k resistor be actively disconnected after detection. One question remained: When exactly to turn it off (disconnect it)?

In fact the way the AT standard's PD state-machine diagram is written, it actually does *not* even permit keeping the detection resistor across the port beyond the exact moment of completion of detection phase (for both Type 1 and Type 2). This can cause backward compatibility issues because the AF standard was quite ambiguous on the PD state machine, and was widely interpreted as allowing this "always present" 25-k resistor. So there are many PDs out there even today, both Type 1 and Type 2 (such as On-Semi's NCP1083) that continue to retain this 25-k resistor across the port always connected. Faced with that market reality, it is not a good idea to design a PSE that doesn't accept this wide PD behavior, whatever the AT standard may or may not say.

It seems the main reason for immediately disconnecting the 25-k signature resistor right after detection was motivated not by a concern for saving power initially, but because of a 2007 presentation at an IEEE Ad Hoc meeting from an engineer from Linear Technology who voiced concern of something that "may" happen during classification (see Chap. 4). This fear seems to have then prompted a PD-state machine within the AT standard that demands immediate removal (disengagement) of the detection resistor right after detection. The consequence is most PD-chip designers nowadays struggle to create a very narrow range above 10 V, say 10.5 to 12 V in which they consciously disconnect the detection resistor, and then slightly above that range, turn on the classification-current-sink. However, combined with the basic problem of variances in diode forward drops and the "guesstimation" of all that beforehand while setting hard thresholds in the PD, this has likely led to many reported PSE to PD interoperability issues.

NOTE *Why do we need any "guesstimation"? The problem is as follows: Whenever we talk of the PSE-side voltages, we are quite right, because the PSE IC can actually read the port voltage accurately on its side of the cable and do the "right thing" at the "right voltage threshold." How about the PD? Can it really read the PD-side PI voltage? We are forgetting that in most PD ICs, the PD has no way of actually reading the PI voltage per se. Because it is on the* right-hand side *of a bridge rectifier, and that introduces* two *forward-diode drops between the PI voltage and the voltage the PD "sees." So a PD IC has to literally guesstimate the diode bridge offsets up front, before setting its thresholds. That can lead to significant errors and reported interoperability issues, since the bridge rectifier forward drops can vary a lot and are also a function of diode current and temperature.*

This disengagement of resistor seems to have been not worth it, because the detection resistor actually serves a useful purpose during both detection and classification as explained in Chap. 4. It is therefore considered OK to leave it in the PD across the port always connected. Typical test suites (such as from University of New Hampshire's Interoperability Laboratory) do not check PDs for this "removal of the detection resistor above 10 V" requirement. If they did, a lot of earlier and modern PDs will fail the test. If we really want to save 100 mW of dissipation from the signature resistor during normal operation, the best option from a systems viewpoint is removing the detection resistor from across the port only *after power-on of the PD is completed,* not before.

Lower Detection Threshold: Practical Concerns in PSEs and PDs

Here is another problem: The lower threshold of the detection resistor is 2.8 V as per the standard. Many have pointed out that this is an oversight on the part of the engineers behind it. The reason is that if the signature resistor is being disengaged, it needs a high-voltage FET inside the PD IC to accomplish that. But the diode-bridge offset is supposed to be about 2 V. In other words, when the PD-side PI voltage is 2.8 V, the PD IC only "sees" 2.8 – 2 = 0.8 V. Therefore, the small, high-voltage detection-resistor FET inside the PD chip has to turn on at 0.8 V. But that is hardly possible, since we need to apply a couple of volts across its Gate and Source typically. Therefore, most commercial PD IC's guarantee the signature is available only *above* 1.5 V at the PD-chip side. From the PSE's perspective, it should therefore not sample for the signature below 1.5 + 2 = 3.5 V at least. This is to avoid interoperability issues, whatever the standard may say.

CHAPTER 4

Classification

What Is Classification?

We start by reacquainting ourselves with the basic handshaking overview presented in Fig. 3.3. After completing a successful detection, the PSE is now "sure" that there is an IEEE-compliant PD on the other end of the cable. It can then proceed to the next stage of the process: *querying the PD as to its power requirement.*

The process by which the PSE queries the PD as to its input power range is called classification. The PD in turn, declares its "class." Alternatively put:

> **Classification is the process by which the PSE asks the PD: *"How big are you (in terms of wattage)?"*** and the PD answers that question, akin to stating its size as indicated in Fig. 3.3.

Types of Classification Methods and Backward Compatibility

Keep in mind that the handshaking process presented in Fig. 3.3 is just exemplary. And it is just one possibility within the broad category of classification. The category is characterized by the use of small currents and voltages for signaling across the copper wire (quite similar to the detection process). Copper is the "physical layer" (or medium) here, and is also called layer 1 or just L1. So this method is generically referred to as physical-layer classification or layer-1 classification or L1 classification. It was the original classification method used in the AF standard.

There is also a very different category of classification based on querying the PD's power rating over the data link (if and when that gets established). This is called layer-2 classification or L2 classification or LLDP-based classification or data link layer (DLL) classification. It was introduced by the AT standard because the AT standard had itself created a new need—the need for *mutual identification* between a (high-power) Type 2 PSE and a Type 2 PD. Classification in the AF standard was traditionally just a one-way street: The PSE would ask the question, and the PD would answer. But the PD never asked anything of the PSE and didn't need to either. The new scenario came about because the new high-power (Type 2) applications that the AT standard introduced were not a subset of existing applications, so (a) a high-power (Type 2) PD could get connected to a Type 1 (AF) PSE, and (b) a high-power PSE could get connected to a Type 1 PD, and so on. The matrix of permutations was so wide that *mutual identification* became the only way to ensure appropriate behavior would result under all permutations. For example, a Type 2 PD could activate its higher-power mode/features, while the Type 2 PSE it was connected to could correspondingly and safely increase its current limits to allow for the higher currents—*provided each realized that the other was Type 2.*

NOTE *LLDP stands for Link Layer Data Protocol. Layers 1 and 2 mentioned previously are a subset of the general 7-layer OSI (Open Systems Interconnection) model. But we need not go into more detail of that here.*

How can mutual identification be implemented? In fact, the most obvious method would seem to be LLDP-based. Which means the Type 2 PSE and Type 2 PD could "talk" to each other over the data link. As simple as that. Except for the fact that both the PSE and PD must at a bare minimum be *capable* of communicating via data. But Midspans are not capable of that. They typically use the spare pairs for PoE, and there are no data transformers inside. So how could mutual identification occur in such a situation? Without going into more details in this section, we just state that the AT standard defined *a second type* of physical-layer classification, pattern in addition to the new LLDP-based classification. The older L1 method is now called a 1-event or 1-finger classification. It corresponds to the voltage being raised by the PSE *once* (a "finger" on an oscilloscope), and is essentially what was described in Fig. 3.3. The AT standard introduced 2-event or 2-finger classification, which, true to its name, raised the voltage twice (with a PD-observable dip in the middle). The idea was simple: If a Type 2 PSE performed a *unique* classification pattern, the Type 2 PD could be designed to "recognize" that pattern, say by means of a very basic flip-flop/ counter, and thereby conclude it was connected to a Type 2 PSE.

Summarizing, as per the standard, mutual identification of Type -2 devices is now possible by either (a) doing a traditional (1-event L1) classification, followed by "further discussion" over the data link (L2), **or** (b) by just doing a 2-event L1 classification (no need for L2 classification). The latter would be conducive to Midspan PSEs but could be used by Endspans too. In that case, the Type 2 Endspan could opt to continue to talk with the Type 2 PD over the data link, and ask further/finer questions about the power requirement of the PD. This "extension" of 2-event classification is not ruled out by the standard.

While framing the AT standard, it was recognized early on that it would be very difficult, if not impossible, for a typical PD to know whether it was connected to an Endspan or a Midspan PSE. That would require an entirely new physical-layer classification pattern. A new one was already in place to distinguish Type 1 from Type 2. Now we would need to distinguish a Type 2 Endspan from a Type 2 Midspan too. Therefore, to avoid needless confusion and complexity, the IEEE standard rightly laid down the *same* classification requirements for both Midspans and Endspans. But in doing so, it also remained conscious of their differing capabilities and limitations as indicated above. That is why the standard does *not* require *any* Type 2 PSE (Endspan or Midspan), to support *both* methods of classification. As indicated previously, a Type 2 PSE can support either 2-event L1 classification (no data link required—good for Midspans and Endspans) or 1-event L1, followed by L2 classification (good only for Endspans).

At this point please consult Fig. 4.1, which reproduces the various PSE-PD possibilities enumerated in Table 33-8 of the AT standard, and also contains an alternative and easy visualization of what is allowed or disallowed by the standard.

This table ensures backward compatibility and "proper behavior" and can be considered the result of the PSE and PD state-machine diagrams contained in the standards (which will be discussed in greater detail in Chap. 8).

At the other end of the cable, Type 2 PDs are *required* by the AT standard to support *both* methods of above-mentioned classification. Why is that? Because a Type 2 PD may find itself connected to a Type -2 PSE that *only* supports 2-event L1 classification, or it may get connected to a PSE that *only* supports 1-event L1 + L2 classification. But they do need to communicate (in the same "language") for successful mutual identification. So since a Type 2 PSE is allowed to speak only one of two languages (1-event + LLDP or 2-event), a Type 2 PD basically has to learn *both* languages to be able to speak to either type of PSE. In other words, a Type 2 PD must support both 2-event classification *and* LLDP.

At least that is what the standard says, and it is logical. Unfortunately, there are some early and/or ultra-low-cost Type 2 PDs out there, which are technically noncompliant because they do *not* support L2 classification (LLDP), as indicated on the lower-right side of Fig. 4.1. We cannot however ostracize them. *Market reality* demands they be supported too. Therefore, most PSE vendors have recognized and accepted very early on, that these low-cost PDs have passed on their burden to the PSE. Since L1 (1- or 2-event) classification is all such a PD recognizes and responds to, to establish communication

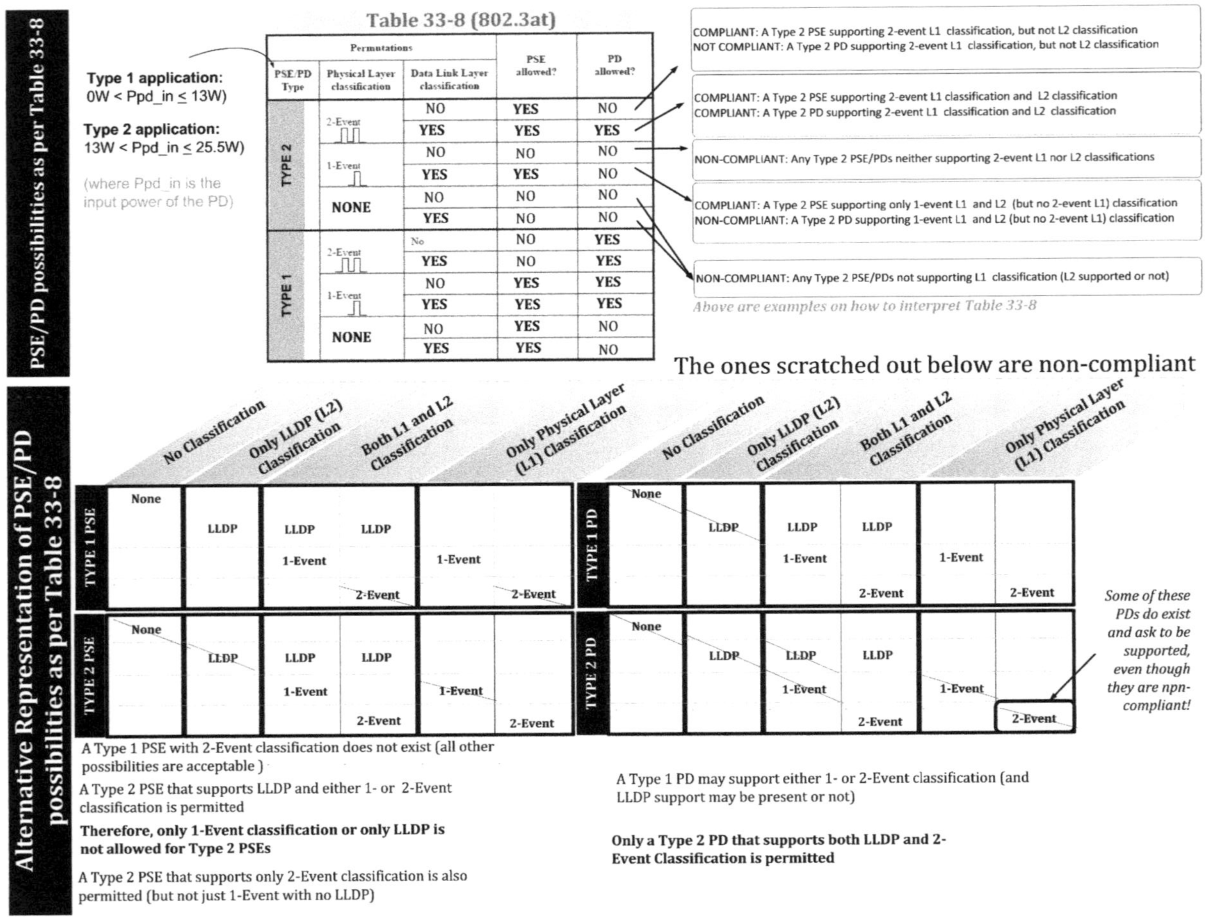

FIGURE 4.1 PSE-PD possibilities are per Table 33-8 in IEEE802.3at and an alternative visualization.

with it, we need to ensure that the (Type 2) PSE (30 W) supports 2-event L1 classification too, even though the standard does not require it to (as long as it supports L2 classification at least). So in reality, the Type 2 PSE has to speak *both* languages: 2-event and LLDP—that's based on market reality, not on the standard.

There is another *practical* reason for insisting that a Type 2 PSE-controller IC support 2-event physical layer (L1) classification. The reason is the IC may be used in PSE Midspans. And so, if the IC doesn't support physical-layer classification, it can only be used/sold in Endspan markets. In other words, though it may not violate the standard, it may just not sell very well.

If a high-power PSE and a compliant or otherwise high-power PD cannot for some reason communicate during classification or after (different "languages" spoken), for all practical purposes, classification gets *bypassed* altogether. In fact, that is allowed too, but only under the "safe" assumption that the application on hand will be treated as a low-power (Type 1) application going forward, not a Type 2 application.

Type 1 PDs, as defined in the AT standard, are largely backward compatible if not identical with PDs defined in the AF standard (though there are some differences in terms of disengagement of the signature for example resistor, as discussed later). So, they *need not* actively support any method of classification. In that case, the PSE will consider them as having no class, or Class 0, and will assign a default power rating to them and proceed to power them up *if power is available.* But this is really not a good idea from an overall systems-management perspective as discussed in the next sections.

Practical Limits of AC-DC Power Supplies

The underlying reason for desiring classification in the first place is that AC power is neither free nor unlimited. The incoming "48V" rail from the AC-DC power supply has practical limits as to its maximum-output power. For example, if a standard-household AC outlet is used as the input to the AC-DC power supply, we must keep in mind that the outlet has a maximum RMS rating of 15-A. Engineers usually derate that by applying a multiplicative factor of 80 percent to get the safe, continuous rating of the AC outlet as 12 A. Then they multiply that (safe) number with the incoming AC (RMS) voltage (say 110 V), and further multiply that by the "power factor" of the power supply (say 0.9), and also the efficiency of the power supply (say 85 percent). The result of all that is the maximum useful power (output wattage) that a single AC-DC power supply unit running off a standard 110 V AC outlet can provide is typically around 1 kW.

Now, 1 kW may be OK for about 24 PoE ports, each drawing 30 W simultaneously, though we also have to consider the overall switch/PHY power requirements (if any), which will also be derived from the *same* AC-DC power supply and will come in from the same AC outlet. If we need to get to the next power level to support more ports, multiple AC-DC power supplies can be paralleled using active load-share circuitry. Paralleling of power supplies is considered expensive unless unavoidable, but it is really not such a bad idea, even for lower powers, because it can add a certain amount of *redundancy*, since at least *some* power is still available after a single power- supply failure.

But to ensure any logic, we need to first start *monitoring* the multiple power supplies. So the state of the output rail of each power supply (good/no-good) is usually checked by the PSE chip, with the help of three or four "power-bank" or "power-good" pins. For example, if one of the paralleled power supplies fails, the available power is obviously no longer enough. The PSE can then "drop" (previously programmed) "low-priority" ports, preferring to *not* interrupt some of the other "high-priority" ports. An example of the latter could be the port going to the IP phone in the CEO's office. The former could be an IP camera in some noncritical location on the premises. Ethernet and PoE are not democratic processes; they are hierarchical by nature.

However, with the power-supply-monitoring data available, to make use of it, we now need power-management software—usually found residing somewhere on the PSE end of the cable. On further thought, we realize that the key input to any power-management software is the result of *classification*, because once the available power is known, it has to be distributed across several devices ("rationed") based on their power-requirements. And that comes from classification. So, eventually, classification enables a whole set of useful features and systems-level opportunities for optimization of the entire power system.

Classification Is Optional for Type 1 Applications But Recommended

For Type 1 devices, classification is *optional*. In other words, knowledge of the *actual* power rating of a given PD is *not* completely essential. But we have just realized it helps a great deal, and should be implemented even for Type 1 devices (it is not optional for Type 2 devices, unless they want to pretend they are Type 1 devices and can be configured to do so by the user).

If the PSE knows power requirements via classification, it can then try to reserve the *right* amount of power for each classified PD, ensuring its uninterrupted operation, and also ensuring that as many other PDs/ports as possible can be powered up later if desired, with any remaining (surplus) power.

If after a given classification sequence, the PSE learns it has insufficient power to provide to the PD, it will not power up that PD but will return to its "idle" state once again. The handshaking process will commence all over again, starting with detection. To prevent the chances of this happening more frequently than necessary, we should try to carefully budget out ("ration") the power to different PDs *after* making accurate estimates of their requirements via classification.

Default Class (Class 0)

The reason we *can* skip classification altogether in some cases is that the IEEE standard specifies a *default value* of power—one that we can assign to *any* unknown PD and would be at least "safe," because it is based on the worst-case assumption of a CAT3 cable. That default value is 12.95 W (on the PD side), which translates to 15.4 W on the PSE-side; see Fig. 4.2. It is the equivalent of "no class," or equivalently Class 0.

But this is also the *maximum* power as per the AF standard. Do we really want to allocate full power to every port unless we are really sure the devices being connected indeed require that much power? It is tempting in a way: The default power value would *suffice* for all Type 1 PDs (up to 12.95 W), and will therefore at least not cause unanticipated shutdowns, of already powered ports (because the set default is already the maximum). Unfortunately, the resulting situation is not optimal. The power-management software will not realize there is power to spare, so it might prevent *new* PDs from being powered up, despite power being available.

The task of performing (accurate) classification, thereby estimating beforehand any leftover power to give to additional ports, leads to a situation where, *statistically speaking* at least, a greater number of ports can be supported by smaller (and cheaper) AC-DC power supplies. We see that even the basic objective of lowered costs can be better realized by performing classification. A default-power rating, as with no classification, is basically the equivalent of "playing safe" in *most* applications—and we should know all too well by now that playing safe (high-design margins/headroom), is the easiest route to over-design, accompanied with all its well-known penalties including escalating costs and diminishing returns.

What if we have a Type 2 PD and the PSE assigns it a default class? In that case, using 12.95 W as its default rating at power-up means the PD will need to run with some of its "power-hungry" features disabled—at least initially. *Otherwise it risks being treated as an overload/short by the Type 1 PSE and being turned OFF*. For example, if the PD is a pan-tilt-zoom (PTZ) camera, in the absence of appropriate classification the pan function may need to be disabled. Or maybe the zoom function, and so on. Not permanently though. Because later, if the data link is available and LLDP is enabled, the PD and PSE *could* "discuss" the actual power requirement and mutually identify

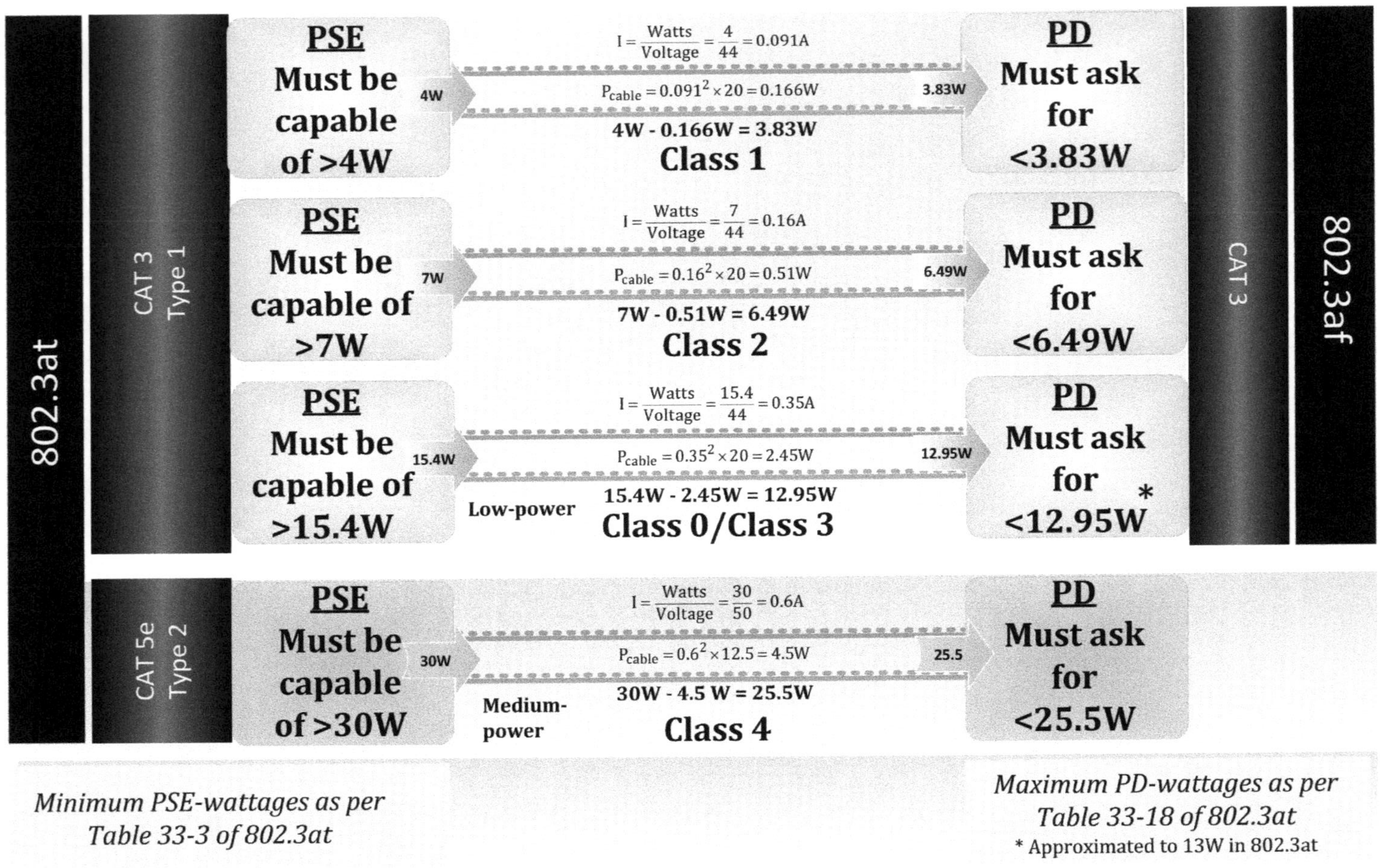

FIGURE 4.2 PoE power-classification levels.

each other as required. However, examine the lower half of Fig. 4.1 carefully. A Type 2 PSE is not allowed by the standard to bypass classification altogether (assigning Class 0 to the PD) and expect to do the entire classification over the data link. For safety sake, a Type 2 PSE must perform 1-event classification at least, and also see Class 4 reported by the PD during that phase. The LLDP part of the classification process can only be confirmatory in nature (how's that for faith in the software guys).

LLDP or Physical-Layer Classification for Type 2 PSEs?

We have learned that *a Type 2 PSE can support either 2-event L1 classification (no data link required—good for Midspans and Endspans) or 1-event L1, followed by L2 classification (good only for Endspans).* See also Fig. 4.1. Colloquially, we just say a Type 2 PSE can do either physical layer classification or LLDP-based classification. What we actually mean should, however, be clear by now. It is actually 2-event, or 1-event + LLDP.

So now we ask: *Which is better* (obviously from the viewpoint of Endspans)? We will discuss this once more at the end of this chapter too.

For the purpose of mutual identification between Type 2 devices, the first method (Type 2 classification completed at layer-2), is much *slower* than the second method (Type 2 classification completed at layer-1). The former can take a minute or more after connecting the PD—the normal time for the data link to be up and available to carry out LLDP-based classification. It therefore leads to significant installation delays and possible inconvenience at times. But despite that, layer-2 classification is often preferred over layer-1 (physical-layer) classification, because the power negotiation and subsequent power allocation can eventually be carried out with much *higher granularity*—up to 100-mW resolution, as compared the several watts of difference between adjacent classes, as offered by traditional physical-layer (layer-1) classification. Layer-2 classification can also be done *continuously* while the equipment is in operation (live), and so power allocations can be adjusted "dynamically." So in high-power/high-port installations where power is at a premium and allocation needs to be done with the utmost care, LLDP is the preferred choice (with 1-event L1 classification preceding it, of course). But for most commonly seen PoE systems, layer-1 (2-event) classification is usually acceptable despite its lack of finer granularity—being preferred for its inherent simplicity, speed, and convenience.

Class Levels in Layer-1 Classification

Now we come to the actual classes in physical-layer classification. When the AF standard was being created, there were only a handful of power classes under consideration, based on existing types of devices in the market. These power levels were called Class 0, Class 1, Class 2, and Class 3. Class 4 was kept as a placeholder for the future but was largely undefined at the time.

1. Class 0 is in effect a "classless default," and it means that the PD (max) power rating is anywhere between 12.95 W (corresponding to 15.4 W measured at the PSE end as per Fig. 4.2), down to almost zero. *Almost zero,* in this case, is just enough power drawn by the PD so as to prevent the PSE from disconnecting it thinking there was nothing connected at the other end of the cable.

   Keep in mind that the maximum PD power of 12.95 W is based on 44-V minimum-operating input and 0.35 A maximum current: 44 V × 0.35 A = 15.4 W. From this we take off cable losses as per Fig. 4.2.

   Class 0 was later rounded up to 13 W at the PD end by the AT standard and is now called "low power."
2. Class 1 is set for 4 W at the PSE end. From Fig. 4.2 we see that this gives us 3.83 W at the PD end. So a Class 1 PD-maximum power rating is anywhere from 3.83 W down to almost zero.
3. Class 2 is set for 7 W at the PSE end. From Fig. 4.2 we see that this gives us 6.49 W at the PD-end. So a Class 2 PD can demand maximum power anywhere from 6.49 W down to 3.83 W (not down to almost zero).
4. Class 3 is called "low power" too, but unlike Class 0, Class 3 corresponds to a PD-max rating between 12.95 W down to 6.49 W (not down to "almost zero" as for Class 0).

   As mentioned, the AF standard was made somewhat future-proof by reserving Class 4 (undefined at the time). The AT standard came along a few years later and used this very placeholder to cover applications up to 25.5 W (corresponding to 30 W at the PSE-end, see Fig. 4.2). So we now add this class to the above list.
5. Class 4 is called "medium power," and corresponds to a PD max rating between 12.95 to 25.5 W.

Keep in mind that Classes 0, 1, 2, 3 are, in the worst case, intended for CAT3 cabling, whereas Class 4 is definitely meant for (better) CAT5/CAT 5e cabling only. The loop resistances are different in each case, so the cable losses are also different as per Fig. 4.2.

The class of a PD is based on its power rating. We have to be clear this is the power flowing into its *input* terminals from the PI. Some of this power is lost in the PD pass-FET, and the remaining enters the input of the DC-DC converter. There will be additional energy lost in the switching-conversion stage, and so finally, perhaps only 60 to 80 percent of the supposed "power rating" of the PD may actually amount to "useful power." It is important to minimize all losses (improve efficiency) in the PD so as to maximize its overall usefulness. One modern technique called "active diode bridge" will be discussed later, and though it improves efficiency, it can cause interoperability concerns because of, typical PD-design architecture. In general, however, efficient PDs are good from the overall *infrastructure* point of view too, since they usually ask for *lesser* input power (for a given useful power), thus allowing more ports to be powered up simultaneously.

NOTE *The continuous power levels mentioned above are actually an average value calculated using a sliding window of 1 s. In other words, the standard does leave some "head-room" for small bursts of power above the average levels as discussed later. It also sets overload thresholds depending on reported class and can terminate power if excessive power is attempted to be drawn.*

NOTE *In reality, the AT standard does not tabulate the minimum thresholds of the power ratings of the Classes. For example, it tabulates both Class 0 and Class 3 with a max power of 12.95 W at the PD-side, with no minimum values stated in the relevant table. So what's the difference? Class 0 and Class 3 seem identical on IEEE-AT-compliant paper! However, to avoid confusion, in all the preceding class descriptions, we made the most reasonable and popularly accepted assumption regarding the minimum thresholds of the power ratings (consistent with the AF standard in which they were stated more clearly actually).*

NOTE *The "almost-zero" power level mentioned previously (for Class 0 and Class 1) is popularly assumed to be 0.44 W, based on a minimum current of 10 mA at 44 V input. It is almost the same number at the PSE end as on the PD end. As per the standard, average PD currents lower than 10 mA can cause the PSE to shut off power to the port for safety reasons, because the PSE can then assume there is no PD present on the other end of the cable. This issue is discussed in Chap. 7. For Class 4, this would be 10 mA* $\times$ *50 V = 0.5 W.*

NOTE *In general, a PSE must draw a minimum 10 mA to avoid being DC-disconnected. This was earlier reported as a minimum power of 44 V* $\times$ *10 mA = 0.44 W for Class 0 and Class 3 in particular, in IEEE 802.3af. However this number is not present in IEEE 802.3at. The reason is a PD actually needs to draw 10 mA for just 60 ms after drawing zero current for 300 ms to avoid disconnection. So the input "duty cycle" is 60/360 = 0.167. Therefore, the average input current is 10 mA* $\times$ *0.167 = 1.67 mA. So the average "keep alive" minimum power draw is therefore 1.67 mA* $\times$ *50 V = 83 mW, not 500 mW.*

1-Event Classification

In Fig. 4.3 we present a basic example of a 1-event (or 1-finger) classification sequence, which is the most basic type of physical-layer classification. The initial part of that process is detection of course. The detection portion just happens to be a two-step ascending staircase in the figure. It could well be a *descending* staircase, and maybe consist of two, three, four or even more steps. It could also be, say, two steps ascending followed by two steps descending, and so on. As mentioned earlier, detection can also be done either with current sources or voltage sources. We see that many possibilities for detection exist within the framework of the IEEE standard.

Key questions through all the initial handshaking sequence is: When does a given phase start and where exactly does it end? And what are its corresponding timings? Unfortunately, this is a source of ambiguity, and there are sometimes contradictory statements and

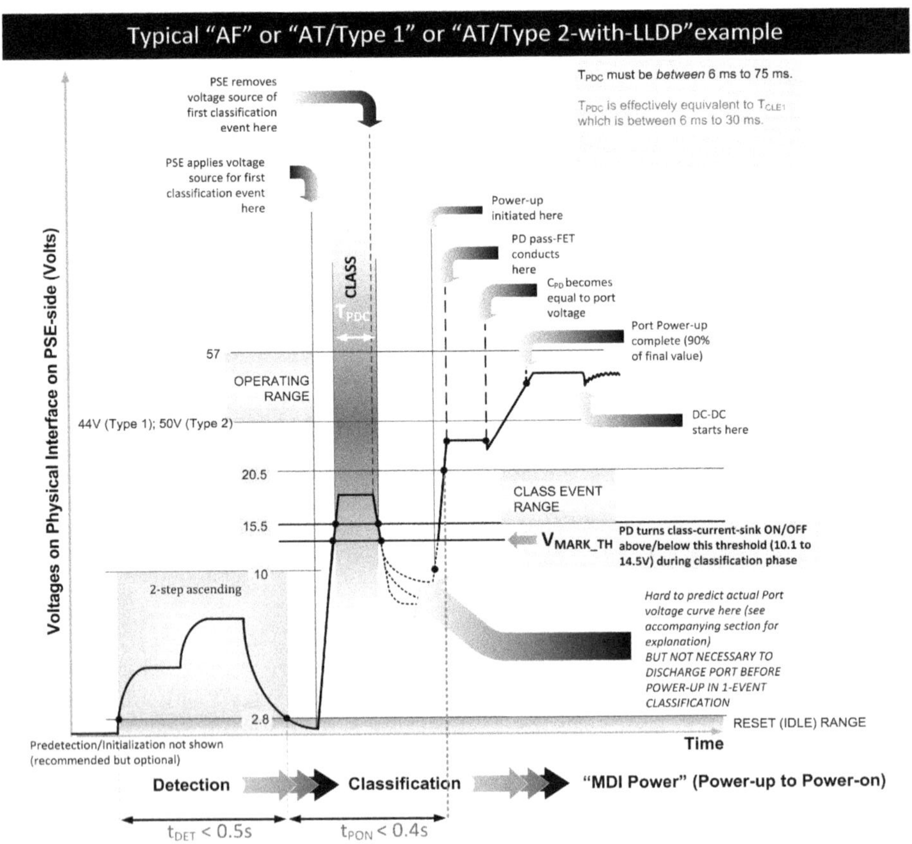

FIGURE 4.3 An example of power-up sequence using 1-event classification.

figures in the AF and AT standards. Therefore, to answer this in the most acceptable manner and to thereby present it in the accompanying figures, we have adopted what appear to be the the most commonly evolving interpretations appearing in PoE IC datasheets and related literature. The guiding perspective is IC *design* and *testability*. For example, both the PSE and PD chips have certain internal-voltage thresholds that are being monitored by them. "Certain things" happen inside these chips at the crossing of these thresholds. So we can conclude that different phases of operation also start and end *at the crossing of certain logically defined thresholds*, and that eventually leads to the associated time intervals.

Based on this approach, as in Chap. 3, we have defined the *end of detection* as the point where the PI voltage (measured on the PSE side) *leaves the detection window*—that is, it goes outside the detection range of 2.8 to 10 V. It can leave that window either by falling *below* 2.8 V (idle/reset threshold) (after which it rises into the classification range as shown in Fig. 4.3), or it could go straight into the classification range without a dip. The latter would be accomplished by the PSE immediately *raising* the voltage above 10 V on detecting a 25-k signature resistor. This is shown in the slightly modified version of Fig. 4.3 presented in Fig. 4.4. Both methods are acceptable and IEEE-compliant 1-event classification patterns. After detection, classification starts. In Fig 4.5, we have a slight variation of Fig 4.4 in this regard, as will become clear soon.

But where *exactly* does classification start? We define that as the moment the PSE applies a voltage source across the port and causes the the PI voltage (at the PSE end) to rise above 15.5 V(> 14.5 V at the PD end)—up to a maximum of 20.5 V (at either the PSE or PD end). This range, 15.5 to 20.5 V at the PSE end, corresponding to 14.5 to 20.5 V at the PD-end, is called the class event or classification-event range.

PSE-chip designers often prefer to say the classification interval starts when the classification voltage source is applied, and ends when that source is disconnected, but from the testability point of view the interval is best defined based on crossing measurable voltage thresholds.

NOTE *We can see that during classification, as per the standard, the PSE is supposed to apply only* voltage *sources across the cable. Early versions of the evolving AF standard had allowed current sources too. In that case, the PD class would be indicated by the measured PI voltage. But this option was removed from the standard long ago.*

The classification-voltage source inside the PSE that brings the PI voltage into the class event range is called V_{CLASS}. It is invariably a linear (series-pass) regulator, sometimes called an LDO for low-dropout regulator, but it does not need to have a low dropout in this application

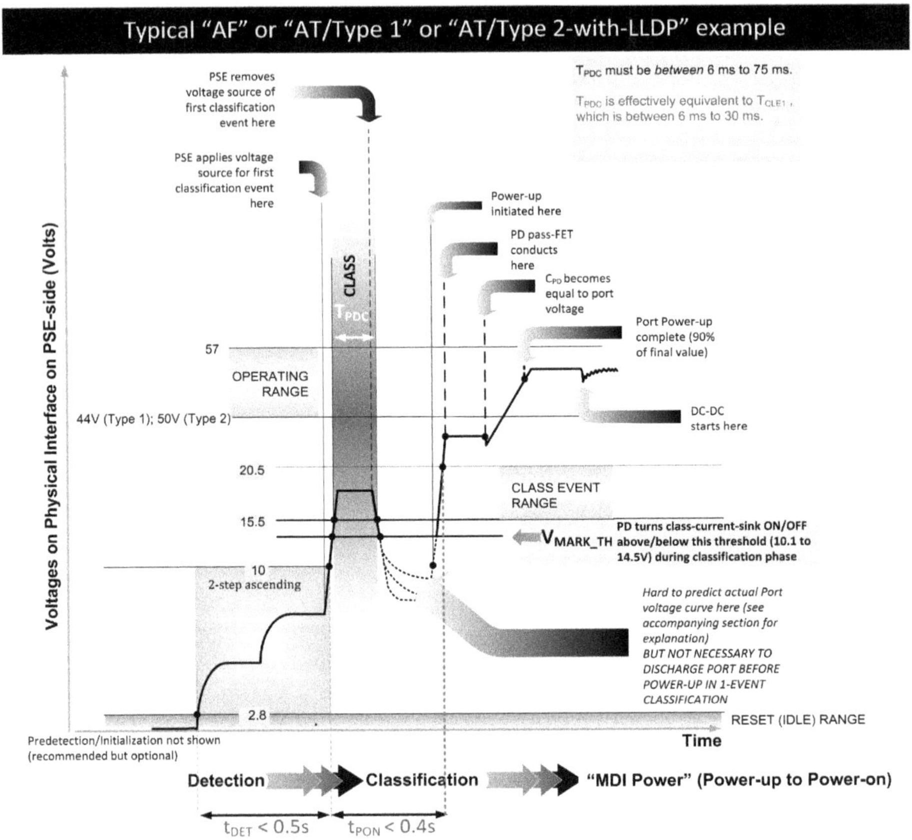

Figure 4.4 Another example of power-up sequence using 1-event classification (straight from detection to classification).

actually (the input to output difference is very large). Its key requirements are that it should be fairly accurately regulated to somewhere in the middle of 15.5 to 20.5 V, and also be able to hold that voltage steady (regulated), with no ringing and so on (be stable), up to at least 50 mA of classification current. For protection, the voltage source should also be current-limited under faults (cable shorts, for example), so that the current from V_{MARK} never exceeds 100 mA. *These are not trivial design requirements.* During PSE-chip design, the stability of the internal LDO (low drop out regulator) providing V_{CLASS}, *must be assured across several possible variations of distributed L and C on the cable,* which could be anywhere from 0 to 100 m in length, and of CAT3 or CAT5e construction.

This PSE-side classification-voltage regulator can be designed such that it is capable only of sourcing current, not sinking it. It is therefore a one-quadrant regulator. Regulators that can source and sink current are called two-quadrant regulators.

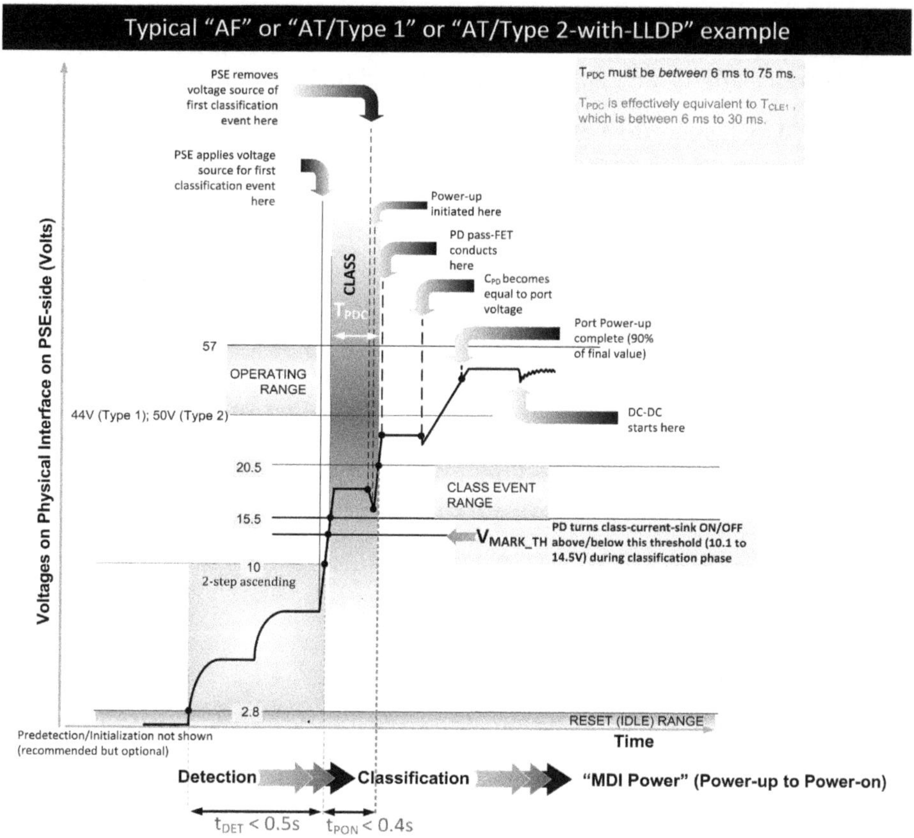

FIGURE 4.5 Another example of power-up sequence using 1-event classification (straight from detection to classification and classification to power-up).

Looking at Fig. 3.3 too, we realize that the very act of raising the PI voltage between 15.5 to 20.5 V amounts to a query from the PSE to the PD. On its part, a properly designed modern PD proactively responds to the classification query by applying a current sink across the port, for the PSE to then read. That current reading tells the PSE the class (input power rating) of the PD.

When exactly does the PD need to apply this current sink? It has to be somewhere between the upper-detection limit and the lower threshold of the class-event range obviously. This was rather *implicitly* stated in the AF standard, but the AT standard formalized it by defining a mark threshold called V_{MARK_TH} between 10.1 to 14.5 V (as measured on the PD side). The PD turns on its class-current sink above V_{MARK_TH}, and turns it off (with some undefined internal-chip hysteresis) as soon as the PD-side PI voltage falls below the V_{MARK_TH} threshold.

Finally, if all goes right, based on the current measurement, the PSE identifies the class of the PD. The PD's designed-in class-current-sink limits, guaranteed over 14.5 to 20.5 V (PD side), are

1. Class 0: 0 to 4 mA
2. Class 1: 9 to 12 mA
3. Class 2: 17 to 20 mA
4. Class 3: 26 to 30 mA
5. Class 4: 36 to 44 mA

As explained in Chap. 3, the PD (design) limits are always tighter (narrower) than the PSE (measurement) limits—to account for imperfect measurement accuracies in the PSE, and other secondary effects like leakage and so on. Based on that reasoning, the PSE-classification measurement limits as specified by the standard, and valid over the extended PSE-side PI voltage range of 15.5 to 20.5 V, are

1. Class 0: 0 to 5 mA
2. Class 1: 8 to 13 mA
3. Class 2: 16 to 21 mA
4. Class 3: 25 to 31 mA
5. Class 4: 35 to 45 mA

Basically, we have just added 1 mA to either side of the PD-design limits (except for the 0 mA lower limit of Class 0), and come up with the PSE-measurement limits.

In Figure 4.6 we have presented a quick look-up chart to connect the classification/power boundaries mentioned above for easy reference, from the viewpoint of the PSE and the PD.

Classification "Gray Areas"

The unspecified regions *between* the above-mentioned class ranges are "gray areas," and the PSE is technically allowed to read either of the adjacent classes. It seems more advisable, however, that the PSE should read the class corresponding to *higher power*, simply to avoid "strange" shutdowns of ports. For example if the PSE reads 23 mA, then as per the standard it is free to interpret the PD as either Class 2 or Class 3. But to avoid an inconvenient port shutdown, a more practical choice seems to be to make the PSE err on the side of higher power.

Prevention is always better than cure. So to avoid this ambiguous situation altogether, the best option is that the PD should itself be *well-designed*—in particular, its current sink should have a tight tolerance (with respect to both applied *voltage and temperature*), and further, its nominal value should be centered fairly accurately between the desired and appropriate class-current limits.

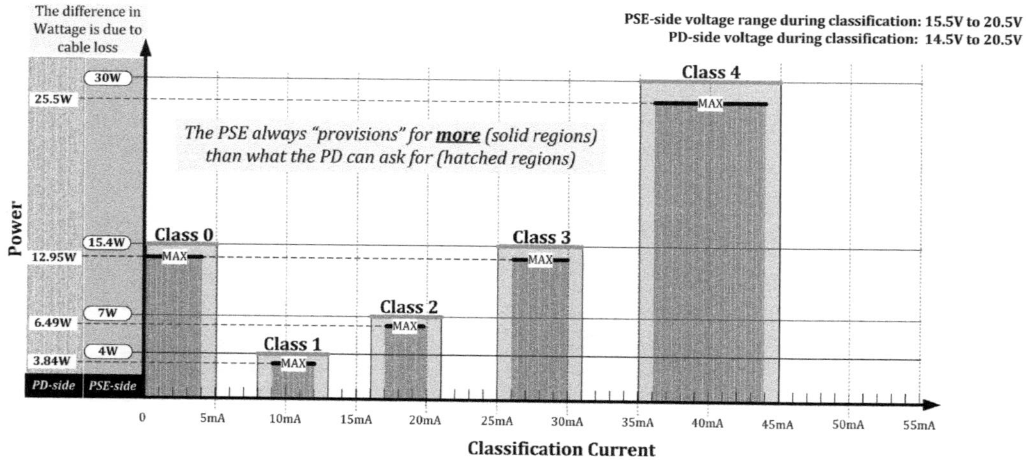

PSE:

The horizontal axis this tells us how a compliant PSE MUST interpret the class based on measured currents. Between the class blocks are don't care regions --- the PSE can declare the class of either adjoining block.

On the vertical axis, we get the range of total (PD + Cable) Watts that is guaranteed supported by the PSE. In other words, the topmost level of each block (solid blue line) represents the minimum power guaranteed to be coming out of the PSE for a given class.

PD:

The horizontal axis this tells us how much class current (its range) that a compliant PD must pull, to indicate to the PSE its class (*"power range"*).

From the vertical axis, we get the max *power range* of the PD. In other words, the maximum continuous (i.e. average) power that a given PD can draw for a certain class, lies within this block, but is also usually higher than the next lower-power class (because otherwise we could have logically declared the PD to be of a lower- power class too).

Note that a Type 1 PSE can ignore classification, and simply report Class 0 as a default value.

Note that in general, a PSE must draw a minimum 10mA to avoid being DC-disconnected. This was earlier reported as a minimum power of 44V x 10mA = 0.44W for Class 0 and Class 3 in particular, in IEEE 802.3af. However this number is not present in IEEE 802.3at. Because in fact, a PD needs to draw 10mA for just 60ms after drawing zero current for 300ms to avoid disconnection.

FIGURE 4.6 Quick reference chart for classification from both the PD's and PSE's viewpoints.

If the PSE measures between 45 to 51 mA, it is free to interpret the PD as either Class 4 or Class 0 as per the standard. Between 51 to 100 mA, the PD definitely needs to be considered Class 0. But power-up is, as always, *optional*. The usual rule/loophole is the allowance contained in the standard which says that a port be powered up only if "enough power is available." The ground reality is no compliance test equipment can ever know for sure, leave aside test, when power is *really available* (to power up a port) or not. So that creates a loophole of sorts to do, say, double-detection beyond the maximum-allowable detection time of 0.5 s, as discussed in Chap. 3.

Where does the 100 mA number mentioned above come from? That is the worst-case (max value) of the current limit of the classification-voltage source inside the PSE. In other words, the standard asks that somewhere between 51 to 100 mA, the PSE should place a fairly accurate current limit on the classification-voltage source, so as to offer protection against circuit faults/shorts. Its worst case should not exceed 100 mA under any situation.

NOTE *It is not explicitly stated in the standard for the classification -current overload, but is implied in general across the standard that whenever overloads are to be measured, a 1 ms settling time is permissible before measurement. In other words, we can assume that if we apply a short during classification, the current must be less than 100 mA measured after 1 ms. Temporarily it may even exceed that. This allowance is good, because no current limit acts instantaneously as we all know.*

Reported "Interoperability" Issues

Poorly designed PDs have historically posed a lot of so-called interoperability issues. One key reason is, however, they have sometimes wrongly reported their class. Now, if they report a higher-power rating than their actual rating, the PSE could provision for that, and the only problem would be that the power supply was not being used optimally. That's not too serious. However, if the PD reports a *lower-wattage* rating than its actual-wattage requirement, the PSE may even cut off power to the PD, thinking there is a malfunction, or that enough power is not available/allocated. As we will see in a later chapter, the PSE sets its maximum-current thresholds based on the reported PD class. So it can terminate power to a PD if the class-based current-limit thresholds are crossed.

Keep in mind that adding any front-end circuitry, say even a splitter/dongle with a bridge rectifier at the input of any PD, can inadvertently change its assumed class. Power class is based on *input* power to the PD, not its output or useful power.

Timings during 1-Event Classification

It seems intuitive at first thought that the *time* for which the PD turns on its class-current-sink, is, or should be, somehow controlled or defined by the IEEE standard. But in fact, we cannot assume any timers are present inside the PD during classification—for the simple reason the PD has minimal power available at this point and is not expected to support potentially power-hungry circuitry (that was at least the reasoning at the time of the AF standard). The PD is therefore designed mainly to *just respond* to the applied voltage levels, both during detection and classification.

On that basis, it is the time for which the voltage source is applied *by the PSE* which is actually governed by the standard, and that *indirectly* sets the duration for which the class-current-sink is turned on by the PD (it tracks the same timing).

In the AF standard, the time spent in the classification range (15.5 to 20.5 V) was stated as T_{PDC} = 10 to 75 ms. The AT standard changed that to 6 to 75 ms for reasons explained later. The minimum-time constraint (6 or 10 ms) gives enough time for the PD to turn on its class-current-sink, and for the PSE to measure that current. The upper limit on T_{PDC} (75 ms) was meant to prevent excessive dissipation both in the PD and in the PSE stages during this rather dissipative classification phase.

Dissipation during Classification

How dissipative is the classification phase really? In fact the dissipation during classification can even be *more* than when the port is powered up and the pass-FET is fully conducting and delivering full AF/AT load current to the PD. For example, if the incoming supply is 48 V, the voltage across the pass-FET of the PSE during classification (class-event range ~18 V) is roughly 48 V – 18 V = 30V. The rest, 18 V, falls across the PD. With a Class 3 current of 30 mA, the dissipation is 30 V × 30 mA = 900 mW (~ 1W) in the PSE pass-FET, and 18 V × 30 mA = 540 mW (~1/2 W) in the PD pass-FET. This is all really significant dissipation. So the maximum classification time during 1-event classification is constrained by the standard.

But in fact, limiting the maximum time for dissipation only affords partial protection. It is not necessarily sufficient under all situations. It may not be enough especially when the ambient temperatures are high to start with. Therefore, many PD and PSE ICs incorporate *thermal sensors* on the chip itself to ensure proper reliability. As engineers, however, we must carefully consider *where* the heat is being generated/dissipated. For PSE ICs with *external* (discrete) pass-FETs, it is not practical to provide thermal protection for the external FETs. That is the reason why PSE solutions with internal FETs are usually considered more reliable. They do, however, have a problem in terms of not being able to spread the heat well across the board, leading to hot spots (thermal constriction effects).

2-Event Classification

As mentioned previously, 1-event classification is a one-way street: The PD is simply telling the PSE (on being queried) what class it belongs to and therefore what power it demands. It does not turn around and seek to identify the power delivery capabilities of the PSE. When the AT standard came about, with an entirely new class, one that was not a subset of existing classes, that situation needed to change. For one, there were already switches/hubs out there that did not support medium-power (30 W). So a new medium-power PD may find itself connected to an older switch/hub—one that does not support its power requirement. In that case, the PD would need to "know" what type of PSE it was connected to, so it could, for example, disable some of its higher-power features if required. Thus came about the need for *mutual identification* during classification—no longer could it be a one-way street as in the AF standard.

The PSE had to proactively signal to the Type 2 PD in some way that it was a Type 2 PSE. And from that came about the two-way street that was appropriately called 2-event (or 2-finger) classification. See Fig. 4.7. Another way to do that is via LLDP, provided that route is available to start with (not so in a Midspan), and provided the PD declares itself as Class 4 during the (1-event) physical-layer (L1) classification. See Fig. 4.1.

Very briefly, if a PD indicates to the PSE that it is Class 4 (medium-power/Type 2), then a Type 2 PSE can proceed to identify itself to a Type 2 PD, by distinguishing itself from an older Type 1 PSE by the simple stratagem of *repeating the classification sequence*. On its part, the PSE is just expecting to read Class 4 once again (and it really needs to read the *same message* twice unless there is an unknown error, at which point the PSE would go back to its idle state and start all over again at detection). From the Type 2 PD's perspective, it sees (and records) a 2-event classification (2 fingers), perhaps keeping count of the fingers with a very basic flip-flop/timer. It thus discovers it is dealing with a Type 2 PSE. If it does not see that special pattern, it simply presumes a Type 1 PSE—at least until the data link is up, when it can talk further to the PSE, and that would be 1-event L1 + L2 mutual identification of Type 2 devices as discussed previously.

There is a key subtlety involved now that we should realize: a Type 2 PD seeing a 2-event classification sequence has the *additional burden* now of "remembering" that it saw a single-classification event (finger) followed by another classification event, and powering up with the preceding mutual identification information *intact*. This demands some additional circuitry inside the PD like flip-flops/

timers, and so on. However, to ensure the PD does not get inadvertently reset ("forgets" the past-class event), the AT standard defines a "mark event range" from 7 to 10 V, and also asks that the PI voltage (at the PSE-end) be kept higher than 7 V during the entire classification process. Because it is assumed that the Type 2 PD's classification memory can get reset if the PD ever sees a voltage less than 6.9 V (at its end).

The introduction of this lowered "mark-event" range also serves to clearly delineate (separate) the two class events (fingers) from each other, so the PD would know for sure it saw *two* class events, not just one big classification pulse.

Finally, to implement all this, the AT standard defines the key classification intervals T_{CLE1}, $T_{CLE2,}$ T_{ME1} and T_{ME2} as shown in Fig. 4.6. The limits on timings are provided within the figure.

Timings during 2-Event Classification

As shown in Fig. 4.7, the PSE is required to hold the PI voltage within the class-event range for a specified period called T_{CLE1} (first class-event duration). T_{CLE1} can be 6 to 30 ms. This duration is considered enough to allow the PD to turn on its class current sink, and for the PSE to measure it (the class current) accurately. T_{ME1} can be anywhere from 6 to 12 ms. So the *maximum* duration for 2-event classification, *not including* T_{ME2}, is 30 + 12 + 30 = 72 ms. T_{ME2} must be at least 6 ms, but is left *open-ended* for reasons explained further below.

In practice, since a Type 2 PSE needs to handle Type 1 cases too with the same chip architecture, T_{CLE1} usually becomes synonymous with T_{PDC} whenever the PSE either has to, or opts to, perform 1-event classification (instead of 2-event classification). That is why a Type 2 PSE datasheet will typically specify its T_{PDC} range to be the same as its T_{CLE1} range (6 to 30 ms). The 75 ms upper limit of T_{PDC} is in effect redundant in modern times.

We now realize why the lower limit of T_{PDC} was reduced from 10 ms to 6 ms by the AT standard. It was to make it the same as the lower limit of T_{CLE1}. The upper limit of 75 ms for T_{PDC} was retained (allowed) for backward compatibility with existing AF PDs. Note that the maximum duration for the entire 2-event classification sequence still targets around 75 ms—except for T_{ME2}, whose *upper-time limit has not been specified in the standard.*

We note from Fig. 4.7 that though the max of T_{ME2} is not specified, it is *indirectly limited* by the constraint on the max value of T_{PON} (see further below). But there is lots of time to spare. Based on max values, a typical 2-event classification cycle can be completed in much less than 100 ms (typically 45 ms), and in fact as quickly as 6 + 6 + 6 + 6 = 24 ms. So why is there an additional time allowance of up to 400 ms? We discuss the key reason for that next.

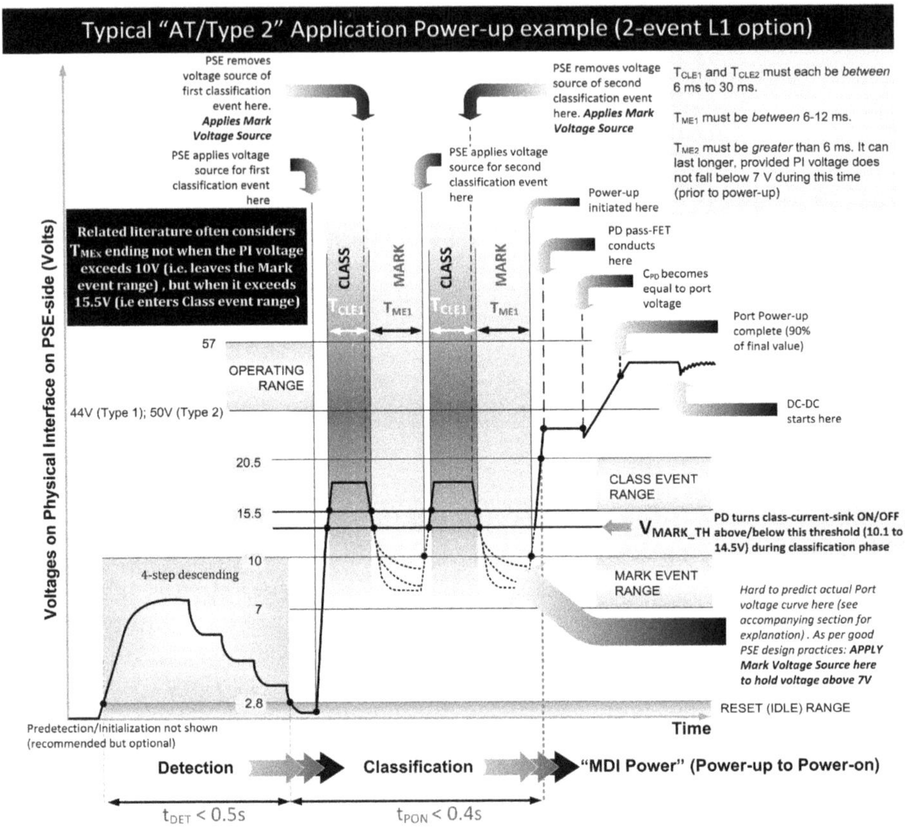

FIGURE 4.7 More detailed example of power-up sequence using 2-event classification.

Overall Timing Constraints

The first timing constraint we need to keep in mind is that we always need to limit the detection time t_{DET} to less than 0.5 s. To confirm compliance, t_{DET} should be measured starting from the moment the PI voltage first climbs above 2.8 V to initiate detection, till the voltage either falls below 2.8 V (again) or goes above 10 V, *whichever occurs first*. See Figs. 4.3, 4.4, 4.5, and 4.7.

Some terms to keep in mind here first: as per the standard, after a successful classification, the PD enters the MDI power states, in which it finds power being released to it and accordingly introduces certain delays before activating the DC-DC converter (as explained in the next chapter). From the PSE's perspective, there are two states/phases involved here. First there is the *power-up* phase, which in terms of testability is best defined as the moment the PI voltage on the PSE-side is raised above 20.5 V (beyond the class-event range). After that, when the PI voltage gets to within 90 percent of its final settling value (very close to the incoming supply rail), *power-on* is said to have occurred (and *power-up* correspondingly over).

We also need to ensure that the power-on time $T_{PON,}$ is less than 0.4 s. Note that despite its name, T_{PON} is not the time until power-on, but is actually the time measured from the end of a successful detection to the start of *power-up* (that is the moment the PI voltage goes above the upper-class event threshold of 20.5 V). For example, see the datasheet of LTC4274 (PSE IC). Usually this time constraint is easy to meet and poses no challenge for a *single-port system*, given the max limits on three of the four classification intervals. But the extra available time and also the "open-ended interval" T_{ME2} come in handy for several other reasons as explained below.

Note that there is still some disagreement about the end of T_{PON}, as discussed a little more in the next chapter.

Multiple-Port Compliance and Systems Issues

From the perspective of PSE-chip designers, they do *not* want to provide the ability to classify several ports simultaneously for the following reasons:

1. The voltage sources used during classification are "expensive" in terms of design complexity and silicon area. They should be multiplexed if possible. So from the cost viewpoint, one classification at a time is the best choice.
2. Further, the dissipation during classification can be very high as we have already seen. Several ports being classified simultaneously can lead to a lot of heat generation, which can take the chip into thermal protection and shutdown, depending on its specific architecture, and where exactly the dissipation is occurring. Also in play are where the thermal sensors (if any) are located, and so on. In any case, one classification at a time is preferred for this reason too.

The problem becomes obvious when several PSE ports are connected to several PDs simultaneously. Assume for now that all ports undergo successful detection simultaneously (though that too is usually staggered to save silicon area and cost). So now, after successful detection, within each port's PSE chip, a T_{PON} port-timer is activated that demands power-on occur in less than 0.4 s. We can now proceed to multiplex the classification hardware (voltage/current sources) for several ports and still keep to within 0.4 s on *all* ports, provided of course the classification of each port is completed quickly enough. Most PSE ICs complete classification of a single port in about 45 ms. So in a multiport situation, there is ample time to multiplex several ports. For multiplexing classification, the best place to keep a port "waiting" for the classification hardware, to get to it, is *between detection and classification*.

There is another issue too. To avoid input rail instabilities and noise, it is also advisable to stagger the actual *power-up* of several ports. Because for each port there is a potentially high in-rush current as it powers up. We do not want the input rail sagging by trying to power up several ports simultaneously. So, in this case, the time between classification and power-up is a good "waiting zone." And it can obviously happen only during T_{ME2}. That explains why T_{ME2} was left open-ended in the standard.

Keep in mind that it is implied that in all cases, T_{CLE1}, T_{CLE2}, T_{ME1} and T_{ME2}, the PSE-side PI voltage should be ensured to stay above 7 V. This is with the help of the V_{MARK} voltage source as explained later.

Also keep in mind that, as per the AT standard, if during the first class event, the PD reports itself to the (Type 2) PSE as either Class 0/1/2/3, *the PSE should not carry out 2-event classification*, but proceed directly to power-up the port. The overall timer constraints remain the same, however. Remember that only if the PD reports itself as Class 4 is 2-event classification allowed.

Discharging Port Capacitances and Actual Voltage "Seen" by the PD

As shown in Fig. 4.5, many Type 1 PSEs will complete classification and go straight to power-up without even trying to discharge the port in between the two phases. In Fig. 4.3, we see there are some cases where the port may discharge somewhat after classification. But none of this is specified because there is no "mark-event range" defined for Type 1 PSEs. The mark-event range was introduced in the AT standard only to handle 2-event classifications.

The need to actively discharge the port to some value between the two class events arose simply to distinguish the class events from each other—*from the PD's perspective*. So it is important to know not just the PI voltage at the PSE-end, but the PI voltage *as seen by the PD*. The latter is unfortunately not the same as the voltage at the PI on the PD-end. The culprit is the *diode bridge*.

This has huge ramifications actually, that affect the PD chip design and PSE design too. Summing those up in two major categories:

1. ***Forward-biased bridge.*** As mentioned in the preceding chapters, whenever we talk of the PSE-side voltages, we are quite right, because the PSE IC can actually read the port voltage accurately on its side of the cable and do the "right thing" at the "right voltage threshold." How about the PD? Can it really read the PD-side PI voltage? We are forgetting that in most PD

ICs, the PD has no way of actually reading the PI voltage per se. This is because it is on the right-hand side of a bridge rectifier, and that introduces two forward-diode drops between the PI voltage and the voltage the PD "sees"—that is, when the diode bridge is *forward-biased* (conducting). The situation can get really ugly if for any reason the diode bridge gets reverse-biased as discussed next. For the case of a forward-biased bridge, we realize a good PD chip design will need to guesstimate the diode drops of the proposed front-end bridge rectifier beforehand, and plan its internal thresholds accordingly. Because the IEEE standard only talks about the PD-side PI voltage (on the *left-hand side* of the bridge), not the voltage that the PD chip will actually see (on the *right-hand side* of the bridge). It is for this reason that, for example, the PD IC LTC 4257 from Linear Technology designates its classification range from 12.5 to 21 V. These are the guaranteed voltages with respect to the IC ground (on the right-hand side of the bridge), not the PI (left-hand side). Basically, the chip designer is assuming that when the IC sees 12.5 V, the PD-side PI voltage is higher by two diode-drops (2 V max), so 12.5 V + 2 V = 14.5 V as required by the standard. This assumption may not be true since forward drop across a diode is a steep function of the forward current and temperature. Besides it varies from one diode vendor to another. There is, unfortunately, no easy solution to this, but that's the way almost all PD chips are designed today. The LM 5072 PD chip from TI defines the chip-side classification range even more aggressively—from 12 to 23.5 V. Based on the same reasoning, it defines the chip-side detection range from 1.5 to 10 V. The On-Semi PD chip NCP 1083 defines its detection range from 1.4 to 9.5 V. If we add 2 V to both the limits, we get a valid detection range between 3.4 to 11.5 V, though the standard requires detection to be guaranteed across 2.7 to 10.1 V. So at low voltages, the NCP 1083 might have a problem. To avoid such "interoperability" issues, we must design the PSE more robustly, to be able to accept all these variations. For example, **the "first sample" the PSE collects for finding R_{DET}, must not be lower than 3.5 V.**

A problem could arise if someone tried to improve efficiency and lower the dissipation by using any of the abovementioned PD chips with a low-drop bridge rectifier, say one made of Schottky diodes, or an "active bridge" (made from FETs). Now the entire PD probably has noncompliant thresholds, and that could cause severe interoperability issues between any PSE and the PD, *unless the PSE was made even more "tolerant"* or "robust." We will discuss this in more detail shortly in the subsection Detection Signature Resistor Disengagement Concerns.

2. ***Reverse-biased bridge.*** To create a "dip" between the two fingers of a 2-event classification pattern, we can depend somewhat on the classification-current-sink associated with Class 4, which has a minimum value of 35 mA. Note that with Class 0, we cannot depend on any classification-current-sink to discharge the port, since its minimum value is 0 mA. But here we are only interested in creating a mark-event for Class 4 in any case. So when V_{CLASS} is disconnected by the PSE on completion of the classification event, this > 35 mA sink remains active at the PD end right until the PD side PI voltage falls below V_{MARK_TH}. That's when the current sink gets turned OFF. The problem is what is the discharge path *after this point*? It is actually quite *unpredictable*, which is why we have represented it with dashed lines in Figs. 4.3, 4.4, and 4.7. In the best case, V_{MARK_TH} may be at 10.1 V, and we would then already be almost into the mark-event range of 7 to 10 V. But in the worst-case, the classification-current-sink would turn off slightly below 14.5 V and *it is unclear how to get the port down to 10 V*.

To make this task easier, the AT standard says that to get the PI voltage to fall, there should be a *mark-event current* that can sink 0.25 to 4 mA. First question is: How can this be implemented? One option is to create a weak-regulated voltage source in the PSE, which actively pulls down the port voltage into the mark-event range (7 to 10 V). This regulator has both current sourcing and sinking capabilities and is therefore a two-quadrant regulator. In contrast, we remember that the voltage source applied during classification is typically only capable of sourcing current (it is a one-quadrant regulator). But does this PSE-side discharge path really help? Does it in fact hurt? In reality, it is of little practical use since the capacitors on the PD-side are usually placed *after* the diode bridge (on its right-hand side), and so the bridge will likely just get *reverse-biased* in this situation. And if that happens, the voltage seen by the PD-controller IC can be *very different* (several volts higher) from what the PSE would be seeing on the opposite side of the reverse-biased diode bridge. Any assumed and expected correlation in PSE-PD behaviors will be put at risk, and assumptions the PSE may make can prove very wrong. In fact, the PD may not even realize there were *two* class events (no dip will be seen by the PD chip). With a reverse-biased bridge, the situation is actually much worse. So what is the solution?

If we think about it we will realize that *any discharge path on the* ***right-hand side*** *of the bridge is far more effective* since it can pull charge out of capacitances on *both* sides of the bridge (for example, the 0.1 µF PSE port capacitance *and* the 0.1 µF PD input capacitance). On the other hand, a discharge path in the PSE only pulls out charge from the port capacitances on the left-hand side of the bridge (for example, the 0.1 μF PSE port capacitance). With a little thought we will realize that with all permissible distribution of port capacitances considered, if we ensure that the *PD-side discharge path is stronger than the PSE-side discharge path,* the diode bridge will never get reverse-biased. That's what we have to ensure.

So, that is why the discharge current of 0.25 to 4 mA that the AT standard specifies is ***specified to be located not on the PSE-side but inside the PD chip***. Summarizing:

- A Type 2 PD must be able to sink 0.25 to 4 mA below V_{MARK_TH}. That is called I_{MARK} in the standard. The classification-current-sink is similarly called I_{CLASS}.
- On the PSE side, we should actually avoid actively discharging the port after the classification event, to avoid reverse-biasing the diode bridge. It is, however, a good idea to place a large-value resistor (> 45 k for reasons explained in Chap. 3) across the port after V_{CLASS} is disconnected by the PSE. This "bleeder resistor" is discussed further below.

NOTE *One potential problem with the assumption of a discharge path after the bridge is that the minimum-classification current for, Class 0 is stated to be 0 mA. That means there will be no discharge path even down to V_{MARK_TH}. Section 33.2.6.2 of the AT standard downplays this by saying: "In a properly operating system, the port may or may not discharge to the V_{MARK} range due to the combination of channel and PD capacitance and PD current loading. This is normal and acceptable system operation. For compliance testing, it is necessary to discharge the port in order to observe the V_{MARK} voltage. Discharge can be accomplished with a 2 mA load for 3 ms, after which V_{MARK} can be observed with minimum and maximum load current." Note that the standard refers to V_{MARK} not as some voltage source, but simply the port voltage in the mark-event phase of classification.*

Detection Signature Resistor Disengagement Concerns

At this point it is a good idea to recall/repeat the discussion in the previous chapter on a very basic issue: *the presence of the 25-k detection resistor*. In almost all cases discussed so far, and also in related literature, this resistor is shown as a stand-alone component just past the bridge on the PD-chip-side. It is "always present." So when the voltage ramps up into the classification range (~18 V), it continues to draw current, of about 18/25 k = 0.7 mA. Not a big deal, and though it adds some current to the constant class-current-sink, its effect is less than 1 mA. When the port is fully powered up, it continues to draw about 50 V/25 k = 2 mA. In terms of wattage this is 50 V × 2 mA = 100 mW.

Not entirely negligible. Keep in mind that we need to draw about 0.5 W to keep the PD-PSE connection "alive" in the first place. Nevertheless, it did lead to some people wanting to shut off this "wastage of power," and they proposed the 25-k resistor be actively disconnected after detection. One question remains: When exactly to turn it OFF (disconnect it)?

In fact, the way the AT standard's PD-state-machine diagram is written, it actually does *not* even permit keeping the detection resistor across the port beyond the exact moment of completion of the detection phase (for both Type 1 and Type 2). This can cause backward compatibility issues because the AF standard was quite ambiguous on the PD state machine and was widely interpreted as allowing this "always present" 25-k resistor. So there are many PDs out there even today, both Type 1 and Type 2 (such as On-Semi's NCP1083) that continue to retain this 25-k resistor across the port always connected. Faced with that market reality, it is not a good idea to design a PSE that doesn't accept this wide PD behavior, whatever the AT standard may or may not say.

It seems the main reason for immediately disconnecting the 25-k signature resistor right after detection, was motivated not by a concern for saving power initially, but because of a 2007 presentation at an IEEE Ad Hoc meeting from an engineer from Linear Technology who voiced the following concern. Consider a Type 1 PSE in the middle of an open circuit detection (before starting detection). The open circuit voltage could be as high as 30 V as per the standard. At that precise moment, if a Type 2 PD was plugged into the PSE's port, the suddenly collapsing open circuit voltage as seen by the PD may be interpreted by the PD as the first of two fingers of a 2-event classification. The PSE would then initiate detection since there was no open circuit anymore, and the PD would be in the mark-event range (since that is about 7 to 10 V and overlaps the detection range 2.8 to 10 V, see Fig. 4.8). So if the PD had a 25-k signature resistor still present during the mark-event phase, the Type 1 PSE would "detect" it as a valid PD, and would initiate a 1-event classification event. That would be interpreted by the PD as the second finger of a 2-event classification, and so the Type 2 PD would wrongly identify the Type 1 PSE as a Type 2 PSE. The PSE would, having received class information from the PD, would then power-up the port. There are some comments that can be made here after deeper thought:

1. So what? If the Type 2 PD wrongly thought it was connected to a Type 2 PSE and demanded more power than the PSE was capable of providing, in no time an "overload" fault would have been detected by the PSE and it would have turned off the port and then started with detection all over again, this time giving no cause for concern. The reverse case of a Type 2 PSE thinking it was wrongly connected to a Type 2 PD and

allowing higher levels of current than necessary would have been of some concern.

2. As shown in Chap. 3, even the time to take one steady sample of port voltage during detection is at least 51 ms (three time constants). Two steps are required for proper sampling, which is the minimum requirement as per the standard to eliminate diode offset. That will take over 100 ms. Whereas T_{ME1} is a maximum of 12 ms only. That is just *not enough time* for any PSE to falsely detect a 25-k resistor were it present during the mark-event phase. There seems negligible chance of the PSE ever thinking that the impedance seen during the mark-event range was a valid PD, and thereby proceeding with classification and power-up. Perhaps that is why there were no waveforms presented by the engineer, just a "graphical depiction."

This fear seems to have then prompted a PD-state machine within the AT standard that demands immediate removal (disengagement) of the detection resistor right after detection.

The consequence is most PD-chip designers today struggle to create a very narrow range above 10 V, say 10.5 to 12 V in which they consciously disconnect the detection resistor, and then slightly above that range turn on the classification-current-sink. However, combined with the basic problem of variances in diode forward-drops and the "guesstimation" of all that beforehand while setting hard thresholds in the PD, this has likely led to many reported PSE to PD interoperability issues.

It seems to have been not worth it, because the detection resistor actually serves a *useful purpose* during both detection and classification as explained further below. It is therefore considered OK to leave it in the PD across the port always connected. Typical test suites (such as from University of New Hampshire's Interoperability Laboratory) do not check PDs for this "removal of the detection resistor above 10 V" requirement. If they did, a lot of earlier and modern PDs will fail the test.

If we really want to save 100 mW of dissipation, the best option from a systems viewpoint is removing the detection resistor from across the port only *after power-on of the PD is completed*, not before.

Note that the AT standard also asks that the *classification-current-sink* be disengaged after completion of classification. Older PDs keep this constant-current sink engaged indefinitely even after power-on. That does help in implementing simple PD front-end designs based on discrete components (no PD chip). Further, this disengagement of classification-current-sink is also never tested by standard PD test suites, so it can be left in if desired. But it is a waste of power given the high class-currents, so most good PD designs will disengage the classification-current-sink somewhere between 20.5 V (the maximum class-event range limit) and 30 V (the lowest "must turn-off" threshold for PDs).

Detection Signature Resistor beyond Detection

The first use of this resistor is in ultra-low-cost PDs. Suppose there are no classification-current-sinks present inside a PD, but there is a 25-k detection resistor across the port always. If so, in the class-event range, the minimum current the detection resistor will draw is about 15 V/25 k = 0.6 mA. This will be naturally interpreted as Class 0 by any PSE. Therefore, power-up will occur, whether optimal or not.

Let us do some more accurate estimates on the dashed discharge curves in Fig. 4.7 for example. We ask how long does it take for a maximum of 670 nF port capacitance (the "must accept" range, see bottom of Fig. 3.5) to discharge from 14.5 V (the max value of V_{MARK_TH}) down to 10 V (the upper limit of the mark-event range)? A simple Mathcad file reveals it takes 6.22 ms. We note that the first mark-event interval T_{ME1} is stated to be within 6 to 12 ms as per the standard. We thus realize that 6 ms is actually *almost insufficient* in this case. In other words, the discharge down to 10 V may still not be complete after the 6 ms marker.

What if we introduce a discharge (bleeder) resistor across the port on the PSE side as suggested above?Assuming it is > 45 k with a max value of 80 k, this bleeder appears in parallel with the 25-k PD signature resistor. The equivalent discharge resistance for the port capacitance is now 80 k × 25 k/(80 k + 25 k) = 19 k. With this lowered-effective resistance, as per the Mathcad file, the discharge time from V_{MARK_TH} to mark-event range decreases from 6.22 to 4.7 ms. And that is much better, since T_{ME1} is > 6 ms. So now we can be quite sure the port *will* discharge into the mark-event range *before* the minimum 6 ms is over.

We have thus learned that in a good PSE design we should introduce a PSE-side weak port-discharge (bleeder) resistor during the classification phase (as between detections), whose min-max range is 45 to 80 k. This particular bleeder should not be present during actual detection as it can seriously affect the detection range of the PSE. It should be switched in only after detection is over.

We note that the AT standard asks that the PD sink 0.25 to 4 mA during the mark-event phase (7 to 10 V). We can calculate that a 25-k resistor on the PD side will draw 7 V/25 k = 0.28 mA at the lower threshold, and 10 V/25 k = 0.4 mA at the upper threshold. So in fact the 25-k resistor *does* meet the AT standard's requirements on this aspect automatically. We do not need a formal mark-event current sink inside the PD chip if we have a 25-k resistor connected across the port. The standard does not explicitly ask for a regulated "current sink" in the mark-event range, it just defines the min to max current values, and we have seen that can be complied with by the basic 25-k signature resistor itself.

We realize that *the 25-k resistor is very valuable during both detection and classification*. The problem described by the Linear Technology engineer seems questionable, and the disengagement of the signature resistor appears to be a bigger problem to implement precisely, given the unknown bridge-rectifier diode drops, and the consequent potential impact on interoperability.

Before we close this discussion we make it clear that we do want to ensure the PI voltage does not keep discharging and ultimately goes below 7 V, especially during the open-ended interval, T_{ME2}. For this we do need a voltage regulator of some sort to hold the PI voltage steady during the mark-event phase too.

For example, the Type 2 PSE, Si3452 from Silicon Labs has the following characteristics as per its datasheet available on the web:

1. V_{CLASS}: Regulated to within 15.5 to 20.5 V, up to 45 mA (current limited to less than 100 mA).
2. V_{MARK}: Regulated to within 7 to 10 V up to 5 mA (current limited a little higher).

Other PSE chip designs guarantee V_{MARK} only to 4 mA (i.e., the right value as explained below). In other words, there is a voltage regulator present in the chip that guarantees the PI voltage is regulated between 7 to 10 V after V_{CLASS} is released. But is it a one-quadrant or two-quadrant regulator? As we have shown above, to avoid reverse-biasing the diode bridge, it is best to depend mainly on PD-side discharge paths, not PSE side. And on the PD side, that can be an active current sink of 0.25 to 4 mA as per the standard, but this can also be fulfilled by a signature resistor provided it is not disengaged (as also required by the AT standard but often ignored). The important thing is the PSE-side discharge path should be *weaker* than the PD-side discharge path. So a bleeder resistor (45 to 80 k) on the PSE-side is the best choice. In other words V_{MARK}, should *not* be a two-quadrant regulator. Its function is *not* to *actively* lower the PI voltage from the class-event range to the mark-event range. Its responsibility is only to *prevent* the PI voltage from falling below 7 V. In a good PSE design it should be able to source, *not sink*. And since on the other end of the cable, the PD may have activated a mark-event current sink of up to 4 mA, this voltage regulator must be stable up to 4 mA.

Besides this, the Si3452 indicates it has a class-event range voltage regulator (apparently also one-quadrant), which is stable up to 45 mA (though the IEEE standard actually asks for 50 mA).

Putting it all together,

1. It is a good idea to have a 25-k resistor on the PD side in an "always connected" mode during the entire detection and classification phase.

2. It is also a good idea to have a bleeder of 45 to 80 k on the PSE side, connected during classification phase only (at least during the mark-event phase).
3. There needs to be a PSE-side class-event range voltage regulator (15.5 to 20.5 V), stable up to 50 mA and current-limited to less than 100 mA. This needs to only be a one-quadrant regulator (sourcing, not sinking capability).
4. There needs to be a PSE-side mark-event range voltage regulator (7 to 10 V), stable up to around 4 mA and current-limited slightly higher. But to avoid reverse-biasing the PD's bridge rectifier, this *should* be a one-quadrant regulator (sourcing, *not sinking* capability).

To reiterate: *to avoid a reversed-biased PD bridge we must ensure the discharge mechanism on the right-hand (PD-chip) side of the bridge is stronger than the discharge mechanism on the left-hand (line) side of the bridge.*

It is becoming clear that for a proper 2-event classification to take place, it is also important to discharge the port capacitance fairly *quickly* into the mark-event range, and for that reason, though a total port capacitance of 670 nF is considered "good" ("must accept" range), and only capacitances above 10 μF are considered truly "bad" ("must reject"), and despite there being such a wide gray area in between, it is actually not a good idea to design systems with total port capacitance exceeding 670 nF at all, even if that is not strictly disallowed by the standard—because as we have learned, more than that value of capacitance can become quite detrimental to a proper 2-event classification sequence in particular. If we have a large capacitance, we have to discharge it too. And that is not as trivial as we may have thought.

IEEE 802.3at Classification Details Summary

The IEEE 802.3at standard makes several key statements concerning classification. These are summarized and bulleted for easy reference as follows, with bold letter/italics introduced to highlight certain words or phrases

General Points

- Section 33.2.6 Page 41: With Data Link Layer classification, the PSE and PD communicate using the Data Link Layer Protocol (see 33.6) after the data link is established. **The Data Link Layer classification has finer-power resolution and the ability for the PSE and PD to participate in dynamic power allocation wherein allocated power to the PD may change one or more times during PD operation.**

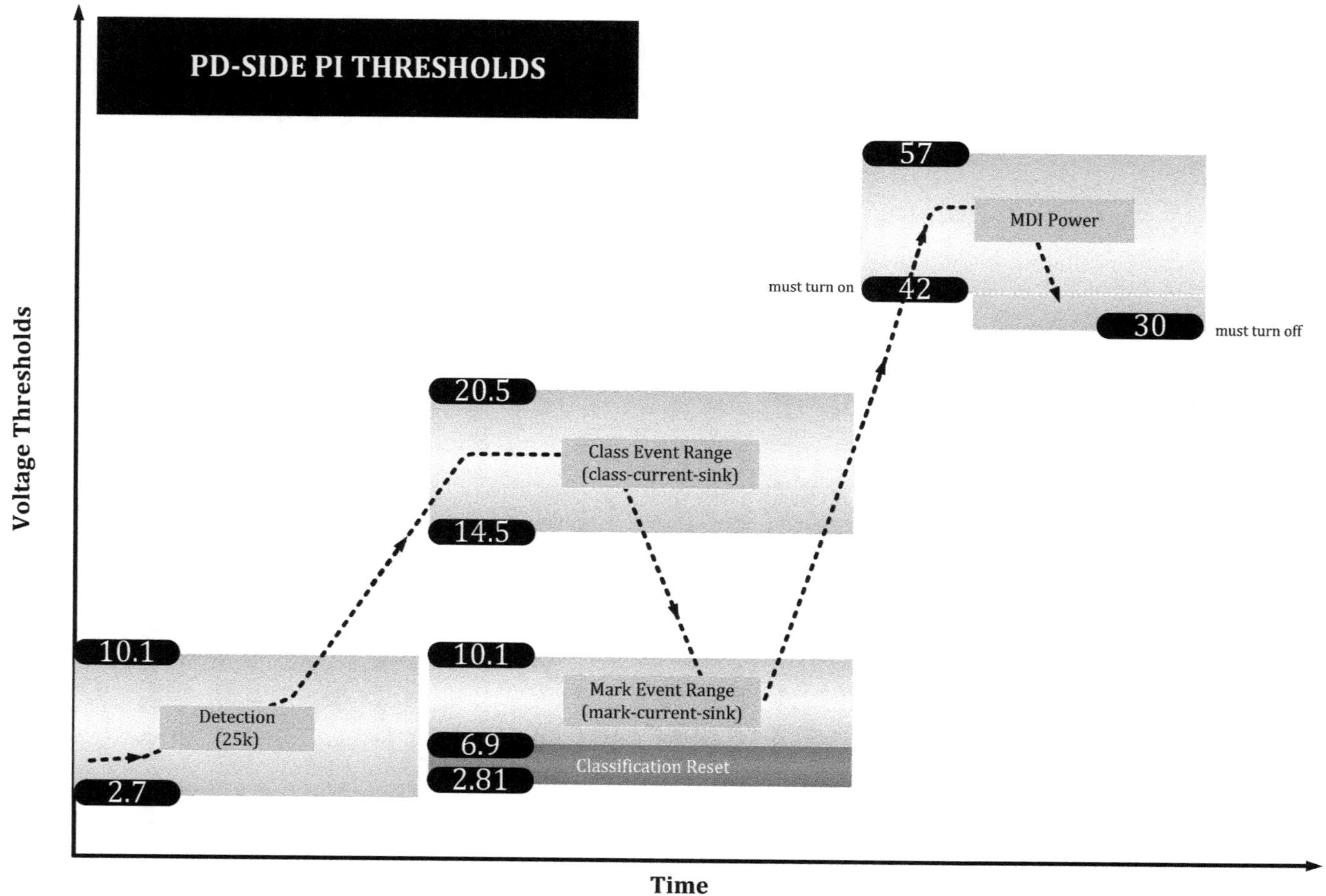

Figure 4.8 PD-side PI-voltage thresholds.

- Section 33.3.5 Page 60: A PD may be classified by the PSE based on the Physical Layer classification information, Data Link Layer classification, or a combination of both provided by the PD.

Author's Note The above statement implies that a Type 2 PD must be able to handle both Layer 1 and Layer 2 classifications—and this is clarified further as follows:

- Section 33.3.5 Page 60: Type 2 PDs implement both 2-Event class signature (see 33.3.5.2) and Data Link Layer classification (see 33.6).

Type 1 Specific Points

- Section 33.2.6 Page 41: Subsequent to successful detection, a Type 1 PSE may optionally classify a PD using 1-Event Physical Layer classification. Valid classification results are Classes 0, 1, 2, 3, and 4, as listed in Table 33-7.
- Section 33.3.5.1 Page 60: Class 0 is the default for PDs. However, to improve power management at the PSE, a Type 1 PD may opt to provide a signature for Class 1 to 3.
- Section 33.3.5.1 Page 60: Type 1 PDs may choose to implement a 2-Event class signature and return Class 0, 1, 2, or 3 in accordance with the maximum power draw, PClass_PD.
- Section 33.2.6 Page 41: If a Type 1 PSE does not implement classification, then the Type 1 PSE shall assign all PDs to Class 0. A Type 1 PSE may optionally implement Data Link Layer classification.
- Section 33.2.6.1 Page 42: If the result of the class event is Class 4, a Type 1 PSE shall assign the PD to Class 0.
- Section 33.2.6.2 Page 43: If the result of the first class event is any of Classes 0, 1, 2, or 3, the PSE treats the PD as a Type 1 PD and may omit the subsequent mark and class events and classify the PD according to the result of the first class event.

Type 2 Specific Points

- Section 33.2.6 Page 42: Subsequent to successful detection, all Type 2 PSEs perform classification using **at least one of the following**: 2-Event Physical Layer classification; 2-Event Physical Layer classification and Data Link Layer classification;

or 1-Event Physical Layer classification and Data Link Layer classification.

- Section 33.2.6.1 Page 42: If the result of the class event is Class 4, a Type 1 PSE shall assign the PD to Class 0; a Type 2 PSE treats the PD as a Type 2 PD but may provide Class 0 power until mutual identification is complete.
- Section 33.2.6.2 Page 43: If the result of the first class event is Class 4, the PSE may omit the subsequent mark and class events only if the PSE implements Data Link Layer classification. In this case, a Type 2 PSE treats the PD as a Type 2 PD but may provide Class 0 power until mutual identification is complete.
- Section 33.3.5.1 Page 60: PDs implementing a 2-Event class signature shall return Class 4 in accordance with the maximum power draw, PClass_PD, as specified in Table 33-18. Since 1-Event classification is a subset of 2-Event classification, Type 2 PDs respond to 1-Event classification with a Class 4 signature.
- Section 33.3.5.2 Page 61 and 33.3.6 Page 61: Until successful 2-Event Physical Layer classification or Data Link Layer classification has completed, a Type 2 PD's pse_power_type state variable is set to '1.'.........The default value of pse_power_type is 1. After a successful 2-Event Physical Layer classification or Data Link Layer classification has completed, the pse_power_type is set to 2.
- Section 33.3.2 Page 54: A Type 2 PD that does not successfully observe a 2-Event Physical Layer classification or Data Link Layer classification shall conform to Type 1 PD power restrictions and shall provide the user with an active indication if underpowered. The method of active indication is left to the implementer.

Author's Note The previous two statements imply that a Type 2 PD must be able to indicate externally that it has detected a Type 2 PSE (or not). The latter statement may call for a visual indicator (like an LED). This is referred to as a "power-status" indicator in the discussion that follows. It needs to be provided irrespective of whether the classification is layer-1 or layer-2 based.

IEEE 802.3at Table 33-8 Explained Further

Table 33-8 on Page 42 of 802.3at-2009 summarizes some of the above-mentioned points cogently. This figure was reproduced in Fig. 4.1 with adjoining *examples* revealing how it should be understood and interpreted. The key conclusions are (for compliant PSEs and PDs):

1. A Type 2 PSE is allowed to depend only on 2-finger classification to seek out Type 2 PDs, with no layer-2 /LLDP classification. But it can also opt to use 1-finger classification followed by LLDP classification. So when the data link comes up, the PSE will use LLDP protocol to communicate with the PD that it is a Type 2 PSE and is thus able to provide more than 13 W to it.
2. A Type 2 PD must be able to accept *both* of the above Type 2 PSE classification behaviors: layer-1 with 2-finger classification *and* layer-1 with 1-event classification followed by LLDP classification.

1-Finger or 2-Finger Classification for Type 2 PSEs?

This is a commonly asked question. We have discussed it previously, but we will highlight and summarize it here.

Assuming PDs are compliant: Despite the added software-based complexity, the LLDP approach (after a 1-finger classification) may be considered preferable, as compared to 2-finger classification for a couple of reasons:

1. When the data link comes up, the PSE will use LLDP protocol to communicate to the PD that it is a Type 2 PSE and is thus able to provide more than 13 W to it. But adding to this statement: LLDP classification actually allows the PSE to tell the PD not only that it can provide more than 13 W, but how much more! Maybe it does not need the PSE to provide full 30 W at PSE-end. And similarly, the PD can in turn inform the PSE (with far greater granularity than layer-1 classification permits), what is the actual amount of power it needs. The granularity can be in steps of 100 mW. In addition, the power requirements can be changed dynamically as operation proceeds. On the other hand, 2-finger (layer-1) classification alone for example, will only allow the PD and PSE to tell each other that *one* of them requires (and the other supports respectively), *more* than 13 W (though less than 25.5 W, of course). In such a case, a PD that operates with even 14 W will technically ask for and be allocated up to 25.5 W by the power-management software. This could end up requiring an over-designed 48 V power supply too, one that caters to up to 30 W on each Type 2 port, irrespective of the actual wattage requirement on the ports.

2. The AT standard requires a *power status indicator* on Type 2 PDs. This status indication helps a user determine if a Type 2 PD is *not* being fully supported in its power requirements. Such a PD is then considered "underpowered" and consequently will not able to provide one or another useful function or feature. This situation occurs when a Type 2 PD is connected to a Type 1 (or "legacy") PSE. An installer/user will typically need to physically connect the PD to a port, look at the indicator, and thereby determine if the PSE port connected to is appropriate for the power requirement. Because, within the "wiring closet," the installer depends completely on the power status indicators. Yes, it is true that if 2-finger classification is done, a PD can more *rapidly* identify a PSE with sufficient (or insufficient) power—in approximately 1 s, as compared to a PSE which uses LLDP classification. In the latter case, the PD power status takes approximately 60 s because this ability is gated by the PD's boot-up time. However, the IP-phone is the most common PD as of 2008. All Type 2 IP-phones can usually power up and provide basic functions even when connected to Type 1 PSEs. Further, IP-phones also tend to use an LCD screen and that also provides status. But that normally requires the IP-phone to boot up fully, which process takes approximately 60 s in any case. Therefore, any visual indication, even of being underpowered, takes roughly 60 s to appear. Quite similarly, the second-most-popular PD as of 2008, the Access Point or AP, typically needs to boot up first, to be able to light up any power-status LED. Therefore, a PSE with 2-finger classification is in reality unable to provide an earlier power-status report. Therefore, 2-finger classification for Type 2 applications is still considered optional in Type 2 PSEs (except Midspans).

CHAPTER 5

Inrush and Power-Up

Overview

The "Power-up" mode (or phase) starts once classification is complete. From the testability point of view, it may be best defined as commencing the moment the port voltage on the PSE-side exceeds 20.5 V, this being the upper limit of the classification range. Several things happen along the way as the port voltage (PI voltage) on the PSE-side rises toward its final settling value. It is clear that this final value is essentially the incoming "48V" rail minus the forward drop across the pass-FET in the PSE. Traditionally, when the port voltage got up very close to its final value, Power-up phase was said to have ended and Power-on achieved. The AT standard, however, laid down one additional qualifying condition before declaring Power-on, as discussed later. In general, Power-up happens to be one of the most difficult phases of PoE operation to understand. There are many prevailing misunderstandings that we need to try and resolve in this chapter.

Inrush Behavior

One of the key events constituting Power-up phase is inrush. As the port powers up, the PD can draw significant amount of inrush current as it tries to achieve steady state. The IEEE standard tries to protect the cabling infrastructure and also the PD by regulating this inrush. But it also wants to achieve steady state in the best possible manner (with no ringing), and reasonably fast too.

The AF and AT specs on inrush current are just a *little* different. But they do *seem* to be very different at first sight. On closer examination, the "differences" are actually only in the first 2 ms of operation. The AT spec is a bit more restrictive in this region. In the discussions leading up to the standards, there were many discussions on the safe-operating region (SOA) curves of typical sense resistors inside PDs and fusing of PCB traces, and the consequent need to ensure that the PSE restricts the allowed currents during the first few milliseconds, to avoid damaging the PD. For such reasons, the AT standard asks that the inrush current be less than 5 A even initially, and that it returns to less than 0.45 A at the end of 1 ms. The limits of the AF and AT standards are compared (overlapped) in Fig. 5.1. For now let us just visually compare the first 2 ms. As we can see, the AF standard is a bit more generous in the initial couple of milliseconds.

Standard AT compliance-test equipment will check for one key condition: *Port current to be less than 0.45 A measured after 1 ms following the application of a test overload/short-circuit during the inrush period.*

Let's be clear: From which instant in time is this 1 ms measured from? Note that all time intervals during inrush (Power-up) are

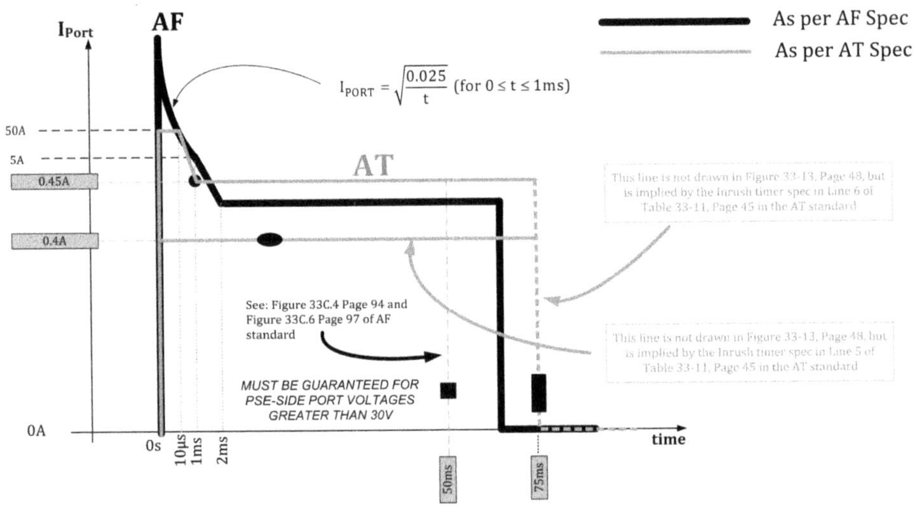

To stay compliant with both AT and AF specs:

- a) 1ms after PI voltage on PSE-side crosses 30V, the maximum inrush current through the PSE must be limited to less than 0.45A
- b) A minimum inrush current greater than 0.4A must be supported by the PSE
- c) The inrush should be supported by the PSE for at least 50ms
- d) The PSE must terminate port voltage before inrush can exceed 75ms

SUMMARIZING:

The PSE must always limit the inrush current to less than 0.45A (settling level)
But the PSE must be able to support at least 0.4A for 50ms: it must not meanwhile foldback or enter thermal shutdown.
Under a fault condition, the inrush/fault current should not be allowed to last more than 75ms

FIGURE 5.1 Understanding the allowed inrush behavior as per AF and AT standards.

commonly interpreted as starting not from the point the PSE-side PI voltage exceeds 20.5 V, *but when it crosses 30 V*, though the differences in the two in terms of time may be very slight. The inrush current profile in Fig. 5.1 is stated to be applicable only for PSE-side port voltages exceeding 30 V. Later, we will also describe the inrush requirements for lower voltages, for example 20.5 to 30 V. But for now we are only talking about what happens above 30 V.

NOTE *30 V was selected as a Power-up phase commencement threshold because it was considered high enough for the PSE to set its inrush current limit slightly below that threshold, and for the PD's pass-FET to be turned ON a little above that threshold, without affecting detection and classification functions at lower voltages.*

To pass PoE-compliance testing, in general, wherever there are differences in the AF and AT standards philosophically speaking, it is always best to try to comply with whichever standard *is more restrictive*. So for the initial inrush period, it is best to follow the AT standard up to the first 2 ms. After that, there is no need to decide: the AF and AT standards concerning inrush are actually identical, *though stated differently*.

While sketching out the inrush requirements graphically, the two standards, AF and AT, ended up presenting the same requirements (after 2 ms) in a rather different manner. In fact, Fig. 5.1 is not exactly what the AT standard sketched out but is *modified* somewhat, based on what is *implied* or stated elsewhere, in the tables of the AT standard for example.

The problem is the AT standard complicated matters somewhat by trying to create an inrush *template*. A template strictly just implies "stay *within* this (outer)boundary" (< 0.45 A, in this case); it makes no mention of an *inner* boundary (> 0.4 A), which actually does exist in the case of inrush for both AF and AT. In other words, the standards demand that the PSE must be able to support at least 0.4 A for 50 ms during inrush. Yes, though *both* the AF and AT standards ask for that, in the case of the AT standard, the lower threshold is stated only in its tables; it does not appear in its inrush template. Hence the possible confusion.

We note that the inrush range of 0.4 to 0.45 A is kept deliberately identical to the normal operating current-limit settings for Class 0/3. That greatly simplifies both PSE and PD chip designs. It also ensures backward compatibility, especially because modern and older devices can have slightly different implementation of inrush *logic* as described later (legacy Power-up).

Based on the allowed inrush range, a nominal of 0.425 A is usually selected as the PSE-chip current limit during inrush.

In terms of *timing*, it is not very clear what the AT standard is saying based on its inrush template. But once again, connecting the figure with the tables, we arrive at the following summary of inrush behavior, as indicated in Fig. 5.1:

1. The PSE must always (even under a fault) limit the inrush current to less than 0.45 A (after the settling period of 1 ms) ← *Protection.*
2. Under a fault condition, the inrush current (0.4 to 0.45 A), must never last more than 75 ms ← *Protection.*
3. The PSE must be able to support at least 0.4 A for 50 ms (it must not during this period either fold back or enter thermal shutdown) ←*PSE Capability.*

NOTE *It is implied that the inrush requirements shown in Fig. 5.1 for the AT standard are the same for AT Type 1 (all Classes 0/1/2/3) and AT Type 2 (Class 4). In other words, even a medium-power application will have the same Power-up characteristics (inrush profile) as a low- or ultra-low-power application. In other words, the higher-, or lower-, current limits, commensurate with the class of the PD (provided of course the PSE can support the class power requirements), are set right after inrush is over. There is also a defined "wait time" for the PD to ask for its class-based operating current as discussed later.*

Purpose of Inrush Limiting and the PD Bulk Capacitance

What is the purpose of inrush current limiting anyway? It is to charge the input cap of the DC-DC converter stage that follows in a *controlled manner*, and also do so in a reasonable amount of time. In the standard, this cap is called C_{PD}. But to make it clearer where exactly it is positioned, we prefer to call it "C_{DC_DC}." Whatever the names, it is just the input bulk cap of the DC-DC stage that follows the PD's front-end (its pass-FET section).

We remember that during detection and classification, this bulk cap is not exposed to the PSE and is kept isolated from the line by the PD's pass-FET. So its initial condition is fully uncharged or 0 V. This FET-isolation feature is basically the difference between an IEEE-compliant PD and a legacy PD as discussed in Chap. 3.

So when the PSE raises the port voltage following a successful detection and classification, at some stage the PD suddenly turns on its pass-FET into a fairly large bulk capacitor, one that is hitherto fully uncharged. Theoretically, that can lead to *infinite* currents for very small cable lengths, because as we know, we should never put a voltage source directly across an uncharged cap. That situation is analogous to trying to apply an arbitrary current source in series with an inductor, which we know, from basic principles of switching-power conversion, can cause huge voltages spikes.

The lesson is that current limiting must be introduced somewhere along the way to limit the inrush current. But *where* should it be? *In the PSE pass-FET or in the PD pass-FET?* Calling it just I_{LIM} for now, that is, not yet deciding where exactly it resides, we do realize that we can now get the cap voltage to rise in a controlled manner as per the basic equation $I_{LIM} = C_{DC\_DC} \times dV/dt$. No infinite currents result anymore.

The standard asks that C_{DC_DC} have a minimum value of 5 μF. The reason for that was this minimum value was seen to be able to support (decouple) typical dI/dt transients lasting up to 30 μs. These transients are, for example, the sudden current demands from the load connected to the DC-DC converter. C_{DC_DC} not only provides bulk decoupling and reduction of noise and EMI, but allows the PD to continue to stay powered, because the presence of this fairly large capacitor can be easily detected by a low-frequency AC probe-signal sent by the PSE down the line in AC disconnect, as discussed in a subsequent chapter. The lowered AC impedance caused by this cap across the port draws increased probing current, and that tells the PSE that there is a PD on the other side, so it does not disconnect the PD (power-off the port) inadvertently. This, too, is discussed further in Chap. 7.

NOTE *In 2007, Yair Darshan from PowerDsine (now Microsemi) identified a corner case at IEEE where the 5 μF seemed insufficient. It was later reported in the minutes as follows: "We determined that there is a corner case where a PSE power transient happens simultaneously with a PD overload event. When this happens, the PD may require a capacitor larger than the minimum allowed (5 μF) to avoid a reboot. In the opinion of the ad hoc, the chance of these two infrequent events happening simultaneously is very small, and the impact on the PD with a small cap is minor (a disconnect followed by a normal reconnect), so the ad hoc committee chose to ignore this case.*

Note that the standard specifies *no maximum* to this DC-DC input capacitance, but the max value is *practically limited* as discussed later in the chapter.

Finally, to decide on *where* the current limit should reside, the standard ascribes the following "responsibilities":

1. If C_{DC_DC} is less than 180 μF, the responsibility of limiting the current rests with the PSE.
2. If C_{DC_DC} is greater than 180 μF, the responsibility of limiting the current rests falls on the PD.

To be very precise, the standard actually mentions C_{PORT} above, not C_{DC_DC} (C_{PD}). In other words, it lumps up the entire port capacitance, which also includes the PSE- and PD-side ceramic port caps.

Why 180 μF? The key reason for choosing 180 μF cap as a golden value of sorts seems to have come from early discussions leading up the AF standard, in particular a paper titled "Proposal for Start-up and Port Line/Load/Cross regulation/Transient Parameters for IEEE 802.3af Standard Power over MDI" by Yair Darshan from Microsemi and Dave Dwelley from Linear Technology. The 180 μF value came from a line regulation concern during normal operation, well after the inrush period was over actually. It involves the case of a sudden "negative line transient" from 57 to 44 V (a delta of 13 V). This is *not* related to a *positive* transient during inrush or even under normal operation—that would lead to current limiting, not a negative transient. But the logic spilled over somewhat it seems. The negative transient could come, say, from a sudden switchover from AC regulated power over to a weak battery backup (in about 1 ms). This is what could happen: Assume the PD is at the time drawing a minimum of 10 mA (to avoid being disconnected as per the standard), then under this transient, the PD's diode bridge would get reverse-biased, so the entire 10 mA will come *not* from the PSE but from the PD's cap C_{DC_DC}. In that case, the PSE would "think" there is no current being drawn, and there is a danger that it might shutdown the port (thinking the cable has been disconnected). However, as per the disconnect specification, if the 10 mA port current resumes within 300 ms, disconnect will not occur. For the 10 mA to come from the PSE eventually, and not from the PD's cap, the cap must get discharged 13 V within 300 ms, with 10 mA current. If we do that calculation for 180 μF, we will get 234 ms. Since this is less than 300 ms, the PSE will start to provide the 10 mA before the 300-ms timer runs out, and so PD disconnect would be avoided. Hence, 180 μF got chosen.

For larger capacitors, the PD may need to be designed to actively discharge the bulk capacitance. Alternatively, the PD should have a much higher minimum current, say 30 mA. That would discharge the PD's bulk cap three times faster than 10 mA, allowing three times larger capacitance than 180 μF.

Practical PSE Design for Inrush Currents

In Fig. 5.2 we present the AT standards's inrush template (worst-case, extracted from Fig. 5.1). This figure now also includes the lower current-limit threshold (0.4 A), and also the minimum value of the inrush timer (50 ms). So this is complete, and it is the AT (more stringent) requirement. A practical PSE design will center its inrush current limit between 400 and 450 mA. So its nominal value will be about 425 mA, but there will be a spread around that, based on various tolerances and drifts. This band of uncertainty is shown in gray in the figure. But keep in mind that the minimum of this tolerance band must be above 400 mA, and it maximum must be

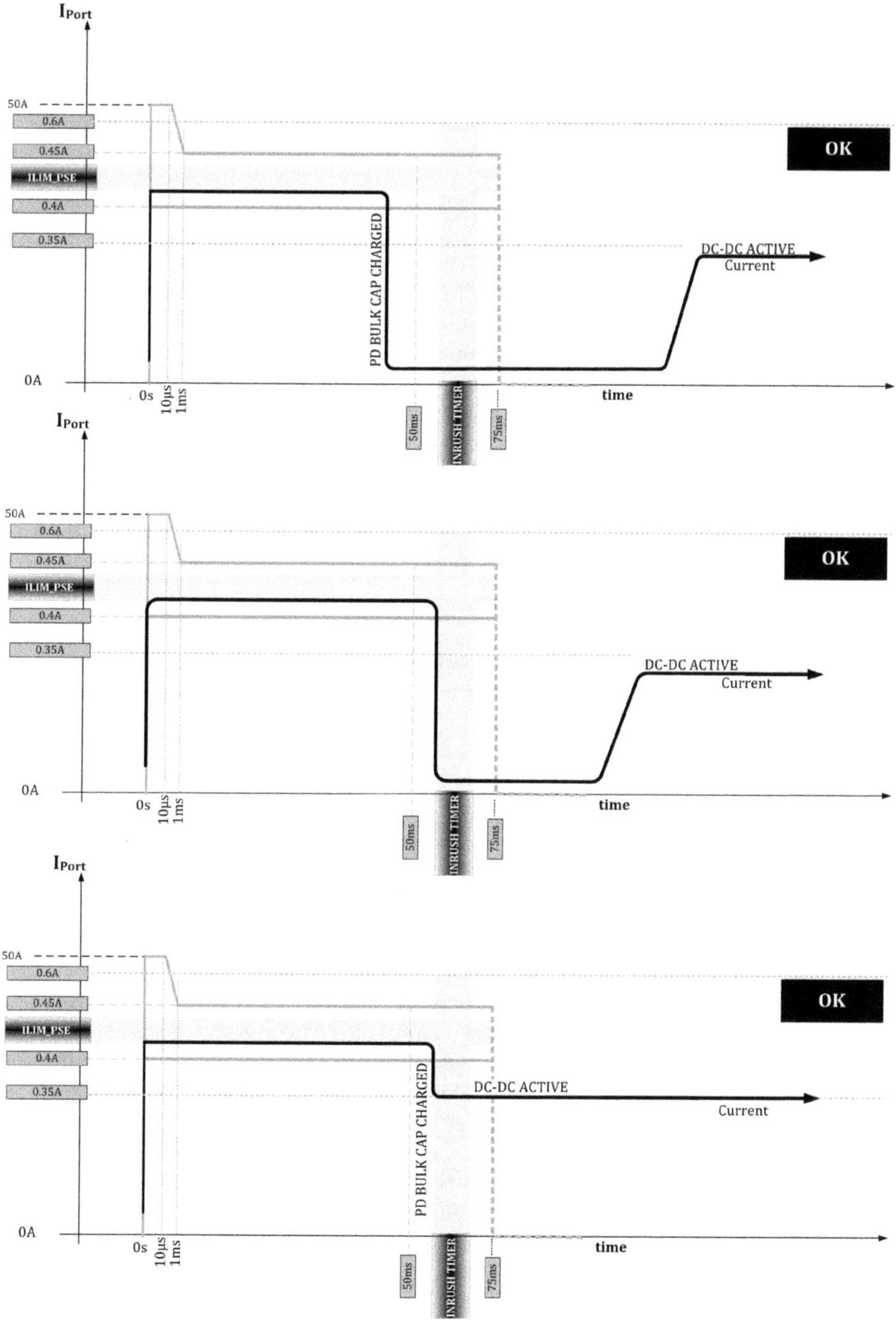

FIGURE 5.2 Sample current waveforms superimposed on inrush template.

below 450 mA. Similarly there is another gray region representing the inrush timer, and we realize its minimum must be above 50 ms and its maximum below 75 ms.

We have shown several more cases in Fig. 5.3. Each of these two figures has three subcases: top, bottom, and middle. Let is discuss each of them in turn.

1. Figure 5.2, top: The PD's bulk cap is charged at the limiting current value (0.4 to 0.45A), and it is small enough to get fully charged before the 50-ms marker. We see that as soon as the cap is charged, the port current ceases and comes down to almost zero immediately. It is thus obvious that the DC-DC converter is *not* active at that moment. The DC-DC however starts drawing current (becomes "active") a little later. In the figure, we have shown that its operating current is a little less than the Type 1 max of 350 mA, so we can be sure the PD continues without being interrupted (assuming the PD was identified as a Class 0/3 prior to Power-up). In any case, the overall behavior during normal operation is not based on the inrush template. We will discuss normal operation matters in the next chapter.
2. Figure 5.2, middle: This shows a borderline case of the preceding case. The inrush lasts a little more than 50 ms because of a large bulk cap but narrowly evades the inrush timer, so port shutdown is avoided once again. This highlights the fact that the standard demands that the inrush current-limiting *not* last *longer* than 75 ms. For example, it does not say the inrush should not last more than 50 ms. It does, however, say the port should *not* get shutdown *before* 50 ms: In other words, the PSE must be able to support a certain amount of inrush at least. So this figure shows behavior that is compliant in all respects. However, if the inrush timer region got just a little wider, the PSE could have shut down the port since the current would be intruding into the vertical (inrush timer) gray region.
3. Figure 5.2, bottom: This highlights the fact that the standard does not ask that the port current fall to near-zero values before the 75-ms marker. The standard only asks that the port current leave the inrush current-limiting region of 400 to 450 mA, by the time the 75-ms marker comes along. So in the case shown in the figure, the PD's DC-DC converter has obviously been active much earlier, perhaps even during the time that the bulk cap was being charged. But it is drawing Class 0/3 currents only, consistent with the preceding classification and the inrush template. *This type of PD behavior, where the DC-DC comes on early, may not be recommended,* but is compliant.
4. Figure 5.3, top: This shows the case where the inrush lasts a little too long and is trapped by the inrush timer. So, the PSE

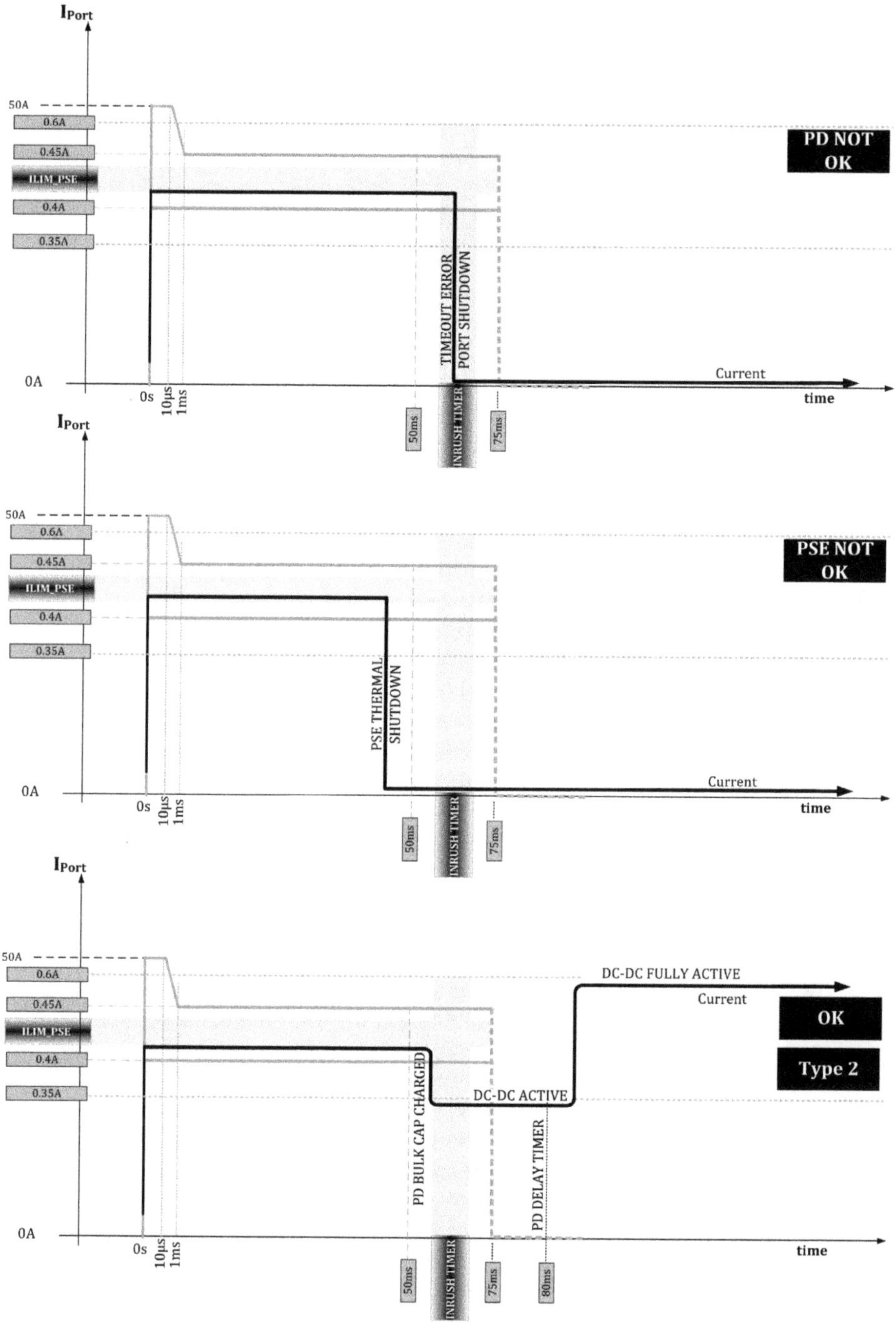

FIGURE 5.3 More sample current waveforms superimposed on inrush template.

identifies the situation as a fault and steps in to decisively turn the port OFF (and also typically place a bleeder resistor across the port to try and actively discharge it). The fault is clearly only the PD's in this case.

5. Figure 5.3, middle: This shows the case of a PSE that just wasn't capable enough to support current limiting for at least 50 ms as required by the standard. The PSE went into thermal shutdown. So this noncompliant, and the PSE needs redesign.
6. Figure 5.3, bottom: This shows a compliant, Type 2 Power-up sequence, though it is somewhat marginal. The PSE does not shut down the port because the current fell to Type 1 levels *just before* the inrush timer was encountered. Note that after a certain PD-side delay timer, as discussed in a later section, the PD's DC-DC converter goes to full power, and the PSE clearly supported that by raising its set current limits from the inrush level of 0.4 to 0.45 A, to the required operating current limits levels consistent with Class 4.

Undervoltage Lockout Thresholds

The standard also creates undervoltage lockout (UVLO) thresholds when it states that the PD power supply must turn ON by the time the port voltage reaches V_{ON} = 42 V. And when the port voltage falls, the PD power supply must turn OFF before the port voltage falls to V_{OFF} = 30 V. Later, the standard makes it even more brief (and rather succinctly confusing), by requiring that the PD turn ON at a port voltage less than V_{ON} and turn OFF at a port voltage greater than V_{OFF}. Question is what part of the PD? Neither is it clarified whether the output of the DC-DC converter is expected to be up (and stable) by that rising port voltage level. It seems questionable if these levels have anything to do with the PD power supply (DC-DC converter stage) at all, at least not directly per se. Therefore, the stated UVLO thresholds are usually interpreted very simply as belonging to the PD interface (pass-FET section). The usual interpretation is *the PD's pass-FET must conduct before the PD-side PI voltage reaches 42 V and must stop conducting before the PD-side PI voltage voltage falls to 30 V*.

But it is not over yet. There is no mention of any "deglitching" in this regard either. In other words, if the port voltage falls below 30 V, how *quickly* should the PD's pass-FET be turned OFF? Nothing is instantaneous, after all.

We can also ask do we really have a *hysteresis* of 42 V – 30 V = 12 V as is often assumed in literature? Actually the standard does not specify any hysteresis at all. Consider a numerical example: Suppose the PD is designed to turn ON at 37.1 V and turn OFF at 37.0 V. That's almost zero hysteresis. Yet 37.1 V is less than 42 V, and also 37 V is greater than 30 V. So they both comply with the stated requirements, yet there is almost no hysteresis really. If the standard demanded hysteresis, it would (or should) have specified the *min* of V_{ON} and the *max* of V_{OFF}. It doesn't.

Despite this confusion, we can perhaps say: the standard *allows for* a Power-up hysteresis up to a maximum of 12 V. Assuming this is true and also that we *could* somehow avail of *all* this 12 V *hysteresis band*, is that even enough? And is it really as *helpful* as we instinctively and intuitively assume?

People generally assume that some hysteresis may be good to avoid ringing and oscillations occurring during start-up when the PD's pass-FET suddenly turns ON. Because, as mentioned, when the PD's pass-FET suddenly turns ON, it presents a fully uncharged capacitor (C_{DC_DC}) to the "48V" voltage source. So were it not for current-limiting somewhere along the line (either at the PSE-end or the PD-end), the current would theoretically be infinite. Therefore at some point, inrush current limiting was introduced, and it was also kept at a level consistent with Class-0/Class-3 normal operation for simplicity. However, the underlying math and distribution of voltages is actually *very dependent on whether the current-limiting function is on the PSE-side or the PD-side*. We will study that in the next section.

There is also some prevailing confusion whether the V_{OFF} and V_{ON} thresholds refer to the PI voltage on the PD-side or PSE-side. The best interpretation is they are on the PD-side and include cable drops, as assumed previously. But that leads to another problem, which some engineers point out amounts to a bug in the standard (See "Power over Ethernet—the reality of designing a Powered Device" by Tony Morgan of Silvertel Ltd available at www.silvertel.com). Assume the PSE-side PI voltage climbs above 30 V. Slightly below 30 V (and above 20.5 V), the PSE would obviously enable its 0.4 to 0.45 A current limit in expectation of providing inrush current to a PD above 30 V as per Fig. 5.1. Suppose the PD was designed to turn ON at exactly 31 V. And suppose the connection from the PSE to the PD was 100 m of CAT3 (20 Ω). Then as soon as the inrush starts, the cable drop will be around $20\ \Omega \times 0.45\ A = 9\ V$, and that would take the PD-side PI voltage down to $31\ V - 9\ V = 22\ V$. That level technically demands the PD turn OFF. In other words, the standard seems to actually overlook the cable resistance, and somehow implicitly assumes that when we have 30 V at the PSE-side, we also have 30 V at the PD-side. We do, but not when inrush starts and when there is a real, long cable in between. What really happens is that the sudden drop takes the port voltage below 30 V, causing the PD's pass-FET to turn OFF again. That would cause the cable drop to reduce to zero again (no current), and that would cause the PD-side voltage to jump up again, so it would turn ON again. Eventually we get repetitive cycling (oscillation). Not pretty! Now if we had *enough* hysteresis to prevent the falling UVLO threshold from being reached at all, and that too with either CAT3 or CAT5e cable, we would perhaps avoid this cycling completely. But unfortunately, as discussed, there is no hysteresis specified by standard. We could try to set V_{ON} very close to 42 V. But even then, in trying to avail of the maximum available hysteresis of 12 V, we can see that with a 9 V drop on 100 m of CAT3 cable, that is cutting it way too close. There is just not enough hysteresis implied, assumed or otherwise. The article mentioned above also suggests raising V_{ON} several volts above V_{OFF} to avoid this. But as analyzed previously, that is not necessarily enough. The other option is to keep the inrush current *much lower* than 0.4 A, *by using a PD with its own (lower) current limiting*. And also setting V_{ON} higher, at about 35 to 37 V. We will discuss this next.

Analyzing the Inrush Phase

Going back to the inrush case, where the C_{DC_DC} cap is fully discharged initially, we ask: How *quickly* can a 180-μF cap be charged? Just the basic calculation only at first! Here we are depending on the PSE-side current limit of 0.4 to 0.45 A, and taking the lowest value of the range (0.4 A) we get the worst-case time as

$$\Delta t = \frac{C \times \Delta V}{I_{LIM}} = \frac{180\ \mu \times (57-0)}{0.4} = 26\ \text{ms}$$

Note that we have retained the starting voltage as 0 V above as a reminder that even though the port voltage may be 30 V or so when the PD's pass-FET turns ON, the capacitor itself is fully uncharged.

A reasonable time is 26 ms. We know from the inrush specification that a PSE must support at least 0.4 A for up to 50 ms, without going into any form of protective shutdown or foldback. So this 26 ms seems well within the required PSE capability, and in principle, the PD should start up with no problem. But does it?

Before we answer that, we need to clarify why it may get difficult to sustain any FET in current-limiting mode for too long. The reason is: *dissipation*. If a FET is fully ON (the voltage across it is almost zero) its dissipation, $V \times I$, is theoretically zero because V is zero. If it is fully OFF (nonconducting), its dissipation, $V \times I$, is again theoretically zero because I is zero. The trouble starts when the FET enters its linear region (triode/ohmic region) with some nonzero voltage across it and some nonzero current through it *simultaneously*. This is exactly what happens when a FET enters current-limiting state (but keep in mind that a FET may enter this region even if it is not current-limiting as we will soon see). Whenever a FET enters its current-limiting state, it is neither fully ON nor fully OFF. It starts behaving as a *variable resistor*. It adjusts its resistance to whatever value is necessary, to restrict the current into the cap as per its set current limit and at the same time have all the circuital currents and voltages consistent with Kirchhoff's laws. That indirectly determines the voltage across the pass-FET (the one performing the actual current limiting).

Now if we introduced PD-side current limiting, and at a level below 0.4 A (the lowest-possible value of the PSE-side current-limit range), then since the PD's pass-FET is in series with the PSE's pass-FET, it would end up dominating the current-limiting function *in most cases*, and be the one operating in its linear region (mostly). And that changes the entire distribution of voltages and currents in the circuit as explained in Fig. 5.4.

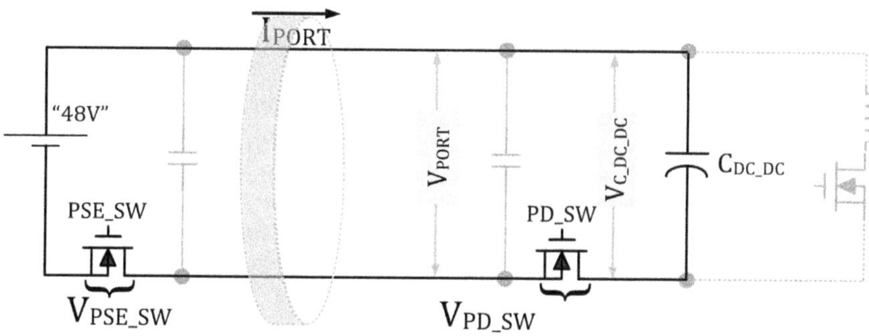

PSE-side is doing current limiting

(0.4 – 0.45 A), no current limit present on PD-side

PSE dissipating (but oscillations)

CASE 1: Significant PD Hysteresis Scenario:
If PSE-SW is current limiting, as soon as PD_SW turns ON (fully conducting), the voltage across PD_SW, i.e. V_{PD_SW} goes to zero. So the *voltage across C_{DC_DC} becomes equal to V_{PORT}*. Since $V_{C\_DC\_DC}$ is intially zero, that means V_{PORT} ***will be dragged to zero***. But before that happens, at some point around 30V, V_{PD_SW} will get turned OFF by the PD control, so I_{PORT} will become zero. Then the whole cycle will start again, *provided the PSE does not turn OFF at some point, "thinking" there is some cable fault.* At best there will be *ugly behavior* as C_{DC_DC} tries to charge up intermittently.

PSE and PD dissipating

CASE 2: Insignificant PD Hysteresis Scenario:
If there almost no hysteresis present in the PD (and no delay/smoothing caps either on the Gate of PD_SW!), this could actually work, because then PD_SW will find itself bouncing very quickly between ON and OFF states, ***which actually amounts to it being quasi-linear***. So now PD_SW will take on the necessary voltage across itself, quite similar to being in current-limiting state, but not exactly. Now, ***the dominant current limit is still 0.4 – 0.45A determined by the PSE***. Both the PSE and PD take some voltage across themselves, and both dissipate in the process. But the start-up behavior would be observed to be ***smooth*** as shown in **Fig. 5.3**. This would work for small or large cable resistances.

PD-side is doing current limiting

(< 0.4 A), because PSE current limit higher than 0.4A

The PSE also has current limiting but the PD has a lower current limit, so it dominates the show *at least initially*. In this case, PD_SW does NOT turn fully ON, but turns on in ***current-limiting mode***. So though the voltage across C_{DC_DC} is initially zero, ***the voltage on the line side of PD_SW does not get dragged down to zero by C_{DC_DC}***. But what voltage it actually stays at, depends on cable length as explained below.

PD dissipating

CASE 3: No Cable Resistance:
Assume that PSE_SW is fully ON, since it is not in current-limiting mode, and has already passed detection and classification. So now the entire steady "48V" appears on the line side of PD_SW, which takes up the difference between that voltage and V_{CDC_DC} across itself— ***till C_{DC_DC} is fully charged.***

PSE and PD dissipating

CASE 4: With Cable Resistance:
PSE_SW is still fully ON initially. So the entire "48V" appears on the PSE-side of the cable. Then there is a cable drop based on $I_{LIM\_PD} \times R_{CABLE}$ and the voltage on the PD-side of the cable is correspondingly lower. The difference between that and V_{CDC_DC} appears across PD_SW *till the two are equal*. After that the PD_SW is no longer current limiting, but is fully ON. C_{DC_DC} *continues to charge linearly*, but now (in this period) *under control of the current limit of the PSE*, i.e. I_{LIM_PSE}. There are thus **two** phases of operation here and the PD current limit gives way to the PSE current limit at an intermediate point which depend on the amount of cable resistance and set current limit of PD_SW.

Figure 5.4 Understanding different start-up scenarios.

In Fig. 5.4, we have shown four possibilities actually. The one with only the PSE doing current-limiting only, and with significant hysteresis is actually the *worst case*, simply because the PD's pass-FET turns on fully into an initially uncharged cap and that voltage then appears across the port, dragging it down, causing the PD to turn OFF, then cycle back-and-forth in an ugly manner. *This is very likely the reason all known commercial PDs perform current limiting themselves, completely disregarding the 180-μF-based current-limiting "responsibility chart" of the AF and AT standards*. Perhaps the PD-chip designers just did some circuit simulations and noticed that only when they put in current limiting inside the PD, did Power-up become smooth. This is actually consistent with the analysis in the right-hand column of Fig. 5.4.

In Fig. 5.4, on the lower-left side, is a very interesting case we call the "insignificant hysteresis scenario." When implemented with a good PD design, it leads to the waveform shown in Fig. 5.5 and is a *good method for also doing bench-testing on a PSE's inrush current limiting capabilities, without the behavior being "masked" by the PD's current-limiting dominance*, or some other strange oscillatory behavior. This is discussed further in the next sections.

In general, it has become increasingly obvious that proper Power-up can only occur if the PD has the current-limiting function inside of it (with a max value of 0.4 A, *must be below* the PSE's range). So the PSE-side current limit just becomes a watchdog (protective upper) limit, one that is encountered only if there are faults in the PD itself, or shorts in the intervening cable, for example. Under normal operation, in effect, the PD takes care of its own reliability, and also ensures smooth start-up behavior.

Ensuring Proper Power-Up Behavior

There are strange interactions and complications during Power-up, involving UVLO-based instability, cable-resistance drops, negative-input impedance of switching converters, and so on, all of which make the entire area of PD-PSE design during Power-up rather complicated and misunderstood. The standard does seem ambiguous itself, if not misleading, at places on this particular aspect. So we do need to fall back on engineering judgment eventually.

As prospective PSE designers, we need to accept market reality too: Commercial PD's are often rightly considered low-tech, and we should expect to find many out there with all sorts of strange behavior during the critical Power-up phase. We should therefore try to make our PSE designs robust enough to at least avoid contributing to such questionable behavior and causing irrecoverable start-up issues.

As prospective PD designers, we realize that there is no minimum hysteresis demanded by the standard during the critical

Power-up phase. For capacitors less than 180 μF, the standard does not ask for the PD to have current limiting. But to mitigate strange behavior, most commercial PDs are designed with several volts of hysteresis and also voluntarily include current limiting.

With all efforts, we may still not get entirely satisfactory Power-up behavior under all conditions, despite our best efforts, simply because as PSE designers, we can't control all PDs out there, and as PD designers, we can't control all the PSEs. But we can usually live with this diversity. Keep in mind that in any case, if there is ringing during Power-up, it is hard to test and measure, and much harder to ascribe fault at the end of it all (such as, was the real culprit the PSE or the PD?). For good reason perhaps, this stability aspect during Power-up is mentioned but rarely, if ever, tested.

Testing the Inrush Performance of PSEs

However, we do need to test the PSE for its overall inrush current capability and protection. The question is *how do we do that if in a properly behaving setup, the PD's pass-FET is always in series with the PSE's pass-FET, and is the one dominating the current limiting*? What we want to do is to be able to test the PSE's current limit, by somehow turning OFF current limiting in the PD. But as indicated in the top-left of Fig. 5.4, as soon as we do that, we usually get the familiar Power-up cycling and oscillations.

As identified in Fig. 5.4 too, *one way out is to run the PD's pass-FET in its linear region (insignificant hysteresis scenario)*. Note that in this state, the PD's pass-FET is *not* performing current-limiting: that function still resides in the PSE's pass-FET only. But the PD's pass-FET is nevertheless allowed to go into its linear region, by suitable Gate control, which keeps it conducting just the right amount. Now the PD's pass-FET can accept whatever voltage it needs to. And since the current-limiting is still coming from the PSE, the PSE's capability can now be tested independently.

To get the PD's pass-FET into its linear (resistive) region, we need to come up with a discrete test circuit with virtually no hysteresis, and no delays either (no FET Gate caps for example). With such a setup, we could for example, design the PD's pass-FET to turn ON fully at 38 V and turn OFF fully at 37.9 V (a very sharp "knee"). Now when the PSE goes into current limiting and the PD's bulk capacitor starts charging up, the excess voltage will naturally and automatically appear across the PD's pass-FET as shown on the bottom-left of Fig. 5.5. Note that this is a viable (smooth) method of operation, since the port voltage does not tend to get dragged toward zero anymore. Instead the port voltage stabilizes very close to 38 V as per the example, while keeping the PD's pass-FET in its linear region and still allowing the PSE to get exerted up to its set current limit. We can thus

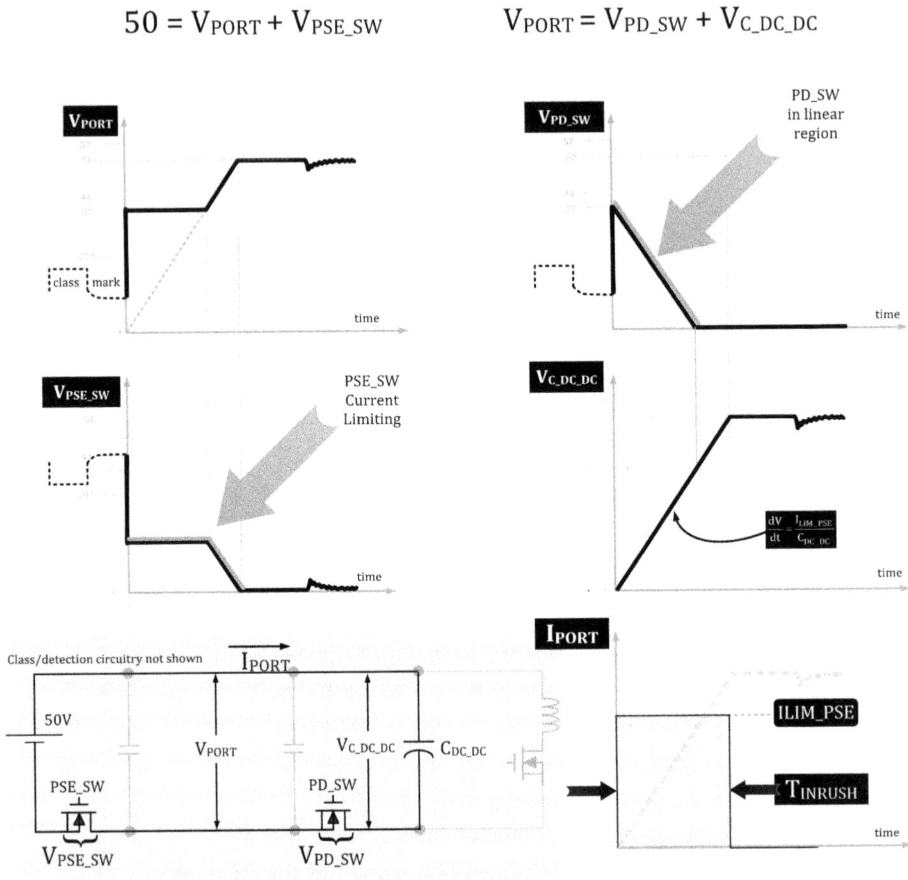

FIGURE 5.5 One power-up possibility using insignificant hysteresis and Inrush current limiting in PSE only. (See also Figs. 4.3 to 4.5.)

test the inrush performance of the PSE. For example, we could, with this circuit, apply larger and larger PD bulk caps (C_{DC_DC}) to study the response of the PSE.

Let us work backward and see what capacitance thresholds we get. We are assuming a port voltage of 50 V in normal operation.

1. Condition: PSE must support at least 50 ms of inrush current as low as 0.4 A.

$$C = \frac{I_{LIM} \times \Delta t}{\Delta V} = \frac{0.4\ \text{A} \times 50\ \text{m}}{50\ \text{V}} = 400\ \mu\text{F (for incoming } V_{PSE} = 50\ \text{V)}$$

In other words, if we put $C_{DC\_DC} \leq 400\ \mu F$, the PSE must *not* turn OFF to be considered compliant.

2. Condition: PSE must *not* allow more than 75 ms of inrush current as high as 0.45 A.

$$C = \frac{I_{LIM} \times \Delta t}{\Delta V} = \frac{0.45\ \text{A} \times 75\ \text{m}}{50\ \text{V}} = 675\ \mu\text{F (for incoming } V_{PSE} = 50\ \text{V)}$$

In other words, if we put $C_{DC\_DC} \geq 675\ \mu F$, the PSE *must* turn OFF to be considered compliant.

A Discrete PD Front-End for Testing PSEs

In Fig. 5.6 we have presented a popular circuit made of discrete components, one that has been used in almost similar forms across the industry for years in PoE testing and in finished products too. Note the LM317 is used as a programmable current source (any three-terminal regulator with a low-bias pin current can be used here). A known problem with this discrete PD front-end circuit is it does not have a very precise and predictable UVLO threshold, nor a very sharp "knee" either. But it does have *almost no hysteresis* and that could explain its wide popularity and usage. It corresponds to the insignificant hysteresis scenario discussed previously (see Fig. 5.4), and thereby, *it enables the PSE current limit to act, while placing the PD's pass-FET into its linear region, thus preventing the port from being dragged toward zero volts by the initially uncharged bulk cap of the PD*. This circuit can be further improved with a little more circuitry perhaps (like PNP-NPN small-signal transistors to enhance its Gate drive and its "knee"). The common feature would, however, continue to remain: *no hysteresis*—because that is why it works so well. It can therefore be used for PSE-testing and PSE-evaluation without the PD masking/dominating the current limiting. Note that based on trying to avoid cycling during start-up, we actually recommend *not* placing any "filtering" capacitors in the Gate of the PD's pass-FET, or elsewhere too, because the caps actually cause delays. That, in effect, amounts to a hysteresis of sorts, which ultimately contributes to visible oscillations, rather than smoothly forcing the PD's pass-FET into its linear region by a quick-acting Gate control.

The Inrush Timer and the Real End of Inrush

Now we consider some *timing* issues. The end of inrush is marked by the fact that the PD's bulk cap is fully charged and neither the PD's pass-FET or the PSE's pass-FET are in current limiting anymore. This supposedly marks the entry into Power-on state (with some qualifications as discussed later).

Discrete Circuit to Simulate PD Front-end

Reference: AN167, Microsemi Corp (PowerDsine)

Table 6: R1 Values

Class	Current [mA]	R1 [Ω]	Minimum Package
1	10.5	127	0805
2	18.5	69.8	0805
3	28	45.3	0805
4	40	31.6	0805

Table 7: R2 values

Class	Current [mA]	R1 [Ω]	Minimum Package
1	10.5	0 (short)	
2	18.5	0 (short)	
3	28	220	1206
4	40	120	1206

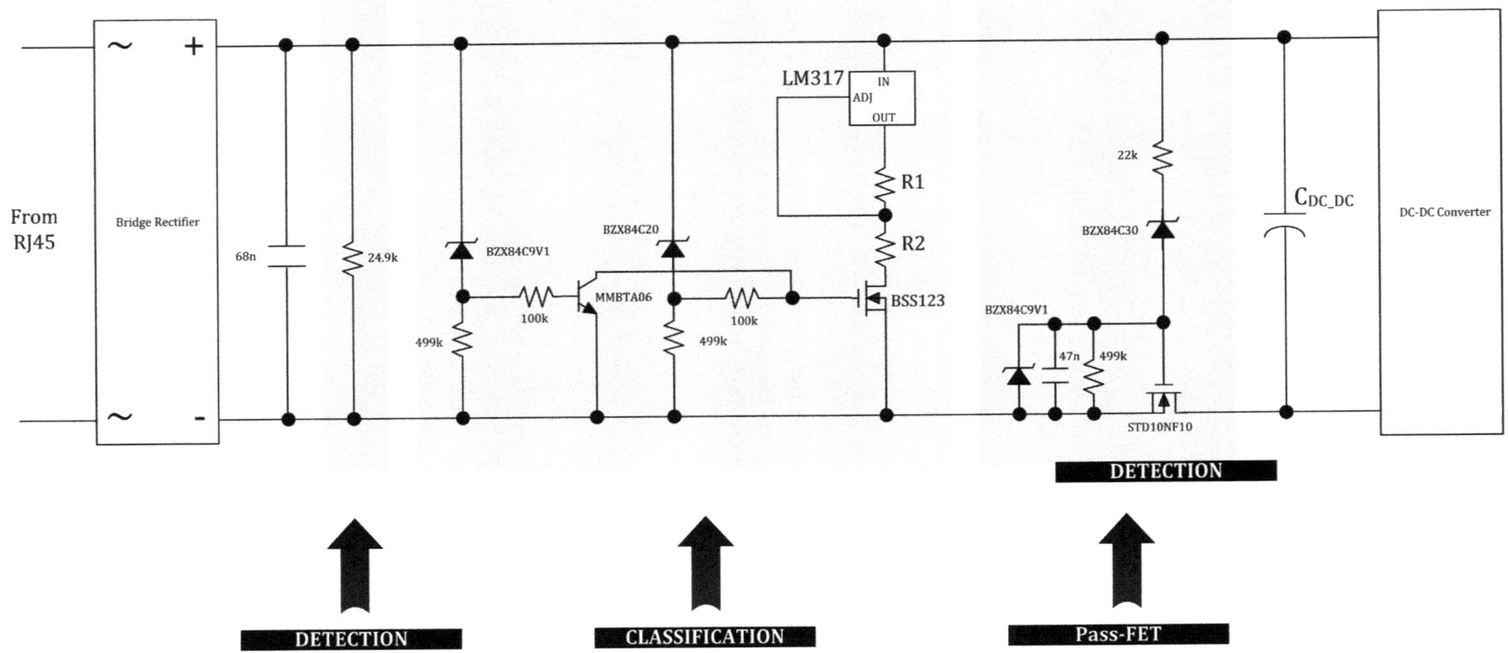

FIGURE 5.6 A PD front-end (one of several discrete circuit versions possible).

How do we know the capacitor has been fully charged? The standard says the inrush is considered complete when the port capacitor voltage is *99 percent of the final settling value*. This number may however prove difficult to measure on a test setup with any precision: 1 percent of 50 V is 0.5 V. The University of New Hampshire Interoperability Lab (UNH-IOL) asks to monitor the port voltage on an oscilloscope to judge when the port is powered up. But with a typical 10 V/div setting for monitoring port voltage, 0.5 V is tiny—rather hard to see.

From the PSE-chip design perspective, the chip usually has an ADC for monitoring port voltage and may be able to know this for its own purposes, with enough accuracy. However, it is dependent on knowing the incoming "48V" rail accurately and also estimating the drop across the PSE's pass-FET. So *many PSE-chip designers just say that inrush is complete when the port voltage reaches the normal operating range, which on the PSE-side is anything above 44 V for Type 1 and above 50 V for Type 2. Or they may just set a fixed threshold of 40 V nominal for both Type 1 and Type 2.* Some PSE chips say inrush is over when the *current* has fallen by say, 10 percent, from its limiting value. By the use of current information, they can also tell whether the PSE or the PD was the series-pass element actually controlling the inrush, and thereby respond accordingly. Some PSE chips may ultimately combine both voltage and current information to ascertain end of inrush.

Quite similarly, PD datasheets (like the TPS2378 from TI) also interpret inrush completion based on the PD-chip's *current*. They specify "inrush period termination" as occurring when the current falls to 90 percent of its current limiting value. This parameter has a min-max range of 80 to 99 percent.

Why is it important to know the exact end of inrush anyway? Some erroneously think it is because the T_{PON} timer needs to be less than 0.4 s. Their error is based on taking the name T_{PON} literally. But as discussed in the previous chapter, the T_{PON} timer starts at the end of detection, and despite its name (T_{PON} stands for Power-on time), it is more correctly interpreted as ending not when Power-on has occurred, or even when inrush has ended, but when the full AC-DC power is first *applied* to the port. Measurably, that is the point where the PI voltage on the PSE-side just exceeds 20.5 V. This interpretation of T_{PON} makes more sense since T_{PON} is in essence a PSE-side specification, and what happens *after* 20.5 V depends a lot on the PD's design too, its bulk capacitor value and so on (which incidentally has no specified upper limit as per the standard). So it is very hard to test a PSE's compliance to T_{PON} if we do not define this interval as ending the very moment 20.5 V is exceeded (or 30 V—that's not much different).

With the correct definition of T_{PON} behind us, the question still remains: Why is it important to know the exact end of inrush anyway? *Because inrush must be completed before the inrush timer times out.* What is this new timer? This starts at the very moment T_{PON} ends (port voltage exceeds 20.5 V). By typical PSE chip design, this T_{INRUSH} timer is set to a nominal value centered somewhere between 50 and 75 ms, say 65 ms. Basically, inrush needs to be completed before this timer runs out (max value of 75 ms as per the standard), *unless there is a fault*. So at the 65-ms marker, the PSE chip is *expecting* to see: (a) no PSE-side current limiting (port current must x percent less than its set value, or below 400 mA by then), and (b) the port voltage must also be very close to its final settling value as explained above. If *both* conditions are true, inrush is deemed over. But if for example, after 65 ms, the PSE either finds it is still in current limit (obvious from the substantial voltage present across it and despite the high current passing through it), and/or the port is not yet close to the "48V" rail, then the port will be shutdown and a fault declared. After a suitable delay, the port will start once again, but at detection.

Types of Power-Up Behavior and Power-On

We now know how to tell when inrush is complete as per the standard, based on the monitored port voltage. We also realize that event can actually happen well before the inrush timer runs out. For example, if we have a very small bulk capacitor inside the PD, the cap will charge up in say only 10 ms, whereas the inrush timer is 65-ms nominal. So the question is: Do we say Power-up is complete at the 10-ms mark or at the 65-ms mark? There is a difference between the AF and AT standards in this respect. The AT standard says: *wait till the inrush timer runs out*, then come in and check the port voltage and current limit conditions, as implied in the previous para. If all is "kosher" at that moment, then consider Power-up phase complete and Power-on having taken place. The AF standard, however, interpreted Power-up phase completion a little differently. It did not *wait* for the 50- to 75-ms timer to go off and then check if port voltage was very close to its final value. The AF PSE was in effect *constantly monitoring* the port, and if the port voltage was almost at its final value at any time before the inrush timer ran out Power-up was considered done. The AT standard, however, rightly recognized that PD designs in particular, were very diverse. Hence it opined: "Using only the PI voltage information may be insufficient to determine the true end of PD inrush current; use of a fixed T_{INRUSH} period is recommended." In the PSE state-machine diagram of the AT standard, for backward compatibility, it allowed for the older type of Power-up (which it called "legacy Power-up"), for both Type 1 and Type 2 devices, but strongly recommended that new PSE equipment be designed based on the fixed 50-75 ms inrush timer. Basically, both Power-up methods, with and without T_{INRUSH}, are *allowed* in compliance testing, but it is not considered a good idea to design any devices with legacy Power-up anymore.

Minimum Inrush Below 30 V

We now know than any typical PSE chip will set its current limit to 425 mA nominal, at about 28 V nominal (< 30 V), as the port voltage ramps up from classification (> 20.5 V) to the commencement of the main inrush phase (> 30 V). But in fact, inrush formally starts just above 20.5 V and the question is: What is the set current limit inside the PSE *between* 20.5 and 28 V?

As mentioned, the inrush template in Fig. 5.1 applies only above 30 V. So what exactly is allowed below 30 V? The standard actually allows/demands the PSE permit a small inrush current even below 30 V. In Section 33.2.7.5 it states

- During POWER_UP, for PI voltages between 0 and 10 V, the minimum I_{INRUSH} requirement is 5 mA.
- During POWER_UP, for PI voltages between 10 and 30 V, the minimum I_{INRUSH} requirement is 60 mA.
- During POWER_UP, for PI voltages above 30 V, the minimum I_{INRUSH} requirement is as specified in Table 33-11.

Table 33-11 mentioned leads to Fig. 5.1 of this chapter. So that is already known and discussed. The question is: What do the first two entries mean? Based on our discussions so far, Power-up does *not* involve voltages lower than 10 V. So why does the PSE need to support inrush currents of at least 60 mA above 10 V? Or more than 5 mA below 10 V? The reason is so far we have implicitly assumed something similar to the sequence in Fig. 4.5, in which classifications leads straight into Power-up. But things could also happen as per Fig. 4.3 and Fig. 4.4. In fact, after 1-event classification, the port could be stationed at ~ 0 V for quite a long time before power was applied to the port. So Power-up actually can start at much lower voltages. Then the PSE has to set current limits as the port voltage rises. That is why it details the inrush levels at low voltages. But note that it only specifies *minimum currents*. So, as per the above three bullets, the PSE could actually set the current limit to 425 mA nominal right at the very moment the Power-up command came in. After all, the first two bullets above do *not* disallow higher currents than mentioned. However, we must remember that the PD's front-end does not expose the bulk cap below 30 V. So it makes no sense allowing higher charging currents for small, ceramic port capacitors. And if we do that, we will get very high dV/dt's. That can cause noise pickup and erratic behavior of control chips (see next section). Keep in mind also that as the port voltage rises, the classification current circuit in the PD sinks current (up to 45 mA), and it will turn ON at some point. We want to be able to provide *more* current than that to the entire PD, so that there is some excess current left to charge up the ceramic port caps too.

Otherwise the port voltage will just not rise. Therefore, most PSE chips very simply set a nominal 75 to 85 mA current limit on the PSE below 28 V nominal. Above 28 V, the nominal current limit is raised to 425 mA as discussed previously. This is all illustrated in Fig. 5.7, which also includes legacy Power-up and modern timer-based Power-up differences. It uses nominal or example values to make things clear.

Type 2 PD Delay Timer

To keep things simple, inrush requirements are classless. They apply to any class. It is only when Power-on starts that the PSE adjusts its current limits based on the reported class. As indicated on the left side of Fig. 5.7, the AT standard requires that a Type 2 PSE await the end of the inrush timer before it raises its set current limit to Class 4 requirements (typically 850 mA nominal). This can take up to 75 ms, which we know is the max value of the inrush timer. Hence, to avoid inadvertent port shutdown, the standard correspondingly asks the PD not to *ask* for higher than Class 0/3 currents before 80 ms. This is the timer T_{DELAY}. We have chosen a nominal value of 90 ms for the curves in Fig. 5.7. All the timings are measured from the start of Power-up as discussed previously.

The delay timer was also present in the lowermost case of Fig. 5.3.

When designing the PD we should remember that the cleanest start-up behavior occurs when the DC-DC converter is not switching initially at all, and gets enabled/activated only *after a* wait-time of T_{DELAY}. That gives enough time (and charging current too) for C_{DC_DC} to be properly charged up. The PSE also has time to set the correct operating-current limits (all within 75 ms) *before* the DC-DC converter tries to start up (after 80 ms).

Note that in Fig. 5.7, though inrush ends at exactly the same time (30 ms as per the chosen example), the end of Power-up is designated in a different manner by the AT standard and by the AF standard (legacy).

To reiterate: the AF standard also *does* define an inrush timer of 50 to 75 ms. The difference with the AT standard is simply that if things look "good" to it, the AF PSE just doesn't wait for the timer to expire before turning on the port (and when that happens, the inrush timer gets obviously reset). In the AF case, the inrush timer only comes into the picture *to* enforce shutdown if the port has still not been powered up when 50 to 75 ms have elapsed since power was first applied to the port.

Rise-Time Limits

We also need to limit the minimum rise time of the rising voltage, to avoid making it too "snappy." That can cause significant noise in the lines, and in the RJ-45 integrated magnetics, affecting other ports too.

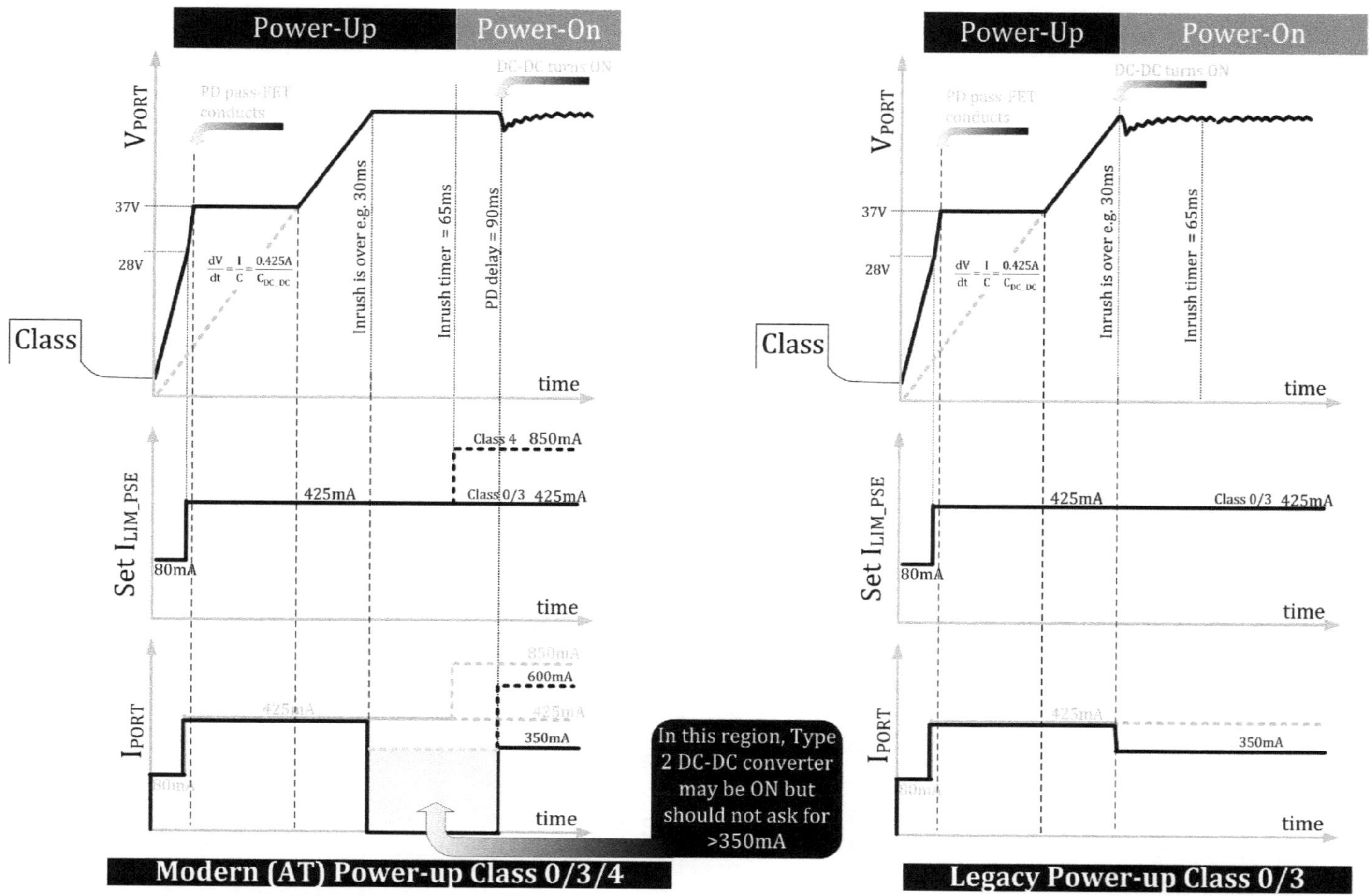

FIGURE 5.7 Currents and current limits during Power-up and different types of Power-ups.

There is also a well-known industry case that caused a Type 2 PD to "forget" it saw two fingers, which would have identified a Type 2 PSE to it. So, despite offering its classification circuit twice, it "thought" it was connected to a Type 1 PSE and did not power up in high-power mode. All because of the steep dV/dt. For such reasons, the standard asks that the rise-time "T_{RISE}" be more than 15 μs. How is T_{RISE} supposed to be measured? From 10 percent of the final port voltage to 90 percent of the final port voltage. For example, if the AC-DC power supply was 52.5 V, and the drop across the PSE's pass-FET was 2 V, we are left with a final port voltage of 50.5 V. So 10 percent of this is 5.05 V, and 90 percent of this is 45.65 V. The difference is 40.4 V. In other words, we need the average dV/dt to be less than 40.4 V/15 μs = 2.7 V/μs.

Now, as we have seen, the port voltage does not rise "monotonically" as sometimes sketched in literature. It has "breaks" and "jumps" in it, on the way to to the top as the bulk capacitor in the PD gets exposed, and so on. So in reality the *average* dV/dt (determined by connecting the 10 and 90 percent points by a straight line), may easily comply. However, we should also look at the rising port voltage curve *piecewise*, to ensure the dV/dt is less than 2.7 V/μs throughout (for the assumed supply voltages). This piecewise examination is not required by the standard but is certainly in the spirit of it all.

PSE compliance test instruments may detect the start of Power-up when the port voltage already higher than 10 percent of the final value. In that case, the instrument will likely internally (mentally) extrapolate the rising curve down to the 10 percent mark, to figure out the total rise time from 10 to 90 percent.

PSE compliance test instruments assume zero bulk capacitance to test the PSE for rise-time compliance. In that case, typically, we have 0.1 μF on the PSE-side port and another 0.1 μF on the PD-side port, that is a total of 0.2 μF. If we charge this with the highest current limit value of 0.45 A, we get a dV/dt of

$$\frac{dV}{dt} = \frac{I}{C} = \frac{0.45}{0.2\ \mu} = 2.25\ V/\mu s$$

This complies with the requirements actually since it is less than 2.7 V/μs. However, it was also once experienced that test equipment manufacturers (Sifos), with their latest test equipment, removed *all* PD-side port capacitance for the rise-time test and then reported a "compliance failure." Naturally, because with only a 0.1-μF port capacitance, we get

$$\frac{dV}{dt} = \frac{I}{C} = \frac{0.45}{0.1\ \mu} = 4.5\ V/\mu s$$

We could argue ourselves hoarse with the test equipment vendor that the minimum PD-side port capacitance as per the standard is

50 nF (0.05 μF), not 0. But the least stressful option is to add more capacitance to the PSE side. For example, we could say that our PSE is designed to work with the minimum recommended value of 0.2 μF. Then, with a total capacitance of 0.2 μF, we once again get 2.25 V/μs and we will be compliant.

Let us do the calculations for 57 V too. Ten percent of that is 5.7 V, and 90 percent is 51.3 V. The difference is 45.6 V. The corresponding dV/dt is 45.6/15 = 3.04 V/μs. So the standard is now saying our dV/dt should be <3 V/μs. This is easier to meet than <2.7 V/μs, so we should continue to do our analysis at 50 V.

Another helpful method for meeting the max dV/dt spec of 2.7 V/μs is to lower the set current limits. So, up to 30 V, we can now choose > 60 mA (say 80 mA nominal) instead of 0.425 A. That will help too. Or we could also think of *slowly* increasing the Gate voltage of the PSE's pass-FET, up to 30 V at least. And so on. We have to be careful, however, that these steps will likely drive the pass-FET into its linear region, so the dissipation can get significant.

Some Practical PD Design Issues

We must keep in mind that once the PSE applies power to the port, it starts an internal inrush timer too and *will* shut down the port if things have not stabilized at the end of that period. Think of this as a race situation, and the start of that inrush interval as a starting gun. We must complete the race in the ascribed time, or be disqualified. For example, if we put a very low current limit on the PD's pass-FET and/or an excessively large bulk capacitance, we will not have stable conditions when the inrush timer times out. So the port will be shut down. This also places limits on the amount of bulk capacitance. For example, if the PD places a current limit of 100 mA, then, to complete the charging within 50 ms (or risk port shutdown), the capacitance must be less than

$$C = \frac{I \times \Delta t}{\Delta V} = \frac{0.1 \times 0.05}{50} = 100\ \mu\text{F}$$

Even with full 0.4 A (PSE-side current limiting) available, the capacitance must be less than

$$C = \frac{I \times \Delta t}{\Delta V} = \frac{0.4 \times 0.05}{50} = 400\ \mu\text{F}$$

Things actually get much worse if the DC-DC converter is *also switching* away while the capacitor is being charged. For example, if the PSE is permitting only 0.4 A, and the DC-DC is drawing 0.35 A from that, we are only left with 0.4 – 0.35 = 0.05 A to charge up the bulk cap. That will cause an inordinate inrush time and we will very likely face an inrush timeout and port shutdown. So PDs should be designed to start switching *after inrush is over*, though the standard does not specify that at all.

To make matters worse, what if the PD was doing the inrush current-limiting, and it was set to 0.3 A, whereas the PD started *trying* to draw 0.35 A while the cap charging was still not complete? Now since the PD is asking for more current than is available, cap charging will in fact never occur. Instead the port will discharge (and perhaps try to charge up again later or enter port shutdown).

Things only get worse because DC-DC converters also exhibit negative input impedance, This means that as the input voltage decreases, the input current tries to increase in an effort to maintain power ($V \times I$) unchanged. So combined with the fact that the incoming current is limited below the requirements to start with, this leads to wild instability and ringing.

A good PD design will enable the DC-DC converter, not based on a port voltage threshold (as wrongly implied by the standard), but after a *certain time delay*. Let's be clear: The standard does not ask for that actually. It only asks for a minimum 80-ms delay before a Type 2 PSE moves over from its initial AF mode to the medium-power Type 2 AT mode. In other words, it implicitly allows the DC-DC converter to switch at low-power levels before T_{DELAY} has elapsed. And that can only take away from valuable incoming current.

To speed up cap charging with the inevitable PD-side current limit, the PD designer should generally avoid placing a PD-side current limit below 0.35 A. It should be positioned as high as possible and above the Class 0/3 operating current limits, that is, between 0.35 and 0.4 A (nominal value 0.375 A, say).

As an alternative to starting the DC-DC converter after a certain time delay, we can also generate a "power-good" signal from the PD front-end to indicate that the capacitor has been charged up (port voltage stable), and use that to enable the DC-DC converter.

None of these practical design aspects seem to have been mentioned in the standard, but it is now clear these are all essential requirements of any practical PD design.

Strictly speaking, we can design a PD with no PD-side inrush current limit, and no power-good flag (or delay) to activate the DC-DC converter, and still be considered compliant. *However, that particular PD may never even start up with a compliant PSE on the other side.*

In another well-known case, it seems the PD designer was being "extra-nice" by incorporating a long delay before turning ON the PD's pass-FET fully (i.e., removing the unnecessarily low inrush current limit)—in fact, much after port voltage had been already applied. It was obviously a waste of time at a bare minimum. All sorts of things could happen in such a situation. The port voltage would go to its final value right away. That could be interpreted by the PSE as the end of inrush. And if it had enabled legacy Power-up, it could even have declared Power-on. But in fact the PD's pass-FET had not even been turned ON yet. So however it happened, this PD was shut down by the PSE. All because of a delay at the wrong moment. Keep in mind that once the inrush timer starts, we just have a limited time to achieve a stable port voltage. It does not help if the PD front-end tries to bring up the voltage too slowly (ultra soft-start), because the DC-DC converter may have a fairly large (but generally acceptable input cap), and might also be simultaneously trying to switch (though in low-power mode), preventing the port voltage from rising up quickly enough or remaining there. The PSE will simply step in and shut the port OFF thinking there is a fault somewhere if it finds the port voltage (even momentarily) below the 90% voltage mark at the specific 75 ms inrush time marker point. That is why we have repeatedly emphasized not to set a very low PD inrush current limit. None of this is really described or governed by the standard, but it will lead to an interoperability issue with no one obvious culprit in sight.

Various nuances of PD design may just cause occasional issues like port reset under a line transient, followed by a normal reconnect, and then everything being okay thereafter. But the most critical design issues, those that we must keep in mind, are very similar to those described above. They are the ones that can cause *irrecoverable* errors. And ironically, they are also the ones most easily avoidable, with a little thought during the PD's design phase. Just don't try only to comply to the bare minimum requirements of the IEEE standard, because it's clear that there is far more to a good PD design than meets the eye.

CHAPTER 6
Operation

Background

Once Power-on state is achieved, there are two very general systems-level concerns: (a) how to keep the port "alive" (avoid PSE-PD disconnect caused by very small PD loads) and (b) how to protect the cabling infrastructure and PSE/PD from faults such as overloads and short-circuits, yet retain an energy delivery system that is robust enough to provide a normally functioning and healthy PD its usual, momentary bursts of power, just a little above its steady, continuous rating.

The first concern listed above is discussed in the following chapter. The second concern is analyzed in this one.

Historically, the AF standard created a triple-layer cake of sorts: a continuous current region whose upper bound was called I_{CONT}, with an overload region right on top, and finally a *short-circuit* region above that (see Fig. 6.1). That was the starting point, however flawed it was in retrospect. Finally, the AT standard came along and changed a lot of the underlying philosophy, and in many senses made the definition of the regions of operation and their related terms more rational. But it also created some odd nuances of its own, so that can also become quite confusing at times. What we can say with confidence is that the AT standard noticeably softened or lowered the bar in terms of PSE-related performance expectations, so achieving AT compliance in this respect is actually easier than it was, and for the same reason does not assure automatic AF compliance.

There are two camps based around how to interpret and handle that odd AF versus AT situation. Some say that the AT standard has superseded the AF standard. So it is sufficient to comply with the overload requirements of the AT standard only. Others still demand AF compliance.

To best understand the evolution of this particular subject, and to successfully debate what is right from an *engineering* viewpoint, we need to start with what the AF standard says, or at least seems to say, and go from there.

Relevant Sections to Refer To in the AF Standard

We will briefly summarize some of the relevant AF sections concerning overload and short circuit, for easy reference, as before we move slowly to the AT standard.

In this list below, sections **common** to overload and short-circuit have been bolded.

For Overload, refer to (in the AF standard):

1. **Table 33-5, Page 42,** which refers to:
2. Section 33.2.8.6 and Section 33.2.8.7 on Page 44
3. Section 33C.1.6, Page 96, which refers to:
4. Figure 33C.2, Page 92
5. Figure 33C.6, Page 97

For Short-circuit, refer to (in the AF standard):

1. Table 33-5, Page 42, which refers to:
2. Section 33.2.8.8 and Section 33.2.8.9 on Page 44
3. Section 33C.1.7, Page 97, which refers to:
4. Figure 33C.4, Page 95
5. Figure 33C.7, Page 98
6. **Figure 33C.6, Page 97**

Reasons for Protection

Here are the basic reasons we want to implement short-circuit protection:

1. Protect the cabling infrastructure
2. Protect the PSE controller—but in fact the dissipation during "short-circuit" may be less than during an overload, because of current foldback (lowering of port current; an optional feature)
3. Limit peak transient currents to control EMI and its effect on data integrity
4. Protect PCB traces of legacy equipment
5. Protect power supply operation in a multiport system
6. Allow reasonable and cost-effective support for port-to-port cross-regulation effects

As indicated, a short-circuit is not defined as 0 Ω, or even close to that. Because that, in fact, is not the most severe condition.

Possible reasons are (a) the AC-DC power supply will likely step in to cause output voltage foldback (lowering of output voltage), and/or (b) the PSE will see the output fall below a certain value and will shut down the port itself.

The most severe condition occurs at the load *just prior to the PSE turning the port OFF*. And that is not 0 Ω.

Brief Overview of Overloads and Shorts as Per AF Standard

In Fig. 6.1 we show what happens as per the AF standard, when the port current I_{PORT}, ratchets upward. I_{CONT} is the maximum continuous current rating and equals 15.4 W/V_{PORT} (default Class 0/3 assumed). In the figure we also see I_{CUT}, the overload current, and T_{CUT}, the overload timer (which later became T_{OVLD} in the AT standard). In addition, the AF standard has a T_{LIM} timer of 50 to 75 ms too, which starts when the current limit is encountered. As indicated, in the figure, we have sketched the actual I_{CUT} and I_{LIM} of a given chip, as well as its actual T_{CUT}. As the port encounters these levels, timers are actuated. If for example, the overload persists long enough that T_{CUT} timer times out, the port is shut down by the PSE.

NOTE *The AF standard sets the I_{LIM} timer exactly the same value as the I_{CUT} timer. So, if the port current rachets up and then hits current limit, the port will always be shut down by the I_{CUT} timer, not the I_{LIM} timer (since the latter started later). So what's the use of the I_{LIM} timer? To have any effect in practice and to dominate the I_{CUT} timer, the I_{LIM} timer should logically have been smaller, indicating that a higher overload, i.e., one amounting to a short circuit (PSE current limit activated), would receive (rightfully so), a swifter response in terms of port shutdown. But that is not the way the AF standard was worded unfortunately. Hence the ambiguity.*

Figure 6.1 is actually a simplified and more accessible version of Figure 33C.6 on page 97 of the AF standard. Observe the triple-layer cake architecture that we had previously mentioned. We have also drawn out the *actual* (set) thresholds in the chip under study. Note, we are *not* talking about the nominal value of the thresholds, or their spreads/tolerances, but on their *actual* values of a given chip. Yes, it is kind of odd that the standard depends on a *specific chip's characteristics* to determine what is considered OK and sent down the line, and what isn't. We will discuss this "device/chip dependency" in more detail shortly and describe how the AT standard correctly removed this particular oddity.

Adding some more detail here, based on other sections in the standard: The general equation for maximum continuous current is

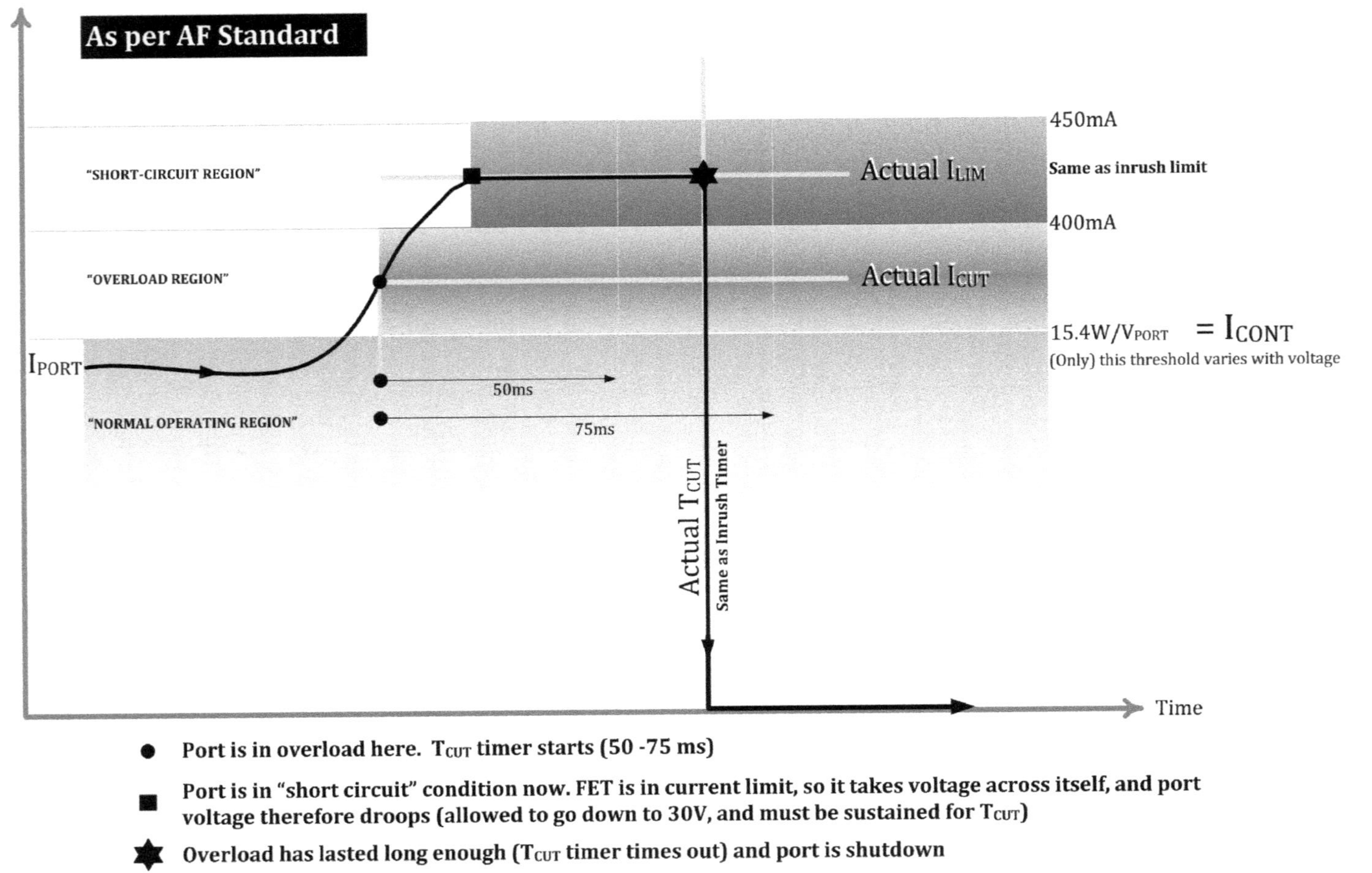

Figure 6.1 Overload to short-circuit as per AF standard.

I_{CONT} = 15.4 W/V_{PORT} (check: 0.35 A = 15.4 W/44 V). So at first sight it seems that the PSE's maximum continuous current is not based on the class of the PD. But luckily, in Section 33.2.8.6 on page 44 it is clarified that I_{CONT} is actually P_class/Vportmin where P_class is the power based on class as measured on the PSE-side.

The continuous current is derated according to port voltage. That is no surprise—it is in keeping with the most basic principles underlying the PoE standards. See the relevant note below as a reminder of the rationale behind that.

What is surprising, however, is that the *upper* bound of the overload region (0.4 A), is *not* dependent on port voltage. Why not? After all, if the continuous operating current falls with port voltage, shouldn't its associated "overload" region track it and have the same profile?

We also notice that this upper threshold of the overload region is implicitly defined only for max AF power (Class 0/3). There is therefore no dependency of the upper threshold of overload region on either the voltage or the class of the PD. That is indeed surprising and seems to amount to a notable omission in the AF standard.

The upper threshold of the short-circuit region (0.45 A) is also not dependent on voltage or class, but that is understandable since the purpose of that threshold is to primarily protect the cable from excessive heating and that depends only on I^2R, not on voltage or class.

NOTE *Most PSE chip vendors went ahead and typically created overload limits based on class anyway, but in reality, these were market-driven and outside the AF standard's stated requirements. PSE chips from Linear Technology for example, even offered the option of selecting either a class-dependent overload current limit (I_{CUT}) or a Class 0/3-based overload current limit. Both options would comply with the AF standard, or rather, the first option would simply not get tested for, by any AF test suite.*

NOTE *The rationale behind derating the continuous current with respect to voltage is explained as follows. The PoE standards are essentially based on power, power being a product of voltage and current. So for example, if we raise the port voltage higher, we need to correspondingly derate the maximum operating current. A 30 W (AT) application, with an I_{CONT} of 600 mA at 50 V, corresponds to a lowered maximum I_{CONT} of 30/57 = 0.53 A at 57 V. In other words, I_{CONT} is voltage-dependent.*

Why was this so defined? Because that is what happens in reality based on the fact that the "last carriage of the PoE train" is a DC-DC converter with a regulated output rail. Let's take a numerical example to explain this. Suppose the input (line) is 50 V and the output rail of the DC-DC converter is 5 V, and that it powers a board which draws a *maximum of 5 A. Assume for simplicity, that the DC-DC converter is 100 percent efficient. So it asks for 25 W from the line too. Now suppose we raise the input voltage to 57 V. How much power does the DC-DC converter now demand? Well, its output is still 5 V and the board it powers still draws 5 A. So it still asks for only 25 W. In other words, the power it draws from the line at low-or high-input voltages remains the same. Therefore $V \times I$ is a constant: if V goes up, I decreases. The standard was written accordingly. But it overlooked the effect of the cable resistance actually. Think!*

Testing PSE's Overload and Short-Circuit Protection as Per AF Standard

In an I_{CUT} test setup, the load on the port is first increased slowly till port shutdown occurs. The measured port current is clearly the actual (set) I_{CUT} of the device (chip) on hand. To comply with the standard, that should be above I_{CONT} and below 400 mA. Once the actual I_{CUT} is known, an overload slightly greater than I_{CUT} is applied for 50 ms. The port should *not* get shut down, indicating the PSE has the required capability to support normal bursts of transient power. Then the same overload is applied for 75 ms and the port *should* shut down to stay compliant.

How do we test I_{LIM}? The AF standard, *at least at some places*, suggests that a short-circuit condition (i.e., where I_{LIM} is encountered) is defined as the maximum current that can be drawn before the port voltage collapses below 30 V. It is implied that if the port voltage falls below 30 V, the PSE will turn OFF the port anyway (by undervoltage monitoring and protection). So the worst-case load is just *before* that happens. To test that we can take the lower threshold of the current limit i.e., 400 mA. Then, assuming the port voltage is just a little above 30 V, the required short-circuit load resistance is $V/I = 30/0.4 = 75\ \Omega$. So we can apply this load for 50 ms to check I_{LIM}. The PSE is expected to *not* shut down the port under this condition. Then, with 450 mA current and 30 V we get $V/I = 30/0.45 = 67\ \Omega$. We can apply this load for 75 ms and we *must* find the port shut down to be compliant.

But to keep it simple, AF compliance test suites usually test short-circuit behavior as follows: A current load in excess of I_{LIM} max, that is, greater than 450 mA (~500 mA) is applied across the port. The test equipment typically expects to see that in a compliant PSE, the port will stay ON for between 50 to 75 ms. This can also be monitored on an oscilloscope. Instead of a constant current load, we can use a resistor. Note that at 30 V, 500 mA requires a resistor of $V/I = 30/0.5 = 60\ \Omega$. This led to the well-known 60 Ω resistor test for AF I_{LIM} compliance.

Later, for a short time during the development of test-equipment hardware for AT compliance, two AF test modules were paralleled with special software to be able to test to higher currents. In that case, the I_{LIM} test became two 60 Ω resistors in parallel, or the 30 Ω resistor test for AT I_{LIM} compliance. This was however too stringent and is no longer required, as explained shortly below.

The Short-Circuit Enigma of the AF (and AT) Standard

It is often said that the maximum short-circuit as per the AF standard is when port voltage droops to 30 V. This is the number indicated in Figure 33C.6, page 97, and also in Figure 33C.4, page 95 in the AF standard.

Does that make complete sense? Consider this: The voltage at the PSE output during current limiting is $I_{LIM} \times R_{OUT}$, where R_{OUT} is the output (load) resistance and I_{LIM} the actual set current limit of the PSE/DUT (it can be anywhere between 0.4 to 0.45 A as per the AF standard). As a side note, observe that Figure 33C.7, page 98 of the AF standard seems to erroneously say that R_{OUT} can be as low as 1 Ω. That would make the output voltage as low as 0.4 A × 1 Ω = 0.4 V during current limiting! However, we also know that no PD is expected to work below 30 V anyway, as per Table 33-12, Line Item 8, page 51 of the AF standard. So it seems clear that Figure 33C.7 is wrong. In reality, it seems this 1 Ω should have been in series with the 33-V zener (and across Rmax, the resistor it apparently erroneously bypassed). So we can ignore Figure 33C.7 and conclude the lowest port voltage really is 30 V. But is that value consistent throughout the AF standard?

We suddenly notice bullet (b) in Section 33.2.8.8, Page 44, which states:

> The power shall be removed from the PI within T_{LIM}, as specified in Table 33-5, under the following conditions:.... Max value applies for any DC input voltage up to the maximum voltage as specified in item 1 of Table 33-5.

Line Item 1 of Table 33-5, page 42 states the applicable voltage range is 44 to 57 V. In fact, the very use of, or reference to, Table 33-5, in the context of power removal due to shorts, implies that the standard requires a PSE to support (maintain the port for the specified duration before shutting it down) only those short-circuits that do not cause the port voltage to fall out of its normal operating voltage range of 44 to 57 V. Therefore, the above quoted statement seems to say that the port can be shut down *right away* (not even awaiting the expiry of the T_{LIM} timer of 50 to 75 ms) if V_{PORT} falls outside its regular range of 44 to 57 V. Finally we ask, is the maximum short-circuit current that the PSE needs to support (for a certain duration of time), one that drags the port voltage down slightly above 30 V, or 44 V? The AF standard is actually inconsistent in this regard.

On this basis, it is often argued that in effect, as per the AF standard, any T_{LIM} requirements when V_{PORT} is outside the range 44 to 57 V (as in an extreme short-circuit condition) are actually not valid or applicable. In other words, the 30 V port voltage droop short-circuit does *not* really need to be supported (for 50 ms, or even *less*). The PSE is *allowed* to shut down the port *immediately* if the port voltage falls below 44 V. Some PSE vendors, whose integrated-FET chip architectures were apparently prone to turning off a bit too quickly under a short-circuit condition (thermal limitations), happily took refuge in the page 44 statement represented previously. But look at it this way: If the port voltage was 44 V to start with, and if there was current limiting, the port voltage would *definitely* have fallen below 44 V, as the pass-FET took on voltage across itself in a current-limiting state. In other words, if we accept that the port can be shutdown immediately just as the port voltage falls below 44 V (instead of below 30 V), then the short-circuit definition itself becomes invalid—because as explained above, if the "48V" rail happens to be a just a little above 44 V, then a short-circuit, as defined by the AF standard (PSE pass-FET in current limiting state), would clearly cause the port voltage to fall below 44 V at which point we can then presumably shut the port down immediately. But that would imply that a short-circuit need not be supported for any amount of time at all (before shutting down the port). So, the AF standard is really quite ambiguous and self-contradictory in this regard. Perhaps the safest design assumption, to avoid failing compliance test suites, is to use the 30 V threshold for short-circuits, not the 44 V threshold. We will soon see that the AT standard unfortunately also does not resolve this issue entirely.

The AT standard actually went ahead and ratified this AF loophole. So, sustaining a 30 V droop for any length of time is not an AT requirement. But what about sustaining 44 or 50 V? Therefore people have likewise argued there is really no short-circuit specification per se in the AT standard—we can turn the port OFF the moment the pass-FET starts current limiting. For AF most people still say 30 V port voltage needs to be supported in short-circuit.

Keep in mind there are still some OEMs who insist on AF compliance and, indirectly at least, ask for this 30 V short-circuit condition (to be supported for 50 ms). It is safer to design the PSE accordingly.

Device Dependency in the AF Standard

Consider the fact that in the AF standard, as per Table 33-5, page 42, I_{CUT} is expressed as a *range* extending from P_{CLASS}/V_{PORT} to 0.4 A. But, looking carefully also at Figure 33C.6, page 97 of the AF standard, we realize that what is really being implied by stating it in this manner is that the systems/IC designer needs to target a nominal value for I_{CUT} centered somewhere roughly in the middle of this range (say 375 mA), and also ensure that the min and max values of this parameter I_{CUT} are always within the prescribed limits—P_{CLASS}/V_{PORT} to 0.4 A, all tolerances, drifts, and so on considered. The understanding is that whenever the actual set threshold of a given PSE device is exceeded, its internal comparator is triggered and its timer activated. Looking closely at Section 33.2.8.7, page 44 and Figure 33C.6, page 97 of the AF standard, we see that if the port current exceeds the set overload threshold of that device for longer than 50 ms, the port is required to be turned off, and certainly should be, before we get up to 75 ms (but *not* earlier than 50 ms either). Let us take two specific numerical examples to understand the implications of this situation, and the strange situation resulting:

1. Suppose we have a device operating at 44 V and at 350 mA with an I_{CUT} that happens to be 370 mA for that particular PSE device. Then, as per the AF standard, we are required to turn off the port completely if the current is say, 380 mA, for more than 50 ms.
2. But now suppose that from another production lot, the device we are using/testing has a set I_{CUT} of 390 mA. Now according to the AF standard, we could actually continue to operate at 380 mA indefinitely.

We can therefore rightfully ask why in the second case, 380 mA was somehow considered completely safe and acceptable to pass indefinitely through the cabling infrastructure, whereas in the first case it was arbitrarily deemed not to be so? This is obviously an incongruous result with a technically unjustifiable expectation—an example of the *implicit device dependency* in the AF standard.

Consider I_{LIM} too. For example, as per the AF standard, the designer needs to ensure the min and max of the set current-limit threshold are always within the limits 0.4 to 0.45 A, the latter being the max of I_{LIM} as per Table 33-5, page 42 of the AF standard. A chip designer will typically set the current limit to 425 mA nominal. But an actual PSE chip under test may have a current limit of 410 mA, while another could be at 440 mA. So there is inherent device dependency in the I_{LIM} response too.

The AT standard solved this inherent device dependency by introducing a *maybe region,* as we will soon see. The chip designer now just needs to put the current limit in the middle of this region and the device dependency then goes out of the picture automatically.

Evolution of Overload/Short-Circuit Perspective

Looking closely at the standards, we will see that terms like I_{CUT}, I_{OVLD}, I_{PEAK}, I_{LIM} and so on, are used sometimes interchangeably, and sometimes rather differently too in subtle respects. Same applies to their associated times like T_{CUT}, T_{OVLD}, T_{LIM}, and so on. It *can* get rather confusing at times. To simplify our basic understanding, we must first note, that though the precise value and meaning of some of these terms may have been changing somewhat over past/present and evolving standards, we can make some commonsense and unifying observations. Besides, there are some commonalties too.

In the AF standard, the normal continuous operating current was called I_{CONT}. In the AT standard it became I_{CON}. But in both the AF and AT standards actually, this current is written out explicitly as P_{CLASS}/V_{PORT} thus clearly taking the class of the PD into account. The latter expression is also helpful because we can clearly see that I_{CONT} depends not only on both the PD's classification level (power) but also the port voltage. Maximum continuous-operating current is *not* a fixed number. The AF and AT standards both agree on that.

Overload (I_{CUT} or I_{OVLD}) is a certain range of currents *higher* than the normal continuous-operating current I_{CONT}, but still not severe enough to be interpreted as a fault condition (which is what a short-circuit is, as discussed next). We also recognize that the primary purpose of any overload specification is to ensure that any compliant PSE possesses a certain amount of overload capability, one which is required to handle normal/typical load transients from valid/compliant-operating PDs without "nuisance tripping" (shutdown) of the port. Yes, if the overload lasts too long (more than 50 ms), the AF standard *does* call upon us to turn the port OFF completely (within 75 ms)—to protect the cabling infrastructure. This actually completely mimics the inrush behavior discussed in the previous chapter and so PSE-chip designs typically reuse the inrush circuitry and blocks.

The AT standard, though likewise demanding a minimum 50 ms time in an overload condition for handling typical PD transients, makes port shutdown after completion of that 50- to 75-ms time optional. So only the minimum-overload time needs to be assured. Other than that, the overload is actually allowed to last forever as per the standard. Which raises another key question: Why should we even restrict the continuous current to a lower level if a higher current can (safely) last forever? We will discuss this further very soon, but this apparently fundamental difference between the AF and AT overload standards can be traced back to the fact that in the AF and AT standards, though the lower limit of the overload region is defined identically (I_{CONT}), the *upper* limit of the overload region is defined very differently. In AF, the upper limit (of the overload range) was a fixed current value of 0.4 A, irrespective of port voltage. It was realized that despite sounding simpler, simple is not always the best. The AF approach was not very rational in a technical sense as discussed previously too. Therefore, in the AT standard, the max of overload was defined more correctly as a certain (though somewhat arbitrary) percentage of *power* capability above and beyond the normal smooth continuous-operating power-level I_{CONT}. That makes it automatically voltage-and class-dependent too.

The (voltage-dependent) upper limit of the overload region in the AT standard is often called "I_{PEAK}." It is the peak current that a normal, healthy PD may demand, and the PSE *will* (must) support. To know this boundary (the value of I_{PEAK}), we have to use a rather complicated equation, one that involves port voltage and cable resistance. (See 33.2.7.4 on page 47 of the AT standard.):

$$I_{PEAK} = \frac{V_{PSE} - \sqrt{V_{PSE}^2 - (4 \times R_{CHAN} \times P_{PEAK\_PD})}}{2 \times R_{CHAN}}$$

Where R_{CHAN} (also called R_{CH} in the AT standard) is the channel-loop resistance. Its value is 12.5 Ω/20 Ω for Type 2/Type 1 for a 100 m long cable for example.

Short-Circuit Range Comparison (AF and AT)

There is a current threshold higher than I_{CUT} in which active/immediate intervention is called for by both AF and AT. This "intervention," however, is implicitly *multi*-level. It does not necessarily mean the port is shut down right away. It means:

1. The current is, first not allowed to go any higher. So the PSE's pass-FET takes on voltage across itself in an effort to restrain the increase of current. The port voltage droops.
2. If the current lasts longer than x ms, the port *must* be shut down to protect the cabling and/or the PD/PSE.
3. If at any time, even before the timer mentioned previously times out, the port voltage sags below y volts, the port is *allowed* to be shut down immediately to protect the PSE.

The placeholders x and y need to be defined for the AT standard in the next section. We should already know what these are for the AF standard.

General Philosophy in Interpreting the AT Standard

We will try *not* to get too pedantic about what a given region or term should or should not be *called*. Because, even in the AT standard, there is some confusion on the regions, as discussed shortly. Instead, we will focus on *what* the AT standard seems to say or imply in terms of recommended and/or mandated PSE *behavior*.

In general, we should always attempt to resolve any behavioral ambiguities on a commonsense basis if necessary, keeping uppermost in our minds the basic engineering intent of the standards.

Expressed from the viewpoint of the PSE, this basic intent is summarized as follows:

1. To protect non-PoE devices and/or terminations from damage (typically due to higher voltage than they are rated).
2. To deliver as much power to a valid PD as possible (but only when asked for or negotiated).
3. To do so without causing any short- or long-term damage to the cable infrastructure (for example, from overheating caused by long-term current overloads).
4. To also be able to cope with any *reasonable but temporary* increase in the PD's (peak) power requirement, without nuisance tripping of the ports, for example.
5. To enhance reliability of the PSE as much as possible, especially under what are obviously "abnormal" conditions—in other words, those not related to "normal" PD behavior. This would usually call for turning off the port altogether unless mandated by number 4.

Overload and Short-Circuit Requirements as Per AT Standard

The key figure to reference for overload and short-circuit requirements in the AT standard is Figure 33-14 on page 49. This is called the Operating Current Template or the Overload Template. In Fig. 6.2, however, we have not simply reproduced Figure 33-14, but added a more information on top of it to help us understand how the AF standard changed over to the AT standard, and how we can comply with both at the same time.

Here are several observations from Fig. 6.2:

1. In general, AT Type 1 requirements are broader or more relaxed than AF requirements. So a PSE that complies with AF standards will almost certainly comply with AT Type 1 requirements. Not vice versa though. We can see that barring the first 10 μs, this is true up to 8.2 ms. The AT standard is generally more liberal. This is actually opposite to what we saw in Chap. 5 in regards to inrush, in which the AT standard was found to be more restrictive up to the first 2 ms (and identical thereafter).
2. The AT standard allows 8.2 ms to return to the settling value of I_{LIM} max (1.75 A), whereas the AF standard demands a return to 0.45 A in 2 ms. But note that there is some ambiguity in the AF standard itself as it allows (in fact, recommends) 1 ms to initialize our current-sense circuitry before we actually

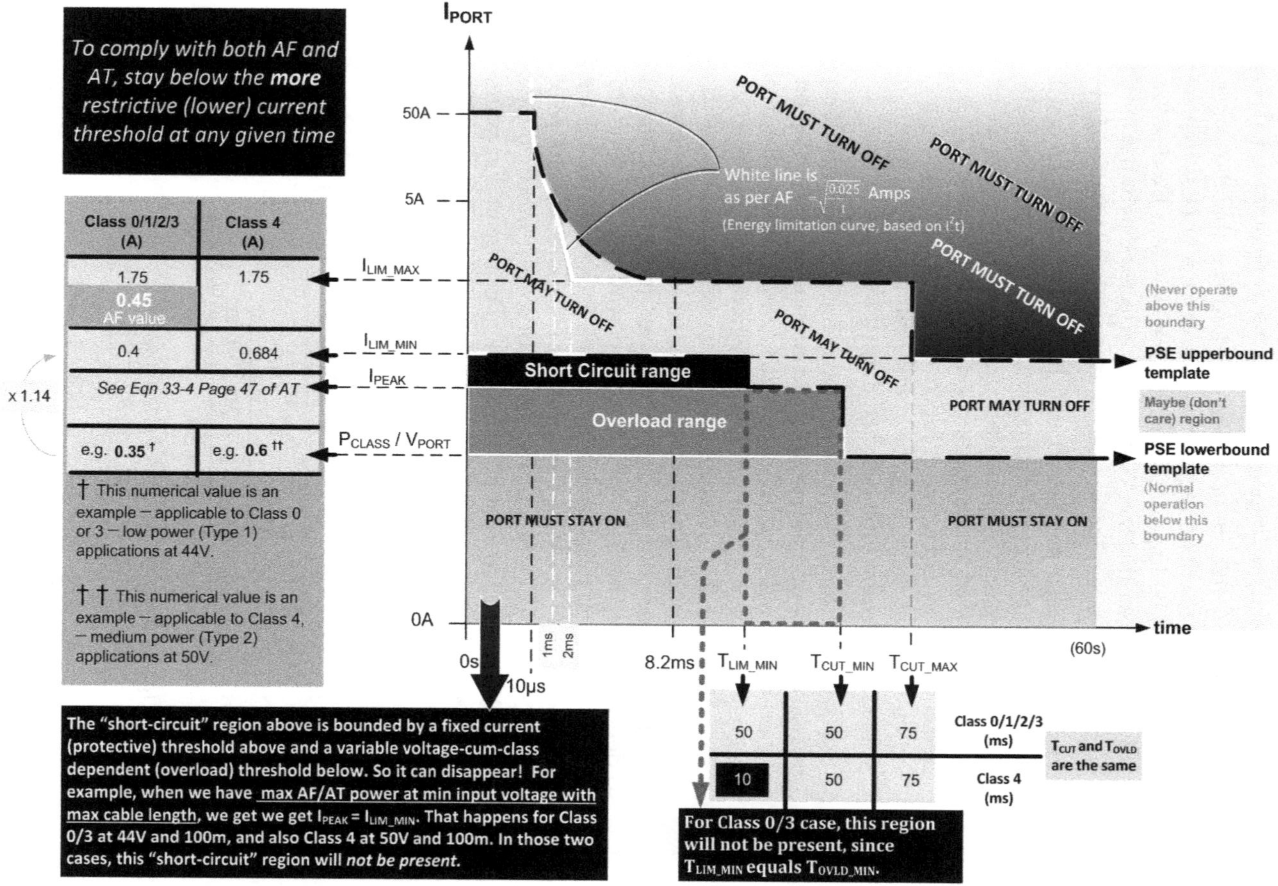

FIGURE 6.2 The PSE operating template as per IEEE 802.3at, with IEEE802.3af limits superimposed.

start measuring the port current. The AF standard also does not limit the max current to 50 A for less than 10 μs as does the AT standard. However, that may be only on paper, as we will in any case take readings only after 1 ms as recommended.

3. One prevalent criticism of the AF standard was that the upper bound of the overload region was not voltage or class dependent as could be logically expected, but was of a fixed value 0.4 A. We can see that the AT standard corrected that by defining I_{PEAK}, which is basically the peak of the permissible overload.
4. However, as also mentioned, the highest region, where we finally seek to protect the cabling infrastructure from overheating, should logically depend only on current, not on voltage or class. The AT standard accordingly defines I_{LIM_MIN} and I_{LIM_MAX}, and the region in between is functionally the current-limiting or short-circuit region. A typical PSE will center the nominal value of its current limit between these thresholds.
5. But we now have a variable voltage-and class-dependent region (overload region) with a fixed current-dependent region (short-circuit region) a little above it. Clearly, between the two there must be a buffer/transition region. And there is, as shown in Fig. 6.2. This region can also rather obviously, stretch or shrink depending on voltage and class—because its upper bound is fixed, but the lower one can move up or down depending on voltage and class. The way the AT standard defined things, this region "disappears" for two cases: if we have Class 0/3 at lowest input voltage (44 V) through 100 m of (CAT3) cable, or if we have Class 4 at lowest input voltage (50 V) through 100 m of (CAT5e) cable. Because at that point, I_{PEAK} equals I_{LIM_MIN}.

   Let us check this out.

   For Class 0/3 at 44 V and 20 Ω, we get

$$I_{PEAK} = \frac{V_{PSE} - \sqrt{V_{PSE}^2 - (4 \times R_{CHAN} \times P_{PEAK\_PD})}}{2 \times R_{CHAN}}$$

$$= \frac{44 - \sqrt{44^2 - (4 \times 20 \times 14.4)}}{2 \times 20} = 0.4 \text{ A}$$

   For Class 4 at 50 V and 12.5 Ω, we get

$$I_{PEAK} = \frac{V_{PSE} - \sqrt{V_{PSE} - (4 \times R_{CHAN} \times P_{PEAK\_PD})}}{2 \times R_{CHAN}}$$

$$= \frac{50 - \sqrt{50^2 - (4 \times 12.5 \times 28.33)}}{2 \times 12.5} = 0.683 \text{ A}$$

Note that above, we have taken the peak power of the Class 0/3 PD as 14.4 W based on Table 33-18, page 62 of the AT standard. The table also states that for Class 4 devices we need to take peak PD power to be 11 percent higher than the continuous power rating. That's how we got $1.11 \times 25.5\text{ W} = 28.33\text{ W}$. We thus see that the stretch region will be nonexistent under the above two cases.

6. In Figure 33-14 on page 49 of the AT standard, *oddly*, this potentially disappearing "stretch" region is called the "short-circuit range." That is inconsistent with the statement in Section 33.2.7.7 on page 48, titled "Output current at short circuit condition":

   > A PSE may remove power from the PI if the PI current meets or exceeds the "PSE lowerbound template" in Figure 33-14. Power shall be removed from the PI of a PSE before the PI current exceeds the "PSE upperbound template" in Figure 33-14.

   The above statement clearly suggests, in conjunction with its title, that it considers the region *above* the lowerbound template to be the short-circuit region. Yet, Figure 33-14 calls the region *below* the lowerbound template as the short-circuit range. That really cannot be so, because (a) this region *can* disappear, in which case would we really say that short-circuit protection is no longer present? And (b) logically, it is the region around the set current limit (i.e., between I_{LIM_MIN} and I_{LIM_MAX} thresholds) which should be considered the short-circuit region. This raises an entirely new question: What should we call this "buffer/stretch" region?

   This discussion only goes to reiterate the futility of trying to get too pedantic about exact names of terms, or what a certain region in the operating template of the AT standard is called. The underlying *intent* should always be kept in mind, as summarized previously. Otherwise even the AT standard will be considered very puzzling at places.

7. In the same vein, we point out that there seems to be an error in the max value of the horizontal scale of Figure 33-14, which states 60 s. In reality, this should perhaps be 1 s, since at several places in the standard, it is indicated that a sliding window of 1 s be used for looking at the currents. So, for example, we feel this template is valid for only one (sliding) second. Which also answers the question we had asked earlier: If the overload current can last indefinitely as Figure 33-14 seems to suggest, what is the difference between an overload current and a continuous current anyway? In fact, the AT standard also states in Section 33.3.7.4, titled "Peak operating power":

> .... the peak power shall not exceed P_{Class_PD} max for more than T_{CUT} min, as defined in Table 33-11 and 5 percent duty cycle. Peak operating power shall not exceed P_{Peak} max..... *NOTE*: The duty cycle of the peak current is calculated using any sliding window with a width of 1 s.

This section makes it clear that a compliant PD's *overload current cannot last for more than 50 ms every one second* (duty cycle is then 0.05 s/1 s = 5 percent). Which also implies that *the entire template in Figure 33-14 is actually valid for 1 s*, not 60 s as stated in Figure 33-14. Going forward, in our figures, we will omit the marker "60 s" altogether, and also ignore any sliding window considerations because of this ambiguity.

8. Looking at some specific details: we see from Fig. 6.2 that there is a "maybe" or "don't care" region: where it is *optional* whether the PSE stays ON or OFF. This is bounded by two lines: the upperbound template and the lowerbound template. Below the lowerbound template, the PSE must be ON. Whereas, above the upperbound template it must *never* be ON (not even momentarily). These clearly demarcated regions effectively removed the "device dependency" of the AF standard that we spoke of previously.
9. We see that any current over I_{LIM_MIN} *must* be terminated by 75 ms for sure. Because at 75 ms, the upperbound template is encountered, and we know we must never encroach above it. Let us break this up carefully: 1) For Type 1, a current of 0.399 A (just below 0.4 A), for example, must be sustained for at least 50 ms, but is also allowed to last indefinitely (that's a *possible* interpretation of the AT operating template, which allows more current down the line than the supposed max of 0.35 A). Whereas a current of 0.401 A (little above 0.4 A) must be terminated by the 75 ms mark. *But it can also be terminated immediately* (not after 50 ms), because it is in the "don't care" region of the AT template. That's a big difference from the AF standard where 0.401 A would have typically lasted at least 50 ms. 2) Similarly, for Type 2, a current of 0.682 A for example, must be sustained for at least 10 ms, but is also *allowed to last indefinitely*. Whereas a current of 0.685 A must be terminated within 75 ms, *but can also be terminated immediately if we so desire.*

NOTE *In the AT standard, certain equations/tables/figures come up with Peak power values for Type 2 at 50 V (or I_{LIM_MIN}) as 0.683 A, or 0.684 A, or 0.685 A. It is not very consistent, so beware.*

10. It is interesting that I_{LIM_MAX} is 1.75 A for *both* Type 1 and Type 2 in the AT operating template. Logically, this seems to be an

oversight. For one, it is not backward compatible with the AF standard *in a big way*. In other words, we cannot hope to achieve AF compliance if we design our PSE in that generous manner. We also have to keep in mind that Type 1 applications are still assumed to be over CAT3 cable, *even in the AT standard*, and since the cable resistance is so much higher in that case, the maximum current through it should also be lower than for a Type 2 application over CAT5e.

Therefore, it is best to restrict ourselves to an I_{LIM_MAX} *of 0.45 A for Type 1 applications*, because that level is *more restrictive*, and if we pass that, we will automatically comply with *both* the AF and AT standards on this count.

11. There was also a lingering question in the AF standard about what really was the difference between the T_{CUT} timer and the T_{LIM} timer: Did we really need an I_{LIM} timer if it had exactly the same value of 50 to 75 ms? The I_{CUT} timer would always timeout first since I_{CUT} timer always started just a little before the I_{LIM} timer. In the AT standard we see that is partially resolved since the T_{LIM_MIN} threshold is only 10 ms for Type 2 devices.

12. But what is the corresponding load current? Looking closely at Fig. 6.2, we see that it is 400 mA for Type 1 (Class 0/3) and 684 mA for Type 3 (Class 4). In other words, 400 mA needs to be sustained for at least 50 ms as per the AT standard, ***for any class*** (Class 0/1/2/3, or Type 1), whereas 684 mA needs to be sustained for only 10 ms as per the AT standard, ***for Class 4*** (Type 2). This lowered (10 ms) duration really helped in the design of high-power PSEs, with integrated FETs in particular.

13. In the AF standard, whenever the current crosses the set I_{CUT} of the PSE chip, the T_{CUT} timer starts. However when the current drops below T_{CUT}, the T_{CUT} timer gets reset. The next overload event it starts counting from zero again. Therefore there could be a situation where several smaller-duration overloads (say 49 ms wide) occurred in quick succession, causing the PSE chip to heat up quickly, but the port would *not* be definitely turned OFF, since the overload needs to be greater than 50 ms (though less than 75 ms), to cause the PSE to turn the port OFF. But now, with a preheated chip, it is increasingly likely that the chip will not be able to support another > 50 ms overload test pulse, and would shut down "prematurely," causing conformance failure.

    The AT standard allows the T_{CUT} timer to *not* get reset every time the current falls below I_{CUT}. T_{CUT} timer thus becomes a *cumulative timer* of overloads, summed over the *preceding second* always. In effect, the timer *retains a memory* of previous, not-too-distant, overloads. So, when the next overload comes along, the T_{CUT} timer will run out more quickly: yes, though it is still counting up to 50 to 75 ms, it starts at a nonzero value now. Thus compliance testing can be more easily passed.

To enable this feature, the AT standard writes in Section 33.2.7.6, titled "Overload current":

"....If I_{PORT}, the current supplied by the PSE to the PI, exceeds I_{CUT} for longer than T_{CUT}, the PSE may *remove power from the PI. The cumulative duration of T_{CUT} is measured with a sliding window of at least 1 second width."*

First: note the "may" above. As we have seen, the AT standard allows the overload to continue indefinitely, though perhaps only with a maximum duty cycle of 5 percent every 1 second window as discussed previously. This is different from the AF standard, which asks for the port to be definitely shut down (due to an excessive overload).

Second: the early draft versions of Section 33.2.7.6 above did not have the italicized phrase: *at least* 1 second width. Its subsequent inclusion however appears to be not well thought through either. Perhaps it is best to just consider the overload timer as being *cumulative over the past one second.*

Peak Power Calculations

Peak power and overload are better defined by the AT standard. We need to understand this issue further now.

First, from the viewpoint of the PD: The peak value expresses a certain overload percentage margin above I_{CONT}. It is also based on fairly *arbitrary* percentages. For Type 2, it allows the PD to demand 11 percent higher power than its continuous value—taking it from a continuous value of 25.5 W to a peak of 28.3 W (temporarily). This value was used in a preceding bullet. For Type 1, Class 0/3, this *overload margin* is set somewhat arbitrarily at 11.2 percent, taking it from a continuous value of 12.95 W to a peak of 14.4 W. The AT standard rounded up 12.95 to 13 W. Based on that, the overload margin is slightly less: 10.8 percent. Note that for Class 1 (Type 1), the allowed margin is 25 percent, thus allowing the PD to go from a continuous value of 4 W to a peak of 5 W. For Class 2 (Type 1), 19.4 percent margin was allowed, taking the PD up from a continuous value of 7 W to a peak of 8.36 W. As per Fig. 6.2 (the operating template), the PSE must allow for this overload (not amounting to a short circuit) *for at least* 50 ms. But beyond that duration, we are free to decide what to do. We can indefinitely sustain the overload current (which is as high as I_{PEAK}), and that is still considered safe for the cabling infrastructure (ignoring sliding windows).

This actually throws up the possibility of a custom PSE-PD operating at much higher power indefinitely, while still complying with the standards. There may even be some "overload margin" (headroom) available on top of that, as explained further in our discussion.

From the viewpoint of the PSE: The full calculation of overload margin is actually very involved because of the intervening 100 m of cable length. And for that reason too, an x percent overload margin at the PD end in terms of power, certainly *does not* translate to an x percent overload margin at the PSE end in terms of power, nor an x percent increase in terms of current. To meet the PD's increased power requirement, we need to increase the current, *more than proportionately*, to offset the drop across the cable. In other words, the increased cable voltage drop at higher currents tends to lower the voltage at the PD end, which in turn tends to diminish the power received there even further. So, to ensure, say 11 percent overload power margin at the PD end, we actually need to increase the overload current/power capability at the PSE end (i.e., the PSE's overload margin), by more than 11 percent.

A Mathcad file was written out specifically for this purpose and its results were pasted into Table 6.1 (at end of the chapter). This table will be discussed in more detail later. Here it suffices to observe that in terms of current (at 50-V input), we had to go up from 0.6 A port current to 0.68 A, an increase of 13.3 percent. At the same input voltage, the power from the PSE side had to be increased by 13.7 percent—from 30 to 34.12 W. This directly indicates the additional power lost in the cable due to the higher current.

Note that above, we had mentioned that we could consider passing a certain level of "overload power" continuously down the cable and still remain "compliant" at the PSE end—though this would require a custom PD, since its power levels will now likely exceed the limits defined in Table 33-18 on page 62 of the AT standard. If we draw that overload current continuously, in general we will not have any margin available on top of this, for handling any transient demands coming from the PD. Therefore, in general, custom PDs drawing overload power on a continuous basis should be designed *not* to exhibit sudden peak demands (plenty of input decoupling and bulk capacitance for example); otherwise the (compliant) PSE could suddenly decide to turn the port off altogether.

A small additional overload/peak margin (in effect, an overload margin for continuous operation in the usual overload region), does in fact become available. This is the emerging stretch/buffer region shown in Fig. 6.2, which in turn depends on Figure 33-14 of the AT standard. This is the region we had argued previously, was somewhat incongruously labeled a "short-circuit" region by the AT standard. However it does exist whatever we call it. To create it, we just have to raise the port voltage by a few volts above minimum port voltage (44 V for Type 1 or 50 V for Type 2). However, to operate continuously in the usual overload region, and to use this emerging region as the overload region for that new higher-power operating mode, we should not derate the port current with port voltage as is commonly done in PoE. Of course we may need to set I_{CUT} higher too. But in this general manner, we can successfully and safely operate continuously at higher powers than the standard specifically allows, though at higher input voltages. We achieve all this with a proprietary (custom) PSE-PD combination, but we can still remain naturally compliant, at least at the PSE end, without further effort. In addition, the proprietary higher-power PD can be designed to appear compliant too, by discovering when it is not connected to its proprietary higher-power PSE partner. Conversely, the higher-power mode can be subsequently established by mutual PSE-PD identification through the data link (LLDP).

Note that for Class-1 and Class-2 devices, the stretch region mentioned above is *always* present, even at the lowest operating voltage.

The Recommended Operating Templates Collected and Explained

In Figs. 6.3 to 6.5 we have finally collected the Type 1 "minimum" operating templates at different voltages: 44 V, 50 V, 57 V, based on our discussions surrounding Fig. 6.2. Figure 6.4 is based on the increasingly common port voltage setting of 50 V, which is used by most AC-DC power supply manufacturers today, to cater to both Type 1 and Type 2 PSEs (they set about 52.5 V as the nominal output, which accounts for the output-rail tolerances and also the drop across the pass-FET of the PSE). These Type 1 templates are essentially an intersection of AF and Type 1 AT requirements. They will help pass an AF compliance test more readily, besides passing an AT Type 1 compliance test.

In Figs. 6.6 and 6.7, we have collected the high-power overload templates based on the governing AT standard and the operating template of AT, and displayed these at both minimum and maximum voltages (50 V and 57 V).

In each case, we have also provided the resistance (load on PSE-side) and the associated power (PSE-side), for easy reference during validation testing. These values are also collected together in Table 6.1 for easy reference.

Looking at Fig. 6.3 for example, we see that if we are testing a Class-0, Type 1 PSE at 44 V, by applying a variable resistive load at its output, we have to sustain *126 Ω and higher* resistor values indefinitely. But any load resistor less than 126 Ω, down to 110 Ω, needs to be sustained only for a minimum of 50 ms. But we can choose to stay ON indefinitely too for load-resistor values larger than 110 Ω. If the resistor is less than 110 Ω, we are required to turn off the port before 75 ms, though we can choose to do so immediately too!

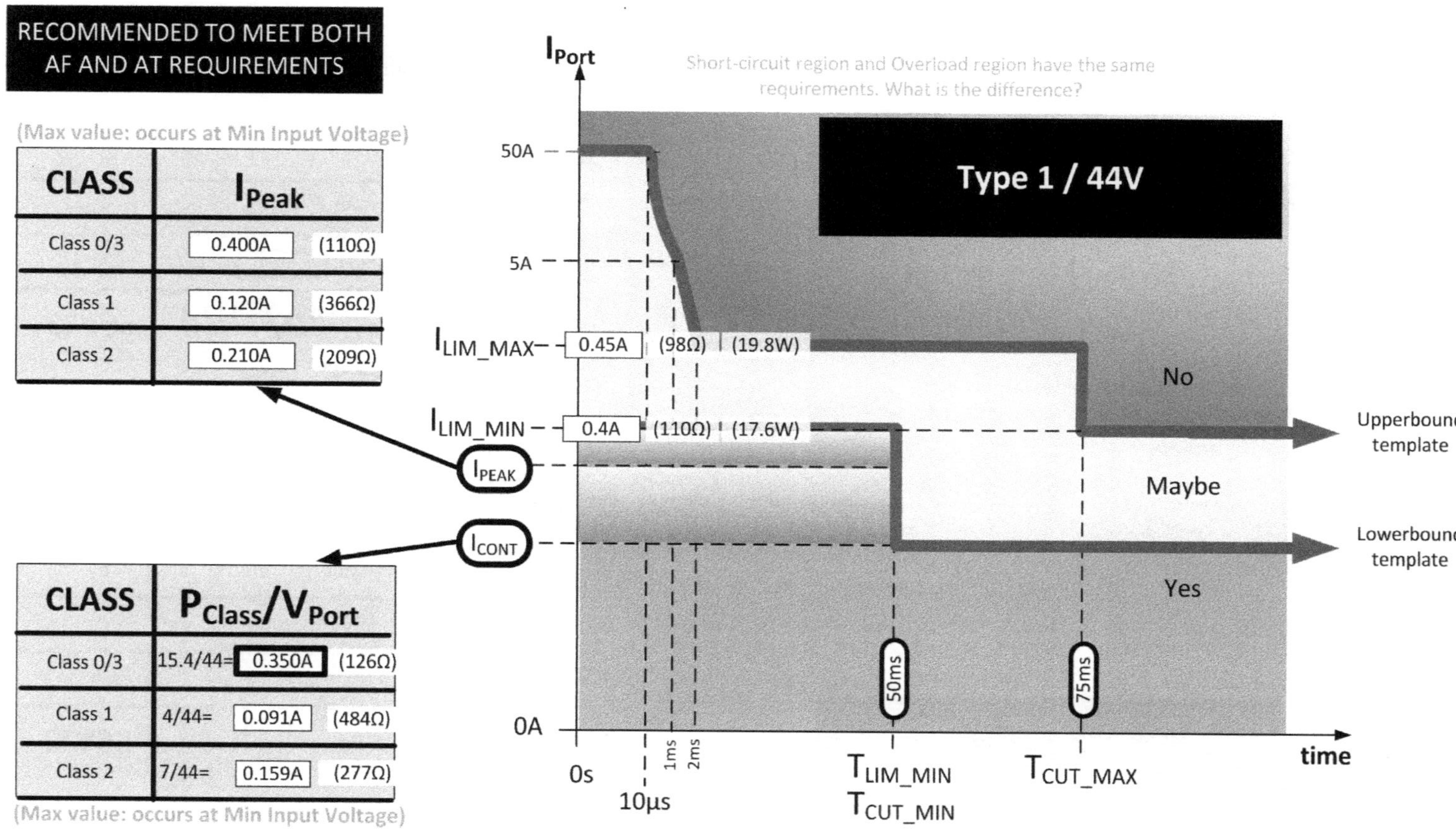

Figure 6.3 The "minimum" template (for AF and AT compliance); Type 1 at 44 V.

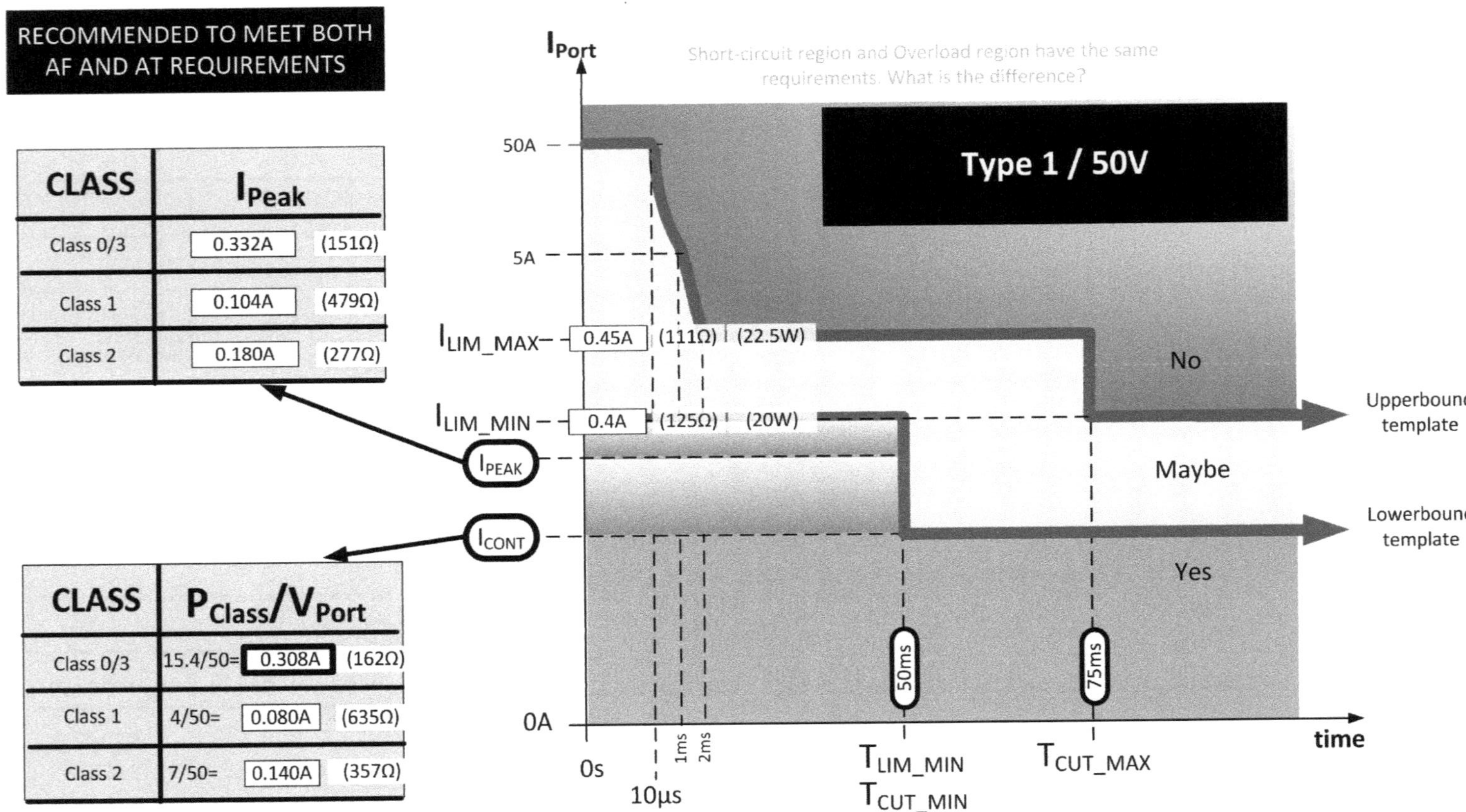

Figure 6.4 The “minimum” template (for AF and AT compliance); Type 1 at 50 V.

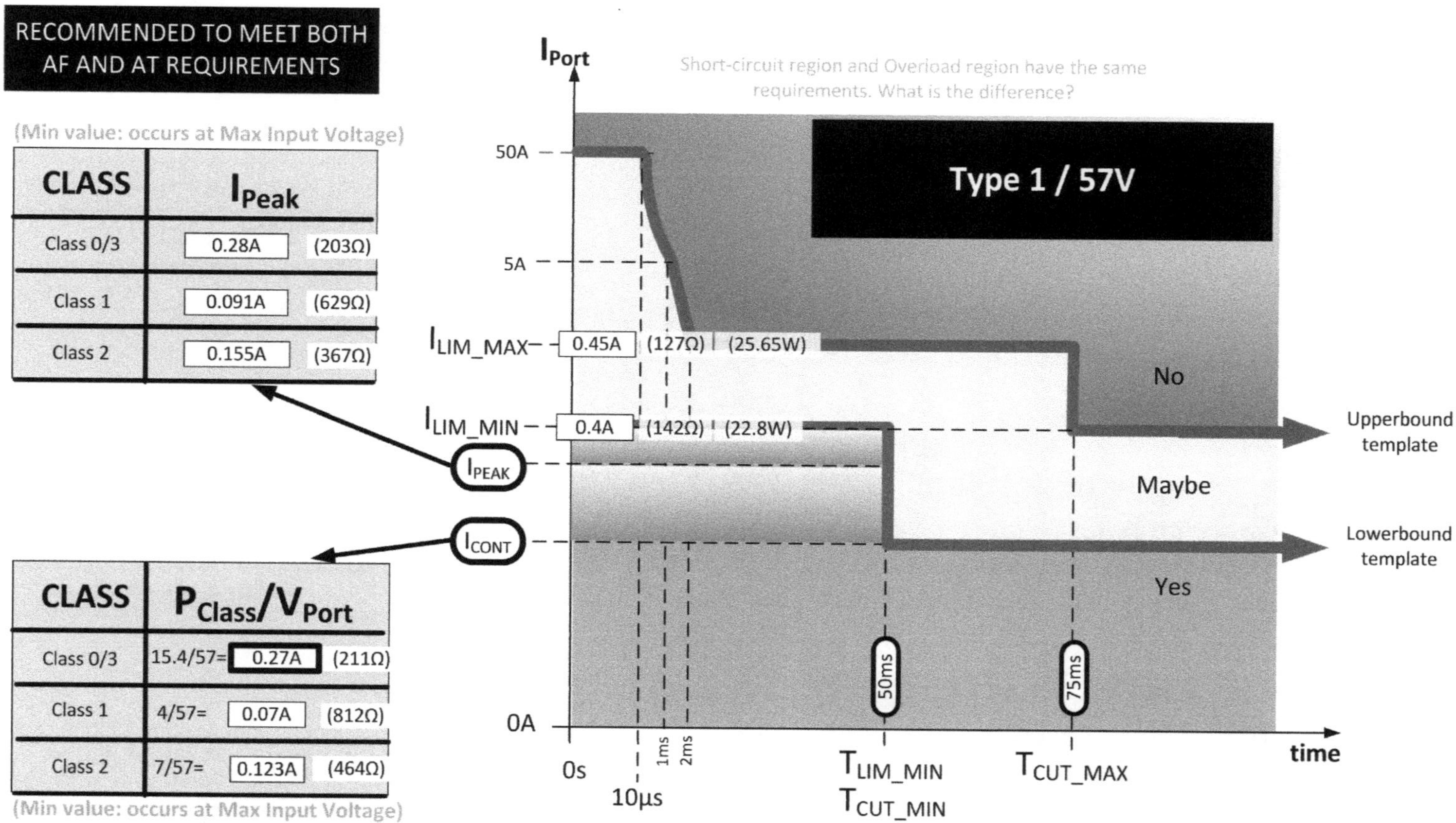

FIGURE 6.5 The “minimum” template (for AF and AT compliance); Type 1 at 57 V.

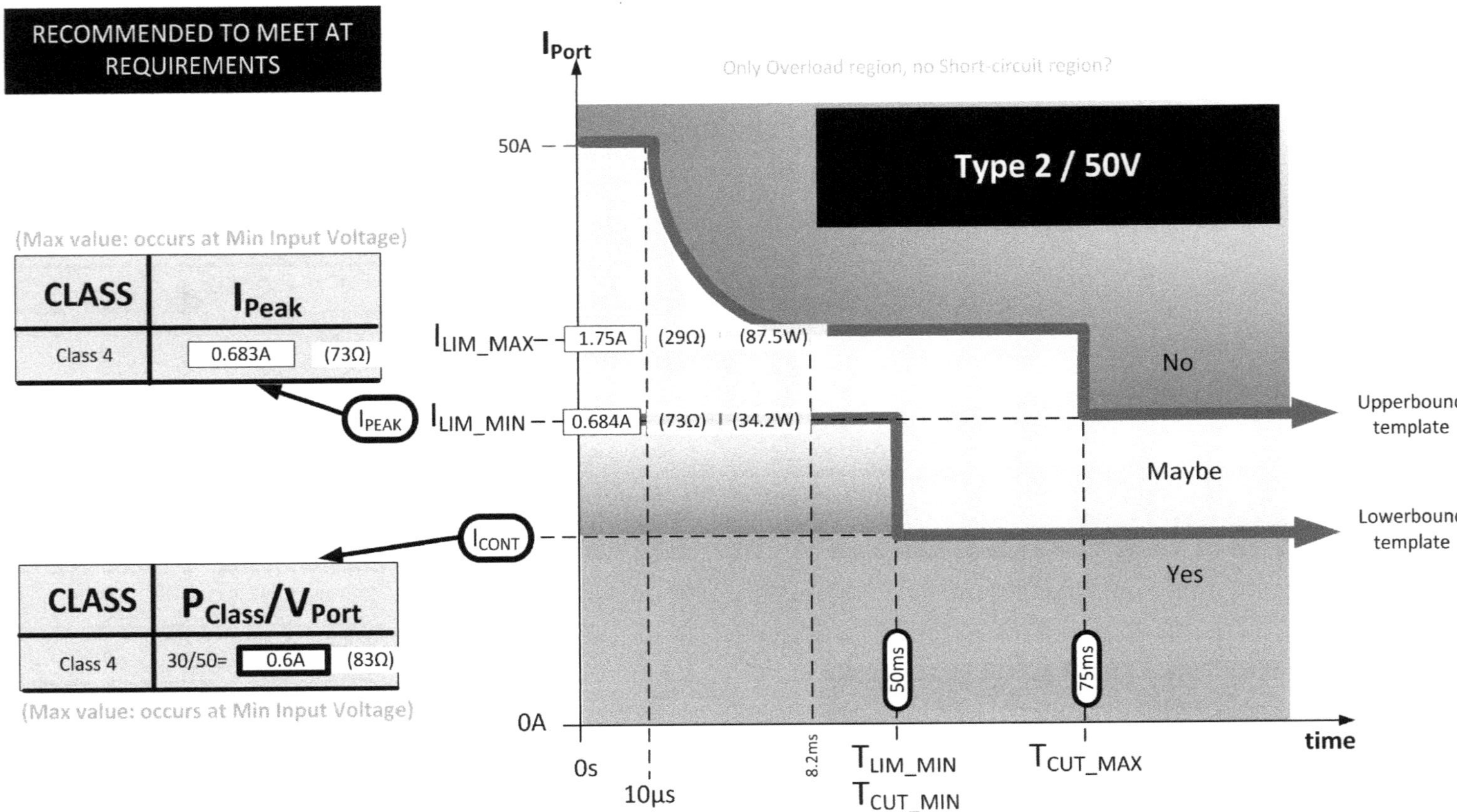

Figure 6.6 The AT template (for only AT compliance); Type 2 at 50 V.

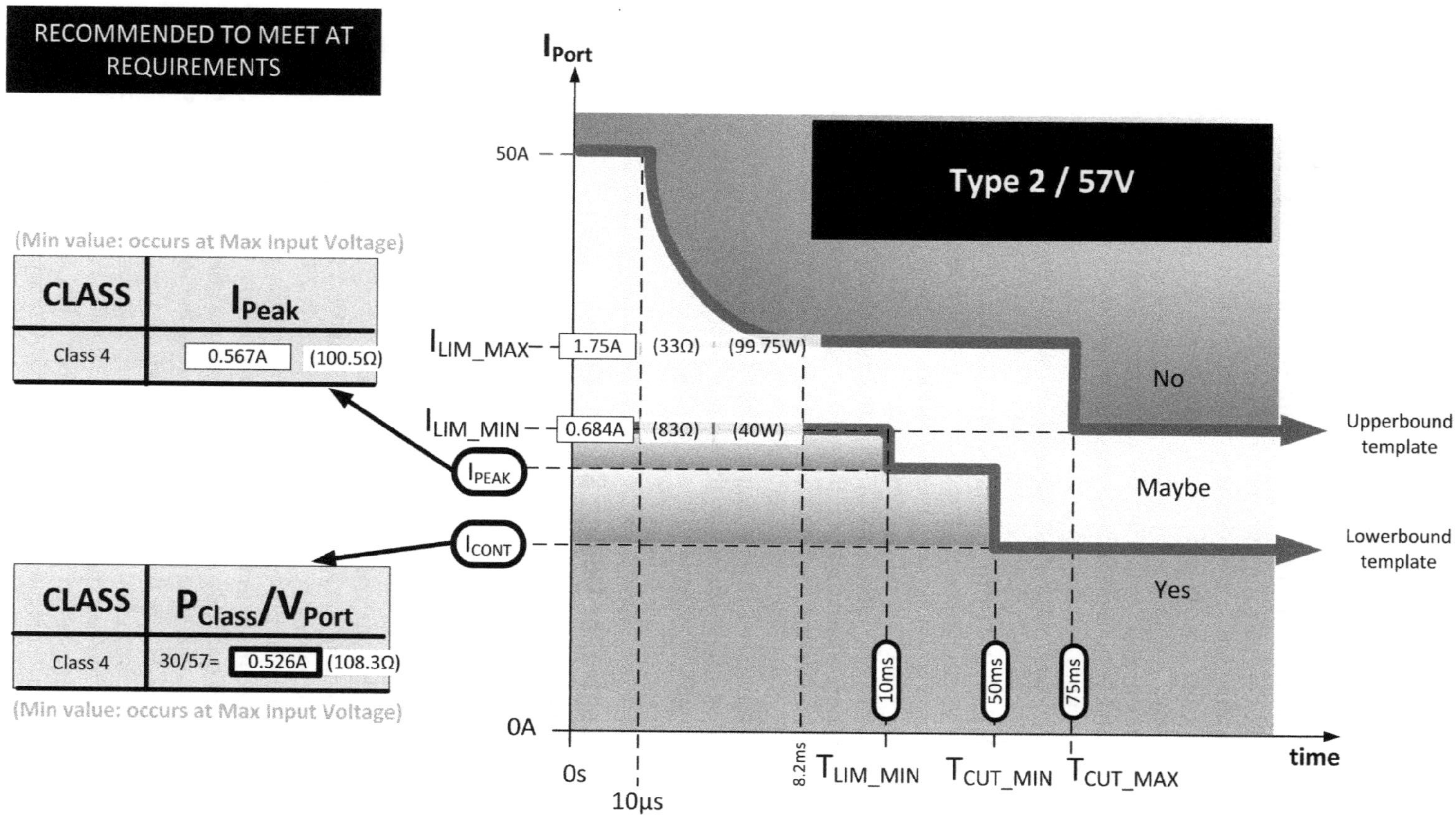

FIGURE 6.7 The AT template (for only AT compliance); Type 2 at 57 V.

Looking at Fig. 6.4 for example, we see that if we are testing a Class-0, Type 1 PSE at 50 V, by applying a variable resistive load at its output, we have to sustain 162 Ω and higher resistor values indefinitely. But any load resistor less than 162 Ω, down to 151 Ω, needs to be sustained only for a minimum of 50 ms. But we can choose to stay on indefinitely too. If the resistor is less than 151 Ω, we are required to turn off the port before 75 ms, though we can choose to do so immediately.

Looking at and comparing the high-power overload templates in Figs. 6.6 and 6.7, we see that the most stringent overload requirement across the entire voltage range is a resistor of 73 Ω applied at the lowest (50 V) input. And that overload must be sustained for 50 ms at least (not 10 ms!). Note that the value of this 50 ms overload test resistor is not 30 Ω (for medium power) nor 60 Ω (for low power) as believed by some in the past, based on the values used by some test-equipment manufacturers in early AF-to-AT transitory test equipment. The reason it is no longer even remotely valid is that the AT standard says nothing about allowing the voltage to droop to 30 V during a short-circuit. In fact if the voltage at the PSE end drops out of its normal operating region (< 44 V or < 50 V for Type 1 and Type 2, respectively), the PSE is allowed to turn the port OFF almost immediately (except for a 250 μs glitch filter as discussed later). This 30 to 60 Ω I_{LIM} test was discussed in the section "The Short-Circuit Enigma of the AF (and AT) standard."

Some PSE-Controller Design Suggestions for AT Compliance

As we can see from all the templates in Figs. 6.3 to 6.7, the logical place to set the nominal value of the current limit (I_{LIM}) of the PSE's pass-FET is roughly centered in the area marked "Maybe," past the 8.3-ms marker. Specifically, this is the area between the upperbound and lowerbound templates. To guarantee the required performance, the min of the current limit spread must be assuredly above the lowerbound template (0.684 A for Type 2 and 0.4 A for Type 1) and its max must be below the upperbound template (1.75 A, but preferably 0.45 A for AF compliance as explained previously). Then we can be sure that the current limit will never get activated in the "Yes" region, and also will certainly get activated before the current treads in the "No" region. Note that so far we are ignoring the subsequent parts of the templates, where the timers take effect.

Now take an actual numerical example. Suppose for an AT application, the actual set current limit (nominal value) of our PSE device is 0.85 A, as suggested previously (between 0.684 and 1.75 A). What happens if the port current is a steady 0.7 A? From the Type 2 templates in Figs. 6.6 and 6.7 we see that any current above 0.684 A *must* be terminated before we get up to 75 ms. But it *may* be terminated *immediately* too, if we so choose. Theoretically, we could therefore just put any timer between 0 and 75 ms and turn OFF the port-whenever the current exceeded 0.684 A. However, we realize because of the current-limit spread, we may end up encroaching on the "Yes" regions prematurely. The safest approach is to allow the timer to timeout somewhere between 50 and 75 ms as shown in Fig. 6.8. That way we can guarantee all the "Yes" regions and also ensure we never enter the "No" regions. Port turn-off will occur only in the "Maybe" region. This is one way of designing our control circuit, and it highlights how we can avoid, or rather plan around, *device dependencies* and chip characteristics, while managing to adhere to the AT template, something that was not possible with the AF standard's way of writing operating requirements.

I_{CUT} Monitoring as Per AT Standard

In Fig. 6.8, we have also shown a narrow sliver called "I_{CUT} monitoring band." As per Table 33-11 on page 45, this monitoring is optional. As per line 7 of the table, I_{CUT} can range from P_{CLASS}/V_{PORT_PSE} to I_{LIM}. But I_{LIM} is itself a range, so that does not sound quite right. Line 7 also has a typo, because the divide-by sign is missing in P_{CLASS}/V_{PORT_PSE}.

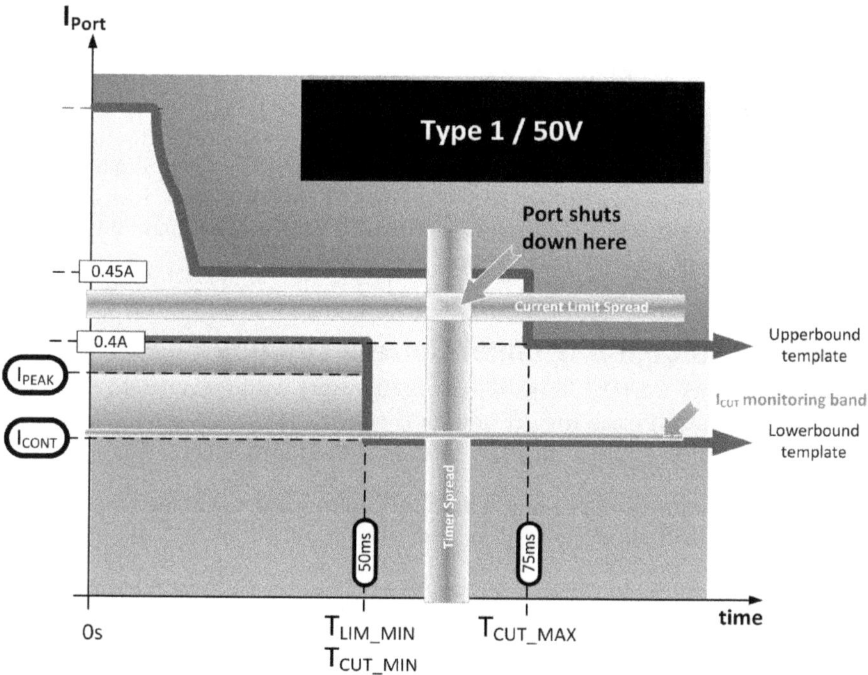

FIGURE 6.8 Setting current limit and timer in typical-PSE chip design.

In 33.2.7.7 on page 48, it is clarified that "The I_{CUT} threshold may equal the I_{PEAK} value determined by Equation 33-4." We take it to mean that while doing this optional I_{CUT} monitoring, we can position this narrow sliver above the I_{CONT} threshold as shown in the figure, and below the I_{PEAK} threshold.

Current Monitoring and Current Limiting Accuracy

A word about current accuracies. All PSEs are very accurate in reading the port current. They need to measure classification currents with just a few mA of accuracy for example. So it is typically stated that the PSE IC from vendor X has better than ±3 percent accuracy. Or another has better than ±5 percent accuracy and so on. A lot of chip-design effort goes into accurate port I-V monitoring. After receiving this information, a specific IC may offer a cumulative timer for I_{CUT} monitoring and so on (see previous section). That is why in Fig. 6.8, we have shown an I_{CUT} monitoring band. The AT standard may not demand it, but such features are often required by major OEMs.

Current monitoring is therefore very accurate but is also done mainly in software, with bits being written and read from chip-memory locations. We also know well by now that software can go horribly wrong. But the safety of the cabling infrastructure cannot be left at the mercy of software. Therefore the (protective) current limit is typically *hardware-based,* which means that it consists of a bunch of transistors that simply get activated by voltage/current thresholds and force port shutdown. Nothing much can go wrong here. At the same time, high-current accuracy is *not* required for this specific purpose. Nor is it required to monitor and report the actual value at which current limiting occurred. Basically, this I_{LIM} hardware-based current limit, as is true of most other hardware-based current limits, is accurate only to about ±10 to ±20 percent typically. That is the reason it has been shown as a rather wide ribbon in Fig. 6.8.

Allowed Port Voltage Sag under Current Limiting

Note that once current limiting occurs, the port voltage starts to fall suddenly. In that case, the AT standard specifically provides further relief. In Section 33.2.7.1 on page 46, the AT standard says:

> A PSE....may remove power... when the PI voltage no longer meets the V_{PORT_PSE} specification.

V_{PORT_PSE} refers to Table 33-11, pages 45. This basically implies that if ever the port voltage falls below 44 V for Type 1 and 50 V for Type 2, the port can be turned off right away.

However, Table 33-11 along with Section 33.2.7.2 (page 46), demands a certain "glitch filtering" for Type 2 (only). Its basic purpose is to make the PSE more robust to withstand shorter, but more severe overload demands from a typical high-power PD (like a PTZ camera, for example, where the motor may suddenly kick in). In effect, this states that if a port-voltage transient (whatever its origin), lasts *less* than 30 μs, it should be ignored completely—even if it ends up taking the port voltage completely out of the valid operating range. But after 30 μs we can turn the port off right away if the voltage is still found to be below 46.2 V. Why 46.2 V? Because that is exactly 7.6 percent below the lowest AT limit of 50 V. This percentage, 7.6 percent below normal for transients, is defined in line 2 of Figure 33-11, and is called K_{TRAN_LO}. Check: $\{(100 - 7.6)/100\} \times 50 = 46.2$ V.

However, if after 30 μs, the port voltage has managed to recover to a level above 46.2 V (i.e., < 7.6 percent below the minimum PSE voltage of 50 V as specified), then that behavior is considered normal PD behavior too, and the port should therefore stay on. However, if the port voltage does not regain its normal operating range of 50 to 57 V at the end of this 250 μs, we are allowed to turn the port OFF immediately (sensing an abnormal condition *not* related to normal high power PD behavior).

In a real case, we know that the current limit for a Type 2 PSE can be set anywhere between 0.684 and 1.75 A. Let us do a calculation for both extreme cases to judge the overall situation when testing compliance to this glitch filter clause.

1. Actual Device Current Limit 0.685 A: Resistance required to cause V_{PORT} to fall below 46.2 V is $46.2/0.685 = 67.45\ \Omega$.
2. Actual Device Current Limit 1.74 A: Resistance required to cause V_{PORT} to fall below 46.2 V is $46.2/1.74 = 26.552\ \Omega$.

From the viewpoint of the test-equipment manufacturer, we realize the manufacturer needs to account both for customers with current limits set high, and also those set low (within the allowed range). So very likely, if this overload parameter is being tested, the most favorable (universally applicable) resistance, of say 68 Ω, may be applied by the test equipment (at any voltage), and the port will be then tested to confirm that it does *not* turn off before 250 μs.

From the viewpoint of the systems designer, we can clearly see that we are only making things harder for ourselves if we set our current limit as high as is allowed by the AT standard. The higher-current-limit case above has to tolerate a much lower resistance. We should therefore try to set the nominal value of the current limit *as low as possible,* just ensuring that the min of the current-limit spread stays above 0.684 A. Then we are fully positioned to take advantage of the clause to turn the port off immediately (after the glitch is filtered out). This amounts to another PSE design for meeting AT requirements as easily as possible.

Resumption after "Error" and Timings

A failed inrush phase (where the current may have exceeded the inrush template for example), or a severe overload or a short-circuit, or a port-voltage droop outside operating limits, are all "error" (fault) conditions that cause the port to be shut down. What happens thereafter?

The PSE first declares "power removal" by turning its pass-FET totally non-conducting ("OFF"), and also activates a timer called the T_{ED} (error delay) timer. Note that after declaration of power removal, the port voltage still needs to be brought below 2.8 V, since port capacitances are charged up. This can be done actively by the PSE, where it places a large bleeder resistor (> 45 k) across the port. In a test setup, the standard allows connection of a 320-k bleeder resistor across the port. The idea is that the port capacitances need to be discharged in a time T_{OFF} whose maximum value is 0.5 s. Once that elapses, or whenever the port actually falls below 2.8 V—because keep in mind, the PSE is not *required* to actively discharge the port, and the 320-k bleeder is inserted in the test setup only to confirm that the PSE's pass-FET can really turn fully OFF, detection can start again. But can it start *right away*? Assume it does and let us see the timings involved. In other words, assume no delay between the port falling below 2.8 V and a fresh detection attempt. We know that maximum detection time is $T_{DET} = 0.5$ s. After a successful detection, classification may be carried out. Finally, Power-up must be declared within T_{PON}, measured from end of a successful detection to declaration of Power-up. This T_{PON} timer has a maximum value of 0.4 s. So adding all these up, we get the maximum delay as $T_{OFF} + T_{DET} + T_{PON} = 0.5\text{ s} + 0.5\text{ s} + 0.4\text{ s} = 1.4\text{ s}$. Of course, along the way there could be a failed detection/classification attempt or power not available, in which case there will be more delays. For example, after a failed detection, an Alt-B PSE must back off for T_{DBO} which has a minimum value of 2 s. But what about *minimum* times involved? T_{OFF} has no minimum time specified in the standard. Nor has T_{DET}, and nor has T_{PON}. So technically, if we could do detection fast enough and so on, then after an error (fault), we could end up powering up very quickly. This is not a good scenario because after a fault condition, time should be allowed for things to settle down, PSE and PD chip temperatures to come down and so on. Besides, we do not want the system to "hiccup" very fast, constantly cycling between error shutdown and power-on. So *the standard basically demands a minimum error delay of 0.75 s*, measured from declaration of power removal to declaration of Power-up (start).

In Fig. 6.9, we show a typical timing chart.

Summary of Peak and Operating Values

Finally, the table of values derived from the Mathcad spreadsheet are presented in Table 6.1 for easy look-up.

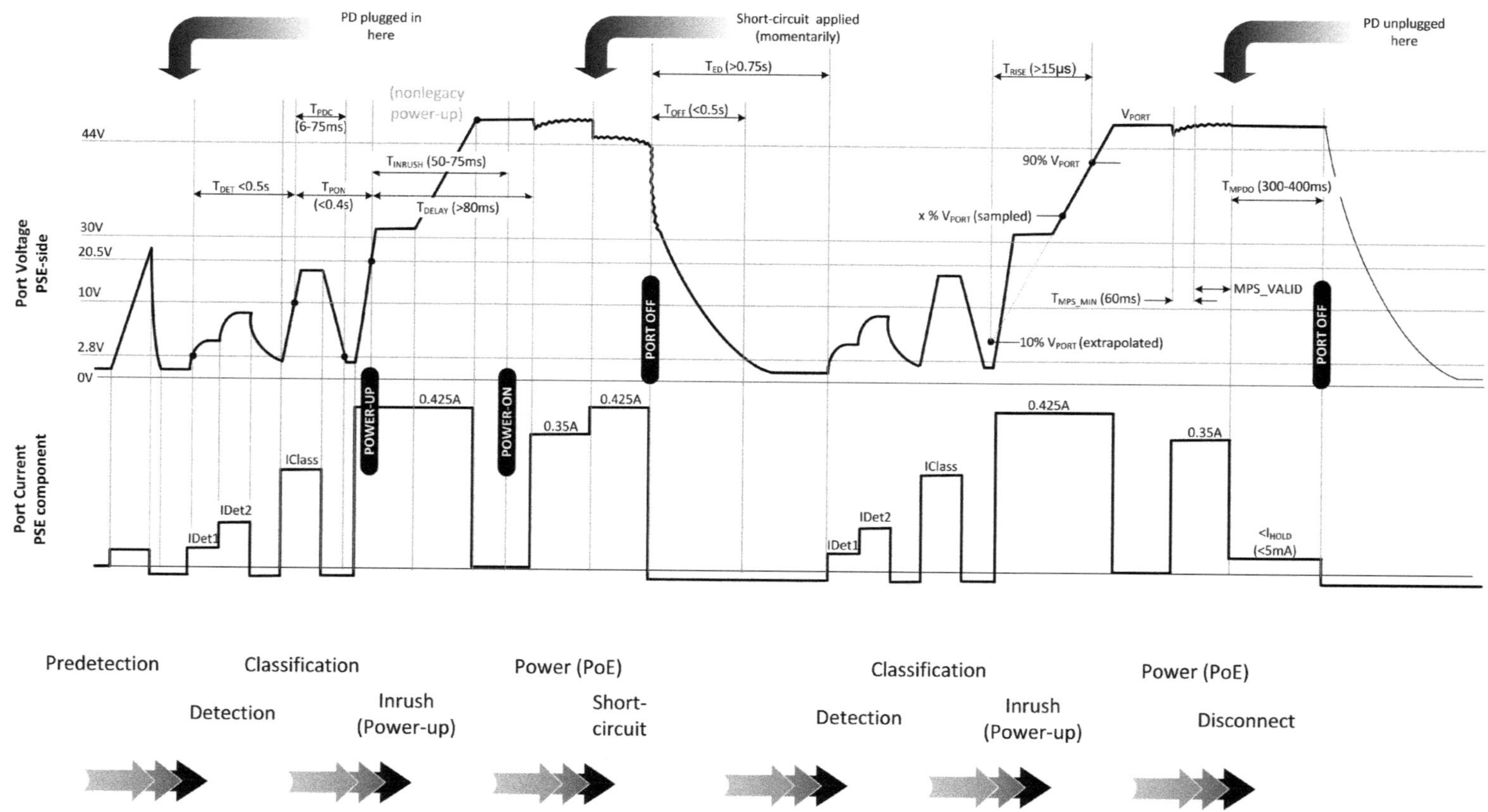

FIGURE 6.9 Typical PSE chart showing timers.

		Operating Values					Peak Values				
		Type 1				Type 2	Type 1				Type 2
		Class 0	Class 1	Class 2	Class 3	Class 4	Class 0	Class 1	Class 2	Class 3	Class 4
Current (A)	at Vpse_min	0.35	0.09	0.16	0.35	0.60	0.40	0.12	0.21	0.40	0.68
	at Vpse=50 V	0.31	0.08	0.14	0.31	0.60	0.33	0.10	0.18	0.33	0.68
	at Vpse_max	0.27	0.07	0.12	0.27	0.53	0.28	0.09	0.16	0.28	0.57
Resistance (Ω)	at Vpse_min	125.71	484.00	276.57	125.71	83.33	110.00	366.04	209.47	110.00	73.27
	at Vpse=50 V	162.34	625.00	357.14	162.34	83.33	150.55	479.13	277.49	150.55	73.27
	at Vpse_max	210.97	812.25	464.14	210.97	108.30	203.44	629.14	367.49	203.44	100.53
Power (Min_PSE) (W)	at Vpse_min	15.40	4.00	7.00	15.40	30.00	17.60	5.29	9.24	17.60	34.12
	at Vpse=50 V	15.40	4.00	7.00	15.40	30.00	16.61	5.22	9.01	16.61	34.12
	at Vpse_max	15.40	4.00	7.00	15.40	30.00	15.97	5.16	8.84	15.97	32.32
Vpse_min is 44 V for Type 1 and 50 V for Type 2. Vpse_max is 57 V for Type 1 and Type 2											

TABLE 6.1 Operating and Peak Values of Current, PSE Load Resistance and PSE Power Output

CHAPTER 7

Maintain Power and Disconnect

Overview

If the Ethernet cable to the PD is suddenly disconnected and then immediately plugged into a "non-PoE-friendly" DTE (like a NIC), the terminations of the new device could get damaged. To avoid this situation two things are required: (a) the PSE must constantly sense the PD to ensure it is still present, and (b) the moment the PD is deemed disconnected, the PSE must remove power from the port. In fact, it is recommended that it actively "discharge" port capacitances to a safe level (2.8 V) within $T_{OFF} = 0.5$ s. This maximum time was set based on how quickly a cable could be disconnected and then reconnected to another RJ-45.

The PSE is, in any case, constantly monitoring the port current, and within its accuracy of measurement can very easily figure out if a PD is present on the other end of the cable or not. For its part, the PD would like to avoid getting inadvertently mistaken for a disconnected PD, and for that, it must continue to draw a minimum current from the cable. This holding current, I_{HOLD}, provides the (maintain-power signature (MPS) to the PSE. It is the first aspect we consider.

Keeping the Port Alive

The AT standard defines a low-current threshold I_{HOLD} with a min of 5 mA and a max of 10 mA. What that means is that a current above 10 mA definitely implies the presence of a PD, whereas less than 5 mA definitely implies no PD. Between 5 and 10 mA is a "don't care" region. Therefore, a typical PSE controller-design puts a nominal I_{HOLD} threshold centered between 5 and 10 mA, say at 7.5 mA. Now the PSE will consider any port current above 7.5 mA as a PD, and any current below 7.5 mA as no PD.

There are two timers to know about here:

1. T_{MPS}: This is the maintain-power signature timer. It starts the moment the port exceeds the nominal I_{HOLD} threshold current, and stops when it falls below that threshold. Table 33-11 on page 46 declares the min value of this timer as 60 ms, with no max value. But how can a timer have a nonzero initial value? That can get confusing. What is really meant is that this timer "amounts to something" only when it exceeds 60 ms. Small excursions above the I_{HOLD} threshold do not count. It is a kind of time-based glitch filter. So, when 60 ms is over, the PSE says that in effect, MPS is valid. And as soon as that happens the dropout timer gets reset, because the dropout (i.e., the excursion below the I_{HOLD} threshold) is effectively over.
2. T_{MPDO}: This is the maintain-power signature dropout timer, or just "dropout timer." It starts counting the moment the port current falls below the I_{HOLD} threshold. It does *not* stop counting automatically if the current momentarily pops up above the I_{HOLD} threshold momentarily. Some conditions need to be in place first as indicated in (1). Table 33-11 on page 46, declares the min of T_{MPDO} as 300 ms and its max as 400 ms. This range indicates that this timer will "time out" somewhere between 300 and 400 ms. If and when that happens, the port *will* be shut down. So the question is, how do we prevent this timer from timing out?

   The answer to that is: If MPS-valid is declared before 300 to 400 ms, the dropout timer will get reset. But it will once again start counting (and immediately so) when the port current falls below the nominal threshold. If MPS-valid does not occur for 300 ms, the PD is in increasing danger of being disconnected at any moment (when the T_{MPDO} timer times out).

This entire scenario is illustrated in Fig. 7.1. As we can see, minor excursions above 10 mA do not count, as it is felt that they could just be noise. To be definitely identified as a PD, the PD must draw more than 10 mA for at least 60 ms. Once that > 10 mA burst is over (it falls below 5 mA), the PD must draw another such burst, > 10 mA, before 300 ms of wait time.

Dropout versus MPS

Let us summarize this for clarity: If for every 300 ms that the port current dips below 5 mA, the PD draws at least 10 mA for more than 60 ms (say the PD's internal burst-timer is 60 to 65 ms), the PSE will

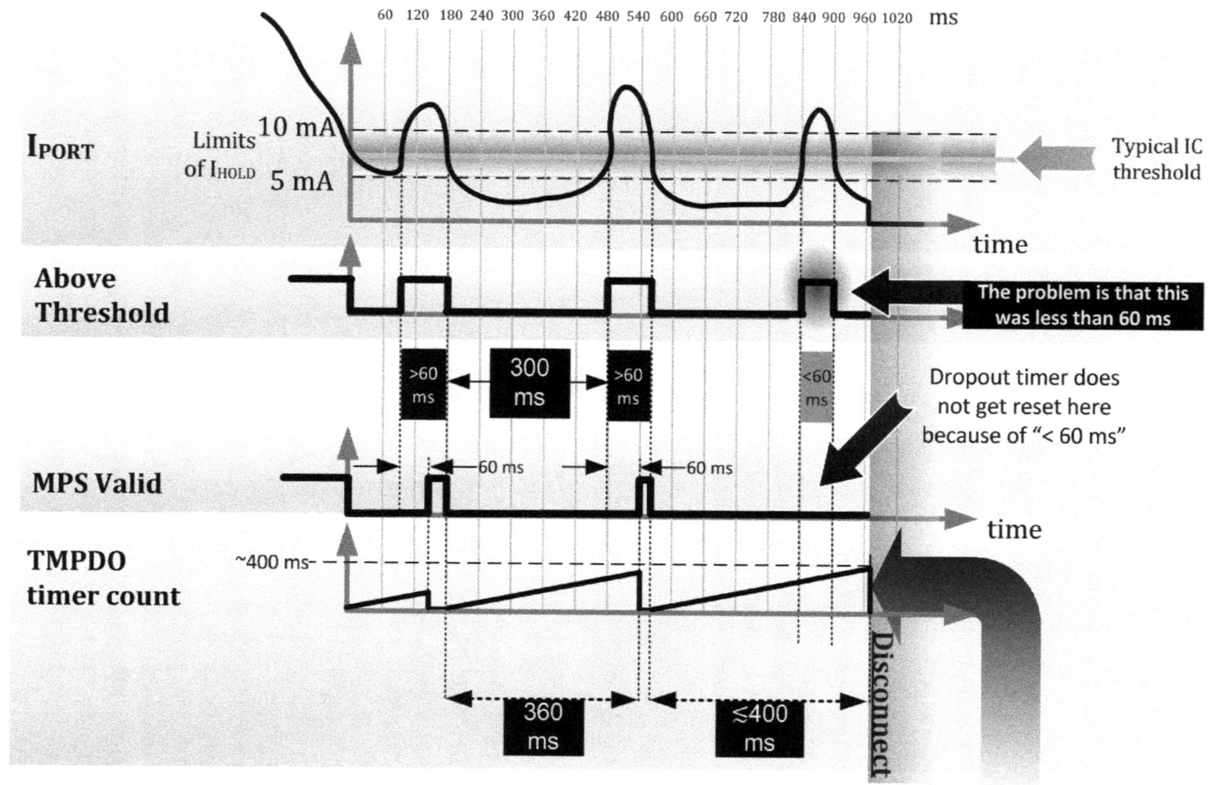

Dropout timer gets reset if MPS is valid. But for MPS to be valid, the port (PD) current has to exceed set threshold for **at least 60 ms**. If this port current is not drawn by PD for at least 60 ms after every 300 ms (i.e., for at least 60 ms every 360 ms), the port **will** be shut down

FIGURE 7.1 A sample case showing how a PD may get disconnected by exhibiting noncompliant behavior.

declare that the signature was maintained and will not disconnect the PD. But if not, the dropout time will be deemed as having exceeded its limits, and the PD *will* be disconnected within 400 ms (the max of the dropout-time interval). The upper boundary of T_{MPDO} is a physical human limit for disconnecting one PD and "immediately" connecting another device.

NOTE *"Power removal" or disconnect just refers to the PSE pass-FET turning off. The actual voltage on the PI is allowed 500 ms more from this point to droop below 2.8 V (at the PSE end).*

One relevant section to refer to is Section 33.2.9.1.2, page 52, of the AT standard, which states:

> A PSE shall consider the DC MPS component to be present if I_{PORT} is greater than or equal to I_{HOLD} max (10 mA) for a minimum of T_{MPS} (i.e., at least 60 ms). A PSE shall consider the DC MPS component to be absent if I_{PORT} is less than or equal to I_{HOLD} min (5 mA). A PSE may consider the DC MPS component to be either present or absent if I_{PORT} is in the range of I_{HOLD} (5 to 10 mA).

Along the same lines, the PSE Test Suite Version 2.4 from University of New Hampshire Interoperability Lab (UNH-IOL) states:

> In order to maintain a valid MPS signature, the PD can draw less than the I_{MIN1} (5 mA) for 300 ms and then draw more than its I_{MIN2} max (10 mA) for the next 60 ms(T_{MPS}) or more.

We realize that the PSE is not allowed to remove power when I_{PORT} is greater than or equal to I_{HOLD} max (10 mA) continuously for at least T_{MPS} (minimum of 60 ms) every $T_{MPS} + T_{MPDO}$ (i.e., 60 ms every 60 ms + 300 ms = 360 ms) as defined in Table 33-11. The duty-cycle "D" is 60/400 to 60/360. This is 15 to 16.7 percent. So the minimum PD power for a Type 1 case is $V \times I \times D = 44\text{ V} \times 0.01\text{ A} \times 0.15 = 0.066$ W. For Type 2, the minimum is $50\text{ V} \times 0.01\text{ A} \times 0.15 = 0.075$ W. Note that this is sometimes stated as $50\text{ V} \times 0.01\text{ A} = 0.5$ W. That is also true, in a way, because reducing the power below this level cannot be done on a random basis. The PD must be designed to be very smart and to carefully exploit this power-saving strategy. It must draw the right amount of minimum power with the right duty cycle, otherwise the PSE may shut down the port. So we can say, theoretically, the minimum power is 0.075 W (for Type 2), but *in practice* it is 0.5 W.

Keep in mind that the standard only regulates the minimum current, not the wattage. So if we raise the port voltage for any reason, we also unfortunately increase the minimum PD power to keep the port alive.

Setting the Timer for "MPS Valid"

The problem is that a min value is given for T_{MPS}, not a design *range*. This has caused some confusion, causing some PSE-chip designers and software folks to set the MPS-valid timer at around 40 ms. They make two assumptions when asked: (a) that this is also IEEE-compliant behavior and 60 ms is actually the max if interpreted "correctly," and (b) by doing so, they end up supporting "green applications," allowing further reduction in minimum power. However, that is not true. We have to understand that the overriding spirit of the PoE standards is always tilted towards *safety*. So it would prefer to turn the port OFF rather than keep it alive. Our reasoning has to support this too. So if the PD is designed to ask for more than 10 mA for less than 60 ms (as shown in Fig. 7.1 too), the PSE will prefer to shut down the port, not keep it ON. *That is the correct interpretation*. We can conclude that we must err on always making it a "little *more* difficult," not easier, for the PD to avoid being disconnected. In other words, the PSE is compliant if it sets its MPS-valid flag at *more* (not less) than 60 ms. We can, for example, set the nominal to 80 ms, with a min of 65 ms and a max of 95 ms. A PD that draws > 10 mA for, say 62 ms, would be disconnected in this case, but that is a better option than a PD that draws > 10 mA for only 58 ms and is wrongly kept connected by the PSE.

So setting the MPS-valid flag a little past 60 ms is compliant behavior, and in line with the spirit of the standard. Of course, the minimum PD power to keep the port alive has gone up slightly, but this interpretation is safer in terms of protection, which is of the highest priority.

PD Preloading

What is the effective PD load-resistance corresponding to 10 mA at, say 50 V? It is $V/I = 50/0.01 = 5\ \text{k}\Omega$. The wattage is $V \times I = 50\ \text{V} \times 0.01\ \text{A} = 0.5\ \text{W}$. If the output of the PD was 12 V, then assuming 100 percent efficiency for the DC-DC converter, we need to preload the PD for 0.5 W minimum load. That leads to a preload resistor of $V^2/\text{W} = 12\ V^2/0.5\ \text{W} = 288\ \Omega$. So we can provide the MPS by connecting a 5-k resistor directly across the port, but that would obviously interfere with detection and classification. So the correct option is to include a resistor of less than 288 Ω across the 12-V output rails.

If the efficiency of the DC-DC stage is, say, 80 percent, then the output power corresponding to 0.5 W at the input is $0.5\ \text{W} \times 0.8 = 0.4\ \text{W}$. So the correct value of preload resistor increases to $12\ V^2/0.4\ \text{W} = 360\ \Omega$.

AC Disconnect and DC Disconnect

So far, we have been implicitly assuming what is called DC disconnect. This involves the PSE taking a simply voltage-current measurement to determine if there is a valid PD on the other side or not.

This manner of implementation, which really was just Ohm's law, unfortunately ran into intellectual property (IP) battles however strange that may sound. So an alternative method came into being, one that involved injecting a low-frequency AC wave on the line. This depends on the fact that when a PD is connected to a port, the AC impedance measured on its input terminals is significantly lower than in the case of an open port (disconnected PD). AC disconnect depends on the PSE being able to "find" the *capacitor* on the PD side. What capacitor are we talking about? *Not* the few nF of port capacitance during detection and classification, but the capacitance when Power-on has occurred and the PD is in normal operation. We know that the minimum value of that cap, called C_{PD_PD}, is 5 μF. That is also in parallel to an effective load resistor, bringing down its net impedance.

For a moment, we ignore the actual implementation details of AC disconnect and worry about the impedances involved. The AF and AT standards tried to maintain backward compatibility with DC disconnect by simply fixing the simple rule: *accept as a valid PD any AC impedance less than 27 k.* Why 27 k? It is not very clear. Some have argued that it was at that value because the detection resistor range is 25 k ± 5 percent, which as we know from Chap. 3, gives us a max of 1.05 × 25 k = 26.3 k (on the PD-side), and about 27 k on the PSE-side. But this is the *detection* resistor range, not the resistance or impedance during normal operation, which is what AC disconnect is actually looking to "spot out." The other interesting thing is that the IEEE standard established a max for the AC detection probing-frequency at 500 Hz and *with no minimum.* It is interesting that there is no min because an AC signal of zero frequency is, in effect, a DC signal. So is that DC detection now? Further, with almost any frequency allowed up to 500 Hz, the range of allowed (valid) capacitances is not clear. What exactly are the limits of AC disconnect in terms of PD construction? That seems to depend on the chosen probing frequency, so it is all not only vendor-dependent, but implementation-dependent and perhaps also device-dependent. Nothing about AC-disconnect specification is really objective. Another strange thing is that in the relevant tables where the number 27 k pops up in the standards, it is mentioned that the probing frequency is 5 Hz. Why 5 Hz?

Finally we ask: How does the standard hope to reject (*disconnect*) a PD? As we see, it is not based on a 5 to 10 mA minimum load anymore. AC disconnect has a very basic rule for that: *Disconnect any PD/load with an AC impedance larger than 1980 Ω.* Why such a unique-sounding number "1980"? That is simply because 1980 Ω is the min of a 2 M ± 1 percent resistor.

So all the AF and AT standards are ultimately saying is any AC impedance below 27 k, keep it connected, and any AC impedance above 2 M, disconnect. "Whatever your probing frequency is, provided it does not exceed 500 Hz!" That is as vague as the UNH-IOL test that tests for AC disconnect in a few simple steps.

AC Disconnect Test Procedure (UNH-IOL)

The procedure used by the University of Hampshire Interoperability Laboratory is as follows:

1. Attach a valid signature to the PI of the DUT such that the DUT enters the POWER_ ON state.
2. Disconnect the PD from the PI of the DUT.
3. Measure the time taken by the DUT to remove power.
4. Observable Results: Verify that 300 ms $\leq T_{MPDO} \leq$ 400 ms.

So this test is a basic OK/Not-OK test. We can work backward to see if we will have any problem with the 5 µF minimum C_{DC_DC} cap (input cap of the DC-D stage inside the PD). And down to which probe *frequency*? We see that with 5 µF and 5 Hz, the impedance is 6.5 k. Since this is well below 27 k, this standard minimum-capacitance PD will stay connected with an AC disconnect feature present in the PSE. Now keeping the same minimum capacitance, we see that the impedance will be just a little below 27 k (accept threshold), if we choose a probing frequency of 1.18 Hz. So that is, in effect, the min of the AC-probing frequency (the max being 500 Hz as we already know). In other words, the AC detection probing frequency range is effectively 1.2 to 500 Hz, even though the standard does not mention it.

The standard does place some additional requirements for safety and low-EMI using AC disconnect. It states that the AC signal-generator be limited to 5 mA maximum. Also, the slew rate must be below 100 V/ms. Both these are actually easily complied with.

Commercial PSE's Interpretation and Implementation of AC Disconnect

Many commercial PSEs, however, use a probe frequency of around 100 Hz. As per their interpretation, AC disconnect should be able to detect even the port capacitor of the PD (typically 0.1 µF). So with 100 Hz, the impedance is $Z = 1/(2\pi \times C \times f) = 1/(2\pi \times 0.1 \times 10^{-6} \times 100) =$ 16 k. Since this is well below 27 k, it will be detected by the AC-disconnect circuitry. But note that this is not the way the UNH-IOL test is written above.

There are other commercial PSEs that use a much lower frequency of about 25 Hz. They are counting on the fact that there is a 0.1-µF cap on the output of the PSE too. Plus some cable capacitance (typically 10 to 50 nF for 100-m length), and so on. So on top of that, adding a PD-side capacitance of 0.1 µF is enough to bring the entire impedance well below 27 k, even with a 25-Hz probe frequency.

Some PSEs are basically using an internal oscillator of low-amplitude (typically less than 5 V) and injecting the signal by capacitor-coupling onto the line. Another way to do that is by first creating a boot-strapped rail by means of a high-frequency charge pump as shown in Fig. 7.2. The reason for using such a high frequency is that other than the output-reservoir bulk capacitor of the charge pump, the small "flying" capacitor within the charge pump can then be of a very small value and can therefore be "on-chip." This boosted rail is max 3-V higher than the port voltage rail. It is then modulated by a low-frequency probe oscillator (try to *not* make this sharp and square, thereby minimizing EMI). This is then directly injected on to the line.

Note the presence of the AC-disconnect diode in the upper rail. Its basic purpose is to block AC signal from going backward and causing the PSE to get confused by the very low 48-V incoming rail's AC impedance (that may have has several bulk capacitors in parallel).

This diode drop is always in series with the port current and causes significant heating and waste of energy. To minimize that, a Schottky is preferred, as shown in Fig. 7.2. It should be rated greater than 60 V actually, since another PSE may be inadvertently connected on the line with this PSE in an unpowered condition. That will apply a worst-case 60-V reverse voltage on this AC-disconnect diode. However, a 40-V Schottky is sometimes preferred in this location because a 40-V diode typically has a much smaller forward voltage-drop than a 60-V Schottky, and so that will greatly lower the heating. But the use of a 40-V diode is based under the following assumptions:

1. A 40-V Schottky typically "avalanches" (has reverse breakdown) at~1.4 times the rated DC voltage, that is 40 V × 1.4 V = 56 V. So it may conduct just a little at around 57 V.
2. If the 48-V rail coming in directly to the anode of this Schottky is from a typically designed AC-DC power supply, then if the Schottky does break down just a little, the output cap inside the AC-DC power supply will charge up quickly, and in a very short time the anode of this Schottky will rise in voltage, and so it will stop avalanching very soon. *If this happens*, the 40-V Schottky *will* survive this abuse condition and can be used as the AC-disconnect diode.

Note that a 2A or 3A diode is all that recommended in this position. A 1A diode will have a much higher forward drop than a 2A diode, so it is not recommended here. On the other hand, Schottky diodes have a relatively high reverse leakage, one that varies significantly from vendor to vendor, even for an "equivalent diode." For example, if we put a 5A diode in this location, it will have a lower drop than a 2A diode and will save some energy, but the 5A diode will typically also have much higher reverse leakage than a 2A diode. *This is known to cause: (a) detection failures, and (b) AC disconnect failures.* Therefore, we have to be very careful in picking this diode. We must also finally put the PSE in an oven and ensure at high temperatures, in particular, the increased diode reverse-leakage is not causing any test-compliance failures.

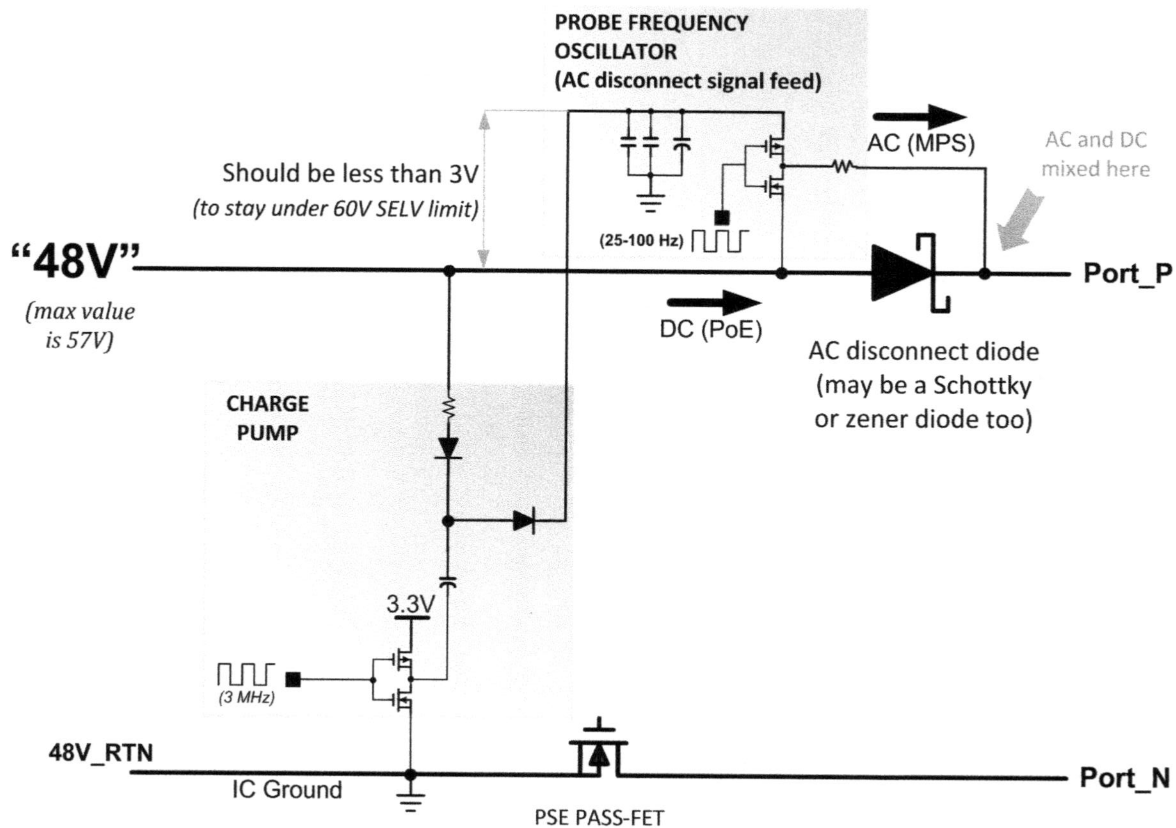

FIGURE 7.2 A charge-pump followed by a probe oscillator for AC disconnect.

Safety in AC Disconnect

There is one requirement that the standard stipulates, but it is widely overlooked and not tested for either: The amplitude of the AC-probe signal. Though that is not directly specified, the standard does say that if the PD is suddenly disconnected, the port voltage (incoming DC with the open-circuited AC superimposed) should never exceed 60 V. The problem with PSEs that use a 5-V probe amplitude is that if the port voltage is at its maximum-DC level of 57 V, adding the 5 V on top of that takes it to 62 V, which exceeds the safety threshold of 60 V. Therefore, any solution, which targets a nominal AC-amplitude of only 2.5 V, with a max of 3 V is compliant, others are strictly not. Though no one may ever know!

One key reason why no test suite or equipment tests this particular condition or violation is they do not take control of the incoming DC voltage. That rail remains whatever it was when the PSE was submitted for testing. That is usually less than 55V in any case.

Reasons to Avoid AC Disconnect

AC disconnect can perhaps skirt some lingering IP issues if required to do so, but it has serious drawbacks, as listed here, and should be avoided.

1. Very careful design of AC disconnect circuitry is required and can cause interoperability issues with some PDs because it is just not precise enough, nor is it very well-defined in the standard itself. It is misinterpreted a good deal too.
2. AC disconnect causes significant waste of energy and overheating.
3. Presence of diode can cause detection and disconnect failures too, especially at elevated temperatures.
4. Because of this blocking diode, when a surge strikes the line, this diode prevents the incoming surge energy from flowing backward into the output caps of the AC-DC power supply and getting safely absorbed there. As a result, very-high voltages may be seen on the port, causing damage to the PSE chip. *DC disconnect, by its very nature, provides much higher surge-withstand capability and therefore higher field reliability.*

CHAPTER 8

PoE State-Machine Diagrams

In this chapter, we have split into several pages and figures what are just a couple of pages in the AT standard. The idea was that each state-machine diagram originally need a lot of going back-and-forth for the novice reader. A "simple" term like "power_applied" in the state-machine diagram may need to be researched all over the standard to see on which exact page it was defined. We were lucky if we had a searchable PDF version. Timers were also defined at various places, but their numerical values could only be found in another referred-to table. It could also be confusing to just see "power_applied" as a *statement*, whereas it actually had *values*: TRUE or FALSE. Similarly, we could find "!power_applied," which was meant to be power_applied = FALSE, but obvious mainly to software designers or experienced persons. For beginners, it could all get very inefficient and confusing rather quickly. That is unfortunate, because it is very important to understand what is truly the heart of the PoE standard: its PD and PSE *state machines*.

This chapter also serves to finally cement all we have learned in previous chapters, including detection, classification, inrush, Power-up and Power-on.

In Fig. 8.1, we have the PSE Initialization phase. It leads to Fig. 8.2, which is Detection. That leads to Classification in Fig. 8.3. Then we have Fig. 8.4, which is Power-on. Finally, we have Fig. 8.5, which is Fault Protection.

The last two figures are for a PD. We have Fig. 8.6, which is basically all of the PD's state machine prior to Power-up and Power-on. This is followed by the rest, in Fig. 8.7, titled MDI Power.

That completes everything about the standard itself that we need to know at this point. There are leftover issues regarding isolation, fuses, and so on, which will follow slowly in the next chapters when we start taking a very close look at *systems-level* concerns.

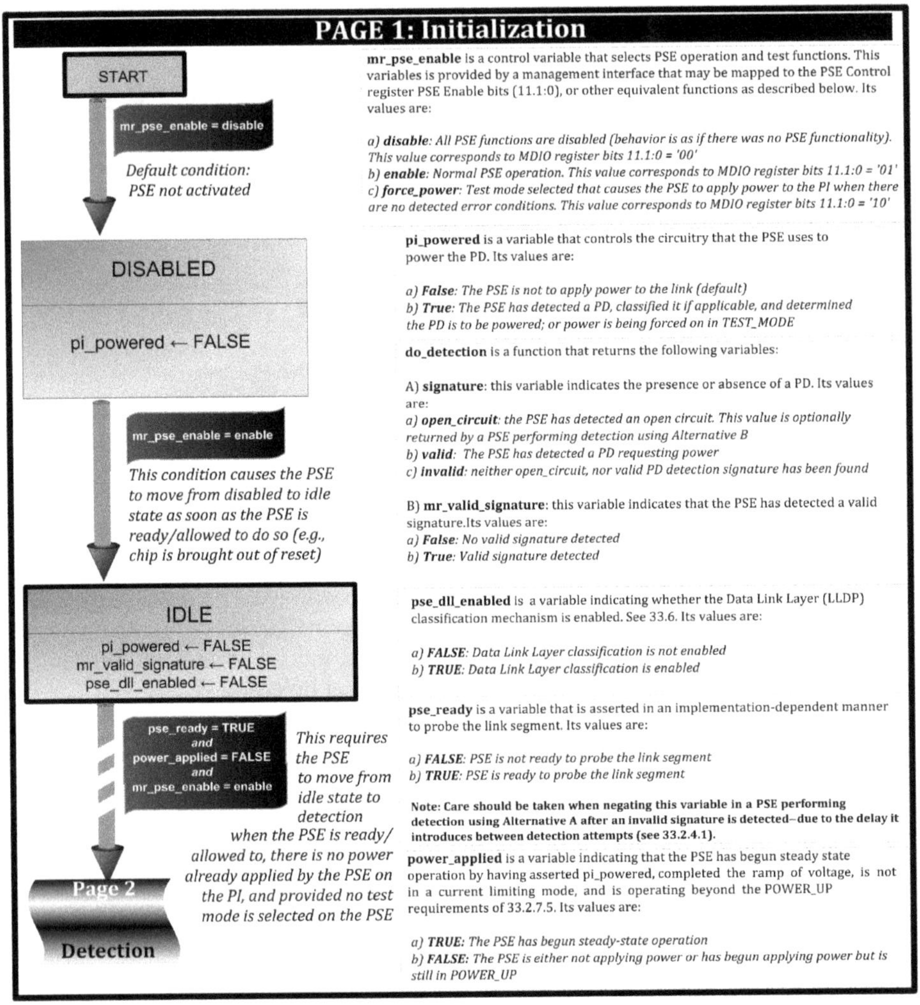

FIGURE 8.1 PSE state-machine diagram—page 1 of 5.

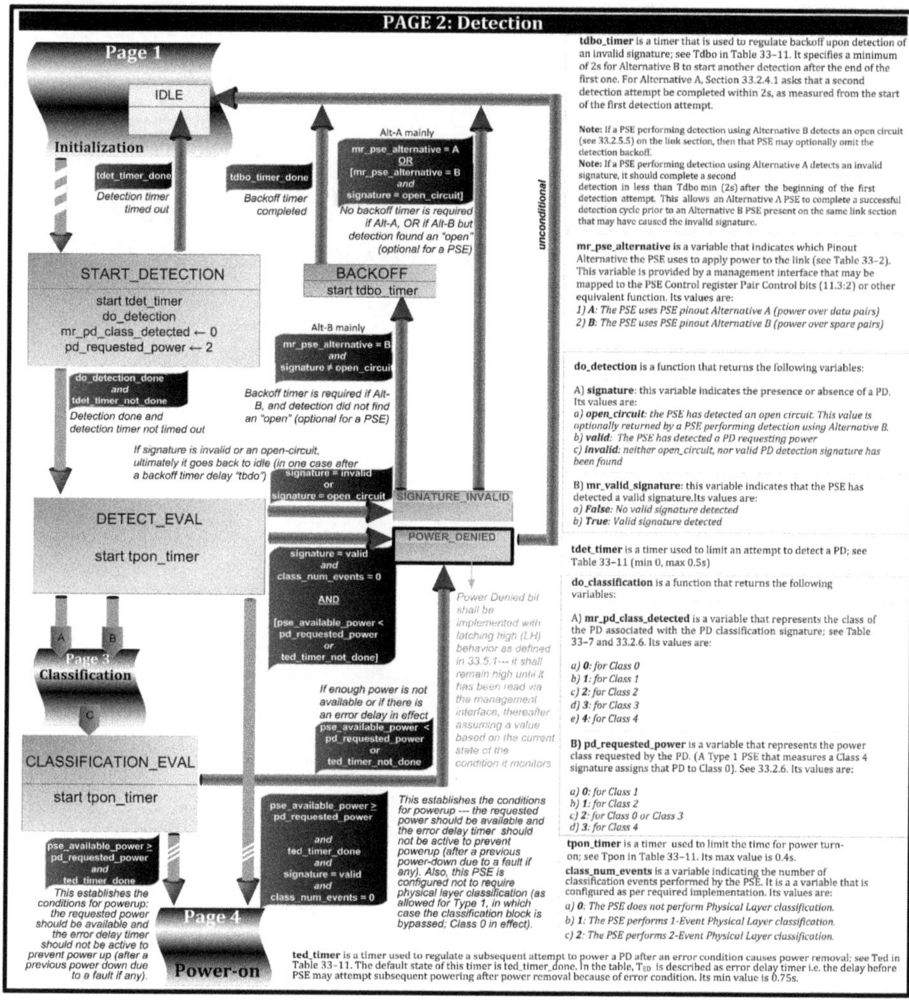

FIGURE 8.2 PSE state-machine diagram—page 2 of 5.

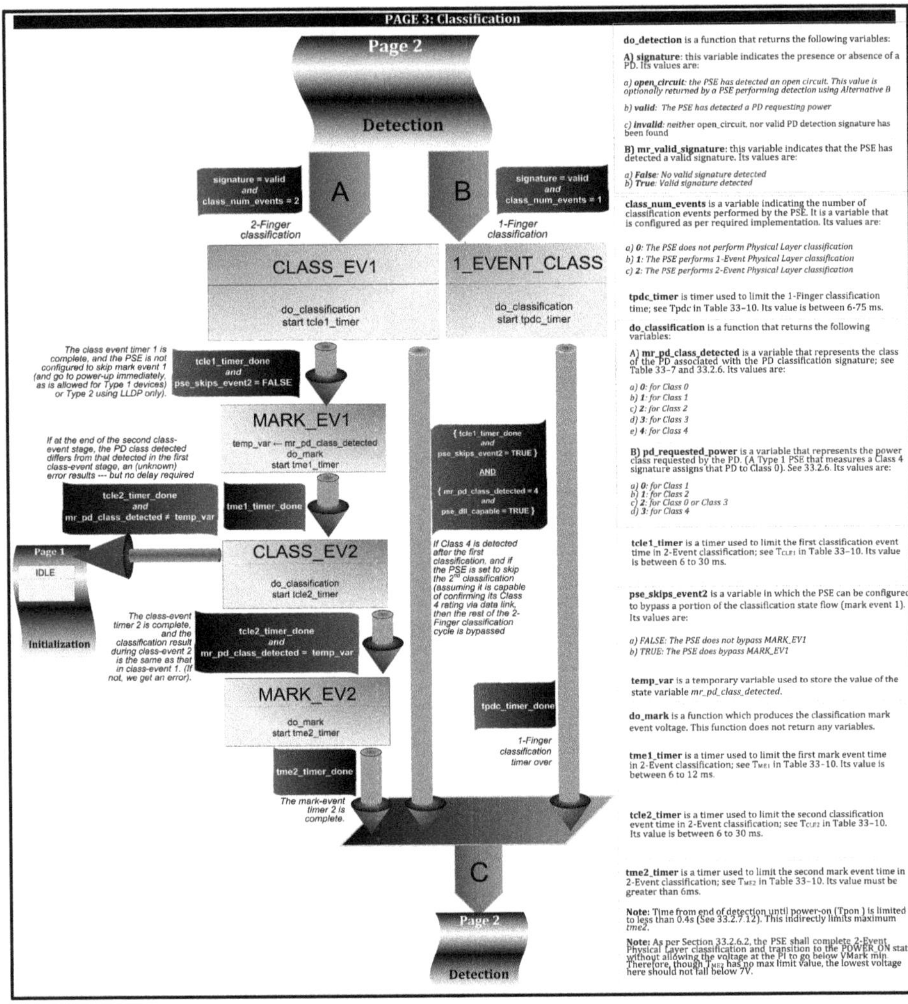

FIGURE 8.3 PSE state-machine diagram—page 3 of 5.

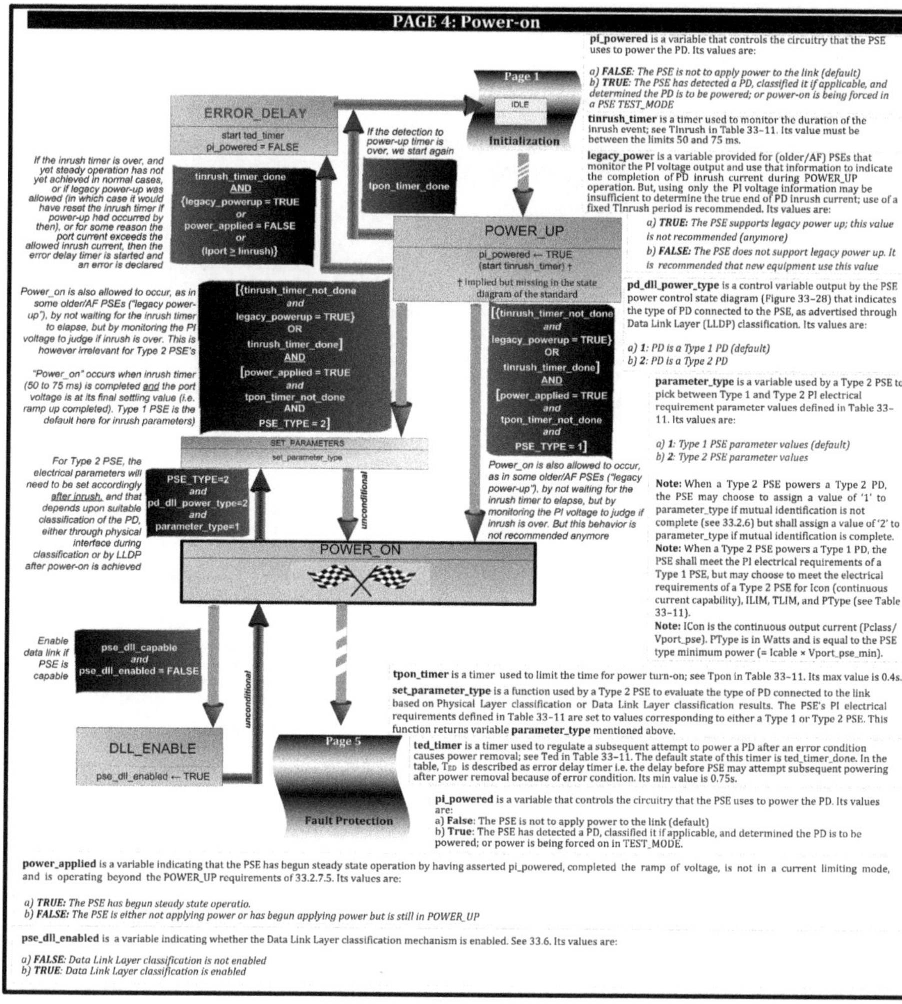

FIGURE 8.4 PSE state-machine diagram—page 4 of 5.

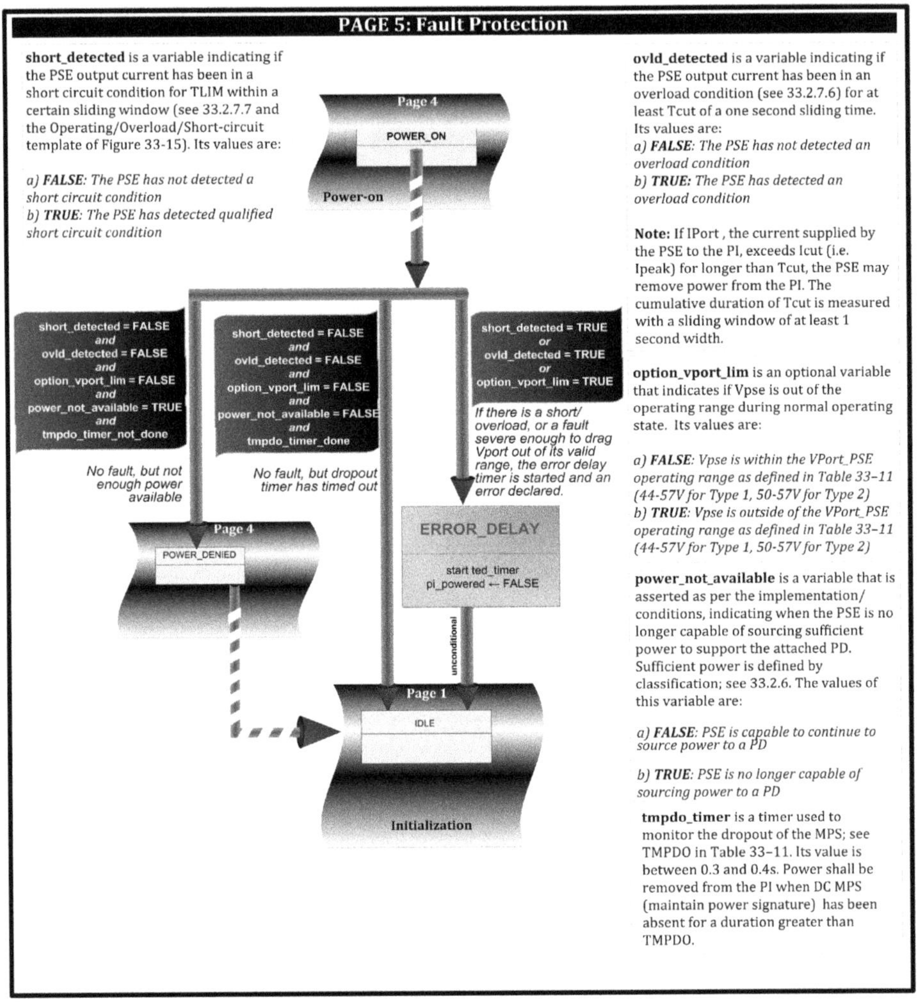

FIGURE 8.5 PSE state-machine diagram—page 5 of 5.

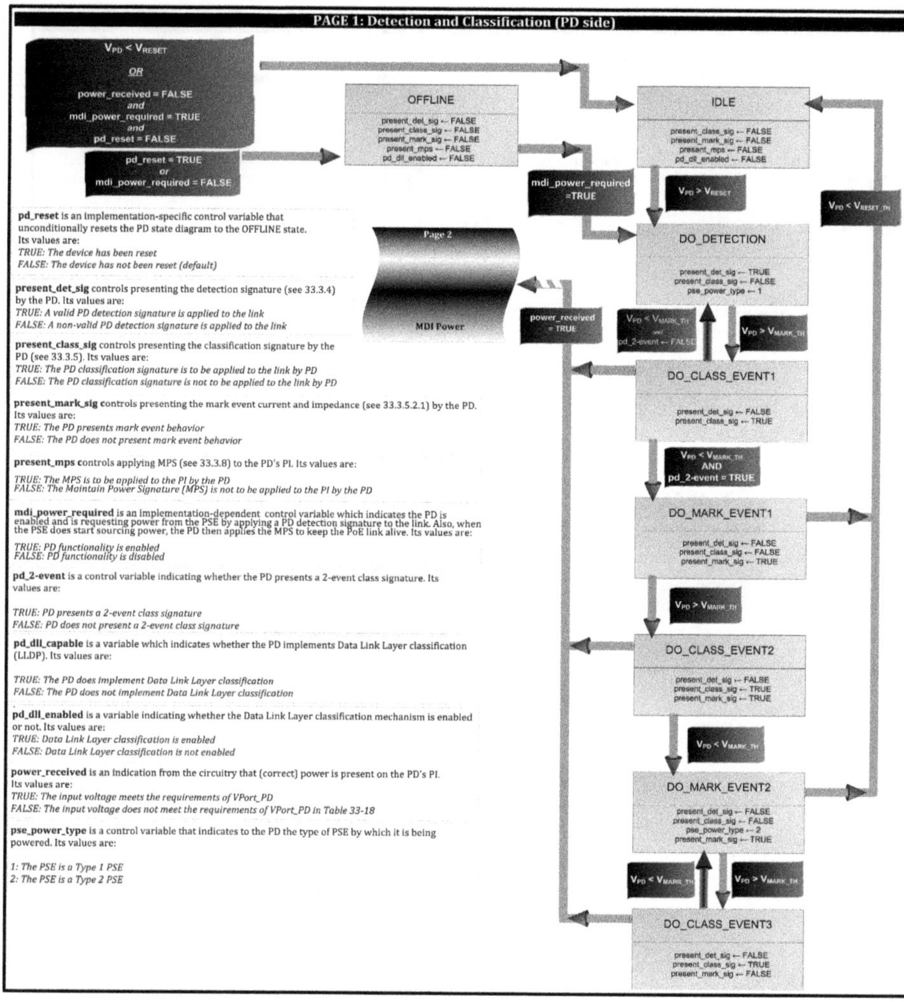

FIGURE 8.6 PD state-machine diagram—page 1 of 2.

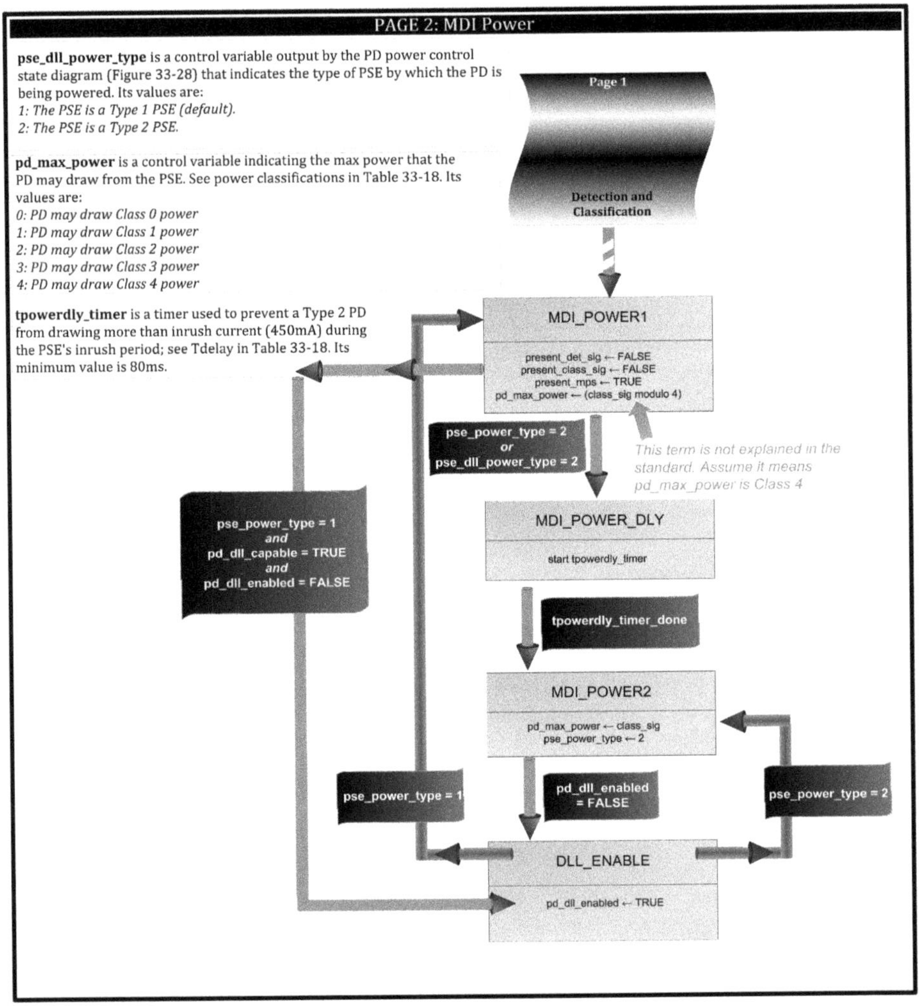

FIGURE 8.7 PD state-machine diagram—page 2 of 2.

CHAPTER 9

Magnetics

Overview

We know from Chap. 1 what an RJ-45 is. Colloquially, it is just a standard Ethernet jack. The word "magjack" is also often used equally colloquially and refers to a jack containing the "mag" in it—that is, its associated magnetics, in particular the pulse/data transformers. We will thus typically find tiny, toroidal transformers inside the metal housing of the magjack. (The housing is always metal for EMI reasons.) We can have various stacked versions of magjacks too: like the 2 × 4 for example. This would be 2 rows containing 4 magjacks per row. Much like a symmetrical 2-floor dormitory, with a total of 8 rooms. Or we can have a 1 × 4 magjack, which would be like a single-floor dormitory with a total of 4 rooms. And so on. We should keep in mind that "MagJack®" is actually a registered trademark of one specific vendor (Bel Fuse). Others are more likely to call it an integrated connector module (ICM), but they have also tried to give it catchy names. Admittedly, none caught on as much as "magjack." Therefore we too will continue to call it a magjack, but it should be understood that that is used in a very generic sense, just the way an ICM is so commonly referred to. It is like Xerox instead of photocopy, to draw an analogy.

A very wide variety of magjacks are available. In general, a single magjack itself (i.e., a 1 × 1) can contain 10 or more components within a single housing. Besides the obvious RJ-45 mechanical connector, these could include port-LED indicators, data transformers, common-mode filters, termination resistors/capacitors, etc. The convenience arising from this integration accounts for the widespread use of magjacks. But that is also the reason why they need to be evaluated carefully for PoE applications, especially for medium-power (PoE+) applications. Space is at a premium inside the usual housing. There is barely any room to cater to PoE/PoE+ by including larger-pulse transformers (to handle any additional energy-storage requirements), or thicker wire gauges. In addition, there are also thermal constraints due to I^2R heating arising from PoE. Also, as per safety requirements, 1500 VAC isolation is required between the PHY side and the line (or PoE) side, so some minimum spacings and clearances may still need to be maintained despite all these additional components inside the housing.

Even without supporting PoE, per se, the basic PoE-friendly magjack will at least have DC-blocking capacitors in the built-in resistor terminations (if the terminations are contained in the magjack of course). As explained in Chap. 2, this cap makes the terminations PoE friendly, and they do not get damaged if 48 V is inadvertently applied to the line.

Focusing on the magnetic components inside the magjack, there are two main points to consider carefully. These are the same key issues we always encounter whenever we deal with magnetic components in general. The two basic questions we need to ask of any magnetic component in any application are:

1. Is the copper of its windings thick enough (in terms of its resistive losses) to handle the required RMS current without overheating?
2. Is its core big enough (in terms of its energy-handling capability) to avoid saturating in our application?

For example, in our specific medium-power PoE case, we must ensure that the coils can handle 600 mA DC current (per pair) without overheating. Also note that in going from a Type 1 load (350 mA) to a Type 2 load (600 mA), the heating goes up by a factor of $(600/350)^2 = 2.94 \sim 3$! The corresponding rise in temperature over ambient also *triples* if the wire gauge (AWG) is not improved. Eventually, a magjack must work satisfactorily not only at room temperature, but at the elevated maximum ambient temperature of our specific application too (typically 45°C). Keep in mind that the local ambient temperature inside the magjack may be 10°C to 20°C higher that the room ambient temperature.

The overriding concern in dealing with magnetic elements is they *do not saturate*. Otherwise, their effectiveness may be significantly reduced because of excessive droop (as explained further in the chapter). The general result will be that the square waveforms transmitted by the PHY will be severely deformed by the time they reach the receiver, thereby increasing the bit-error rate (BER) and causing transmission delays and much-reduced speeds (lower bandwidth).

The size of a magnetic component in a power application is almost completely determined by the peak energy we need to store in it. That energy is, by simple physics, equal to $\frac{1}{2} \times L \times I^2$. To know the core size we therefore need to know (a) what L is, and then, (b) find out what exactly is the I (to plug into $\frac{1}{2} \times L \times I^2$).

Open-Circuit Inductance (OCL)

In Fig. 9.1 we show how a transformer behaves when we apply a square pulse to it, as opposed to applying a sine wave to it. We remember that a transformer only passes *changes* in voltages to the other winding, as per the induced EMF (Faraday's/Lenz's) law, which leads to what we call transformer action. So if there is only DC voltage present across the main (primary) winding, on the other (secondary) winding (placed on the same core), we will get no (zero) induced voltage.

With that perspective, consider a step pulse, which goes from a certain DC level (say, -1 V) to another level (say, 1 V). Both -1 V and 1 V are, before and after some time, just steady DC levels, so there cannot be any induced voltage across the secondary winding, either *initially* or *eventually.* But we do get a sudden induced-voltage step across the secondary winding *at the exact moment* we go from -1 to 1 V (the step), since at that very instant we have a large and sudden change in voltage—which has, in effect, high-frequency AC components. Therefore, the *initial voltage* across the secondary winding is 0 V, and the final voltage is also 0 V. But along the way, there is first a sudden step rise, followed by an exponential decay back to 0 V as shown in Fig. 9.1.

A square-pulse train (data stream) is essentially a series of successive up-and-down steps. So the waveform on the other winding (line side) has several successive droop sections as shown in Fig. 9.1 too. The time constant of this droop is L/R, where L is the inductance of the winding (OCL) and R is the resistance *across* the winding (in effect

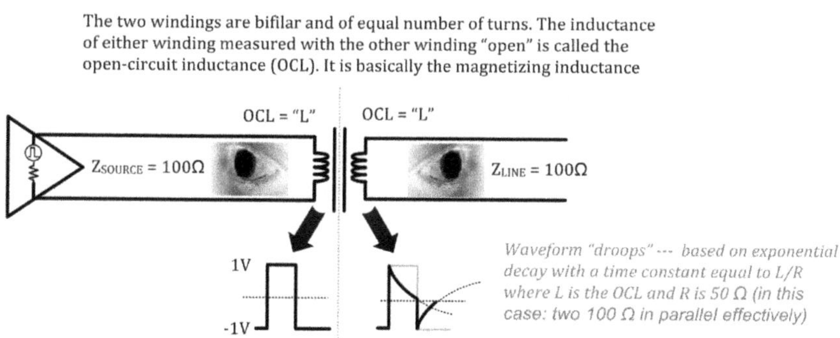

For example:

As per Gigabit Ethernet:

We can accept 73% droop in 500ns. So

$e^{-Lt/R} = 0.73$. Therefore, solving,

t=1.6us (time constant).

So L =80uH

Full calculation using Mathcad:

Droop := 0.73 Pulse_Width := $500 \cdot 10^{-9}$ R := 50

$\tau := \frac{Pulse\_Width}{-\ln(Droop)}$ $\tau = 1.589 \times 10^{-6}$

$L := R \cdot \tau$

$L = 7.944 \times 10^{-5}$ Henry (OCL = ~80uH)

FIGURE 9.1 Sample calculation for calculating minimum OCL in Gigabit Ethernet.

50 Ω in this case). If we first fix the permissible droop, based on the design of our receiver (its tolerance to waveform distortion), we can calculate the minimum acceptable OCL. A sample calculation is shown inside the figure.

Note that core saturation, not considered so far, makes this "droop" even worse because by definition, saturation basically implies a large fall in inductance (typically 20 to 50 percent). And if the inductance falls by say 50 percent, the time constant L/R will halve too, and the droop will be more severe in a given time interval.

DC-Bias Current Caused by Baseline Wander

In the underlying magnetics equation: Core_Volume ∝ Stored_Energy = $\frac{1}{2} \times L \times I^2$, we have discussed the L (OCL) above. Now we need to know what I is.

Well before PoE came into the picture, the size of the transformers was indeed very small, because the signal-related currents sent down the line were really small. However, there was still a concern about several 1s, or several 0s (actually -1s), happening to fall in succession. This basically amounts to a big and rather long step, with an excessive DC component. It leads to baseline wander (BLW). BLW occurs because the ground or reference (DC) level is not transmitted through a transformer. It can be explained as follows: Only if we have exactly symmetrical waveforms do we get equal areas of that waveform above and below its geometrical center, which then forms the baseline for the signal. Therefore, if we do not have exactly symmetrical waveforms, as in the case of a steady one-sided voltage (a stream of 0s for example), the baseline literally wanders. See Fig. 9.2. On the other hand, the receiver does not get a proper "fix" on the signal's wandering baseline with respect to its hitherto fixed hi-lo decision threshold. This can lead to it interpreting what is really an incoming "high" as an incoming "low" and vice versa. In other words, we get a bit error.

To avoid this scenario, various scrambling techniques emerged in Ethernet, such as Manchester coding, as mentioned in Chap. 1. Yet there was still a lingering chance of some wander, based on successive 0s or 1s. Therefore, well before PoE even emerged, the requirement was that Ethernet transformers (for 10/100-Mbps applications) have a minimum guaranteed OCL of 350 μH at a minimum DC bias of 8 mA. The BLW was estimated to be able to cause up to 8 mA of effective DC bias, and that may be enough to cause a really tiny transformer to saturate just a little, as compared to zero DC bias.

Modern PHYs try to deal with some amount of droop and transformer saturation by what is called BLW correction circuitry. In effect, these are DC restoration techniques.

Using PHYs with BLW correction, we can use lower inductances and/or let the transformer saturate somewhat, putting the burden of

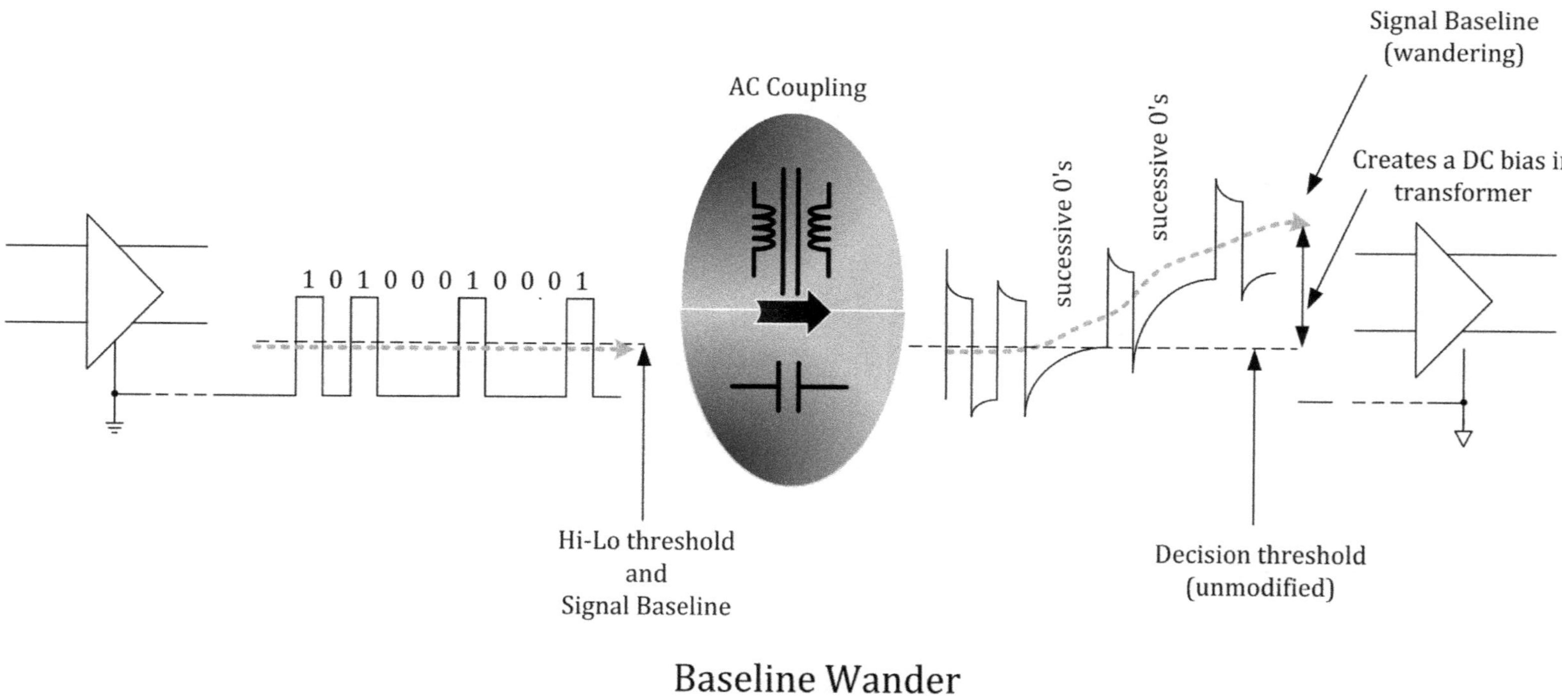

FIGURE 9.2 Baseline wander and DC bias in transformers explained.

signal-droop correction and its correct interpretation (hi-lo decision) on the BLW correction circuits of the PHY chip. This is of great help, especially in PoE, because PoE *adds* to the 8 mA BLW-related DC bias as we will soon see. Even though not all PHY chips working around us today have BLW-correction circuitry, the most recent ones all seem to have it.

Summarizing, historically, for 100 BASE-TX applications, the original requirement of 350 μH at 8-mA bias (as per ANSI X3.263: 1995), was included to mitigate BLW, which could cause incorrect decoding of signals with long runs of steady-state signals (+1 or −1). However, in 1997, PHY vendors began implementing BLW-compensation circuitry in their PHY transceivers. PHY vendors later conducted tests and confirmed that PHYs released after 2003 (with BLW-correction circuitry), using transformers having an OCL of just 120 μH, interoperate and achieve low BERs of 10^{-8} or better. These results were obtained with different cable lengths, up to and greater than 100 m.

Stored Energy and Core Saturation

If PoE currents start causing core saturation of the data transformers, the inductance (OCL) will fall, and that can greatly affect signal integrity and cause data corruption. We had started to explain all this in Chap. 1 (see Fig. 1.12). In PoE, because of center-tapping, the current splits up equally in the two winding halves—in an ideal case. If the two windings were *exactly* equal (and the rest of the circuit identical too), we will get zero DC-bias contribution from the PoE currents.

However, we do need to be very clear that even if that were true, in an Ethernet transformer, the current as "seen" by the core may in fact be very different from the current as "seen" by the *copper* (windings). For example, if the currents in two windings (having the same number of turns) are equal and in opposite directions (which means current in one winding goes clockwise, and in the other it is counterclockwise), the net flux, which is proportional to the net ampere-turns, is *zero*. So the core does not "see" any flux (ideally). Nevertheless, the windings obviously do get hot—they do "see" the current. The copper wire gauge (AWG) must be selected based on suitably reducing this self-heating.

In the following discussion, we actually repeat the very same key issues we always encounter whenever we deal with magnetic components in general. As mentioned, the two basic questions to ask are invariably: (a) is the copper of the windings thick enough (in terms of its resistive losses, to handle the required RMS current without overheating)? And (b) is its core big enough (in terms of its energy-handling capability), to avoid saturation?

In the high-power PoE case, we must ensure that the coils can handle 600-mA DC current (per pair) without overheating. A magjack may therefore work satisfactorily at room temperature, but perhaps not at the elevated maximum ambient temperature of our specific application. Again, that depends on our expected operating and environmental conditions.

Still we have only discussed an ideal situation. In reality, there are differences in the two halves. The resistances are not equal for one, so the PoE current does not really split up exactly equally at the center-taps. Rather than 600 mA becoming 300 mA and 300 mA, we may for example get 308 mA in one wire and 292 mA in the other wire of the twisted pair. The currents and flux do not cancel anymore. In fact, the effective DC bias ("seen" by the core) is the *difference current*: 308 mA – 292 mA = 16 mA in this example. And to that we need to add the 8 mA coming from BLW. So all in all, we now need the transformer to guarantee 350 μH with 16 mA + 8 mA = 24 mA of bias current. In this example, the DC-bias rating of the transformer has gone up by a factor of 24/8 = 3, because of PoE-related inequalities. In terms of volume, since core volume is proportional to current squared, the effective core volume has gone up nine times! It is very likely this core can no longer be accommodated in a standard magjack housing. We may have to reduce inductance and run the risk of having higher droop, and consequent higher BERs. Or we need better magnetic design and materials with higher-energy storage capability per unit volume, and so on.

For PoE systems engineers, this highlights the importance of striving to minimize resistance inequalities in the two halves. But we also need to know how resistance imbalance is expressed to start with, and how it relates to the resulting current imbalance. It is not as obvious as we may have thought. We start to examine that next.

Resistance Imbalance

Suppose we have two resistors: one is 10 Ω, the other is 10 Ω × 1.03 Ω = 10.3 Ω. So it seems natural to intuitively opine that the resistance imbalance is 3 percent (because that's how we got the factor 1.03). But this statement is not true always, certainly not in PoE. There are actually two ways of expressing resistance imbalance. One is as per the TIA-568B wiring standard mentioned in Chap. 2, which itself is based on the standard ASTM D4566. (ASTM stands for American Society for Testing and Materials.) So this is basically the American way of expressing resistance imbalance (or unbalance):

$$R_{\text{UNB}} = \frac{R_{\text{MAX}} - R_{\text{MIN}}}{R_{\text{MIN}}} \times 100\%$$

This coincides with our intuitive interpretation above. However, the PoE standard refers to the international (European) standard called

IEC 61156. The AF standard mentions it explicitly, whereas the AT standard just uses its version of resistance imbalance (or unbalance):

$$R_{UNB} = \frac{R_{MAX} - R_{MIN}}{R_{MAX} + R_{MIN}} \times 100\,\%$$

Note that this equation did *not* average the two resistors in the denominator, but just summed them up. So going back to our resistor example, as per PoE standards, the unbalance is

$$R_{UNB} = \frac{10.3 - 10}{10.3 + 10} \times 100\% = 1.5\%$$

This difference in expression is very important to know because when the PoE standard allows a resistance imbalance of 3 percent between the two wires of a twisted pair, it is in effect 6 percent as per the ATSM/TIA way.

For example, the TIA-568B allows the two wires of a twisted pair to have a resistive imbalance of 5 percent. But that is actually only 2.5 percent as per the PoE standard's way of expressing resistive imbalance. The PoE standard asks the PoE system to tolerate up to 3 percent of resistance imbalance in the two halves, which clearly exceeds the imbalance caused by the actual wiring (2.5 percent), and includes imbalance contributions from other sources too, however insufficient as discussed later.

Common sources of DC bias are:

1. Differences in the wire resistances of each twisted pair
2. Differences in the DC resistance (DCR) of the windings of the two halves of the data transformer
3. Differences in connector contact resistances along the way (where mating of male and female RJ-45s occur)
4. DC bias caused by baseline wander (BLW)

We will discuss and quantify all these in the next few sections, and we will also realize that the PoE standard's limit of 3 percent is rather understated (optimistic). The actual worst-case imbalances (DC bias) can be much higher, especially for short cables.

"Imbalance" as Per PoE Standards

A certain "resistance imbalance" (or "unbalance"), creates a corresponding "current imbalance." We need to know how they are connected in PoE specifically.

The previous PoE standard (802.3af) says in Section 33.2.8.12 titled "Current unbalance":

> The specification for I_{UNB} in Table 33-5 shall apply to the current unbalance between the two conductors of a power pair over the current load range. The 10.5 mA value is based on a simulated output current unbalance of 3 percent.

In Annex 33E (informative), the AF standard implies that this 3 percent current unbalance is caused by a 3 percent resistance unbalance:

> The 3 percent cabling resistance unbalance is specified for the ISO/IEC cabling channel illustrated in Figure 33-18. At the maximum current allowed, this resistance unbalance equates to a 10.5 mA difference between the two paths. . . .
>
> Using a transformer that can only tolerate this amount of DC bias reduces the maximum current the PSE can deliver without saturating the transformer to: 350 mA × (8 mA/10.5 mA) = 267 mA.

The math behind the 10.5 mA number is not so obvious and is provided in Fig. 9.3. Note that one resistance is 6 percent higher than the other, yet the resistance imbalance is 3 percent (not 6 percent) as per the IEC method. The AF standard also introduces current unbalance I_{UNB} of 3 percent, and it defines it just the way it defines resistance unbalance. So once again, the ratio of the currents in the two branches is actually 6 percent, not 3 percent. This can all get somewhat confusing. But we will follow it up with several examples to clarify. The most important thing to keep in mind is the connection between resistance imbalance and DC bias. We can summarize that this "connection" is as follows:

1. If the ratio of resistances in the two halves is X, the resistance imbalance *as per IEC* (and PoE standards) is $X/2$. The DC bias is $I_{BIAS} = X/2 \times I_{PORT}$.
   For example, if the ratio of resistances is 6 percent, the resistance imbalance is 3 percent. The DC bias is 3 percent of 350 mA (for AF), which is 10.5 mA.
2. If the ratio of resistances in the two halves is X, the resistance imbalance *as per ASTM* (and TIA standards) is X. The current imbalance is $I_{UNB} = X$. The DC bias is $I_{BIAS} = X/2 \times I_{PORT}$.
   For example, if the ratio of resistances is 6 percent, the resistance imbalance is 6 percent and the DC bias is 3 percent of 350 mA for AF, which is 10.5 mA.

Either way, we get the same final results, but the intermediates are different.

The AT standard retained backward compatibility to the AF standard by allowing 3 percent resistive imbalance and asking that the imbalance in current be 3 percent of the maximum port current.

Example: 3% resistance imbalance (AF case)

$$\text{CHECK: } R_{UNB} = \frac{1.06-1}{1.06+1}\times 100 \approx \frac{0.06}{2}\times 100 = 3\% \text{ (as per IEC method)}$$

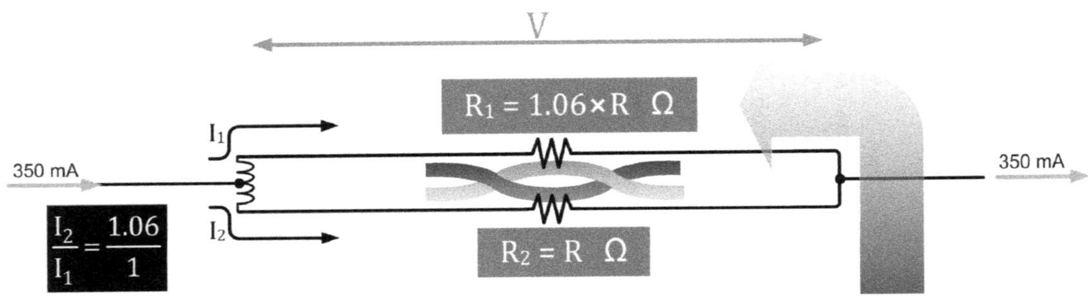

$$\frac{I_2}{I_1} = \frac{1.06}{1}$$

NOTE: Resistance is 6% more, but this is still a 3% Resistance Imbalance as per IEC

$$I_1 = \frac{V}{1.06\times R};\ I_2 = \frac{V}{R}$$ *Individual branches*

$$V = (I_1 + I_2)\times\frac{(1.06\times R)\times(R)}{(1.06\times R)+(R)}$$ *Effective Resistance and total current*

Solving (substituting V):

$$I_1 = (I_1+I_2)\times\frac{(1.06\times R)\times(R)}{(1.06\times R)+(R)}\times\frac{1}{(1.06\times R)} = (I_1+I_2)\times\frac{(R)}{(1.06\times R)+(R)}$$

$$I_2 = (I_1+I_2)\times\frac{(1.06\times R)\times(R)}{(1.06\times R)+(R)}\times\frac{1}{(R)} = (I_1+I_2)\times\frac{(1.06\times R)}{(1.06\times R)+(R)}$$

Ratio of Currents

$$\frac{I_2}{I_1} = \frac{(1.06\times R)}{(R)} = 1.06$$

Current splits up in inverse proportion to resistance in branches

Using componendo and dividendo rule

If $\frac{A}{B}=\frac{C}{D}$ then $\frac{A-B}{A+B}=\frac{C-D}{C+D}$ we get

$$\frac{I_2-I_1}{I_2+I_1} = \frac{1.06-1}{1.06+1} = \frac{0.06}{2.06} \approx \frac{0.06}{2} = 0.03$$

Since $I_2 - I_1 = I_{BIAS}$

(if a transformer with zero DCR is present with equal turns in its halves),

and $I_{PORT} = I_1 + I_2$

$I_{BIAS} = 0.03\times I_{PORT}$

So DC bias is equal to resistance imbalance ratio times the port current

But ratio of split currents is twice the resistance imbalance ratio

$$I_{BIAS} = 3\%\text{ of } I_{PORT} = \frac{3}{100}\times 350\text{ mA} = 10.5\text{ mA}$$

I_2 = 175mA + (10.5mA/2) = 180.25mA

I_1 = 175mA - (10.5mA/2) = 169.75mA

CHECK: Ratio I_2/I_1 = = 180.25/169.75=1.06

FIGURE 9.3 Three percent cable resistance imbalance leads to three percent of I_{PORT} transformer DC bias.

It retained 350 mA for the maximum port current for Type 1, but for Type 2 it changed the max current to I_{PEAK}, which is 684 mA, as we know. So, in effect, it asked for the PSE to tolerate

1. $I_{BIAS} = 3$ percent $\times$ 350 mA = 10.5 mA for Type 1
2. $I_{BIAS} = 3$ percent $\times$ 684 mA = 20.5 mA for Type 2

Clearly, for Type 2, the DC-bias current is twice that of Type 1, so the transformer will need to be four times larger, unless better magnetic materials are used.

Actually, we should also add the 8 mA coming from BLW to each case above. Because the above numbers clearly do not include that. It seems while writing the AT standard, commercial compulsions came to the fore, because clearly, if worst-cases were taken into account, the transformers would become too big to accommodate in the existing housings.

However, in all the discussions, a term called current-imbalance I_{UNB}, seems to have been poorly defined in the PoE standards, and that can cause much confusion today. So we need to clarify that now.

Current-Imbalance I_{UNB}: What Is It Really?

Let us recapitulate what we learned in Fig. 9.3. We defined resistive imbalance in a slightly nonintuitive way. If the resistance in one branch was 6 percent larger than the resistance in the other branch, the resistive imbalance was 3 percent. Since currents entering a parallel network distribute in inverse proportion to resistance, the current in one branch was also 6 percent more than the current in the other branch. However, *since the split current halves were roughly half of the total port current*, this 6 percent translated into 6%/2 = 3% of the total port current. That was the DC bias. So DC bias was calculated to be 3 percent of the total port current.

Now we make an observation: What exactly is the imbalance *current*? Unlike *resistive* imbalance, it is not defined in either the AF standard or in the AT standard. But it is used very freely nevertheless. Hence the possible confusion. The AF standard made several references to it. These were mentioned in a preceding section titled "Imbalance as per PoE Standards." By saying that a 3 percent resistive imbalance led to an exactly equal 3 percent current imbalance, it implicitly assumed that I_{UNB} ratio was a ratio, much like the resistive-imbalance ratio, and it too was defined in an IEC-like manner as

$$I_{UNB} = \frac{I_2 - I_1}{I_2 + I_1} \times 100\% = \frac{I_{BIAS}}{I_{PORT}} \times 100\%$$

For example, for a low-power case, we get 10.5/350 = 3 percent. And this is exactly what was stated in the AF standard and in the AT standard (at places). The problem is that the AT standard makes several self-contradictory references to I_{UNB}, some of which are:

- In Table 33-11, it says that I_{UNB} is 3 percent of I_{CABLE} for Type 1 and 3 percent of I_{PEAK} for Type 2. That is consistent with what the AF standard implied in the very few places it mentions I_{UNB}. The confusion arises in the following places:
  - Section 33.8.3.4, page 105, under "Electrical specifications applicable to the PSE and PD," it states that a 100 BASE-TX Type 2 Endpoint PSE and PD will meet the requirements of Clause 25 in the presence of ($I_{UNB}/2$). All of a sudden, I_{UNB} has gone from a dimensionless *ratio* to an *absolute current*.
  - In Figure 25-1, it states "I_{BIAS} is the current $I_{UNB}/2$."
  - In Section 33.2.7.11, page 51, it states "Type 2 Endpoint PSEs shall meet the requirements of 25.4.4a in the presence of ($I_{UNB}/2$)."

So it implies that I_{UNB} is now not only *not* a ratio, but for some reason is twice the bias current. That would also imply it is asking for Type 1 transformers to be able to handle a bias current of only 10.5 mA/2 = 5.25 mA. That does not make sense in terms of backward compatibility with the AF standard.

We need to ignore $\boldsymbol{I_{UNB}}$ *throughout the AT standard and directly connect from resistive imbalance to DC bias. For that is what really matters eventually anyway.*

Keep in mind that typical PSE-PD compliance suites do not test for unbalanced situations anyway.

Worst-Case Imbalances and DC Bias

The TIA-568B cable spec we had mentioned in Chap. 2 specifies up to 5 percent imbalance *between conductors of a given pair*. Note that *between pairs*, the measured imbalance was around 8 percent as seen on the bench. But clearly, *interpair* imbalance does not affect the DC bias of transformers, only *intrapair* imbalance does. We also know that as per the IEC method, the resistance ratios stated above need to be halved.

There are other contributions to total imbalance that we now try to consider. For example, a typical male-female connector can have a contact resistance typically 20 mΩ (but can be as high as 40 mΩ after aging, though we will ignore the aging aspect here). We can have a case where the contact of one wire of a pair is very good, at 0 Ω, whereas the other contact is 20 mΩ. We can also have multiple such connectors on the way, as the connection is made from the PSE to the PD via patch panels and so on. It is often assumed that up to five connectors can be placed en route, though a more practical case is actually four connectors.

NOTE *There can be typically up to five connectors (male-female combinations) in going from the hub to the PD. The contact resistance for each pin of each connector is allowed to vary from 0 (perfect contact) up to 20 mΩ as per IEC512-2. So in the worst case, we could theoretically have* $5 \times 0.02\ \Omega = 0.1\ \Omega$ *total contact resistance for one particular strand (branch) of a given pair, and 0 Ω on the other strand (of course, that may not be likely, but theoretically it is possible).*

Older transformers could have winding halves of DCR as high as 0.5 Ω with 5 percent imbalance. As per PoE+ magjack datasheets, today the winding halves can be about 200 mΩ and maximum 3.5 percent different. However, we can speculate that if the halves are wound bifilar (two strands wound simultaneously as a single strand) with the same wire gauge, the halves can be, in fact, very well-matched as shown in Fig. 9.4. We will therefore make that assumption in the calculations that follow, but note that if this bifilar winding strategy cannot be confirmed from the vendor, we should assume 3.5 percent difference (resistance imbalance). That will skew the results even more (higher-DC bias).

In Figs. 9.5 and 9.6, we finally show some hopefully "not overly pessimistic" imbalance calculations. The attempt is to see the worst-case DC bias, but still retain a certain measure of reality. The maximum imbalances in current (the maximum DC bias), occur with *small cable lengths*. The reason is large-cable lengths have more cable resistance. There is a significant ballasting (equalizing) effect coming from the component of that resistance which is exactly equal in both branches. The component that is different creates further inequality. It all depends on which effect dominates, but with a maximum 2.5 percent resistive imbalance, the ballasting effect obviously dominates. So long cables help, whereas short cables don't. And that exposes another shortcoming of most automated test suites—just as they have no control of the "48V" supply and can therefore not explore weaknesses in the PoE setup as the input varies, these test suites have no control over the length or resistance of the cable resistance either. But even if they did, a full test suite will need to include both data and power, to see if for example, short cables are causing excessive BERs (bit error rates). In other words, this critical aspect of PoE design—whether the data transformer is sufficient with PoE present on it—is, in reality, rarely tested.

Derating Power Based on DC-Bias Capability

In the AF standard

> Using a transformer that can only tolerate this amount of DC bias reduces the maximum current the PSE can deliver without saturating the transformer to: 350 mA × (8 mA/10.5 mA) = 267 mA.

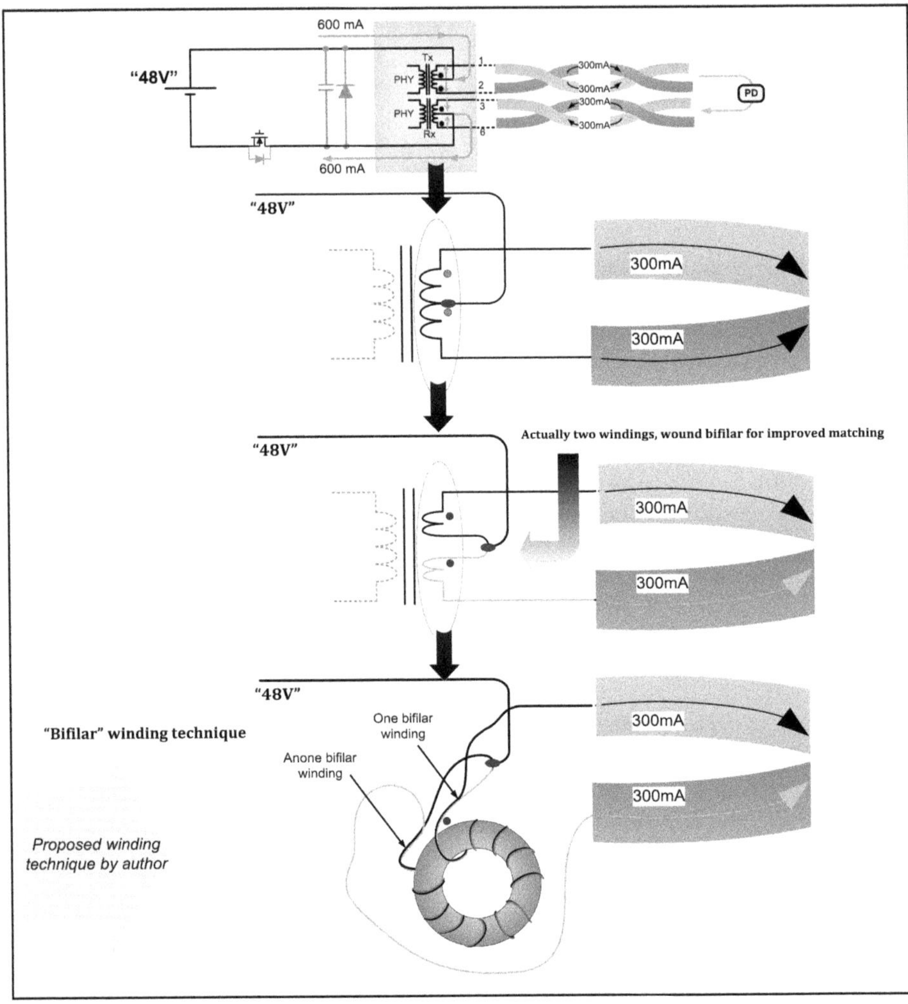

FIGURE 9.4 If winding halves are wound as shown, their relative matching is very good.

We can see that the standard is derating the maximum power so that the DC bias from PoE stays within the capability of the transformer. It is not adding BLW-based DC bias (8 mA) to the number from PoE (10.5 mA). For supporting 350 mA of PoE with 3 percent imbalance, theoretically, we would need a transformer with a DC bias capability of 10.5 mA + 8 mA = 18.5 mA. So a transformer with 8-mA bias is theoretically no good for any amount of PoE at all (zero power). The only reason we get away with PoE (if not PoE+) on a transformer with just 8 mA of bias-capability transformer is that modern transceivers have built-in BLW correction circuitry as discussed previously,

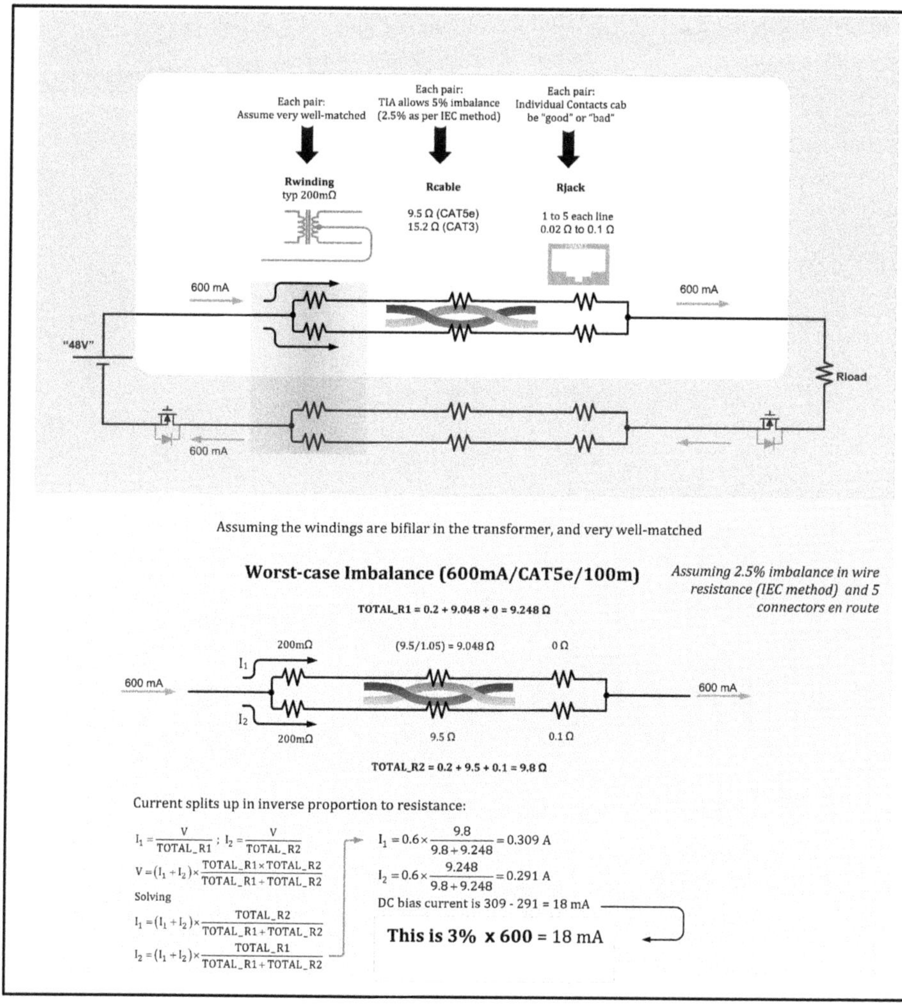

FIGURE 9.5 Calculating worst-case DC bias for 100-m cable.

and can therefore handle the effect of a certain amount of successive 0s or successive 1s. In other words, the 8-mA bias capability of the transformer gets somewhat freed up and available to handle the imbalance coming from PoE currents.

Coming to the effect of resistive imbalances, it can be shown that in the worst case, with five interconnects and just 5 m of cable, the current imbalance can exceed 42 mA even with just 350 mA of PoE current. It may not be practical to look for transformers that can handle this DC bias current—certainly not in a magjack. One way of trying to reduce this DC bias is to introduce *ballasting resistors*. This is discussed next.

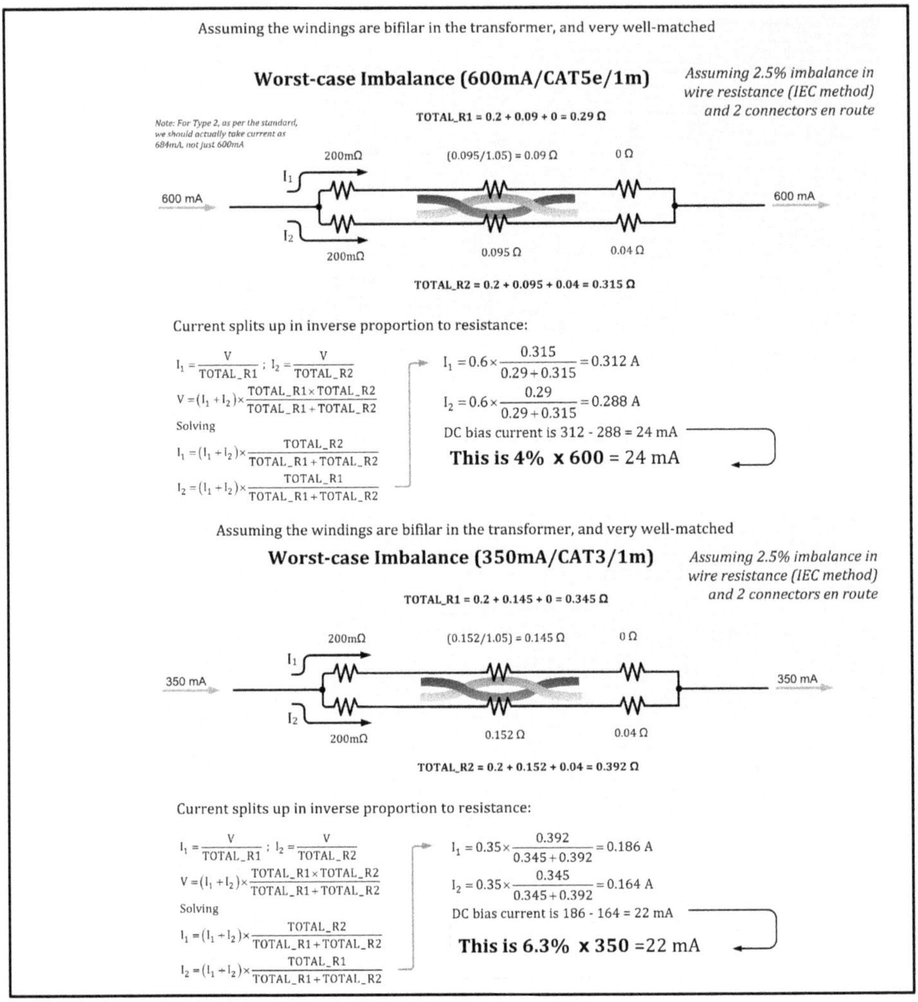

FIGURE 9.6 Calculating worst-case DC bias for short cables.

Ballasting Resistors

These will take the form of two equal resistors (a few ohms each), inserted in series with each half-winding. Though this is often recommended, despite the small associated power loss, and is in principle useful, especially for short cables, the truth is nobody is really using ballasting resistors. In Midspans, however, there is a greater need to use such techniques, since there is no ballasting effect from the DC resistance of the (equal) winding halves of a transformer. Note that common-mode filters are often inserted in series with the PoE rails, and these also help in this regard, often unknowingly. Though the filter is primarily present for EMI filtering and common-mode noise rejection, its windings provide ballasting too. We will discuss EMI filters later. For now we try to write out the equations for ballasting resistors. See Fig. 9.7, wherein we have derived a closed-form equation for setting the ballasting resistance and calculating the net bias current. We see that even two 1.5 Ω resistors inserted in series significantly reduce the bias current from 44 to 18 mA. The latter is within the capability of most PoE+ transformers, so there is then no concern when using short cables.

We have used 2.5 percent resistance imbalance for the cable, but note that is only as per IEC/PoE method. By the ASTM/TIA method, the same imbalance is expressed as 5 percent, which is why we have used the factor 1.05. Also, since the ballasting resistor may have tolerances too, assuming 1 percent resistors, that is actually a delta of 2 percent between min and max, since it is in reality ±1 percent, not just 1 percent.

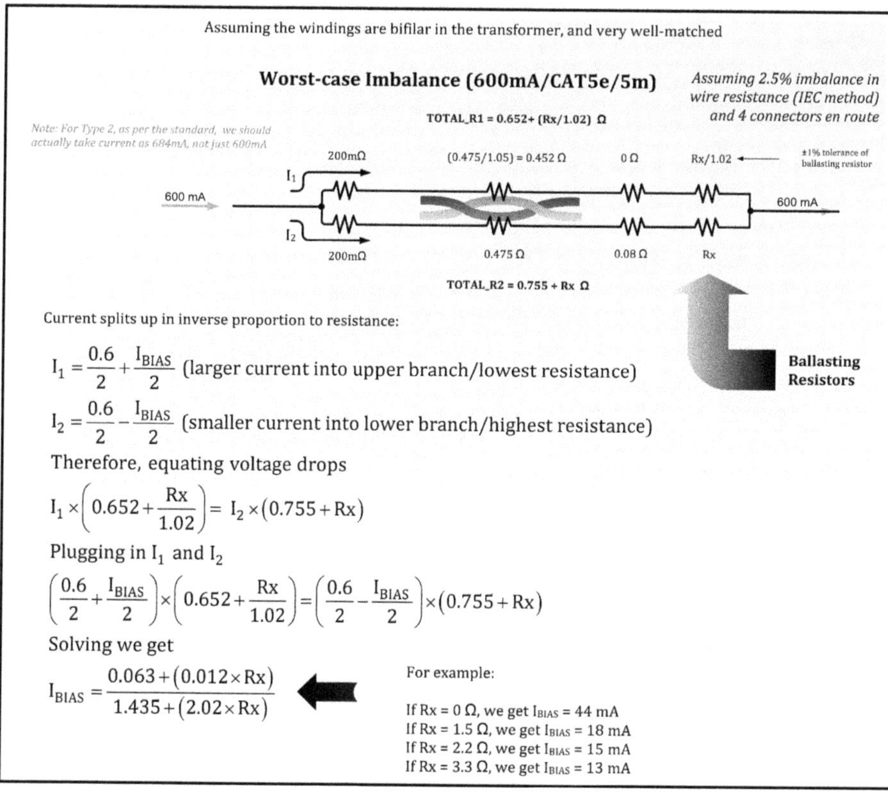

Current splits up in inverse proportion to resistance:

$$I_1 = \frac{0.6}{2} + \frac{I_{BIAS}}{2} \text{ (larger current into upper branch/lowest resistance)}$$

$$I_2 = \frac{0.6}{2} - \frac{I_{BIAS}}{2} \text{ (smaller current into lower branch/highest resistance)}$$

Therefore, equating voltage drops

$$I_1 \times \left(0.652 + \frac{Rx}{1.02}\right) = I_2 \times (0.755 + Rx)$$

Plugging in I_1 and I_2

$$\left(\frac{0.6}{2} + \frac{I_{BIAS}}{2}\right) \times \left(0.652 + \frac{Rx}{1.02}\right) = \left(\frac{0.6}{2} - \frac{I_{BIAS}}{2}\right) \times (0.755 + Rx)$$

Solving we get

$$I_{BIAS} = \frac{0.063 + (0.012 \times Rx)}{1.435 + (2.02 \times Rx)}$$

For example:

If Rx = 0 Ω, we get I_{BIAS} = 44 mA
If Rx = 1.5 Ω, we get I_{BIAS} = 18 mA
If Rx = 2.2 Ω, we get I_{BIAS} = 15 mA
If Rx = 3.3 Ω, we get I_{BIAS} = 13 mA

FIGURE 9.7 Connecting reduced-bias currents with ballasting resistors.

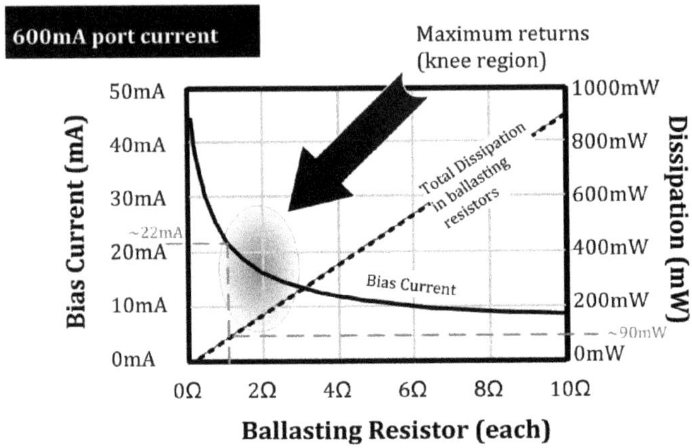

FIGURE 9.8 Plotting bias currents and additional dissipation versus ballasting resistance.

Finally, the equation is

$$I_{\text{BIAS}} = \frac{0.063 + (0.012 \times Rx)}{1.435 + (2.02 \times Rx)}$$

where Rx is the ballasting resistance in each branch. This is plotted out for easy reference in Fig. 9.8. We see that there is a "knee" where the bias current falls dramatically. After that we get diminishing returns. In addition, the dissipation from the two ballasting increases proportional to Rx, so the knee is the optimum design region. As an example, we see from the plot that even just 1 Ω resistors brings the bias down from 44 to 22 mA (half), adding less than 100-mW dissipation. A ballasting resistance of 2 Ω seems well-centered within the shaded-gray "optimal area" of Fig. 9.8. It will bring the bias current down to almost 15 mA, adding barely 200 mW to the total dissipation. So this seems to be the most ideal design point.

EMI Filtering and Common-Mode Filters with PoE

As discussed in Chap. 1, there are two EMI-related concerns that make twisted pairs so attractive: emissions and susceptibility. Noise is picked up in common-mode manner and it is important to not let it affect the signal, which is transmitted in a differential-mode manner. Remember that in Chap. 2 (see Fig. 2.10), we provided the rationale behind adding two small ferrite beads in each line, to prevent the center-taps of the send and receive transformers from getting connected together from the viewpoint of the high-frequency noise signals.

In addition to that, common-mode filters are generally recommended in series with the PoE and data lines too.

In many cases it is OK to just put a small common-mode filter on the PHY-side of the data transformer. That does not interfere with PoE and can be very small. However, some signal-integrity engineers want to put the common-mode transformer on the line (PoE) side. We have to be very careful that the PoE currents do not saturate this common-mode transformer. So clever tricks to create flux-cancelation are used. In Fig. 9.9 we have shown two such methods. The top half shows how to do that with two inductors, each with three windings.

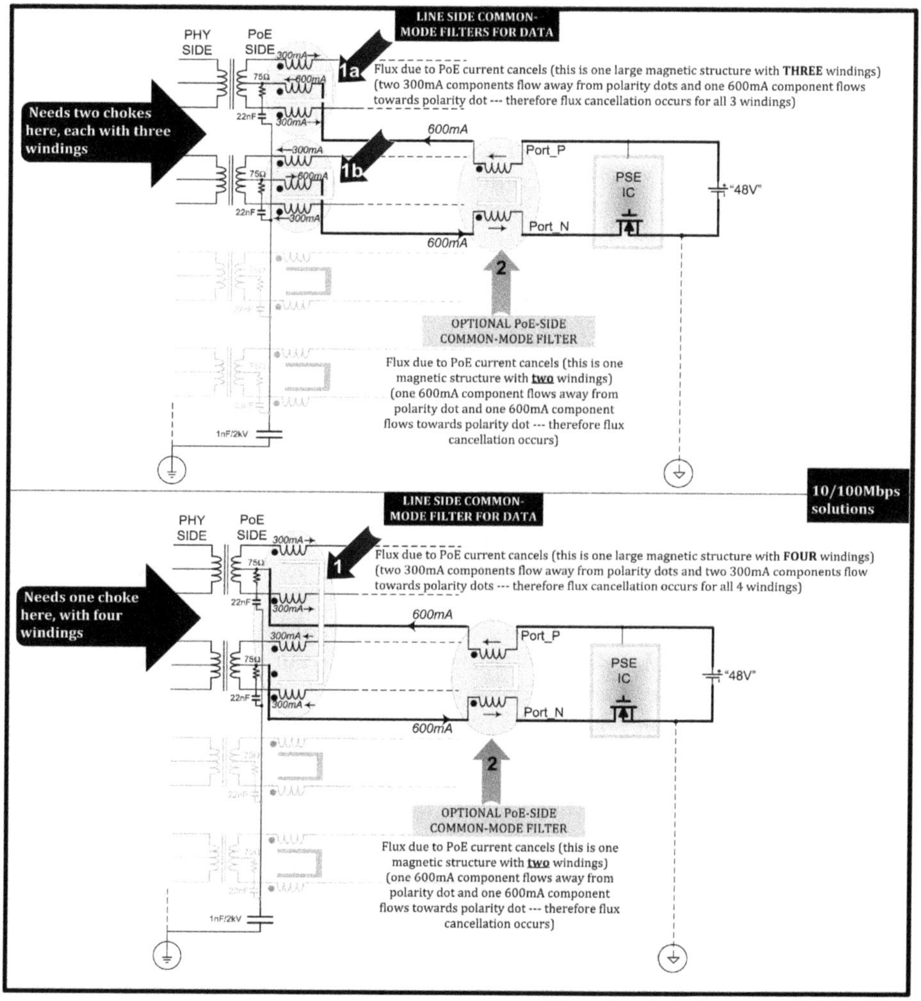

FIGURE 9.9 Two methods of adding line-side common-mode chokes with PoE present.

The bottom half shows how to do the same with a single magnetic structure, with four windings.

Note that in both cases, the main aim was to provide line-side common-mode filtering for keeping the noise and signal separated from each other. The "tricks" were just a way to prevent PoE from affecting that primary function. But as shown in the same figure, an optional common-mode filter is sometimes inserted in series with the rails from the PoE stage. There may be small capacitors on either side of these common-mode inductors, for better filtering (not shown). The purpose of this common-mode filter is not really understood, even by its practitioners. Here are the points to consider in the decision-making process.

1. The PSE and the "48V" supply hardly suffer from any susceptibility problems to justify this common-mode filter. So that is not the purpose of these filters being discussed.
2. If the filters are for meeting emissions requirements, that may, in fact, be of some use. But where exactly is the noise coming from? It is coming from the "48V" switching-power supply. The PSE is just a gate that opens or closes, it can hardly add to any noise coming from elsewhere. So shouldn't we just have better filtering at the output of the AC-DC power supply (i.e., inside it), instead of provisioning for a separate filter stage after the PSE?
3. The common-mode filter windings do have some finite resistance. In which case they will act as ballasting resistors to equalize currents, especially over short cable lengths. But to reduce dissipation, we need to keep them small.
4. There is, in fact, another reason to keep them small. We will see in a subsequent chapter, whenever a surge event occurs (as from lightning far away), a huge amount of energy is unleashed on the lines. The normal way to achieve reliability during this critical situation is to have the surge energy be quickly absorbed in a fairly large electrolytic bulk capacitor—like those normally present at the output of the "48V" switching power supply. We can visualize that in such a situation, adding resistances between the lines and the output caps of the AC-DC power supply can only inhibit the quick absorption of surge energy, causing high-voltage differentials across the PSE, ultimately leading to failure of semiconductors in the PSE. So we should be very guarded about adding large values of ballasting resistances, and also high-resistance common-mode filters for the same reason.

Once we understand all these issues, we realize why some vendors separate the transformer from the common-mode filter altogether.

For example, Coilcraft suggests that for this power level (30 W) the magnetic components do not fit into a conventional Ethernet jack. The Coilcraft solution separates the isolation transformers from the common-mode chokes. We thus have the HPX2126L, which contains two separate data transformers, one for the Rx pair and one for the Tx pair. And this is to be combined with the HPF2187L which has common-mode chokes wound on one core.

Isolation Requirements in Magnetic Components

We will discuss isolation requirements in general in much greater detail in the next chapter. Here we will go through some concerns related to magnetics.

Because the data lines travel through the building, they are prone to picking up rather large, though transitory, voltage spikes, either capacitively or inductively. These spikes often come from motorized equipment operating nearby, because inductors are known to be able to create huge voltage spikes. Or the spikes can be the result of relatively rare but much more severe surge events, like lightning striking far away or atmospheric discharges in general. Besides product reliability, a key concern, in fact the most pressing concern, is that the spikes should not harm the unsuspecting user. The spikes are not considered lethal, but can certainly deliver a nasty electrical shock. Therefore the standards seek to separate the lines from the user by calling for isolation between the line (MDI) and any user-accessible metallic surfaces (the frame/chassis ground).

We should be clear what the AT standard says in particular. In Section 33.4.1, page 68, under "Isolation," it says

> PDs and PSEs shall provide isolation between all accessible external conductors, including frame ground (if any), and all MDI leads including those not used by the PD or PSE. Any equipment that can be connected to a PSE or PD through a non-MDI connector that is not isolated from the MDI leads needs to provide isolation between all accessible external conductors, including frame ground (if any), and the non-MDI connector.
>
> Accessible external conductors are specified in subclause 6.2.1 (b) of IEC 60950-1:2001.
>
> This electrical isolation shall withstand at least one of the following electrical strength tests:
>
> (a) 1500 V rms at 50 Hz to 60 Hz for 60 s, applied as specified in subclause 5.2.2 of IEC 60950-1:2001.
>
> (b) 2250 V DC for 60 s, applied as specified in subclause 5.2.2 of IEC 60950-1:2001.
>
> (c) An impulse test consisting of a 1500 V, 10/700 is waveform, applied 10 times, with a 60 s interval between pulses. The shape of the impulses

shall be 10/700 (10 is virtual front time, 700 is virtual time of half value), as defined in IEC 60950-1:2001 Annex N. There shall be no insulation breakdown, as defined in subclause 5.2.2 of IEC 60950-1:2001, during the test.

The resistance after the test shall be at least 2 MΩ, measured at 500 VDC.

Note that (a) and (b) are just variations of the familiar "hi-pot" (high potential) test because 1500 VRMS has a peak of $\sqrt{2} \times 1500$ V = 2121 V. No one really seems to know why the AT standard picked 2250 V above, instead of 2121 V. Yet they are very close at least.

In (c) we see the mention of an *impulse test*, as opposed to the hi-pot test, which is actually a steady-DC level. Most reputable OEMs, concerned about user safety, disregard (c) completely. But admittedly, there are many smaller vendors that take refuge in (c), disregarding (a) and (b) completely. In fact, (c) is really just cheating, and many have argued it should be removed from the AT standard. We will shortly explain why (c) is so easy to meet.

The hi-pot test most people typically use consists of a 2500-V (2.5-kV) source. As in most product-qualification programs, the hi-pot needs to be applied for one full minute. In production, that can be reduced to 1 s for speed. Keep in mind that the DUT is fully unpowered for a hi-pot test. Also, the current is monitored as the voltage is slowly and steadily increased to 2500 V, and the high-voltage source ("megger") is typically set to trip if the current exceeds 5 to 10 mA, since that indicates a dielectric breaking down somewhere (a failed hi-pot test results).

Is the hi-pot test mandatory? Yes, even though the PoE standard is not mandatory, product safety is constantly monitored in all countries. So Ethernet/PoE equipment in the United States for example, will need to pass Underwriters Laboratories (UL) testing. The entire equipment will be put through safety testing, *but not individual components like capacitors or magjacks*. In other words, if the entire unit survives a hi-pot test, the safety agency will be happy and will likely not dig deeper. They will not, for example, question why a particular component is rated only 2 kV for a 2.5-kV hi-pot test, provided it survived the 2.5-kV test. In other words, when it comes to TNV-1 testing, the mindset is a little relaxed, along the lines: "the proof of the pudding is in the eating." Which is exactly why many low-end manufacturers of data transformers do not do any 1500 VRMS testing, arguing that the end-responsibility rests on the OEM/ODM. But at the same time, PoE in particular is an evolving standard, and there is a possibility that matters are still not fully thrashed out. A key question is how best to do the hi-pot test and ensure safety? We will discuss this further and see how it impacts the design of magjacks in particular.

NOTE *Spikes aside, the basic (underlying) voltage level in PoE, which is a maximum of 57 V, is below the SELV (safety extra-low voltage) upper threshold of 60 V and is therefore considered relatively safe. PoE falls under the safety standards category TNV-1 (for telecommunication-network voltages). And this is what leads to the more relaxed mindset.*

In Fig. 9.10, we show the most common method of doing the hi-pot test in PoE, and calculate the voltages across the caps when we apply a steady DC of 2500 V. We finally take a numerical example based on 22 nF (blocking) termination caps and a 2-nF Y-cap. The math is very simple actually. The two Y-caps of the data pairs come in

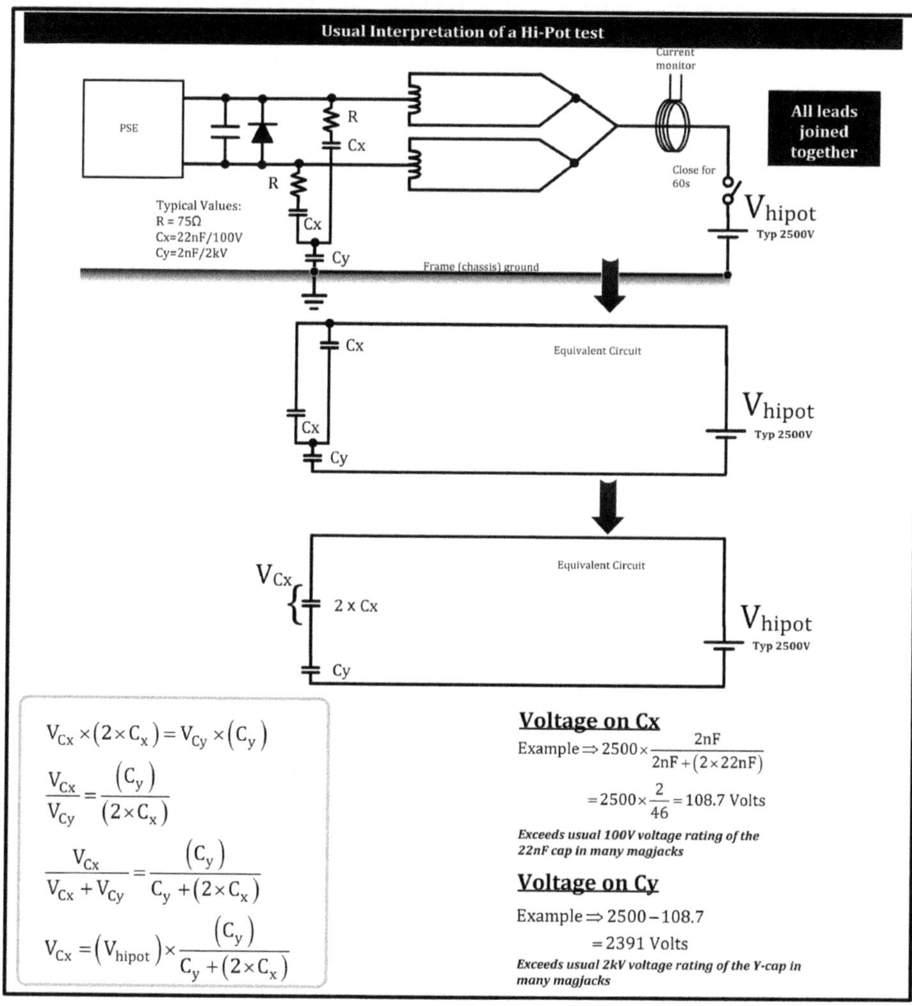

FIGURE 9.10 The usual method of doing hi-pot testing in PoE and the effect on voltage ratings.

parallel and become 44 nF in series with 2 nF. In a series network of caps, the cap voltage divides up in inverse proportion to the capacitance. So since the 44-nF cap is 22 times larger than the 2 nF cap, it takes on 22 times lesser voltage. Since the 2-nF cap gets almost all of the 2.5 kV across itself, the voltage across the two 22-nF caps is about 2.5 kV/22 = 113 V. This is very close to the results of the exact calculation (108.7V). We see that a 22 nF/100 V cap is actually not enough at this location. In fact most magjack vendors continue to put 2-kV-rated Y-caps, knowing fully well that the minimum hi-pot test is 2250 V as per the AT standard. They are, in effect, depending on the fact that most quality ceramic caps can handle 20 to 40 percent higher voltage than their rating. Same for the 22-nF cap selection. This voltage distribution, especially across the termination caps (Cx), *gets worse* if (a) the value of Cx is further reduced, and or (b) the Y-cap value is increased. On the other hand, we can use, for example, 47-nF/100-V caps instead of 22 nF, and/or 1-nF Y-caps. That will help.

In Fig. 9.11, we do the same hi-pot test but without physically connecting all the wires from al the ports together as in Fig. 9.10. Now we do not get the two 22-nF caps in parallel, and in fact we can see the 22-nF/100-V cap is terribly underrated, as it is taking on 208.3 V across itself. Yet, there are magjacks using these sample values. No wonder they can really not pass *this version* of the hi-pot test, and may just say they pass 1500 V test but applied using the 10 μs/700 μs hi-pot profile. Such a fast application of high voltage (though only 1500 V, not 2500 V or 2250 V anymore), does not give enough time for the voltages to become steady across the caps, as they charge up through the 75-Ω termination resistors. That is why isolation test (c) in the AT standard is truly a "tip of the hat" to commercial interests, at the expense of product and user safety, whether intentionally or inadvertently so. It needs to be removed.

This also leaves the lingering question open—which hi-pot test procedure is correct, Fig. 9.10 or Fig. 9.11? We briefly try to answer that now.

Hi-Pot Testing for PoE

Figure 9.10 seems to be based on the old method of testing multioutput AC-DC silver-box power supplies. In hi-pot testing, all the output cables were physically "tied" together, and a high-voltage was applied between that clump of output cables and the metal enclosure of the power supply. The underlying philosophy was that all the paths between output and earth ground were now in parallel, and so the weakest link, whichever it was, would at some voltage level, break down and pass current. *This seems to have been carried over to PoE testing without much thought*. Because we now realize that connecting all the possible breakdown paths in parallel works, but only provided those paths are truly independent. If there is a "sneak" parallel path

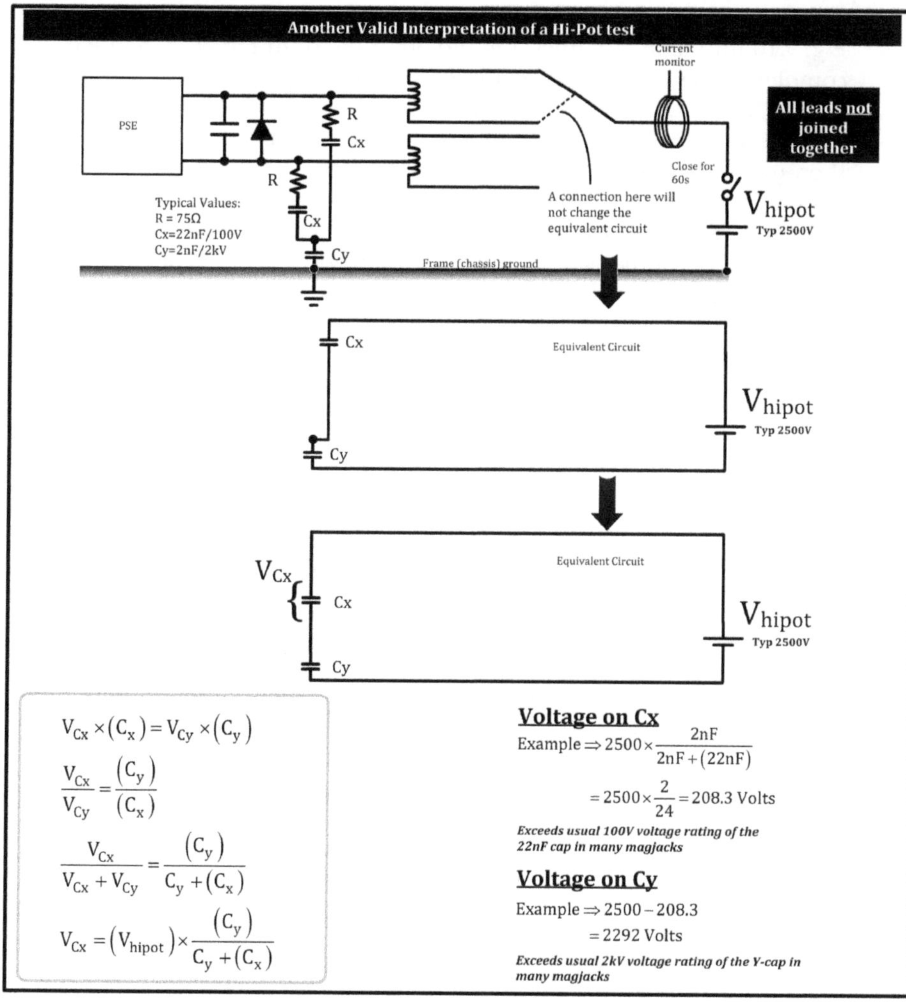

FIGURE 9.11 A better (worst-case, but realistic) method of doing hi-Pot testing in PoE.

present between them, which somehow alters their respective voltage distributions and thus helps pass the hi-pot test, then the underlying assumptions are wrong and we should actually test each breakdown path separately, as we did in Fig. 9.11.

Let us relate this to the basic intent behind safety testing too, that it should relate closely to what can happen in real life. In real life, as we argued in Chap. 1, noise spikes and disturbances are more likely to be picked up equally on the two wires of a given twisted pair—because they are twisted. Unfortunately, the four twisted pairs in an Ethernet cable are not twisted among themselves, so how can we assume that the voltage spike is delivered equally on all of them as Fig. 9.10 implicitly assumes. On the other hand, in Fig. 9.11, had we completed the connection to the other wire of the twisted pair as shown, the equivalent circuit would have remained the same. And the voltage across Cx would still have been over 200 V. It would stand thoroughly exposed in terms of its inadequacy.

Our tentative conclusion is that Fig. 9.11 is actually realistic and a better method for testing for isolation. The "usual" method shown in Fig. 9.10 is perhaps too optimistic and does not replicate a real situation.

Limits on the Y-Capacitance in Magjacks

The basic purpose of Y-caps is EMI suppression. It is a path for common-mode noise to flow into the chassis and away, thus preserving signal integrity, since we know that data signals exist in the differential domain, not in the common-mode domain.

Should we make the Y-caps big or small? Perhaps for EMI reasons we would want to make them as big as possible, but that is not a good solution overall. For example, for ensuring isolation, we have already realized that if the Y-cap is too big, its impedance is very small, so a larger voltage will appear across the termination (blocking) caps (Cx). If we made Cy equal to 22 nF in Fig. 9.11, the voltage on Cx would, by simple divider action, be 2500 V/2 = 1250 V! So, the only reason we can use 100- or 200-V blocking termination caps is that the Y-cap in series is so small, and thus the Y-cap takes on most of the applied voltage across itself.

We will see in Chap. 11 that in surge and cable ESD (CESD) testing, excessive Y-capacitance can severely impact the survivability of the equipment. Here the key criterion is reducing the *net Y-capacitance* of the entire system. In a multiport switch, *all the Y-caps of all the magjacks/ports aggregate, and effectively come in parallel. This is a surprising result to many*. But the logic behind it is shown in Fig. 9.12. A small-AC source, representing an LCR meter, sends out an AC signal. Depending upon through which caps it manages to flow, the total Y-capacitance is measured. Because the other caps are relatively large and behave as shunts, the effective capacitance is very simply

$$Cy_{\mathrm{TOTAL}} \simeq n \times Cy + Cy_{\mathrm{PORT}} + Cy_{\mathrm{AC\_DC}}$$

In other words, the total (net) capacitance from any conductor on the MDI side to the chassis ground is the algebraic sum of (a) the Y-capacitance connected to that specific port, (b) the Y-capacitance inside the AC-DC power supply, and (c) the sum of all the Y-caps in all the magjacks. The latter term is usually $n \times Cy$, where n is the number of ports, since each port has one Y-cap inside it (shared by all

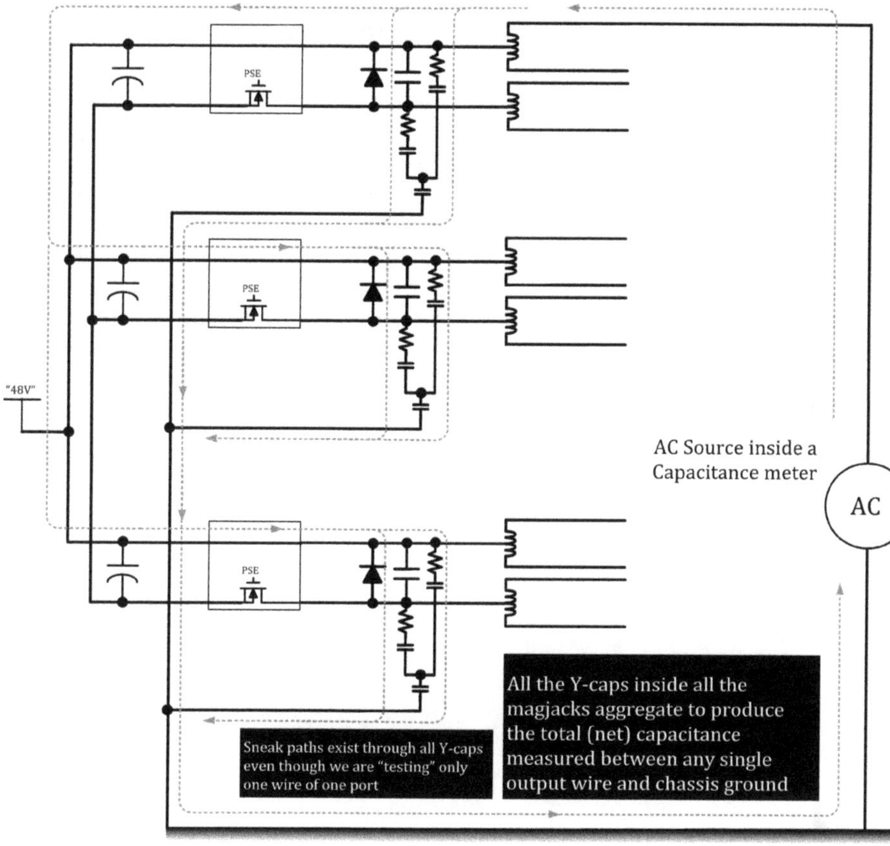

FIGURE 9.12 How all the Y-caps in all the magjack aggregate together.

the twisted pairs of that port). Note that it does not matter whether the Y-cap is connected to the upper PoE rail ("48V") or its return.

It has been determined by several bench tests that this total capacitance is a figure of merit of sorts, and must not exceed

1. ~22 nF for AC-disconnect PSEs, for ensuring their reliability up to about 2-kV surge testing as per EN61000-4-5.
2. ~ 200 nF for DC-disconnect PSEs, for ensuring their reliability up to about 2-kV surge testing as per EN61000-4-5.

We will discuss surge/CESD testing aspects in Chap. 11. Here it is important to know that *if, for example, we have a 48-port PSE with AC disconnect, we must pick a magjack with only 470-pF Y-capacitance inside it, because if we choose 1-nF, for example, we will get a total Y-capacitance of 1 nF × 48 nF = 48 nF, which exceeds our target of 22 nF substantially and will provide reduced surge capability (only up to about 1 kV).*

Vendors Cheating on Y-caps—to Our Advantage

We will now explain how some magjack vendors cheat a little here. We know, or at least assume, that each port should have one Y-cap. But in configurations like 2 × 4, some vendors take the top and bottom ports of this stacked configuration and have all their eight twisted pairs share a single Y-cap. What is worse, their datasheet makes no mention of this.

You will need to cut into the housing to catch this. Or do an LCR measurement as explained previously and you will get half the net capacitance you were expecting. This sharing can cause unwanted interactions between top and bottom RJ-45s and is not considered a very good idea for that reason. But because the net capacitance is significantly reduced, the survivability of such a switch/PSE in surge testing, *especially during cable ESD* (*CESD*) *testing*, is so much higher.

So this is one rare example of someone cheating us to help themselves—and us. Perhaps all they needed to do was *declare* it in their datasheets, and it wouldn't have looked so bad.

CHAPTER 10

Isolation, PCB Design, and Safety

Safety Standards Overview

The international safety standard applicable to information technology equipment (ITE) and telecom (including Ethernet and PoE) is IEC 60950-1. Its first edition was released in 2001. That is the early version we will refer to in this chapter because it is the version referred to by the PoE standard IEEE 802.3at-2009. We note that the second edition of IEC 60950-1 followed in 2005. Yes, there are some changes and clarifications in that over the first edition, but none to cause any major impact on the conclusions and/or recommendations of this chapter.

It is normal practice for countries to "rebadge" the parent IEC standard as their own national or regional safety standard. In this manner, international safety requirements have become increasingly "harmonized." In the United States, the standard is called UL 60950-1, in Canada it is CAN/CSA C22.2 No. 60950-1-03, in Europe EN 60950-1, and so on.

Note that the prefix UL stands for Underwriters Laboratories, the key safety agency in United States. We also note that IEC 60950-1 (which is currently in its second edition) is the successor to IEC 60950 (which went up to its third edition before becoming obsolete). Also note that UL 60950-1 (first edition) was released in 2003, not in 2001, as was its parent document, IEC 60950-1 (first edition).

There is another safety standard, relatively unknown, that some people refer to, and which is actually quite useful. This is called ECMA-287, and it can be located on the Web, at www.ecma-international.org. This website is the home of the "European association for standardizing information and communication systems," or Ecma International as it is more commonly known.

As declared on this website:

> The advent of multimedia products has blurred the borderline between different classes of products, like IT equipment, audio-video equipment,

communication equipment, and the environment within which the equipment is used.

This changing situation has generated a new set of conditions that are to be taken into account when designing new equipment. In order to take into account these conditions Ecma had prepared the first edition of Standard ECMA-287.

The philosophy applied to this new Standard has been to define hazard-based requirements, using engineering principles and taking into account relevant IEC product standards and pilot safety documents. Where technical discrepancies between standards emerged, a conclusion was based on engineering principles.

In other words, ECMA-287 is not distinct from the IEC standard and in fact, is largely based on that. However, it is also more directly an *engineering document*. This also makes it very readable to most engineers. It can certainly be used to better understand the concepts behind IEC 60950-1 and also to clarify various interpretation and/or test issues when they arise. We have referred to it in this chapter quite often.

PoE and Safety

The IEEE 802.3at-2009 standard refers to IEC 60950-1 at two places as shown in Fig. 10.1. We are initially focusing on the first row of the figure, Section 5.2.2, which is the test procedure subsection under Section 5.2 titled "Electric Field Strength." That is colloquially referred to as the telecom/PoE hi-pot (high-potential) test requirement. We first introduced this in Chap. 9 with reference to magjacks, and the reader is advised to read that first.

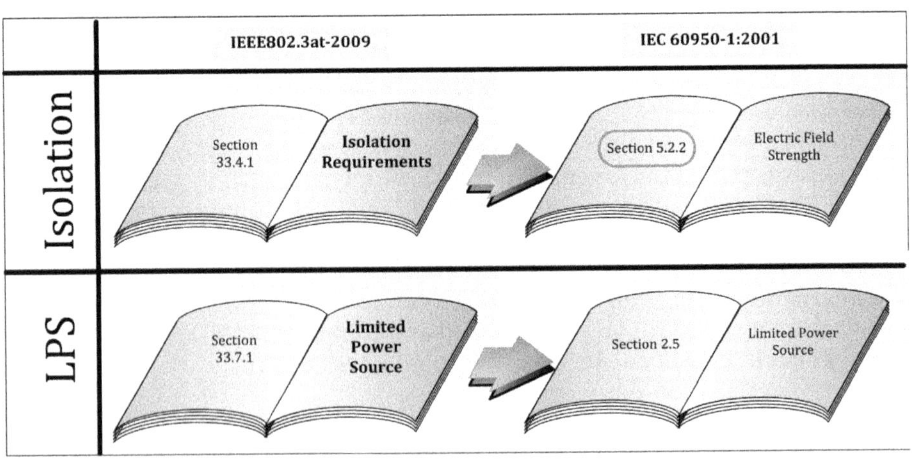

FIGURE 10.1 IEEE 802.3at references to the International Safety Standard.

Steady and Transient Voltages

The first thing we need to understand is that voltage is by definition a potential *difference*, and despite being stated as an absolute number, actually implies an electrical gradient between two points. In a typical circuit, we implicitly measure and state the voltage at any point with respect to a certain reference rail, which we have designated as the system ground. However, in safety, we are concerned with voltages measured with respect to *earth ground*. That reference determines whether a voltage is considered hazardous (capable of causing electrocution) or safe. We ask why on earth is earth ground so important in safety? Because that is the *safest level*—we stand on it every day.

So from a safety perspective, "grounding" means connecting something to the earth ground. For example, we typically ground the metal enclosure of a device, with a view to making it safe to touch, but it also acts as a shield for EMI.

We can visualize that an unsuspecting user/operator can get "zapped" or electrocuted, if any easily accessible exposed metal surface (e.g., the enclosure or an output terminal or lead) has a dangerously high voltage appearing on it—with respect to earth ground. If a user touches such a surface, the potential difference between that point and earth ground will drive a current—through the user. That is not pleasant at the minimum.

The safety standard identified 60 V as the upper threshold of what is referred to as SELV, for safety extra low voltage. This voltage is completely safe to touch and can therefore be allowed on exposed metal surfaces, such as the device enclosure, or output cable ends or connectors.

A high, possibly hazardous voltage level, can be either steady (DC) or it can be temporary/transient in nature. A steady high-voltage level is certainly dangerous, but we can also have relatively safe *momentary* overvoltages riding *on* what is basically a safe SELV baseline level (0 to 60 V). This voltage profile is still considered quite safe. It can cause an unpleasant sensation to a user if somehow accessed and touched, but it is not considered lethal. Safety standards allow this type of voltage onto exposed metal areas *which are not easily accessible*—like the metal contacts in an RJ-45. Such devices come under the safety category called TNV-1 (TNV stands for Telecommunication Network Voltage).

Many telecom lines such as Ethernet (with or without PoE), fall into the TNV-1 category. The reason for the overvoltages (spikes) is that the data cables pass very close to noisy mains wiring, picking up voltage spikes inductively or capacitively as they snake through the building.

Fault Conditions

A basic underlying concept in safety testing is that of single-fault conditions. Safety standards require that the user of any equipment or device be protected, not only under normal operating conditions, but also under overloads and single component failures. In general, abnormal conditions include overloads and single-faults. A single-fault for example, could be just one discrete component anywhere on the PCB, failing either open (perhaps missing altogether in assembly or production) or shorted (perhaps because of its failure mode). For example, the P-N junction of a zener device can simply melt under excessive energy dissipation and finaly fuse in a shorted condition on cooling. Or we could have an inadvertent/intermittent bridging caused by excess solder. Or perhaps some screw/hardware coming loose inside the enclosure.

Safety testing requires that the user remain protected under any *single* fault condition. It is implied that two *simultaneous faults*, both breaking down at the same spot in time, are highly improbable. Like, for example, not one but two layers of insulation, somehow getting weak or developing a defect at exactly the same spot at the very same moment, *despite the fact that each layer was supposed to, or was designed to be, sufficient to bear the entire voltage differential across the barrier on its own*. Or consider two components in series, both failing open or shorted, at the same exact moment, though each was supposed to be robust enough to handle the entire situation (current and voltage) on its own, and so on. The safety standard essentially ignores any *non* single-fault conditions as *highly improbable*.

The most hazardous voltage for us everyday is obviously the incoming AC (mains) line. But, to generate a SELV rail from that, not only does the AC-DC switching power supply need to step down from a high voltage to a voltage below 60 V, but several other conditions need to be met too, to guarantee that no single fault condition anywhere can cause a hazardous voltage to appear on designated SELV areas (enclosure, output leads, and so on).

The general aim of safety is that there should be at least *two* layers (levels) of protection present between any exposed metal surface (accessible to a normal user) and the shock hazard (the incoming AC mains for example). So, if one gives way, there is another level of protection still present, fully capable of protecting the user. For example, inside a typical step-down transformer of a switching AC-DC power supply, we will find at least two layers of insulator (safety agencies actually expect three layers in this particular location for ensuring safety). These three layers are placed between the primary (mains) winding and the secondary (SELV) winding.

In this chapter we will often refer to a "layer" of insulation/isolator in a more abstract sense—as a *level* of protection. It could still just be an actual *layer* of insulator with appropriate ratings. Or, it could be, for example, air, which is actually also an insulator—the thickness of this particular insulator is in effect the *distance through air*, or the *clearance*.

Colloquially, it is often said that in AC-DC power supplies, we need internally 4 mm of clearance. But that is actually just *one* level of protection, out of the two required between AC mains and SELV. Yes, 8 mm of clearance would constitute two such levels of protection and would suffice.

In grounded (earthed-enclosure) equipment, the grounding constitutes one level of protection. We can therefore go ahead and *combine* protection levels. For example, a metal enclosure which is separated from hazardous circuitry by 4 mm of clearance and is *also grounded* is considered SELV (safe to touch). But if we wanted to make our circuit very compact and did not have the luxury of providing 4 mm of clearance, we could instead use a suitably rated insulator. If the ground is not assured, we need *two layers* of insulation, since in that case, the grounding really does not count. Let us explain this a little better.

Grounding (earthing) is usually considered to be one possible level of protection. But for it to really be counted as one such level, it needs to be assured. For example, there should be a firm connection to a water pipe somewhere in the facility. On the equipment itself, there must be grounding screws on the enclosure, and so on. If all the requirements of the safety standard are met in this regard, grounding of the metal enclosure is considered "real" from a safety perspective, and only then does it count as a valid level of protection. But in addition to that, we still need one layer of isolation/insulation. That layer is called basic isolation or basic insulation. Note that in this configuration, lack of grounding may still occur, but it is *considered highly unusual*. It is therefore treated like all other possible failures—as a single-fault condition. And that still leaves us one of the two levels of protection present to protect the user.

Equipment that uses grounding as a valid level of protection falls under Class 1 category in the safety standard. Portable equipment with exposed metal surfaces, but with only a two-prong AC plug for example, falls under Class 2, since it does not rely on grounding. A Class 2 device will have two physical layers of isolation/insulation between exposed metal surfaces and hazardous circuits. These two layers are then called basic and supplementary isolation/insulation. Together they constitute "double" isolation/insulation. Or we could have just one very sturdy insulator that counts as *two* layers, in just the same way as 8 mm counted as two levels of protection. This single insulator, that does the job of two levels of protection, is called *reinforced* isolation/insulation.

Note that Scandinavian versions of the IEC safety standard incorporate stricter Nordic deviations. For one, they *do not accept grounding* as a valid level of protection. The underlying reason is most buildings there do not even have a three-prong AC outlet (no grounding). So lack of ground is not considered a rare event, but *the norm*. That is the reason, why equipment being made for worldwide use, ends up using reinforced or double insulation anyway, whether grounding is intended or not.

Keep in mind that grounding is not just for safety, but for EMI suppression too. So grounding may or may not count as a safety level, but it is certainly valid (and necessary) for meeting EMI limits.

Another way of looking at safety is that we need to provide 100 percent "headroom" for single-fault conditions. For example, an SELV output, which as we know is 60 V or less under normal DC-operating conditions, is actually allowed to go up, under single-fault conditions, to twice that level: 120 V (albeit for less than 0.2 s). The point is that 120 V is still an essentially safe level, and is therefore called ELV (extra-low voltage). Yes, the user is still safe. But this particular voltage level has no further headroom left. In other words, ELV (120 V) is the equivalent of just one level of protection.

Summarizing, we realize that SELV outputs must have two levels of protection from hazardous levels (like the incoming 240-V AC mains supply). If one layer/level breaks down, as under a single-fault condition, we still have one layer/level left to protect the user from the hazardous voltage. On the other hand, an ELV level is separated by only one level of protection from the AC mains. Under a single-fault condition occurring on an ELV surface, the user would be exposed to the full-hazardous incoming-AC mains. That is the reason under normal use, ELV levels are *not* allowed on any easily accessible metal surfaces,and usually not even on the relatively inaccessible metal contacts of an RJ-45 for example. Only SELV levels are allowed on such metal surfaces.

PoE is not really SELV, but SELV-derived, and is therefore allowed on the RJ-45. We will explain that better in the next section.

PoE Rails, Ethernet/Telecom Systems

In PoE, as per the governing IEEE 802.3at standard, we typically use a "48V" rail coming from a silver-box AC-DC power supply positioned at the input of the PoE circuitry. Note that the PoE voltage rail does not exceed 57 V, so it is by choice an SELV-voltage level.

But is it really SELV just because it is less than 60 V? Alternatively we can ask is it really completely safe? Not necessarily so! Because as indicated earlier, a low-voltage level by itself does not constitute an SELV rail right off the bat. A SELV rail is a SELV rail not only because of what it is, but what it will be after a single fault! For example, a sub-60-V rail must also have two levels of protection present, between itself and any hazardous voltages. These protection levels need to be present inside the power supply and everywhere else too, wherever required, for it to be considered SELV. For example, if we had a 48-V output from an AC-DC power supply that had only one level of insulation/isolation between the AC mains circuitry and the output rails, its output would be considered ELV, *not SELV*. Such a power supply would *not* be suitable for use in PoE applications. Because PoE voltage rails always need to be based on SELV rails as per IEC 60950-1 (see Fig. 10.1).

In the described case, we could however start with an ELV rail of 48 V, and follow that up with a 48 to 48 V isolated DC-DC converter module/brick (based on the buck-boost/flyback topology). That module then needs to have only one level of protection from its input to its output. But cumulatively, from the AC mains, we would now have two levels of protection, and also a low final output voltage (< 60 V). So the output of the module would certainly qualify as an SELV rail, suitable for powering up PoE circuits. It is safe.

We now need to recognize that long data cables (with or without PoE present on them) can pick up transient overvoltages, typically as high as 1000 to 1500 V. These spikes can be the result of capacitive or inductive coupling as the cables pass through the building. Or they could also arise from deliberate injection (say by telephony ring tones or Ethernet "link pulses" superimposed on the "safe" DC level). Or by atmospheric discharges (surge events), and so on. They can then appear on either end of the cable, where if a user contacts them, he or she can get "zapped." So yes, though PoE circuitry is SELV to start with, once it is connected to a telecom network, it is no longer completely safe, or at least not very comfortable to touch. Hence the need for isolation in telecom circuits, with or without PoE present.

As mentioned, PoE is SELV-derived (TNV-1): its baseline voltage is less than 60 V, and it has two levels of protection up to the mains, and there are also temporary (transient) overvoltages riding on it. Just for information, other telecom categories are TNV-2 (no transient overvoltages present but steady voltage exceeding 60 V), TNV-3 (DC voltage exceeding 60 V and with transient overvoltages present), and also RFT. RFT stands for remote feeding telecommunication, (steady voltage greater than 120 V, XDSL applications, not covered by IEC 60950-1). We will ignore RFT completely here.

What exactly are TNV circuits? By definition, all TNV circuits—TNV-1, -2, and -3—share one important characteristic: *user access is not easy to any exposed metal connected to them.* The general rationale behind TNV circuits is that an untrained, unsuspecting person is not going to get (easily) zapped when working around TNV, especially when only TNV-1 is present.

Of the three TNV levels, TNV-1, in which category PoE falls, is the safest of all, because even if access is somehow gained to any exposed metal surface where it is present, the steady voltage there would be found to be less than 60 V (under normal operation). And the only high voltage present on TNV-1 is *transient* in nature—enough to cause discomfort, not kill. Note that a "test finger" is defined in the safety standard to replicate a human finger and check if someone can unknowingly or unwittingly touch any exposed metal with TNV present on it, such as the copper inside an RJ-45.

Nancy Araway of NAF Consulting, expresses the situation rather succinctly on the Web at http://www.mail-archive.com/emc-pstc@listserv.ieee.org/msg31257.html:

> Telecom line overvoltage spikes are the primary reason for ensuring separation of TNV and accessible SELV, since they are considered to be a "normal" environmental condition of the telecom line, not an abnormal or fault condition.
>
> That said, ring voltage will still knock you off a metal ladder, (or do serious harm to an infant in wet diapers on a conductive surface) and are required to be "not readily accessible" as per the rounded test finger, including under single fault conditions.

Isolation Requirements

Under any typical overvoltage spike, such as we may see on TNV-1/PoE/MDI circuits, we do not want the insulation between the PoE/MDI side and the host side to break down, because *that would allow the transients to come on to the easily accessible metal surfaces*. In other words, we want the interposing dielectric/insulators to be able to withstand a typical overvoltage spike (~2 kV). That is the purpose of the hi-pot test conducted in PoE and telecom in general (discussed in Chap. 9), and the reason for the first row in Fig. 10.1.

We must point out that in PoE, there are actually no *specific* requirements in terms of any insulators, their thicknesses, their voltage ratings, physical separations, or grounding concerns. The only mandatory requirement is the result itself—that of passing the 1500 V RMS hi-pot test. If we pass the test, however we do it, we are through. That is all that is expected of us by the PoE standard and the safety standard. Because, the voltage profiles we are dealing with (SELV and TNV-1) are both *essentially safe*.

Please refer to Figs. 10.2 and 10.3 to understand more clearly what a typical Ethernet system looks like. Note that there are two key domains in effect: the PoE/MDI domain, which is "hot" (has overvoltages, TNV-1), and the "host side," which is "cold" (safe to touch, SELV). The latter domain connects to the chassis/enclosure either directly, or through a small 0.15-μF/100-V Y-cap typically. This cap does not count in terms of voltage rating or isolation and is there for EMI reasons primarily (not to shunt *too much* common-mode noise on to the metal enclosure, perhaps causing it to radiate).

In Fig. 10.3 observe the hazardous voltages related to the AC mains (labeled Primary Circuit). That is actually the *third domain*. As mentioned, per the safety standard, the Primary Circuit is separated by two layers from both the SELV sections and from the TNV-1 (PoE/MDI) sections. That is why, in the power transformer of the high-voltage switching power supply shown, we have indicated several

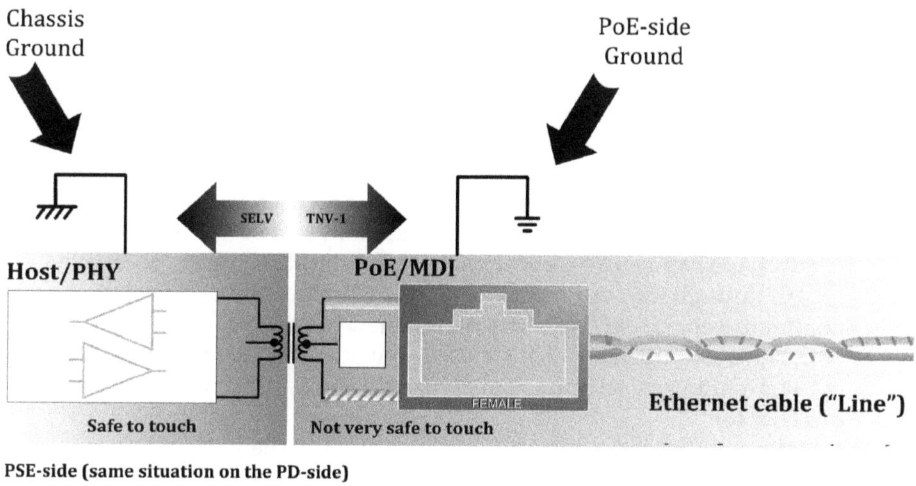

Figure 10.2 Where SELV ends and TNV-1 starts.

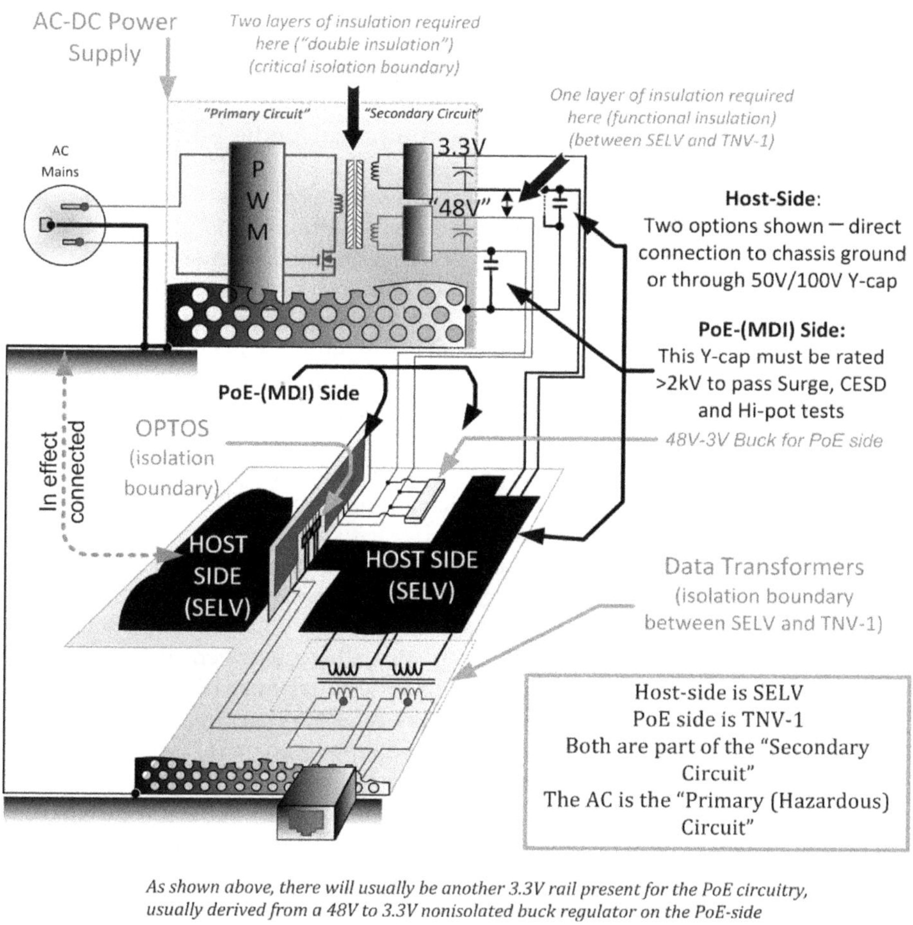

Figure 10.3 A typical single-port Ethernet switch showing isolation boundaries and "domain."

layers of isolation inside—in practice, three physical layers of insulating tape are used between primary to secondary windings, since it is implicitly assumed that one layer of tape can have pin-holes or prior defects to start with. Since there is no specified test to check for that, it is assumed one of the three layers may not be valid at all. So in effect, we have two levels of protection from three layers of tape.

All through the system, *including the PCBs*, there must be a certain minimum separation/isolation maintained between the hot (MDI/PoE) sides and the host side. Otherwise we will likely fail the hi-pot hi test. Defining those minimum PCB separations is the main focus of this chapter.

Some of the high-voltage Y-caps ascribed to the PoE/MDI side, may in fact be "hidden" inside the AC-DC power supply. It is very important to check this out, since there have been cases where power supply manufacturers made telecom power supplies just the way they were making used to making them for ITE, and really did not understand that in PoE, there are domains that needed to be kept separated. If they fail to keep these domains separate, they end up not only violating PoE and safety standards, but seriously compromise the reliability of the equipment under surge/CESD testing. Mysterious failures may occur and they could be the result of flashovers occurring inside the power supply itself. The first step when troubleshooting mysterious field failures is to rule out the AC-DC power supply! PoE systems engineers must open the power supply and confirm connections and ratings as shown in Fig. 10.4. Note that to pass the hi-pot test, any Y-capacitor present between PoE/MDI side and the chassis ground, or from PoE/MDI to host-side, must be able to withstand at least 1500 V RMS. Since 1500 V RMS has a peak value of $1500 \text{ V} \times \sqrt{2} = 2121 \text{ V}$, we actually need a Y-cap rated at least ~2500 VDC. As pointed out in Chap. 9, magjack vendors put 2-kV caps and manage to pass the hi-pot testing at the last stage when the equipment is submitted for testing to safety agencies.

Strictly speaking, since we are not relying on grounding as a layer of protection, and we are assuming that the incoming "48V" rail is SELV with two layers of protection to hazardous mains voltage, there are no mandatory isolation requirements in place here between SELV and TNV-1 domains. In other words, we could use any standard 2.5-kV capacitor on the PoE circuitry (even inside the AC-DC power supply on the secondary side). It does *not* have to be a UL-approved or TUV-approved safety capacitor, with all the well-known approval stamps/markings on it. However, many vendors do not realize this and place an expensive safety-agency approved Y-cap. That is acceptable if not unnecessary. But it does help meet much higher surge and hi-pot ratings.

Though its marking may indicate 250 V~ or 250 VAC (referring to the mains voltage), or even 2 kV, by IEC standards, any approved Y-cap for off-line (mains-related) applications must undergo 100 percent

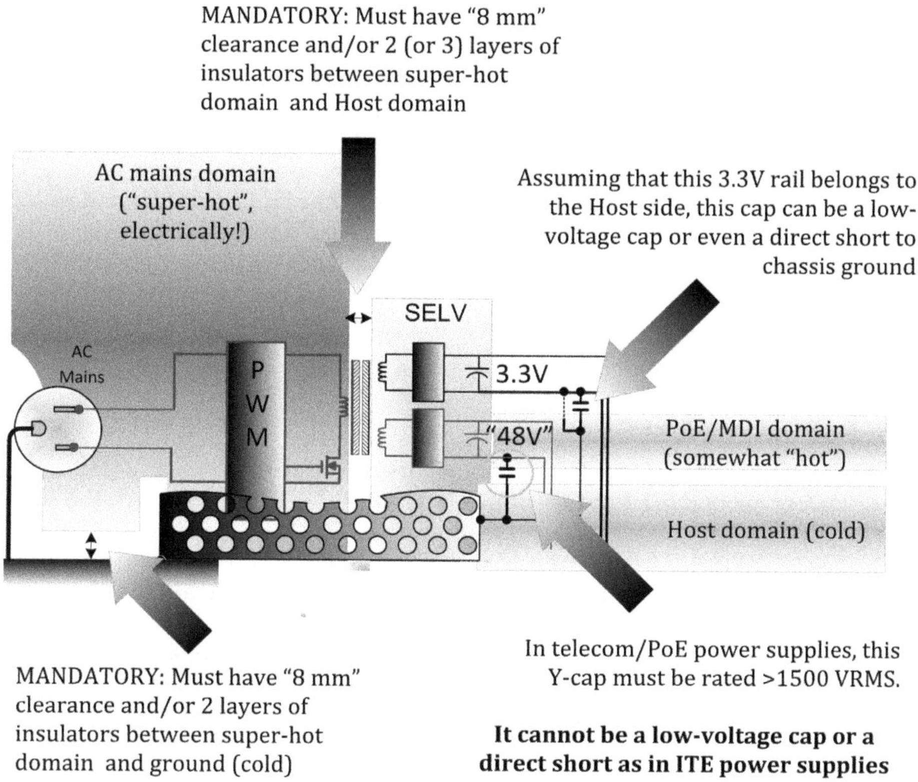

FIGURE 10.4 Internal construction for AC-DC power supplies for PoE applications.

production testing with an applied impulse voltage anywhere between 2.5 and 8 kV. So a safety-approved Y-cap more than suffices in this location. However, a plain 2-kV capacitor instead of a plain 2.5 kV capacitor, as used by some magjack/ICM vendors, is actually questionable, and may lead to hi-pot, surge, and cable ESD (CESD) failures. Incidentally, agency-approved X-caps (placed across the line), are also freely used as Y-caps (to earth ground). And though that is not really recommended, it can work in PoE applications. The X-caps are usually marked 250 VAC, but are impulse tested up to 2.5 kV—since they are expected to deal with severe voltage spikes that can occur differentially across the live and neutral wires of the incoming AC line. Keep in mind that AC wiring in the house is certainly not composed of nice twisted pairs, so differential noise pickup can be severe unlike in telecom. X-caps need to handle those differential spikes.

Even if PoE was not present, the data cables would obviously pick up transient overvoltages. That is one reason why the host-side is always separated from the MDI side by a data transformer (isolation plus common-mode rejection). Therefore, all Ethernet-data transformers, with or without PoE present, need to pass a hi-pot test of their own. The hi-pot test for data transformers when PoE is present is actually a little more stringent (1500 V RMS) than when PoE is *not* present (1000 V RMS).

Another reason why data transformers are used in the first place is to break up ground loops as discussed in Chap. 1. Because if we do not do that and connect equipment separated hundreds of meters apart directly, we will get strange ground bounces and huge ground loops with common-mode currents flowing through the earth ground. That cannot be good for EMI or for signal integrity.

The PoE Hi-Pot Test

We discussed this in Chap. 9 too. Here we will get into more detail. As per Section 5.2.2 of the safety standard IEC 60950-1, to which the PoE standard IEEE 802.3at refers (see Fig. 10.1), we need to apply 1500 V RMS AC (50 or 60 Hz) for 60 s across the boundary that needs to be tested. This is the high-potential or hi-pot test. We are thereby testing the voltage-withstanding capability of everything that separates the following two circuits—TNV-1 (PoE/MDI side) and SELV (host side). Note that "everything" includes circuit components (like the data transformers and the high-voltage Y-caps we talked about previously), the interposing air (leading to the concept of minimum physical separations), and the PCB itself where TNV-1 and SELV traces can run adjacent to each other.

Basically, 1500 V RMS has a peak value of 1500 V × √2 = 2121 V. We are allowed to use either an AC voltage during the hi-pot test (AC hi-pot test), or its peak equivalent DC (DC hi-pot test). The latter is obviously preferred if there are capacitors bridging the isolation boundary since they will have low impedance and mimic a dielectric "giving way." For the same reason, the applied-DC voltage must be ramped up slowly to its final value.

We note that IEEE 802.3at alternatively refers to the test voltage as 2250 V DC. Everybody ignores the slight difference—because there is really no practical difference between a device/equipment designed to withstand 2121 V versus one that can handle 2250 V. But in fact, this test level is often stated to be 2.5 kV peak—see Section 6.2.2.1 of IEC 60950-1. For that reason, 2.5 kV is actually our basic design target in component selection, PCB design and so on, even in PoE.

Note that the device under test is not operational during the hi-pot test—it has no input voltage. To pass the hi-pot test, the basic criterion is that the current passing across the boundary by application of the prescribed applied voltage source, *remains controlled* during the entire test time. No upper limit for this crossover current is actually specified. But typically in all hi-pot tests, the trip level of the hi-pot tester (often called a Megger) is set at 5 or 10 mA. If during the test, the tester does not trip, it implies there was no insulation breakdown. But the resistance *after* the application of this high voltage must also be measured, and needs to be greater than 2 MΩ when measured with applied 500 V DC—that is, a current less than 500V/2 MΩ = 250 μA. If that is met, the device under test passes the hi-pot test. Otherwise it fails.

Note that perhaps a better way to carry out the hi-pot test is to actually increase the trip level of the Megger starting from a few tens of microamperes, in small steps, during the initial setup phase. That way we can account for "normal" currents—those that do not represent any breakdown of insulation. This characterization is especially important if we do an AC hi-pot test as opposed to a DC hi-pot test, since the value of capacitance across the safety barrier can vary, causing corresponding variation in the normal current flowing. In general, for PSE testing, we may prefer to set the trip level of the Megger much lower, say between 0.2 and 1 mA, instead of the standard 5 mA/10 mA. That way we can perhaps avoid inadvertent and serious damage to the boards under test.

The safety standard also makes several recommendations about the number of times we need to *repeat* the test, and also how to alternate its polarity and so on. Though we realize that if we prefer to use 1500 V RMS AC instead of 2250 V DC, the alternating of polarity happens automatically! However, as indicated, AC-waveform testing has its own problems. For one, the trip level may have to be set much higher if we find large caps are allowing significant AC current because of their lower AC impedance—this being normal current flow, certainly and clearly not representing insulation breakdown.

Using AC, or even using DC and alternating polarity, may have other inherent problems. For example, electrolytic capacitors may get reverse biased in the process of the test. We may need to remove them to do the hi-pot test safely. In other words, hi-pot testing can be more complicated than seems at first sight. We also need to refer to Sections 6.1.2 (subsection 6.1.2.1), 6.2.1, 6.2.2 in the safety standard for more details on testing safety of insulation of telecom networks in particular. It is in these sections we will also see that if PoE is not involved, the typical hi-pot test for data transformers in telecom networks is 1000 V RMS, not 1500 V RMS.

In the case of AC-DC power supplies, the hi-pot test needs to be done at 3000 V AC or 4242 V DC, corresponding to two mandatory levels of protection, each level being 1500 V AC or 2121 V DC. In production, since it is not practical to wait for 60 s for each device, we are allowed to reduce the test time to 1 s, so long as we pass the entire test suite during the initial product validation phase. The hi-pot test needs to be done during production, on every single outgoing manufactured device. Safety agencies are also allowed to walk in any time to check compliance. But in PoE, as mentioned in Chap. 9, this is not mandatory, and a lot of corners are skipped even by magjack vendors.

Failing the Hi-Pot Test

We can fail the hi-pot test for any number of reasons. With special attention to PoE, some of these reasons are

1. Capacitors bridging the two domains may have inadequate ratings. In PoE these caps are the Y-caps from PoE/MDI side to the enclosure/host. They should be 2.5 kV, but most vendors use 2 kV.
2. We also showed in Chap. 9 that the 10- or 22-nF blocking caps in the terminations can fail because of the voltage distribution.
3. The same could happen with the port capacitors, especially if they are very small. We can calculate how much voltage will directly fall across the port capacitors (both the PD and PSE sides) with 2500-V applied hi-pot voltage.
4. The data/pulse transformers can be substandard. We mentioned in Chap. 9 that some vendors just "pass the buck" on this aspect to the systems designer, since it is not mandatory for the pulse transformer per se.
5. The optocoupler/isolator ratings may not be adequate. For implementing I^2C bidirectional communication nowadays, monolithic inductive or capacitive isolators are being increasingly used (from Analog Devices and Silicon, Labs, for example). But these may have lower voltage ratings than traditional-technology bidirectional optoisolators (from Clare/Ixys). Designers need to check the ratings.
6. The clearances and creepages between the PoE and host domains may not be enough. PCB design is critical here.

In Chap. 9 we made a case for *not* collecting all the wires of all the ports, tying them, and conducting a hi-pot test. We argued that the voltage distribution changes and does not fully expose dielectric weaknesses that can show up when the high voltage is applied. Please reread that at this point.

When we get down to it, *we will notice a very strong correlation between failing a hi-pot test, failing a surge test, and failing a cable ESD test.* All these different tests, in effect, *probe/test the boundaries of the isolation barrier between PoE and host domains.* In all cases, these tests eventually end up applying a high-current/voltage between the data lines and the earth ground. So any weakness in this respect will have widespread consequences on reliability. It is very important to consider all the points mentioned in this section very carefully indeed.

We now start looking carefully at point 6: clearances and creepages.

Separation Anxiety

It should be obvious, but nevertheless it is worth emphasizing that hi-pot testing is always performed with the equipment powered off. But that's just to avoid damaging it from a functional viewpoint. After all, we can't safely apply a steady 2500-V on the output or across any power supply while it is working!

Some engineers question whether it makes any sense to consult the clearance and creepage tables of the safety standard for deciding how good the separations need to be to pass the hi-pot test. They argue that the hi-pot test is conducted with the device powered off, whereas the clearance/creepage tables in the safety standard assume that the equipment is actually powered up.

To answer that, however, the *purpose* of the hi-pot test must be clearly understood. Its basic purpose is to test the *dielectric-withstand capability* of the barrier between SELV and TNV-1 circuits as it will present itself while the equipment is *operational.* The eventual purpose is to avoid "zapping" a user who may touch the equipment *while it is working* (usually). Therefore, we are justified in consulting the above-mentioned tables of the safety standard. Further, there are no *mandatory* requirements specifying a certain *X* mm of separation between SELV and TNV-1. So all we seek is guidance. In the next few sections, the IEC tables will be consulted for advice.

Causes of Isolation Breakdown and Recommended Minimum Clearance

When we apply 2500 V DC, the first and most obvious, though not likeliest cause of breakdown is a flashover *through air*. Air is an insulator, with a certain dielectric constant and also a certain breakdown voltage. Its breakdown voltage is often stated in literature to be 33 kV/cm, or 3.3 kV/mm, and since there are 394 mil in 1 cm that is equivalent to 33,000/394 = 84 V/mil. Therefore, to hold off 2500 V DC, we need a minimum of 2500/3300 = 0.75 mm or 30 mil. To account for variations in the breakdown voltage based on atmospheric variations, we should leave a little more margin here. We thus fix this minimum distance as *40 mil or 1 mm for meeting 2.5 kV*.

Note that the distance through air is called "clearance." So this is what we are saying: If we keep a minimum of 40 mil clearance between PoE and host sides, the interposing air will not flash over and we will pass the 1500 V RMS hi-pot test. This 40-mil minimum spacing is therefore the minimum distance written in bullets 3, 4, 5, and 7 of Fig. 10.5.

In Fig. 10.5, the stated distances are valid only provided both traces across the interface are "coated." If not, we have to consider the fact that a flashover can flow along the *surface* of the PCB instead of through the surrounding air, in which case we may need to increase

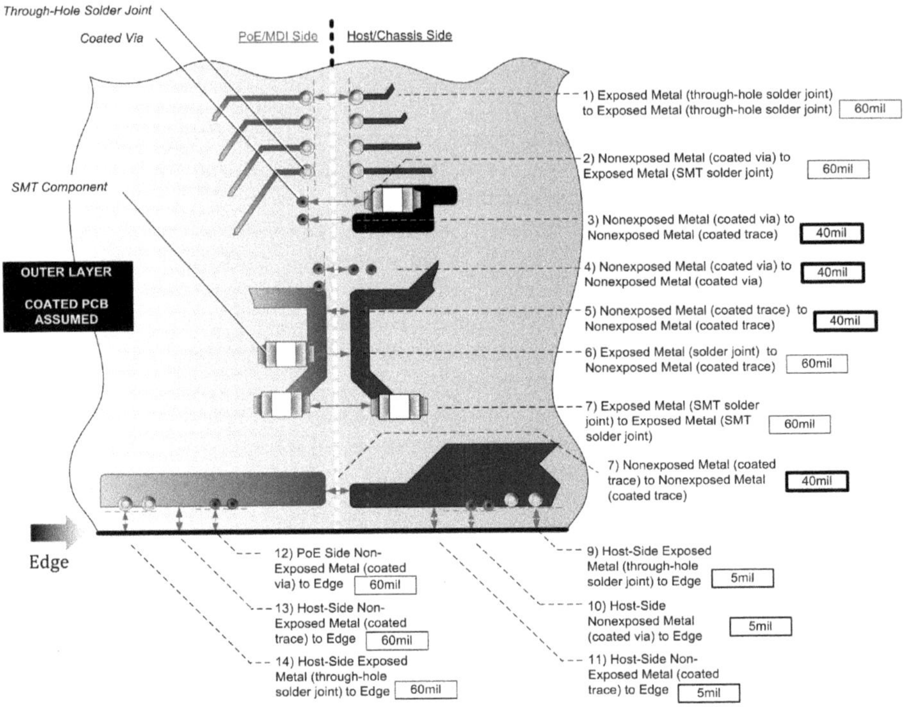

OUTER LAYER SEPARATION RULES:

1) Minimum 60mil separation required between any two points from PoE to Host side, unless BOTH these points are coated. ONLY in the latter case distance can be reduced to 40mil.

2) In bullets 9 to 14, we have implicitly assumed that the edge of the above daughtercard could be mounted on to a motherboard (Host/PHY/Switch domain), or it may lie flush against the chassis, or even against the bare casing of the ICM (which connects to Host ground). In other words, the edge of the above PCB is assumed to effectively be the host-side domain for the purpose of calculating separations.

Figure 10.5 Recommended separations on outer layers of PCB between PoE and host sides.

the minimum separation to almost *twice* as much. This minimum separation along the surface of an insulator (the PCB in this case), is called "creepage," as discussed shortly.

Returning to clearance one again, some people argue that the minimum distance for a given voltage should be increased substantially to account for humid conditions. In doing so, they wrongly assume humid air has a lower breakdown voltage than dry air. In fact, under humid conditions, air can even have *a higher* breakdown voltage than dry air. The reason for that is, very briefly that water vapor does not have the same (conductive) properties we expect from water!

However, the breakdown voltage of air does depend on atmospheric pressure, which can vary from place to place around the globe, according to altitude, and also from day to day, depending on prevailing weather. Therefore it is *usually* safe to choose 40 mil (1 mm) as the recommended minimum clearance (through air) for meeting PoE hi-pot requirements, irrespective of humidity or altitude.

As mentioned previously, IEC 60950-1 has no mandatory-clearance requirements for separation between PoE (TNV-1) and host (SELV) sides. However, indirect-distance recommendations can certainly be extracted from its tables because, mandatory or not, the tables are based on physics and a wealth of experimental data, which can help us pass the hi-pot test unambiguously.

Though not strictly applicable, we now look at Table 2K in IEC 60950-1 reproduced in Fig. 10.6. This refers to minimum clearances in Secondary Circuits. Note that both TNV-1 and SELV are on the secondary side in Fig. 10.3, so this is the most appropriate table to consult for guidance here. We should also keep in mind that IEC 60950-1 assumes a certain default operating environment in terms of incoming voltage spikes. In fact, it assumes Overvoltage Category II, which basically refers to equipment placed inside a room drawing electricity from a normal AC-mains outlet via the internal wiring of the building. It factors in typical 1500-V spikes riding on top of the incoming mains supply. IEC 60950-1 also assumes spikes with a default value of 1500-V peak riding on TNV-1 circuits. Therefore, in our case it makes sense to look directly at the row corresponding to *DC voltages less than 71 V DC* (1500-V spikes are already factored in). In this row, since one layer of protection corresponds to basic insulation, in effect, we read off 1 mm (40 mil). That is in line with our previous conclusions too (in Fig. 10.5).

Note that ECMA-287 also has a table that we can refer to for guidance. That is Table 3.9, applicable up to 2000 m (6562 ft) altitude. The table has not been reproduced in this chapter but can be found on the Web at http://www.ecma-international.org/publications/files/ECMA-ST/Ecma-287.pdf. It directly provides minimum clearance versus peak voltage withstand requirements. Consulting it, we will see it recommends a minimum of 1 mm clearance for peak voltages up to 2000 V and 1.5 mm for 2500 V, permitting interpolation between rows. In other words, as per ECMA-287, we should actually maintain almost 50 mil (not just 40 mil) spacing to meet the 2250 V DC hi-pot test requirement.

Summarizing Recommendations for Minimum Clearance

Here is a list of recommendations for ensuring minimum separation through air to meet typical hi-pot and cable ESD tests:

- Between PoE side and host side we need at least 40 mil (see bullets 3, 4, 5 in Fig. 10.5), but that may need to be increased to 50 mil (as per ECMA-287 too), depending on PCB-production tolerances and slightly higher actual-peak voltages than those assumed by Table 2K shown in Fig. 10.4.
- Staying completely within the PoE/MDI side, or completely within the host side, traces can actually run as close together as 5 mil from each other (same domain, see bullets 9, 10, and 11 in Fig. 10.5), but that may need to be increased to 8 mil based on PCB-production tolerances.

Table 2K – Minimum clearances in secondary circuits

		CLEARANCES in millimeters																	
WORKING VOLTAGE up to and including		Nominal AC MAINS SUPPLY voltage ≤ 150 V (transient rating for SECONDARY CIRCUIT 800 V) see 5)						Nominal AC MAINS SUPPLY voltage > 150 V ≤ 300 V (transient rating for SECONDARY CIRCUIT 1500 V) see 5)						Nominal AC MAINS SUPPLY voltage > 300 V ≤ 600 V (transient rating for SECONDARY CIRCUIT 2500 V) see 5)			Circuit not subject to transient overvoltages see 4)		
Voltage peak or d.c.	Voltage r.m.s. (sinu-soidal)	Pollution Degrees 1 and 2			Pollution Degree 3			Pollution Degrees 1 and 2			Pollution Degree 3			Pollution Degrees 1, 2 and 3			Pollution Degrees 1 and 2 only		
V	V	F	B/S	R	F	B/S	R	F	B/S	R	F	B/S	R	F	B/S	R	F	B/S	R
71	50	0,4 (0,2)	0,7 (0,2)	1,4 (0,4)	1,0 (0,8)	1,3 (0,8)	2,6 (1,6)	0,7 (0,5)	1,0 (0,5)	2,0 (1,0)	1,0 (0,8)	1,3 (0,8)	2,6 (1,6)	1,7 (1,5)	2,0 (1,5)	4,0 (3,0)	0,4 (0,2)	0,4 (0,2)	0,8 (0,4)
140	100	0,6 (0,2)	0,7 (0,2)	1,4 (0,4)	1,0 (0,8)	1,3 (0,8)	2,6 (1,6)	0,7 (0,5)	1,0 (0,5)	2,0 (1,0)	1,0 (0,8)	1,3 (0,8)	2,6 (1,6)	1,7 (1,5)	2,0 (1,5)	4,0 (3,0)	0,6 (0,2)	0,7 (0,2)	1,4 (0,4)
210	150	0,6 (0,2)	0,9 (0,2)	1,8 (0,4)	1,0 (0,8)	1,3 (0,8)	2,6 (1,6)	0,7 (0,5)	1,0 (0,5)	2,0 (1,0)	1,0 (0,8)	1,3 (0,8)	2,6 (1,6)	1,7 (1,5)	2,0 (1,5)	4,0 (3,0)	0,6 (0,2)	0,7 (0,2)	1,4 (0,4)

1) The values in the table are applicable to FUNCTIONAL (F), BASIC (B), SUPPLEMENTARY (S) and REINFORCED (R) INSULATION.

2) The values in parentheses are applicable to BASIC, SUPPLEMENTARY or REINFORCED INSULATION only if manufacturing is subjected to a quality control programme that provides at least the same level of assurance as the example given in annex R.2. In particular, DOUBLE and REINFORCED INSULATION shall be subjected to ROUTINE TESTS for electric strength.

FIGURE 10.6 Abridged Table 2K, recommended clearances as per IEC 60950-1.

The Concept of Creepage

In Fig. 10.5 we established minimum 40 mil for the case of completely coated traces/vias ("nodes") with a voltage differential of 2.5 kV between them. We also said that it needs to be 60 mil *if either or both nodes are not coated*. The assumption is that the coating on one node may not be perfect (small pinholes for example), and so if the other node is by design open and exposed, there can be conduction between the two nodes along the surface of the PCB. Current flow along the surface is aided greatly by contaminants (as in condensed water too) present on the surface of the board, and it usually becomes much easier for current to arc over the surface of an insulator rather than through air. Therefore, to avoid this new flashover mode, we need to increase the physical separations further from what clearances dictated (40 mm for every 2.5 kV).

In other words, the 40 mil minimum separation was based on flashover through air. The 60 mil minimum separation was based on flashover across the surface of the insulator. The former is called clearance, the latter is creepage. Whichever calls for *greater* distances (separations) at a given point, dominates the show (weakest link). Here are more formal definitions.

Creepage

Creepage is the shortest path between two conductive parts, measured along the surface of the insulation. One of those conductive parts could, for example, be the metallic bounding surface of the equipment (enclosure). Proper creepage protects against "tracking." Tracking is a process that produces a partially conducting path (localized deterioration) on the surface of an insulating material, as a result of electric discharges on or close to the surface of the insulator. The degree of tracking depends on two major factors: the comparative tracking index (CTI) of the material and the degree of pollution in the environment (e.g., condensation on the PCB). The CTI of a material is the numerical value of the voltage that will cause failure by tracking during standard testing. Tracking reduces the efficacy of insulating material because of one or more of the following reasons:

1. High humidity
2. Surface contamination (flux residue too)
3. Corrosive chemicals
4. High altitude

IEC 112 provides a fuller explanation of tracking and CTI.

Clearance Distance

Clearance is the shortest distance between two conductive parts *through air*. One of those conductive parts could for example be the metallic bounding surface of the equipment (enclosure). Sufficient clearance helps prevent dielectric breakdown between electrodes caused by the ionization of air. The dielectric-breakdown level is further influenced by relative humidity, temperature, and degree of pollution in the environment.

Yes, we could also have a dielectric breakdown path *through* the insulator. So, in general, conduction across/through any safety barrier can take place in the following ways:

1. Through air
2. Along the surface of the PCB
3. Through the PCB material/insulator discussed later in the section titled "Minimum Vertical Separation in PCB"

Note that each of these three possibilities will dictate a minimum separation distance to pass the hi-pot test (and to avoid flashover/conduction). In practice, a hi-pot failure will occur if any of the three minimum-separation distances are not maintained. We need to identify the weakest link and ensure minimum-separation for that mode of conduction. And logically, the weakest link is clearly the mode that asks for the greatest minimum-separation distance. Usually, this is possibility number 2, in other words flashover/conduction along the PCB surface (creepage currents). We need to pay closest attention to that.

The difference between clearance and creepage is illustrated in Fig. 10.7, which shows an optocoupler placed between two separate domains, say PoE and host domains. The slot on the PCB is usually seen only in high-voltage (AC-DC) applications, not in PoE, but it does illustrate the basic concepts here more clearly. As we can see, creepage may be measured along the PCB or along the IC package. Usually the latter is of lesser concern, so we can ignore it from this point. Though we should keep in mind that if and when we deal with really small SMD optocoupler packages in high-voltage applications, that does become of serious concern, and often the only answer is to just move to a larger package since a PCB slot can't help here.

As we can see from Fig. 10.7, if the slot was not present, as is the case in PoE, then the actual physical distances involved, the clearance, and also the creepage along the PCB surface, are almost the same.

Many recent experiments conducted on PCBs at UL indicate that modern PCBs typically have a withstand capability of 40 V/mil, compared to 80 V/mil for air as discussed earlier.

See: http://www.ul.com/global/eng/pages/offerings/industries/hightech/printedwiringboards/silver/.

In other words, the minimum-distance along the PCB (creepage) for a given voltage requirement, will be *double* the minimum distance

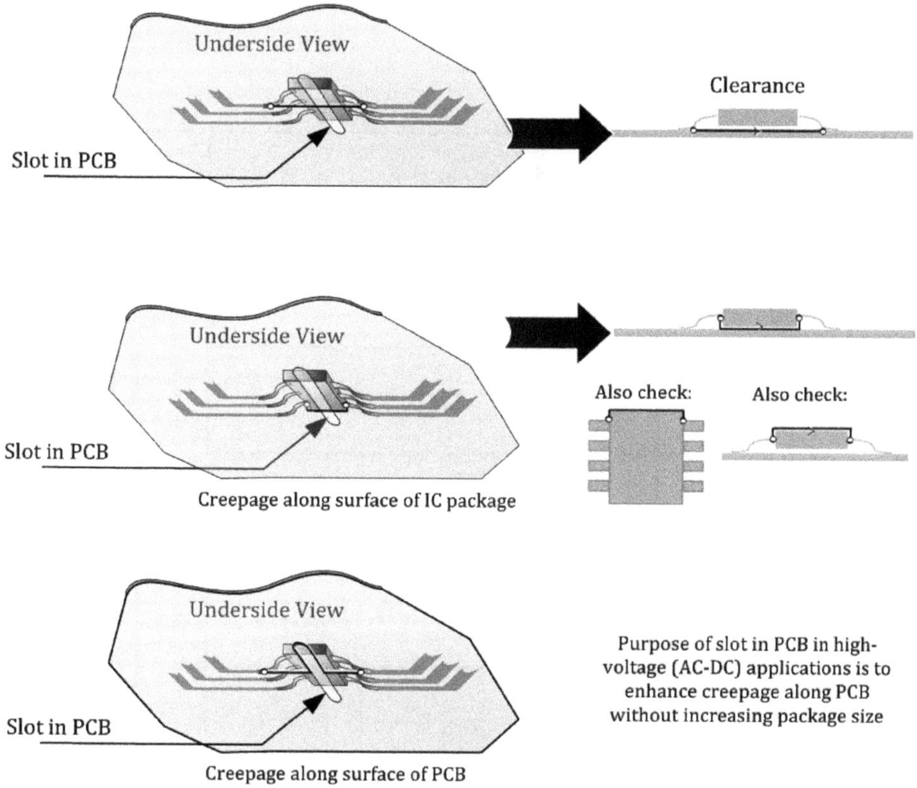

FIGURE 10.7 Comparing clearance and creepage.

through air (clearance). And so by complying with minimum-creepage requirements, we automatically comply with minimum-clearance requirements.

NOTE *As per IEC, the slot in Fig. 10.7 must be wider than 1 mm for it to count at all. If it is less than 1 mm wide, we have to go over it when measuring thee creepage, just as we do when measuring the clearance. In effect, any slot, or indentation, or actually any granularity on the insulator that is less than 1 mm, becomes "invisible" to creepage—commensurate with a deeper understanding of the phenomena of tracking as outlined in the URL given previously.*

Even though there are no-mandatory-creepage requirements in our case, it is useful to see what the IEC mandatory-creepage distances are in particular. We refer to Table 2L in the safety standard, reproduced in Fig. 10.8. If we abide by that, we should be in fairly good shape to pass the hi-pot test.

Table 2L – Minimum creepage distances

CREEPAGE DISTANCES in millimeters							
WORKING VOLTAGE V	FUNCTIONAL, BASIC and SUPPLEMENTARY INSULATION						
	Pollution Degree 1	Pollution Degree 2			Pollution Degree 3		
	Material Group	Material Group			Material Group		
r.m.s. or d.c.	I, II, IIIa or IIIb	I	II	IIIa +or IIIb	I	II	IIIa or IIIb
≤ 50	Use the CLEARANCE from the appropriate table	0,6	0,9	1,2	1,5	1,7	1,9
100		0,7	1,0	1,4	1,8	2,0	2,2
125		0,8	1,1	1,5	1,9	2,1	2,4
150		0,8	1,1	1,6	2,0	2,2	2,5
200		1,0	1,4	2,0	2,5	2,8	3,2
250		1,3	1,8	2,5	3,2	3,6	4,0
300		1,6	2,2	3,2	4,0	4,5	5,0
400		2,0	2,8	4,0	5,0	5,6	6,3
600		3,2	4,5	6,3	8,0	9,6	10,0
800		4,0	5,6	8,0	10,0	11,0	12,5
1 000		5,0	7,1	10,0	12,5	14,0	16,0

Linear interpolation is permitted between the nearest two points, the calculated spacing being rounded to the next higher 0,1 mm increment.

Annotations: 60V — PoE/TNV-1 set to 60V; 1.24 mm (Pollution Degree 2, IIIa or IIIb); 2 mm (Pollution Degree 3, IIIa or IIIb)

FIGURE 10.8 Recommended creepage distances as per IEC 60950-1.

Note that a typical PCB material falls in the worst Material Group, in other words, IIIa or IIIb, and that calls for the maximum separation. The Pollution Degree default level in IEC 60950-1 is actually Pollution Degree 2 (*some* mild condensation possible). As per the table, and the interpolation clause underlined in the lower part of the figure, we actually first interpolate to get 1.24 mm at 60 V. But then, since we need to round it up to the next higher 0.1 mm, it becomes 1.3 mm or 51 mil. Including a safety margin for PCB tolerances, that is 60 mil, consistent with our previous discussions.

If we consult ECMA-287, we will learn from Table 3.10 (not reproduced here) that for voltages less than 63 V, the recommended minimum creepage is 1.25 mm, close to the numbers we have listed. This is why in Fig. 10.5 we had two basic separation distances: 40 mil and 60 mil.

A more conservative estimate for creepage is not 60 mil (1.5 mm), but 80 mil (2 mm). That is based on the 40 V/mil finding of UL, mentioned previously. It also corresponds to the recommendation for Pollution Degree 3 in Fig. 10.8 corresponding to very wet conditions.

Let us do the math. The recommended minimum clearance was 40 mil, which corresponded to flashover through air, and we know 40 mil of air can hold off 40 mil × 80 V/mil = 3200 V. Whereas 80 mil of PCB surface can hold off 80 mil × 40 V/mil = 3200 V. That is the same—2500 V with some additional safety margin.

Therefore, our recommendation for minimum creepage based on this analysis is: *Between PoE side and host side the minimum is 60 mil (see Fig. 10.5).Though in some cases, this may be needed to be increased to 80 mil.*

Coating versus Noncoating

In IEC 60950-1 (first edition), there is a useful table called Table 2N, meant for coated PCBs. This coating is the usual PCB "green-masking." The table seems to say to some engineers out there that we can use PCB coating to minimize distances dramatically, even with high-voltage differentials between them. Quite obviously, since the board is separated from the elements, the recommended separations in the table no longer depend on pollution degree or material group. The coating, is supposed to "kill" tracking altogether, and no current conduction can occur along the PCB surface.

Further, the IEC standard, and also ECMA-287, imply that if *either one* of the two conductive surfaces is coated, even the possibility of flashover through air gets nullified, because the coating acts as an insulator between the nodes and it has a much higher withstand capability (measured in terms of V/mil) than air. In effect, Table 2N says that we can bring the traces as close as 0.1 mm (4 mil!) for voltages less than 63 V. That seems to be an improvement by a factor of 10! Too good to be true? Yes indeed.

The problem is that engineers do not realize that this table is applicable only if the PCB is subjected to several tests to ensure the *quality of coating material and its application*. That is not easy to implement in most cases, in a commercial environment. So for all practical purposes, we should actually *assume the coating does not even exist*, except perhaps for long-term protection from environmental conditions and consequently lower field-failure rates and longer life. In brief, *we cannot use Table 2N* for the purpose of defining minimum clearance or creepage. We can, however, draw a compromise somewhere, since there are no mandatory requirements for clearance and creepage in PoE. We explain that as follows.

In Fig. 10.5, the recommended separations on the outer layer of the PCB falls somewhere between assuming the coating helps *completely* (over-optimism), and assuming it does *nothing at all* (over-conservatism). Since there are no mandatory requirements in PoE as mentioned, we are actually free to interpret data in a reasonable manner. Our ultimate aim is to pass the hi-pot test of 2.5 kV, yet stay commercially viable. So unlike the IEC standard, which assumes the (tested) coating is so good that it has a hugely beneficial effect even if it covers only one of the two conductive nodes present, we are assuming it is good enough provided it covers (or at least seems to cover) *both* the nodes. In effect, we are actually leaving margin for one of the "covered" nodes to have tiny pinholes (imperfect quality of coating).

Separations in Inner Layers

Let us first understand basic multilayer PCB construction. There are two basic elements: (a) copper foil of varying thicknesses, and (b) FR-4 ("fire-resistant 4" category, as per IEC). The FR-4 may be in the form of prepreg (preimpregnated) material, or in the form of a core. The core material is *cured* fiberglass-epoxy resin (i.e., hardened to start with). Prepreg however consists of thin sheets of fiberglass impregnated with *uncured* epoxy resin. Several layers of prepreg may be combined to achieve a desired final prepreg thickness. Finally, the entire assembly is pressed together, and under heat, the uncured epoxy also gets cured (it hardens).

We can have many variations of thicknesses, and finally, the distance between copper layers may be widely different. In fact, since high-frequency impedance depends on the distance between parallel copper traces or planes, these interlayer distances are carefully adjusted on high-frequency boards. Therefore, if the PoE sections are placed on the same motherboard, they will have as many layers available to them as are there for the switch/PHY. And they will also have the same "odd" interlayer spacings. An alternative method is to have the PoE sections on a separate "daughter card"—which is a small board mounted on the motherboard with a multipin solderable connector.

This solution is more flexible since PoE can now be added on or removed at will from the switch/hub. PoE is just an optional feature now. Further, second-sourcing of PoE also becomes viable. Different PSE chip vendors usually have unique pin-outs. However, the *daughter-cards can now have the same pinout*, and so the daughter card, using, say, PoE chips from vendor A, can be simply swapped out with a daughter card with the PSE chips from vendor B. Thanks, of course, to the open PoE standard and the simple laws of interoperability.

The daughter card on motherboard concept also helps reduce cost for two reasons:

1. We can now have a cheap four- (or even two-) layer board for the PoE circuitry rather than waste, say, 12-layer real estate just for a simple PoE board.
2. We can also now design the board with appropriate prepreg and core thicknesses so as to finally achieve *equal distances* between copper layers. This type of board still happens to be low-cost and very popular as a four-layer board, especially in Asia. Also the cheapest option is to have *equal* copper thicknesses on all layers too. This is often referred to as 1/1/1/1 or 2/2/2/2, referring to the thickness being "1-ounce copper" or "2-ounce copper."

Having understood the construction, we should know that IEC 60950-1 (first edition) interpreted inner layers of PCBs as cemented joints. In effect, for inner layers, there was *no concept* of clearance or creepage, just distance through insulation. The insulation, in this case, was the cement/prepreg. In the second edition, the IEC standard qualified this, and now several tests are required on cemented joints for PCBs operating over 90°C to prove the quality of the joint. Only if we pass that testing can we logically say that the joint is good enough and clearance and creepage distances don't apply. This aspect is discussed on page 21 of the document at this link: http://www.ul.com/global/documents/offerings/industries/hightech/informationtechnology/new/60950-1_2ndEd_Analysis_rev_May%2021%202010.pdf.

ECMA-287 also makes the same point on page 42 of the standard. Basically, the overall view is that without tests to confirm the quality of the cementing, we need to assume there is no cement. And that is the real-life situation we have. In that case, *we need to apply our normal concepts for clearances and creepage in outer layers, now to internal layers too*.

In effect we are assuming significant voids in the cementing are possible within inner layers, so the PCB is really not one big air-tight cemented joint. Conduction currents *can* travel along the PCB surface over inner surfaces too. Whether they do or not, that is the assumption we need to make in the absence of tests to confirm the integrity and quality of the cementing.

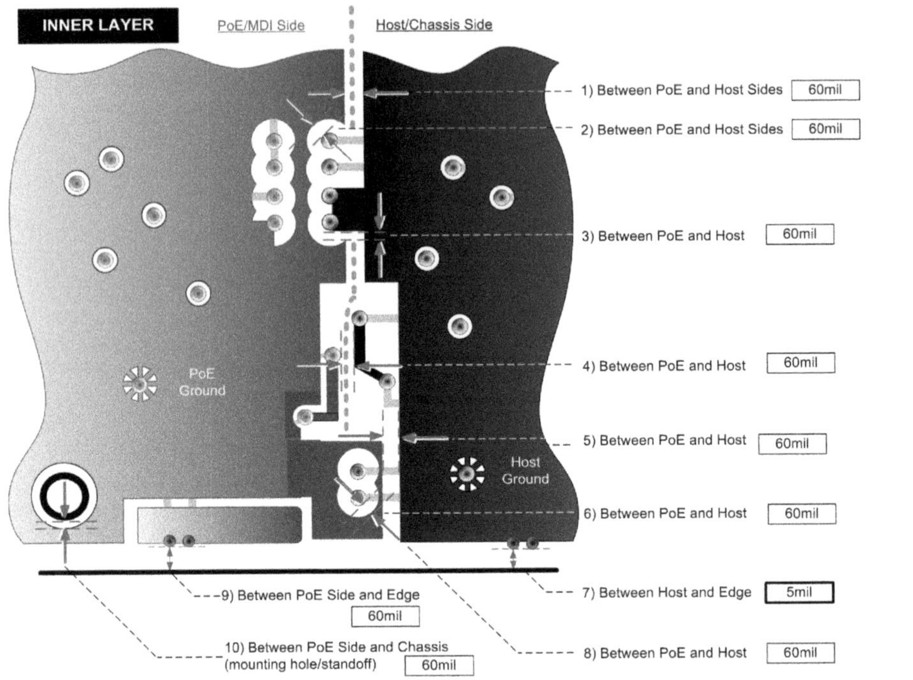

INNER LAYER SEPARATION RULES:

1) Minimum 60mil separation required between any two points from PoE to Host sides. Look at all the key places (examples above) where this minimum separation requirement may get overlooked. In particular, establish the identity (i.e., PoE or Host side) of various bordering vias/through-holes with respect to the identity of any surrounding copper-fill. Maintain required separation.

2) It is also typically assumed that the edge of the board may lie flush against the (uncoated) chassis, ICM, Motherboard etc – i.e. the PCB edge is effectively the host side, and therefore corresponding distances to PoE side must be maintained (i.e. > 60mil). No separation is therefore necessary from edge to host side, except perhaps for functionality, In that case > 5mil is recommended.

Figure 10.9 Recommended separations on inner layers of PCB between PoE and host sides.

There are some who argue that, in fact, moisture and contaminants can get trapped in internal voids much more, and that can in fact *worsen* the insulating performance as compared to outer layers. However, to stay commercially viable, we have only disallowed clearance-based separations (40 mil) and have asked for maintaining minimum separations of 60 mil in internal layers too, as shown in Fig. 10.9.

Minimum Vertical Separation in PCB

As mentioned previously, flashover can occur not only through air, or over the surface of the PCB, but also *through* insulation. Our final concern is therefore avoiding flashovers through the PCB material inside the PCB the *between copper layers*. Note that typical FR-4 PCB material has a dielectric strength of 700 V/mil or 2756 V per 0.1 mm. So to hold off 2500 V, we basically need a minimum distance through

PCB material of only 0.1 mm (4 mil) The rule is: *Minimum distance through insulation (PCB) is 0.1 mm (4 mil) for 2.5 kV.*

Note that standard PCB thickness is 63 mil or 1.6 mm. With a four-layer PCB, assuming the four copper layers are equidistant from each other, the distance from one layer to another will be 63/3 = 21 mil. This is just fine since it is well above the minimum of 4 mil through dielectric. If we have a 12-layer board (once again assuming copper layers are equidistant), the distance between layers is 63/11 = 5.7 mil. We are still *almost* OK. Not beyond that. We can still deal with more layers though, by *not* having the copper of different domains to be directly stacked one over the other on different layers. We need to separate them *sideways* (laterally) too.

Lateral separation of domains is actually-highly recommended, not only for creepage and clearance, but because noise from the "48V" rail, which is derived from a switching-power supply, can get coupled easily to the host side through interlayer capacitances if the traces of the two domains overlap. This can cause strange system upsets and signal-integrity (reach) issues. For the same reason, *do not use a full copper plane for the 48 V as is common practice*. That is really not a good idea since the 48-V plane now becomes a huge radiator of EMI too. It is very advisable to only sticking to *ground* planes throughout the PCB, in any domain. These ground planes are quiet and do not radiate.

It is good to keep in mind that though IEC 60950-1 has no specific recommendations on the number of layers; if compliance to more stringent telecom standards like Telcordia GR-1089 is required, the minimum distance through insulation is not the only concern. To comply with GR-1089, we must ensure that we also *maintain a minimum of two distinct PCB layers between host and PoE domains* (i.e., akin to the double-insulation concept mentioned previously).

In Fig. 10.10, we finally give an illustration of various recommended separations in a PoE PCB, based on all our learning so far. Pay close attention to what happens around the standoffs and vias and also where the traces of the two domains overlap. Note also that for fabrication-quality reasons, it may be generally recommended that all traces (PoE and otherwise) be at least 8 mil apart from each other. We also need to pay close attention to the *thickness* of the traces when calculating the actual distance between traces. Note that usually buried vias are not commonly used. They are shown in Fig. 10.10 just to illustrate the overall concept behind vertical separation through insulation.

Secondary Discharge

This is a common occurrence in troubleshooting field returns. For example, PSEs with their 100 V ceramic port capacitors literally charred and cracked have shown up. How could that happen? To crack open a 100-V ceramic-port cap would perhaps require 2 or 3 times its rated voltage delivered instantantaeously on it. Surges could account for

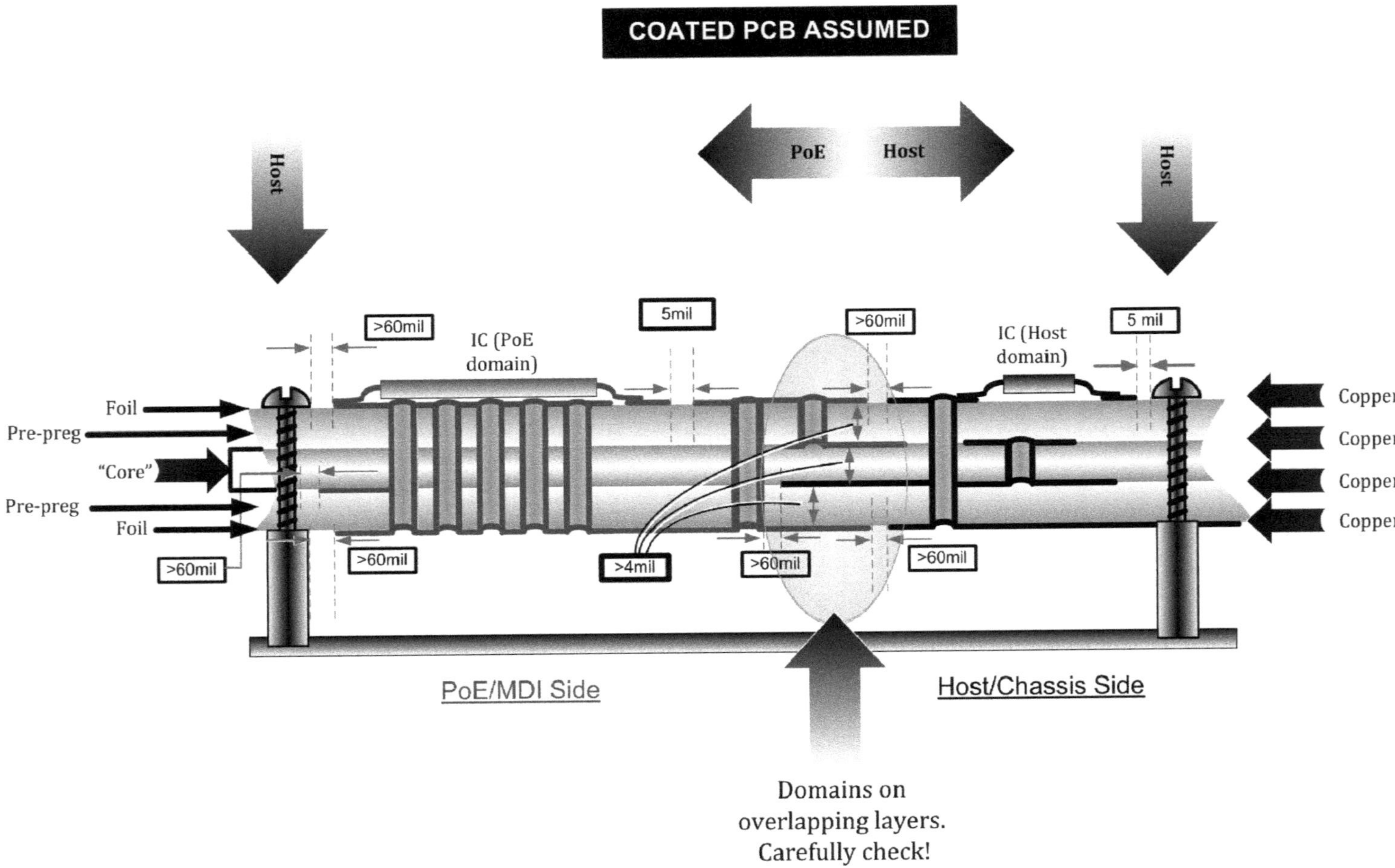

FIGURE 10.10 Recommended internal separations in a typical PoE PCB.

that, but the unit had correctly rated Y-caps and so on, so the differential voltage across the port could really never be that high. This could very well be a failure from secondary discharge. This is best explained by Teseq (formerly Schaffner Test Systems) on Page 9 of the Application Note at http://www.teseq.com/com/en/service_support/technical_information/01_Transient_immunity_testing_e.pdf:

> A secondary discharge can occur within equipment if the discharge current through the product attempts to take a path which includes an air gap. The voltage across the gap increases until the gap breaks down, and this secondary breakdown can be more stressful for the circuits than the initial event, because it is closer to them and probably involves a lowerpath impedance. The breakdown can occur simultaneously with the applied primary discharge, or it can result once several primary discharges have occurred such that an isolated conductive part has built up sufficient voltage to discharge itself. Secondary discharge is best dealt with by ensuring that no sneak air gaps exist, or by making them large enough not to break down, or by bonding across them, and by avoiding sharp edges which encourage high field gradients. Floating (isolated) metalwork or copper areas on PCBs must be avoided.

In the author's experience this is what happens: If there is a sudden arcing on the board, the applied voltage suddenly *redistributes* as the voltage across the air gap collapses. This rapidly collapsing voltage leads to a rapidly increasing voltage elsewhere—for example, across *the port output caps*. So we get a huge current through these with a large voltage too. The port ceramic cap just cracks open.

It is very important to establish good clearance and creepage, not leave floating copper or metal anywhere (in the PCB and in the enclosure), and also avoid sharp edges of metals, since "pointy"-charged surfaces have much higher electric fields associated with them.

PD Isolation Requirements

In Fig. 10.11, we present the isolation domains end to end. We have also now explicitly shown the PD's domains. The requirements for the PD are typically the same as discussed so far. With one exception: *the PD may be fully housed in plastic*, with no potentially unsafe exposed metal areas. That can help reduce the cost of the PD. As stated in the figure:

> If PD enclosure is completely plastic (no exposed metal), there is no need to have an isolated DC-DC converter. There is no SELV requirement, and no need for isolation anywhere. We need not even use a data transformer, except for common-mode noise rejection (and better reach).

We can also drastically reduce the Y-caps. They serve little purpose here. Some Y-capacitance will still be present inside the magjack for

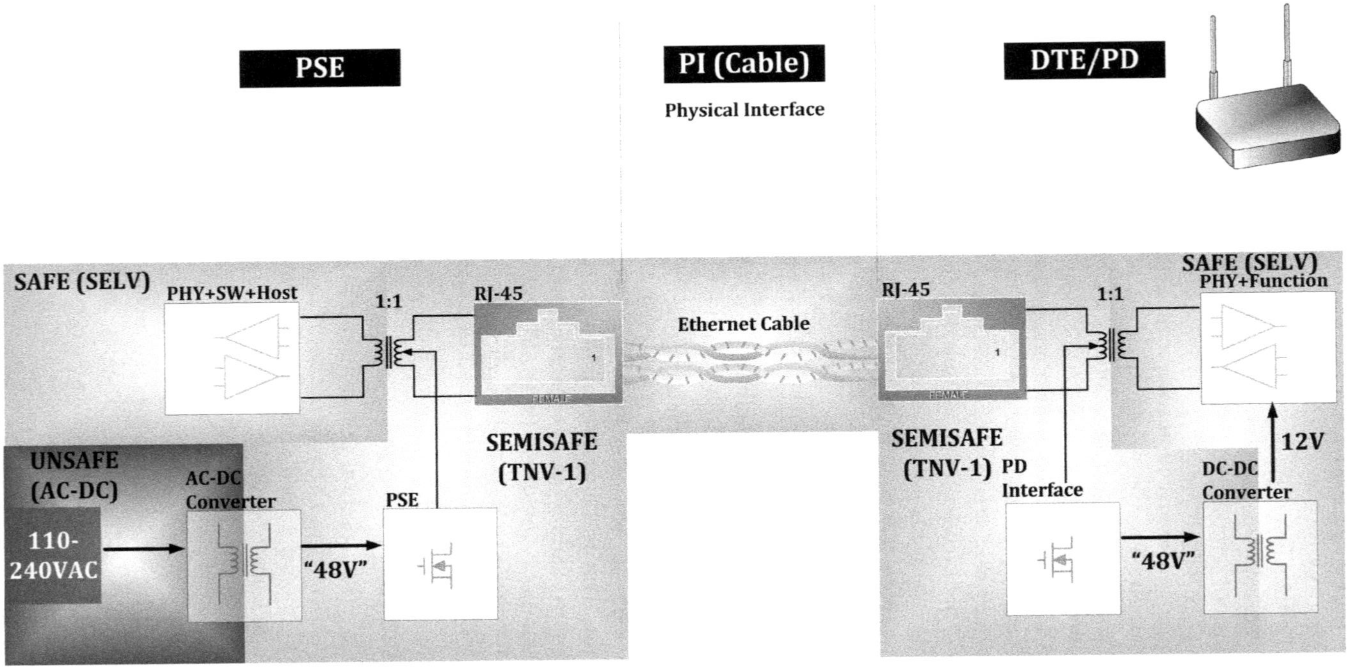

Line-side ("TNV-1") is considered "hot" (semisafe) because of possible spikes picked up on the line, which could zap the user. Therefore, we need >1500VAC isolation from TNV-1 to "SELV". "Host side" is safe (SELV). On the Primary-side of the AC-DC we have unsafe (hazardous) regions. We need >3000VAC *certified* isolation between "UNSAFE" and SELV or TNV-1.

Note: If PD enclosure is completely plastic (no exposed metal), there is no need to have an *isolated* DC-DC converter. There is no SELV requirement, and no need for isolation anywhere. We need not even use a data transformer, except for common-mode noise rejection (and better reach).

FIGURE 10.11 The Isolation domains, end to end.

EMI suppression, but we may not need to add any more Y-capacitance on the PD board itself. And since the Y-cap provides the path for surge/hi-pot/CESD currents to flow, reducing it can have very beneficial impact on the field reliability too, as pointed out more clearly in the following section.

Higher Surge, Cable ESD, and Reliability

We will discuss surge and cable ESD events in Chap. 11 in more detail. Here we will just cover it rather briefly for completeness sake.

In surge and cable ESD testing, we apply the test pulses between PoE/MDI side and chassis ground. So, in effect, there is a high-voltage differential between PoE and host sides—just as in a hi-pot test. The dielectric withstand capability of the interface between domains (including PCB separations), must be assured at least to the level we are testing to. In fact, by bench testing we can confirm that the hi-pot test is *less severe*, volt to volt, than a surge test. For example, we will see that a unit that fails hi-pot prematurely at just 2 kV, will fail the surge test at about 1500 V. If it fails hi-pot at 1.5 kV, it will likely fail surge test at just 1 kV, and so on. Therefore, passing a 2.5 kV hi-pot test is a prerequisite for passing a surge test of 2 kV. If the customer asks for 2 kV surge survivability, the first thing that needs to be rechecked is the 2.5 kV hi-pot test. As mentioned in the previous chapter, there are not just magjack vendors, but even switch vendors who take refuge in the 1500-V, 10-µs/700-µs hi-pot test allowed by the AT standard.

Why is a surge test more severe? Because in a surge test there is not just a V, but also a dV/dt, and that creates a higher electric field. Not to mention that the hi-pot test is done with the equipment powered off, whereas the surge (or cable ESD) test is administered while the equipment is actually in operation. Cable ESD (CESD) is not as severe as a surge test, since it has significant limiting impedance. So it is possible to have a unit that passes only a 2-kV surge, but passes 4-kV CESD.

Today, there is an increasing trend towards higher and higher surge voltage and cable ESD tests. So if, for example, we want to ensure we can survive a 4-kV surge test, we must double all our distance recommendations so far. All distances shown in Figs. 10.5, 10.9, and 10.8 must be doubled for achieving 4-kV surge capability because we first need to achieve 5-kV capability in a DC hi-pot test. *That is a prerequisite.* And besides doubling the voltage ratings of all Y-caps, and also of the termination-blocking caps, we need to have much higher-minimum separations on the PCB to avoid any flashover.

We conclude that to enhance field reliability, *isolation* between PoE and host domains must be made very robust. Eventually, the strength of any barrier or isolation interface is judged by its *weakest* link. Whichever gives in first, ends up determining the field reliability.

The recommendations of this chapter are therefore extremely relevant to achieving that goal.

Limited Power Source

Finally, looking at the lowermost row of Fig. 10.1, we have the LPS requirement. This stands for limited power source. What this basically asks from us is that we place a fuse in series with either Port_P or Port_N to limit the total power to 100 W (VA) in case of failure of the PSE (like shorted FET). So most people calculate: 100 VA/57-V = 1.75 A, and put either a 1.75-A (but preferably a 1.5-A) fuse in series with the port.

The IEC safety standard also places a limit of 250 VA on the power coming into the PSE chip, but typically, this is open to multiple interpretations today and is being somewhat ignored. However to be unequivocal, we need to use either a limited energy AC-DC power supply (called a SELV-EL power supply, where EL stands for limited energy), or in larger systems where this input power restriction is not practical, we may need to place, say, one 4-A additional fuse for every four ports on the input side of the PSE, in addition to the 1.5-A or 1.75-A fuse normally placed on every outgoing port.

A problem occurs in four-pair operation, when despite the higher normal-output power (60 W), under fault conditions we cannot exceed 100 W (measured after one second). Fuses are not so precise or fast-acting. Therefore, several PSE-chip vendors have started approaching safety agencies like UL to certify their PSE chips themselves as overcurrent protectors, much like USB solid-state current limiters. With that approval, they are able to sell fuseless solutions, even for four-pair (60 W) applications. Worldwide safety approvals without fuses is still a question mark as the field of PoE evolves. If PSEs (with integrated FETs) do receive safety approvals as valid current limiting devices, it is usually understood that the usual 1.5-A or 1.75-A fuses are no longer required on their outputs. However, the question remains, whether the 4-A input fuse (or equivalently, the 250 W incoming energy limit) is required or not. This situation is still not fully resolved or universally accepted. However, the International safety standard IEC 60950-1 recently went into its Second Edition, First Amendment, incorporating Annex CC, which contains specific tests for solid-state current limiters (like PSE ICs with integrated FETs). Once the PSE chip is certified, in principle, this allows removal of all fuses at the input and output side of the PSE chip (with no 250 W incoming energy limit either). However, the First Amendment is still being gradually accepted by member countries worldwide. Note that the PD69104/69108 from Microsemi became the first PSE IC in the world to be tested and certified as per the First Amendment (Annex CC).

CHAPTER 11

Surge Testing and Protection

Overview

High-energy transients (surges) can appear on data ports and can cause anything from system upsets to hard failure accompanied by charred boards.

Surges are known to have caused data transmission errors, memory scramble (mess-up), process interrupts, program lockups, latchup/failure in SCR/ICs, power-supply failures, hard-disk crashes, and general circuit-board failure. Add to that now, more specifically, burnt PHYs, PSEs, and PDs.

Interest in surge events has literally peaked in recent years, because manufacturers of switches/hubs, in particular, have noticed a very strong correlation between products that had been previously observed to have relatively lower, or somewhat compromised, surge-survival thresholds in internal product qualification testing, and their escalating number of field returns, especially after storm activity in a certain area. Data/telecom cables snake around a building, past noisy mains wiring, often going outside too, and thus become great antennae, not only for picking up noise, which we struggle to reject by the use of twisted pairs, data transformers, and so on, but *surges* too. Surges have a lot of residual energy, which can hardly be rejected, or wished away, by just twisting cables and other small things like that. We can't afford to ignore surges either. PoE sections being "front-end," from the viewpoint of an incoming surge on data lines, are relatively more vulnerable as compared to the data circuitry on the host side. Therefore, to avoid a rash of product returns, a keen understanding of what exactly happens during a surge event is required and is the key to enhancing product reliability and brand reputation.

Approximately 80 percent of recorded surges are caused by internal switching transients, in turn caused by turning on/off motors, transformers, photocopiers, and so on. Externally generated surges caused by induced lightning, grid switching, or from adjacent buildings account for the remaining. In particular, those surges related to *lightning strikes* can produce surge energies of hundreds of joules. These surges can be the result of a direct lightning strike (very rare, almost impossible to survive), or more frequent cloud-to-ground and cloud-to-cloud discharges. All these events can create powerful electric and magnetic fields, which can then capacitively or inductively couple on to the mains wiring and onto LAN/telecom cabling.

ANSI/IEEE C62.41 is a relatively modern standard titled "IEEE Recommended Practice on Surge Voltages in Low-Voltage AC Power Circuits." Along with UL 1449, it has become the de facto standard for characterizing and implementing surge protection. Keep in mind that the older version of C62.41 was called IEEE 587-1980, and that was for years the go-to reference document on this topic. The C62.41 standard lists different waveforms a surge suppresser is to be tested with. It has three categories (A, B, and C), each having three subcategories (1, 2, and 3). For example, it has created the B3 ringwave category, and also the B3/C1 *combination wave* to represent higher-energy internal surges. It also has a category C3 combination wave (20 kV, 10 kA), which represents very high-energy surges caused by lightning. A surge-suppressor device (SPD) gets an Underwriters Laboratories (UL) 1449 "listing" when it is tested with the C62.41 waveforms and declares its let-through voltage. Voltage let-through refers to the amount of transient voltage passed through, or past, a power conditioning unit to the load. SPD ratings range from 330 to 6000 V.

The guiding international (European-origin) standard for surge waveforms and surge protection of equipment is CISPR 24 titled "Information Technology Equipment—Immunity Characteristics—Limits and Methods of Measurement." CISPR stands for Comité International Spécial des Perturbations Radioélectriques, or loosely translated into English as: "Special International Committee for Radio-electric Perturbations." This standard is advisory in nature, much like the IEC standard for safety, IEC 60950-1. Ultimately all these international standards need to be ratified by local governments and are then accepted as law in that region. So for example, in Europe CISPR 24 became the (mandatory) European Norm EN 55024.

EN 55024 lays down the requirements for surviving surges. But it also refers to another pan-European standard, EN 61000-4-5, for the actual test methods and procedures.

For PoE, based on the EN documents, we arrive at the following summary of requirements:

(a) The mandatory pass level is Level 2, corresponding to ±1 kV surge (see Tables 11.1 and 11.2).

(b) The surge waveform applicable here is the "1.2/50 μs" open-circuit *voltage* waveform, which is the same as the "8/20 μs" short-circuit *current* waveform (see Fig. 11.1 and Table 11.2).

Level	Open-Circuit Test Voltage ± 10% (kV)
1	0.5
2	**1.0**
3	2.0
4	4.0
x	Special

Note: x is an open class. This level can be specified in the product specification.

TABLE 11.1 Definition of Levels as per EN 61000-4-5, with Mandatory Level for Ethernet/PoE Bolded

(c) The surge should be applied *common mode* (equally, and precisely at the same moment on two or more data lines, with respect to Earth ground).

(d) The mandatory minimum to pass, is Performance Criteria B. (See Tables 11.2 and 11.3.)

(e) The total source impedance for the surge waveform, as applicable to PoE/Ethernet testing, is 42 Ω (that includes 2-Ω source impedance inside the surge generator).

(f) Five zaps of positive polarity followed by five zaps of negative polarity are required (but *do not alternate the polarity*: That can create up to twice the voltage swing than necessary).

(g) The interval between successive zaps is 1 min or faster (can set exactly to 60 s).

Many of these points will be further elaborated upon in this chapter. Here, an important closing remark is that the surge test should preferably be done with the PoE link "up and alive" (with a small PD on the other side). This ensures that the pass-FET of the PSE (or PD) is fully *conducting* during the surge testing. If the FET is nonconducting, the incoming surge may actually impose higher stresses on the front-end of the equipment under test and also on the FET itself, *especially with AC disconnect*. In any case, by keeping the PoE link alive during the surge test, helps us monitor the full performance of the equipment during surge testing. Note that this aspect is *not* clarified in the standards, but it is the way being increasingly done today. A fluid situation always results when a new area or discipline, such as PoE, develops and evolves. There is some natural, temporary ambiguity. Not everyone agrees with all procedures. But what they all do seem to agree upon by now is that the mandatory requirements are not enough, as discussed next.

	Environmental Phenomena	Test Specification	Units	Basic Standard	Remarks	Performance Criterion
2.1	Radio-frequency continuous conducted	0.15-80	MHz	IEC61000-4-6	See (1) and (3)	A
		3	V (unmodulated, RMS)			
		80	% AM (1 kHz)			
2.2	Surge, Line to Ground	1	kV (peak)	IEC61000-4-5	See (2)and (4)	B
		1.2/50 (8/20)	Tr/Th µs			
2.3	Fast Transients	0.5	kV (peak)	IEC61000-4-5	See (3)	B
		5/50	Tr/Th ns			
		5	Repetition Frequency kHz			

NOTE 1: The frequency range is scanned as specified. However, when specified in Annex A, an additional comprehensive functional test shall be carried out at a limited number of frequencies. The selected frequencies for conducted tests are: 0,2;1; 7, l; 13, 56; 21; 27.12 and 40,68 Mhz (1%).
NOTE 2: Applicable only to ports which according to the manufacturer's specification may connect directly to outdoor cables.
NOTE 3: Applicable only to cables which according to the manufacturer's specification supports communication on cable lengths greater than 3 m.
NOTE 4: Where normal functioning cannot be achieved because of the impact of the CDN on the EUT, no test shall be required.
Table 2 of BS EN 55024:1998 CISPR 24:1997 (Incorporating Amendments 1 and 2—dated June 2003.

TABLE 11.2 Mandatory Requirements as per EN 55024

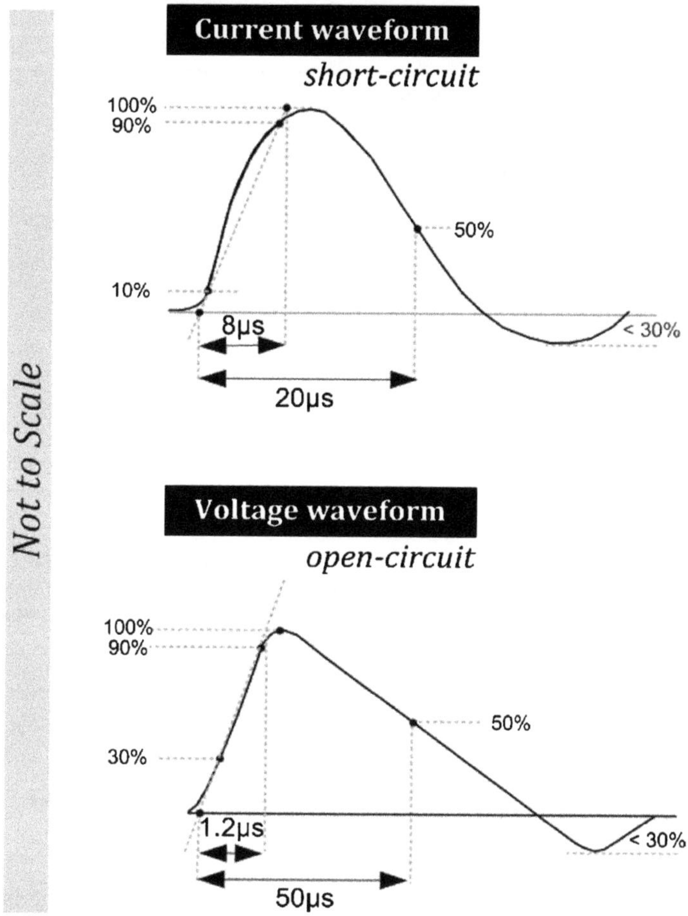

FIGURE 11.1 Surge waveform (expressed as open-circuit voltage or short-circuit current).

Mandatory versus Custom-Driven Requirements

OEMs have seen a strong correlation between field reliability and surge survivability. The mandatory EN level is just 1 kV, but as we can see from Fig. 11.2 (sourced from IEEE 587), we can get hit by one surge of 5 kV once a year, and 3 to 4 kV thrice a year (in the United States at least). That can be a lot of field returns. So there is an increasing demand for passing higher and higher surge ratings. Several issues arise in deciding the best target to chase in an economical manner.

How high should we aim for? From Fig. 11.2, we see that spark-overs (flashovers) almost naturally protect most equipment locations for surges above 6 kV. This is explained in the figure too. So 6 kV is the

Performance Criterion	Description
A	During and after the test, the EUT shall operate without: • error rate beyond the figure defined by the manufacturer; • requests for retry beyond the figure defined by the manufacturer; • speed of data transmission rate beyond the figure defined by the manufacturer; • protocol failure; • loss of link.
B	Error rate, request for retry and speed of data transmission rate may be degraded during the application of the test. During testing degradation of the performance as described in criterion A is permitted provided that after testing, the normal operation of the EUT is self-recoverable to the condition immediately before the application of the test. In these cases, *operator response is* permitted to reinitiate an operation.
C	During testing, degradation of the performance as described in criteria A and B is permitted provided that after testing the normal operation of the EUT is self-recoverable to the condition immediately before the application of the test or can be restored after the test by the operator.

EUT stands for Equipment Under Test
Source: CISPR 24 Second Edition, 2010.

Table 11.3 Performance Criteria as per CISPR 24

upper limit we should be concerned with. Note that poor or mediocre wiring insulation materials ironically help protect equipment better than excellent and expensive insulator materials.

There is increasing talk about China requiring 6 kV surge protection. We now realize why 6 kV was picked. However, it should be kept in mind that, so far, Chinese requirements are voluntary. Even if they do become mandatory, note that they already include designated "levels" of surge withstand capability: 2.5, 4, and 6 kV, similar to CISPR 24 in Table 11.1—so it is probable that much like EN 55024, we may need only to comply with a lesser-than-max level. It is not clear what the surge impedance referred to is either in these Chinese requirements. But we do note that the surge waveform specified for data ports is actually much less severe in terms of its profile (10/700 μs)—its "attack time" (rise time) is sluggish compared to the 1.2/50-μs waveform, and as we will shortly see, the stress on the actual device is therefore much *less* (under conditions of limited Y-capacitance). Incidentally, CISPR 24/EN 55024 is also now proposing testing data ports with only the (softer but wider)

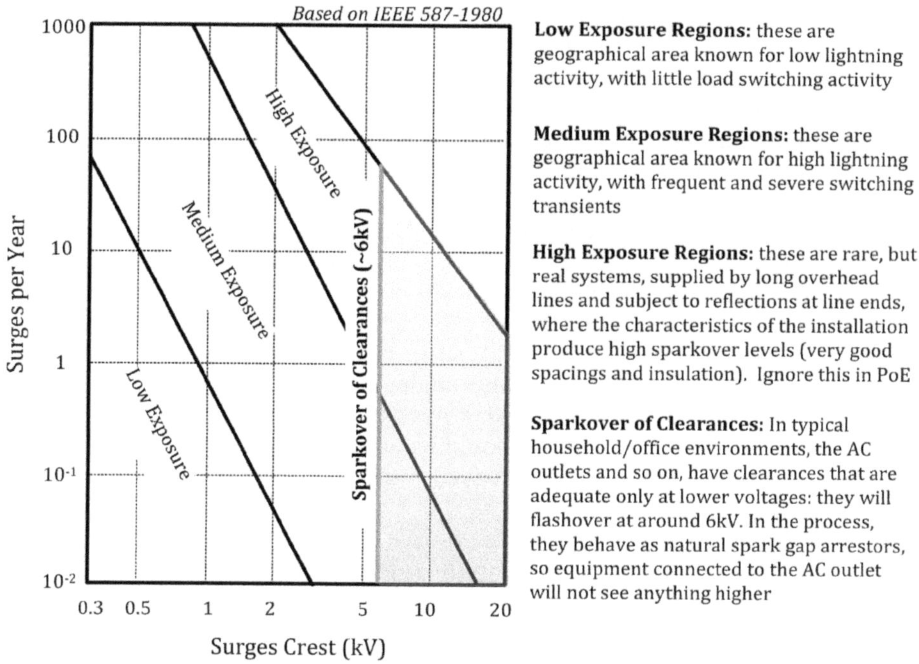

FIGURE 11.2 IEEE-587 study of surge events per year in North America.

10/700-μs profile. But it does have a loophole to revert back to the usual 1.2/50-μs test. (See Table 2 on page 17 of the standard, footnote g: "where the coupling network for the 10/700 μs waveform affects the functioning of high-speed data ports, the test shall be carried out using a 1.2/50 (8/20) μs waveform and appropriate coupling network.")

NOTE *The proposed Chinese standard does in fact require the harsher 1.2/50-μs (8/20-μs) profile at the mains input of AC-DC devices. That is, indeed, hard to meet, but out of our scope here. Also for testing to such high-surge levels, buildings actually need to have "primary protection" against lightning. That is front-end protection (SPDs), which allows much reduced "let-through" voltages (< 4 kV) to reach equipment/appliances inside the building.*

We should remember from Chap. 10, that not only must clearance/creepage be virtually doubled to meet these high-surge levels, but component ratings have to be changed too. For example, the data isolators/optocouplers need to be checked carefully. But the biggest culprit or stumbling block is the 2-kV Y-cap inside every magjack port. That takes the brunt of the entire surge voltage across itself.

How can we hope to pass a 4-kV test with any certainty with an *existing* magjack? Yes, with discrete magnetics, it is possible to achieve, but eventually, no one is going to accept that solution in high commercial volumes readily. Remember the original battle in PoE was to keep the magjacks in the same old housing, with or without PoE. So why would discrete magnetics suddenly become attractive or even acceptable to anyone now? It seems with all these questions, the Chinese requirements will likely die a natural and clumsy death.

With that said, how should *we* qualify equipment (a switch/hub for example) for reliability during the development phase? We will soon realize we can actually push existing designs under development to 2 kV, or even up to 2.5-or 3-kV surge capability quite easily—with minor redesign and some essential engineering prowess. But untill we get into those details, we first create a template for testing. Typical PoE equipment under test (EUT) should be able to pass all the stated test levels, provided the cost-effective recommendations that we will soon make are adhered to.

Template for Testing during PoE Design Qualification Phase

Here are some simple recommendations for testing surge reliability during the qualification phase. This particular test regimen was actually implemented by at least one high-profile switch manufacturer, who later reported good correlation with increased field reliability.

Test Level 1

Pass ±1 kV (1.2/50 μs; 8/20 μs) with 42-Ω surge impedance. With 1 min (or less) between successive zaps, apply 10 positive zaps followed by 10 negative zaps. Note that in terms of number of zaps, this is more severe than the mandatory-minimum requirement of 5-positive and 5-negative zaps (as per EN 61000-4-5 and EN 55022).

Test Level 2

Increase the surge voltage in steps of 200 V. Eventually pass ±2 kV (1.2/50 μs; 8/20 μs) with 42-Ω surge impedance. With 1 min between successive zaps, apply 5 positive zaps followed by 5 negative zaps at each step.

Test Level 3

Increase surge voltage in steps of 100 V. Eventually pass ±3 kV (1.2/50 μs; 8/20 μs) with 42-Ω surge impedance. With 2 min between successive zaps, apply 2 positive zaps followed by 2 negative zaps at each step. Note that the additional time allows components to cool down between stressful voltage applications.

Performance Criteria

In all cases, we can expect to pass with the following *performance levels*:

1. After each zap, we must check if the *data link* is working. Unfortunately, the surge test cannot be easily carried out with the data link up at the time of delivering the zap because the coupling capacitors (the coupling decoupling network, or CDN) used to inject the surge waveform onto the data lines, effectively short-out the lines from the viewpoint of the signal. So we must disconnect the CDN after each zap, connect an oscilloscope to the port and check if we can see it sending out a link pulse at least. That would indicate there was no damage to the switch/PHY sections by the surge test.

   There are a few coupling networks available in the market that do allow the data link to be up and running *during* the surge test (such as from Fischer Custom Communications, Inc). These are noninvasive from the high-frequency signal viewpoint. Therefore, some major OEMs did demand, at least in the past, that the surge test conform to Performance Criterion A in Table 11.3 (data being unaffected). But we also understand that in almost all other cases, it is impossible to either confirm or not-confirm Performance Criterion A for data because there is no active data connection during a typical PoE surge test. Maybe that will evolve at a later stage. As of now, what we can try to do at this stage is to achieve Performance Criterion A *for PoE* (in the sense or spirit of the standard, even though the standard itself does not offer any granularity between data and PoE).

2. We should therefore demand that the PoE link does not get interrupted at all—either in the actual port being zapped, or a nearby port. If we use a small LED-based PD to keep the port alive during zapping (see sample schematic in Fig. 11.3) we can see from the lack of flicker whether the PoE link remained firm—or not. Keep a few neighboring ports alive too with small PDs connected to them, as explained below.

Recommended Surge Test Setup

In Fig. 11.4 we show an overview of the surge test setup.

Four PDs: We should preferably use not one, but four 10-mA/LED PD load cards based on Fig. 11.3. We should monitor not just the port being zapped, but nearby ports too for possible interruption or damage. Further, out of the three unzapped ports, at least one should come from the same multiport PSE chip on which the port being zapped comes from (we should use multimeter continuity

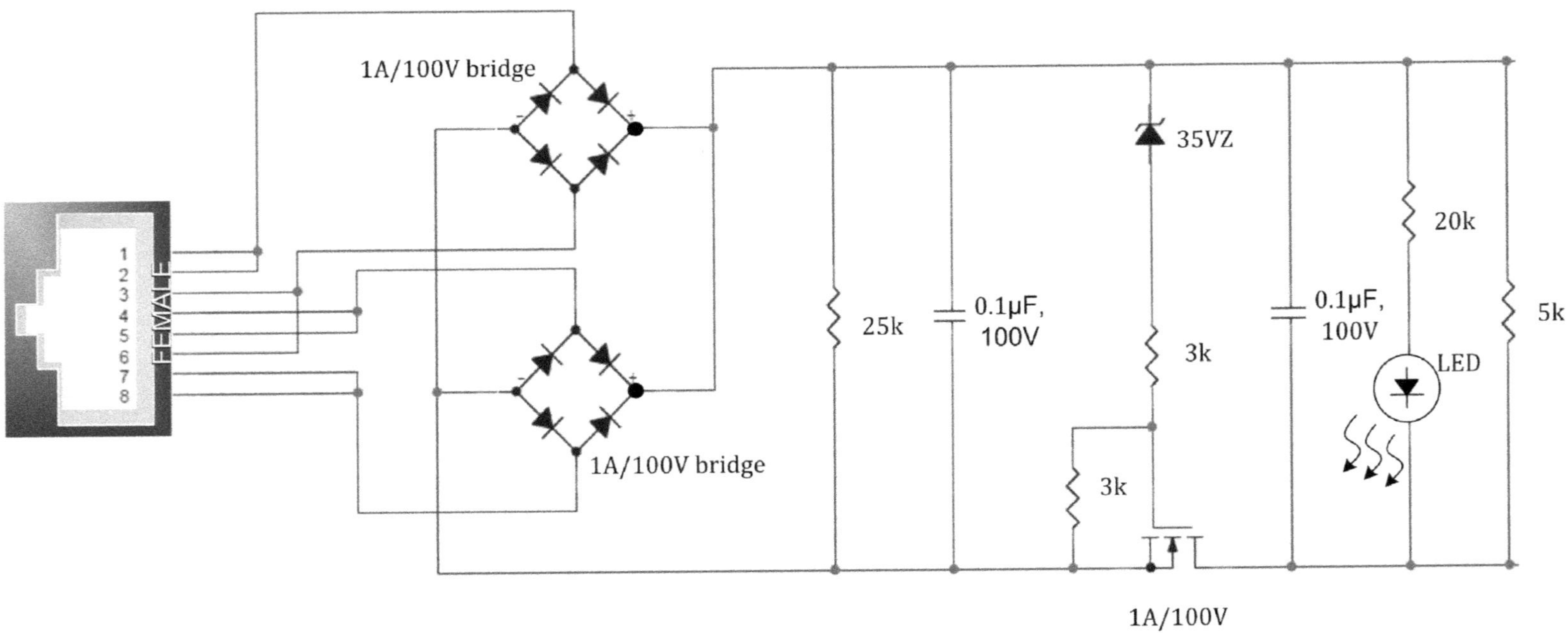

Figure 11.3 PD load for surge testing.

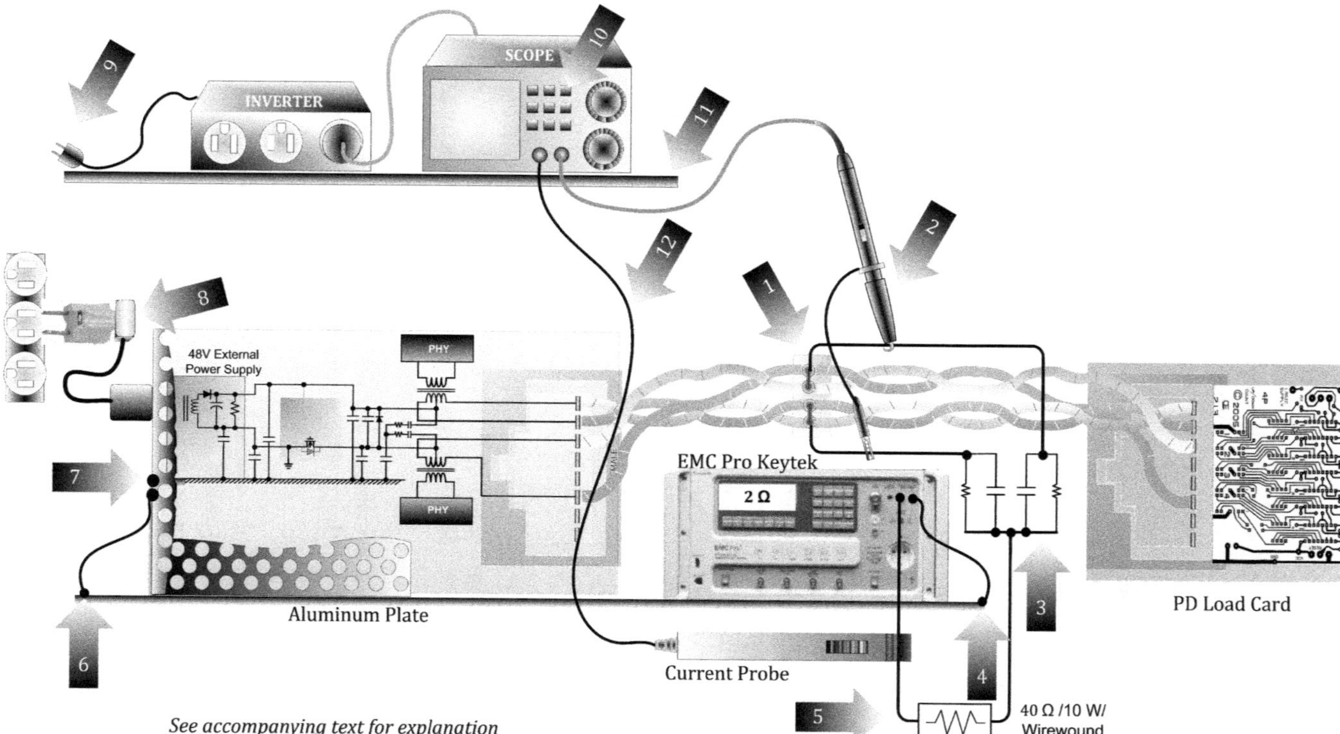

1) A small "breakout board." 2) A high-voltage scope probe. 3) Two metalized polyester or preferably polypropylene high-voltage caps, each of 0.5 μF/4 kV, parallel to a bleeder resistor chain rated 1 M/4 kV. 4) Screw firmly to aluminum plate: this serves as the chassis (earth) ground for the surge test. 5) A 40 Ω wirewound resistor combined with 2 Ω from the surge generator (e.g. EMC Pro Keytek) provides a total of 42 Ω surge impedance. 6) Screw the enclosure of switch/hub firmly to this metal plate. 7) The connection on the enclosure which connects to the aluminum plate. 8) 3-pin to 2-pin AC adapter plug to break the connection of AC-DC power supply and Earth (plug this in during surge test) 9) Unplug AC cord of inverter during the surge test. 10) Scope is floated (on battery power). 11) Wooden shelf 12) Current probe to measure surge current.

FIGURE 11.4 Recommended surge test setup.

checks from the breakout board to the pins of the PSE chips, to identify that chip first).

Length of Cable: The breakout board (with exposed prongs connected to the wires of the cable) on which we will inject the surges into the RJ-45 of the PSE should be connected by a short (1-m) cable to the PSE. The PD should be somewhat separated from the PSE for this test, so it should be connected to the breakout board by a relatively long cable (5 to 10 m).

Isolation from Mains: The most crucial aspect of the surge test is to avoid sneak paths through the AC (mains) wiring. We want to deliver surge energy into the EUT not chase it around the building. So, many steps have been taken to eliminate these stray paths. In particular, lab-power supplies should be avoided for delivering the "48V" rail. Such power supplies surprisingly often have varistors inside, between the output terminals and the mains input terminals, and that creates sneak paths that compromise the validity of the test. The actual silver-box AC-DC power supply being used in the finished product is usually the best one to use here. Also avoid so-called "isolation transformers." They are rarely as completely isolated as we may believe. There may be capacitors present between the primary and secondary windings of the transformer. To reduce capacitive coupling, the transformer may employ shields, but these are usually earthed, and so they will typically carry any common-mode AC currents/noise (such as related to surges), straight into the mains wiring. Further, we also know that AC currents in one winding couple inductively into the other winding. So the only thing it may guarantee is *DC* isolation. But a 1.2/50-μs waveform is hardly DC.

Caution: Do not touch any metal surfaces during the duration of the surge test as the entire setup floats up with respect to Earth ground (on which we stand). Step far away using protective glasses.

In Fig. 11.4 we have placed labeled arrows at various places. Here are some pointers based on those. The numbers correspond to the labels on the arrows.

1. This is a small breakout board. Two RJ-45s are mounted on this (not shown). Assuming Alt-A, we have soldered pins 1 and 2 together and also pins 3 and 6. Two high-voltage caps are connected to each of these two links to inject the surge. The data pairs can either just pass through or get terminated (it doesn't matter here).
2. This is an ordinary high-voltage scope probe. We see that we are running the scope on battery power for preventing sneak paths through the wiring of the building. We can therefore just

use a simple high-voltage probe. Another option is to *not* run the scope on battery power, instead using a high-voltage *differential* probe.

3. Two metalized polypropylene caps, each 0.5 μF, rated at least 2 kV. We need *very high dV/dt* capability here, and these caps are hard to find. They will very likely need to be paralleled for higher capacitance, and "series-ed" for higher voltage to make up an equivalent 0.5 μF/2 kV with very high *dV/dt*. We can start with three 0.15 μF/2000 V (Series 943C20) from CDE Cornell Dubilier. That will give us 0.45 μF/2 kV (close to 0.5 μF/2 kV). To test to higher voltages (if required), we may need to make a series combination of two such "equivalent" caps for each "single cap" shown in the figure. The principle is as follows: For example, to achieve 4-kV rating we can take two caps of 0.5 μF/2 kV in series. That gives one cap of 0.25 μF/4 kV. We parallel two of these strings to give us 0.5 μF/4 kV. And so on.

   For the resistors across the caps, these are just bleeder resistors of 1 MΩ each. Actually there are not two as shown, but *twenty* resistors here. Why? Because it is not possible to meet high-voltage ratings with just one commonplace resistor. It is therefore suggested here, to use ten 1-W leaded metal-oxide resistors in series, each of 100 kΩ (example RSF 200 from Yageo). Finally we should wrap heatshrink tubing around the chain of 10. Note that we need 10 resistors for *each* of the 2 capacitors.

   This entire coupling-decoupling network (CDN) should be finally placed in a tough polypropylene container to protect the user in case the capacitors or resistors develop a fault and explode.

4. Screw this terminal coming from the surge generator firmly to a large aluminum plate underlying the switch/hub (EUT). This plate serves as the local chassis (Earth) ground for the surge test.

5. This is a minimum 40-Ω/10-W wirewound resistor, which adds to the preset 2-Ω surge-generator impedance to give the required 42-Ω surge impedance as per EN 61000-4-5.

6. We should screw the enclosure of the switch/hub firmly to this metal plate.

7. The connection on the enclosure of the EUT connects to the aluminum plate.

8. Use a 3-pin to 2-pin AC adapter plug to disconnect the AC-DC power supply from the Earth wiring. *Plug this in!* It is just shown unplugged to reveal the adapter plug's construction.

9. The scope is isolated from the mains supply and run on battery power from this "inverter." But the inverter itself is not battery-powered—it has an AC cord to the charger inside the inverter. So we must make sure we unplug this AC plug for the duration of the surge test.
10. This is the oscilloscope to measure the surge voltage/current. It must be floated on battery power. If we do that we can use ordinary high-voltage probes, otherwise we need *differential* high-voltage probes.
11. Use a wooden shelf for the measurement equipment, to prevent sneak paths.
12. The connection to the current probe to measure surge current.

What Happens during the Surge Test

In Fig. 11.4, we monitor the voltage and current associated with the surge. Let us do some simple math around this. If we apply 2 kV across a dead short (say the almost-uncharged caps at the input or output of the PSE), we will get an instantaneous current of $I = V/R = 2000/42 \approx 47$ A. Well, this is the *simplified* upper limit. In reality, we can get much less. There are two related reasons for that. First, the voltage does not reach 2000 V immediately. It takes a few microseconds to get there. And things can change quite dramatically during that brief instant. Which brings us to the second reason: All the surge current must pass through the Y-caps (in series, thus acting as a bottleneck). So if we reduce the Y-caps, the current will obviously decrease too. In fact, a good test to check the integrity of our surge setup is to disconnect/remove all the Y-caps that we know of, including in the terminations, and inside the "48V" power-supply unit (PSU). If we still see any current at all, using the current probe in Fig. 11.4, there is something very wrong. We should not proceed with the surge test till all "sneak paths" have truly been eliminated. The two reasons above are related as mentioned, because a relatively small Y-capacitance will charge up at least partially during the few microseconds that the surge waveform takes to peak. Eventually, the current at the peak is less than 47 A. Because, now the current is based on the *difference* between the instantaneous surge voltage and the voltage of the partially charged Y-cap because that is the actual peak voltage appearing across the limiting surge resistance. So if the cap had meanwhile gotten charged to 800 V, the current when the surge waveform peaks would be $(2000 - 800)/42 = 28.5$ A instead of 47 A.

If we reduce the total Y-capacitance, we will significantly reduce the stresses in the PSE circuitry arising from the surge. But if we indiscriminately increase the Y-capacitance, all the applied 2 kV may eventually appear across the PSE circuitry and will certainly destroy it—unless we can *absorb that energy safely* (we will come to that shortly).

The PSE certainly can't get damaged by a completely common-mode surge, more so if there are no Y-caps present, unless the disturbance is somehow "differential-mode" across the data pairs (positive-and-negative-port rails). Differential-mode surge waveforms are typically not a requirement for PoE, though some more stringent standards like GR-1089 may require it, as discussed later. Yet we deal with differential-mode surge effects, because though the applied surge voltage is common-mode, because the impedances are so different on the two PoE rails on which it is injected, this so-called common-mode waveform produces a significant differential-mode component across the positive and negative data lines. Eventually this overvoltage can cause the port pins of the PSE chip to burn out first. Then something else behind the PSE chip can also burn out, like the 58-V TVS, and so on. In fact, it is because of this common-mode to differential-mode conversion that we are having to take so many steps to understand the problem and find a solution. GR 1089 explicitly asks for differential-mode surge testing, but as we can see, standard common-mode surge testing produces the very same effect, and the solutions apply equally.

Having understood this, the obvious temptation is to entirely dispense with all Y-caps on the PoE board. Unfortunately, we need Y-caps for EMI suppression. We do not want the cable either to emit excessive EMI or pickup excessive EMI (susceptibility for the data traffic). For the latter, we have the Y-cap C3 as shown in Fig. 11.5. But many more Y-caps are typically mounted on the PoE board, such as Cpsu, C1 (2), and C2 (2). Their main purpose is to prevent noise from going out on to the data lines (EMI emission). So we need to stop and ask: where is the noise coming from? The PSE is, philosophically speaking, just a gate that opens or closes for the incoming "48V" DC rails. It can rarely contribute to noise/EMI itself. Yes, the PSE may have some high-frequency onboard clocks for timing or data-communication functions, or for on board data processing, but usually the EMI from all that is insignificant. Basically any outgoing EMI from the PSE is very likely coming from the PSU, *not the PSE.* So, rather than use brute-force methods like sprinkling Y-caps everywhere, it is better to ask the vendor of the PSU to ensure their PSU has very low EMI on its output cables too. Most ITE power supplies are only tested for EMI on the AC mains input lines, but power supplies for telecom applications should actually be tested for EMI on their outputs too.

Summarizing, *reduction of Y-capacitance is one of the ways for ensuring surge survivability, especially in AC-disconnect cases.*

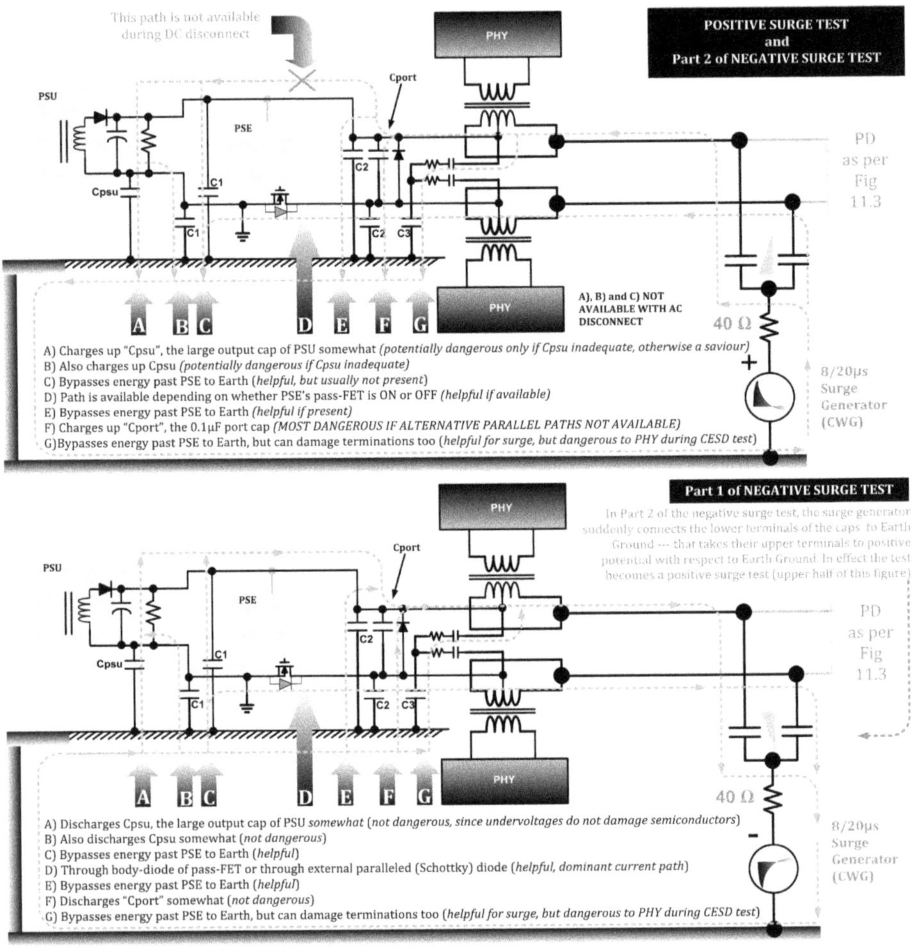

FIGURE 11.5 Explanation of currents during surge test.

The reason for this statement emerges more clearly from a quick study of the current paths under a surge event, as sketched out in Fig. 11.5. Let us discuss this figure now.

1. In the top half of Fig. 11.5, we are applying a surge of positive polarity from the CWG generator to the PoE board. The current paths are shown.
2. In particular, paths A, B, and F are marked "dangerous," because they charge the input and/or output caps of the PSE.
3. F is certainly the most dangerous, because we can imagine that 47-A (or 28.5-A) of surge current, were it solely passing through C port (0.1 µF), would destroy it immediately by overcharging it.

4. A and B are paths that come in parallel with F and are usually the reason we *don't* have to worry about path F at all, because A and B go through a very low impedance, Cpsu, and therefore divert most of the current through this path rather than F. Since the value of the capacitance Cpsu is typically very large, it can easily handle even 47-A of surge current with aplomb, not getting excessively charged. In brief, path F, despite being marked dangerous, is actually the savior with DC disconnect. We can fix the amount of bulk capacitance required, as shown further below.
5. With AC-disconnect, paths A, B, and C are blocked by a diode in the positive port rail (not shown in Fig. 11.5). So now all the stress can appear on path F, which charges up the port capacitor. Clearly, such a huge surge current cannot be handled with such a tiny cap, and we need to take drastic steps to reduce the total Y-capacitance to cause the surge current to slow down or stop completely, within the few microseconds it takes for the surge waveform to peak.
6. In the lower half of Fig. 11.5, we have explained that the *negative* surge test actually creates very similar waveforms as a positive surge, but that is only during part 2 of this test. In part 1 of the negative surge test, the CWG pulls in current from the PSE. Most of this current comes either through the body-diode of the PSE's pass-FET, or through a Schottky diode (typically 2 A/100 V) in parallel to the FET (placed pointing in the same direction as the body-diode, *placed very close to the pass-FET and connected to it by short and thick copper traces*). See Fig. 11.6. The reason for asking for a paralleled external diode across the PSE's pass-FET is to actually divert the surge current away from the body-diode, *to protect the bond wires of the PSE/FET pack.* A Schottky-bypass diode is used since the drop across this diode must be much less than the drop across the body diode of the FET it seeks to bypass, so that the surge current will "prefer" this external diode over the internal body-diode. From Fig. 11.6 we see an alternative method can be used, not requiring a Schottky, but it is much more lossy. It is, however, less sensitive to layout or diode characteristics. But *neither of these external bypass-diode methods is typically required unless a surge capability of more than 2 kV is sought (using 8/20-μs waveform and 42-Ω surge impedance).* It is however necessary to survive GR-1089 (discussed later). Therefore a placeholder for these diodes is recommended.
7. If the PoE circuit survives part 1 of the negative surge test (with the help of a parallel diode if necessary), then, as explained in Fig. 11.5, we move to part 2 of the negative surge test. This reversal happens when the CWG suddenly raises the voltage

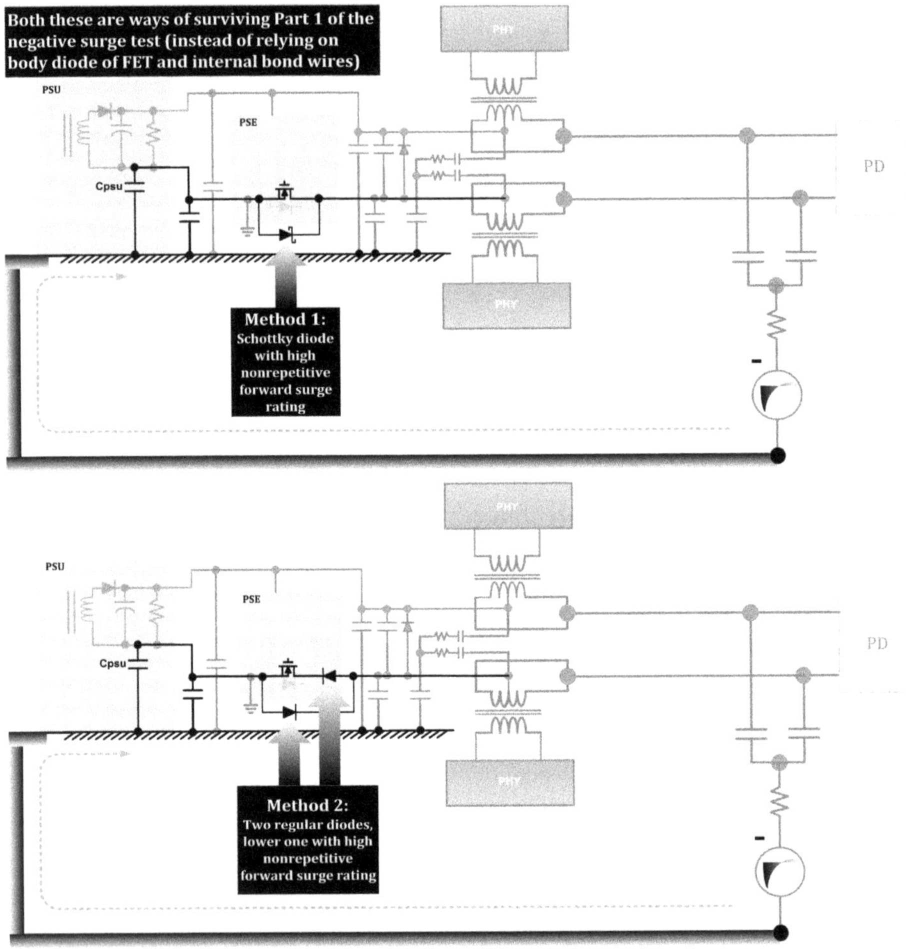

FIGURE 11.6 Surviving high-voltage negative surge tests.

at its end of the injection cap to Earth ground. Since the injection cap is charged in one direction and cannot discharge immediately, the voltage difference across it is maintained. So its other end suddenly gets raised, and that in effect conducts a positive surge test on the PSE, albeit with somewhat diminished amplitude. In other words, part 2 of the negative surge test is almost identical to the regular positive surge test. And it too has the potential of creating *overvoltages* (not undervoltages as in part 1). The PSE chip can fail exactly the same way, and so, the fixes we propose for the positive surge test, apply equally to the negative surge test. The failure modes are almost identical too.

The equivalent circuits that explain rather intuitively what happens during a surge test are presented in Fig. 11.7, based on our

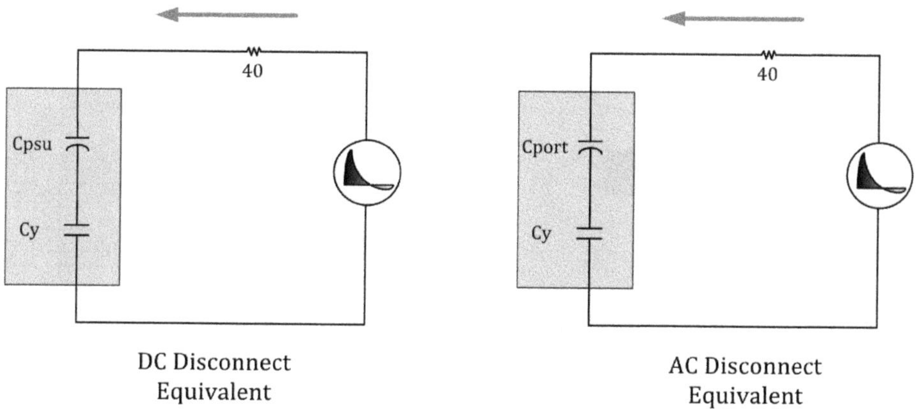

FIGURE 11.7 Equivalent circuits during surge test.

understanding of Fig. 11.5. To start putting some numbers to this, we need to first understand the characteristics of the surge generator (CWG), as discussed next.

Other Setups for Surge Testing

In Fig. 11.8, we describe several surge setups commonly seen. The topmost one is the same one we were discussing previously in Figs. 11.4 and 11.5, except that we have made the setup universal, to test either Alt-A or Alt-B (for example a switch or a Midspan). The middle setup is also OK as per IEC 61000-4-5, but may give slightly different (worse/lower) surge survival thresholds compared to the topmost schematic. However both top and middle setups expose the vulnerability of AC disconnect, and that has nothing to do with the setup really. But it is important to test with a PD connected to keep the pass-FET ON. Also, if we use a 40-V Schottky as the AC-disconnect diode, it typically avalanches (behaves as a zener), at a little over 60 V, so it will then carry the surge energy into the bulk cap of the power supply output. Yet, if the surge lasts too long, as in the case of a very large Y-capacitance or a low-resistance preload resistor inside the power supply, the Schottky/zener can get damaged. So we are walking a fine line. *The best option is to use DC disconnect* as explained previously.

The lowermost setup in Fig. 11.8 is surprisingly often seen being carried out in certifications. The lab technicians connect either all 80 Ω or all 160-Ω resistors to each pin of the RJ-45, based on some equation in the IEC 61000-4-5 standard. But they don't realize that the relevant figure in the standard (Figure 14) does not apply since it was originally intended for symmetrical lines (equal impedances). It would apply if there was no PoE present, but with PoE riding on the lines, the impedances are not symmetrical anymore.

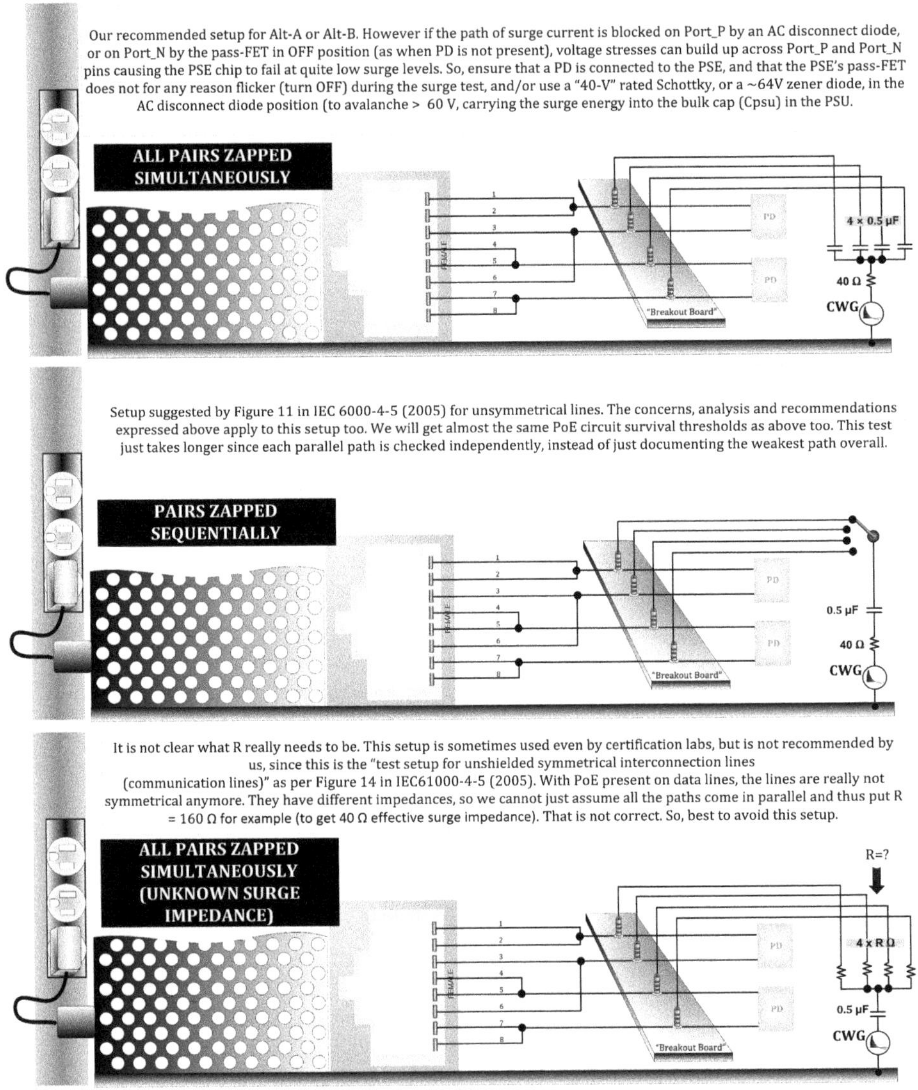

FIGURE 11.8 Other setups for surge testing.

Modeling the Combination Wave Generator (CWG)

The standards describe the waveshape associated with a lightning surge. But they make no mention of how to design the piece of equipment (the CWG) which will have the required characteristics. The required waveshapes were described in Fig. 11.1. The problem is that one of them is an open-circuit voltage waveform, one a short-circuit current. Neither is realistic because we intend to connect a nonzero, noninfinite circuit impedance across the terminals of the CWG.

We have to account for the 42-Ω resistive impedance in series with the DUT/EUT. In this particular application, 2 Ω out of the total 42 Ω is expected to come from the CWG.

The actual rise time of the waveform in our application is neither going to be close to 1.2 μs (open) or 8 μs (short). It is somewhere in between, but what exactly is it? In the appendix of this chapter, we have presented the derivations for the design of the CWG shown in Fig. 11.9. Here we just report the results.

The output voltage (across *Zdut*) in Fig. 11.9 is

$$Vo(t) = \gamma \times e^{-\alpha t} \times \sin(\beta t)$$

where

$$\alpha = \frac{1}{2} \times \left(\frac{Rm}{L} + \frac{Zb}{L} + \frac{1}{R1 \times C} \right)$$

$$\beta = \sqrt{\left[\frac{1}{LC} + \frac{Rm}{R1 \times LC} + \frac{Zb}{R1 \times LC} \right] - \left[\frac{1}{4} \times \left(\frac{Rm}{L} + \frac{Zb}{L} + \frac{1}{R1 \times C} \right)^2 \right]}$$

and where,

$$\gamma = \frac{R2 \times Zdut \times Vdc}{\beta \times L \times (R2 + Za)}$$

$$Za = Zdut + Zext$$

$$Zb = \frac{R2 \times (Zext + Zdut)}{R2 + Zext + Zdut}$$

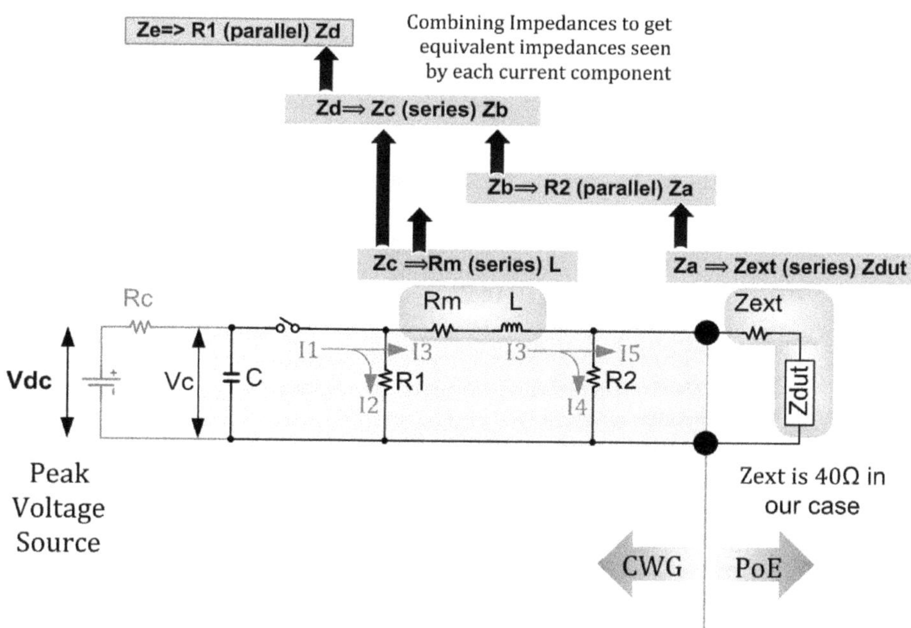

FIGURE 11.9 Analyzing the CWG.

Hint: *Setting Zdut as 1Ω, the voltage across this 1Ω resistor numerically equals the current out of the surge generator. Thus we know surge current too.*

NOTE "Cpsu" earlier is same as "Cbulk" here

Both set of curves below are essentially the same (though with AC or DC disconnect), they use different time scales.

AC Disconnect calculations (using Mathcad plots) (bulk cap of PSU not available to absorb surge energy)

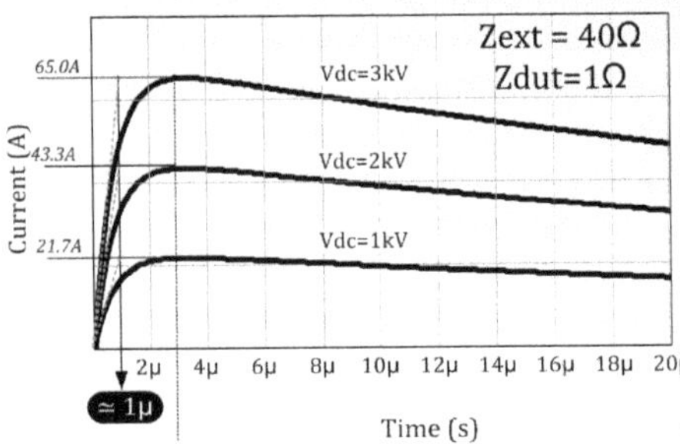

The total Ycap "Cy"should charge up exactly at the rate of rise of surge waveform. Initial rate of rise is V/t where V is the max voltage Vdc, and t is 1 μs (extrapolated). So

$$Cy = \frac{I \times \Delta t}{\Delta V} = \frac{I \times 1\mu s}{Vdc}$$

and I ≈ V/R=Vdc/Zext where Zext = 42Ω. So

$$Cy = \frac{\cancel{Vdc} \times 1\mu s}{42 \times \cancel{Vdc}} = 24nF$$

All peak at 3.2μs

So if we set Cy to be less than 24 nF, it would have charged up fully by the time the applied surge waveform peaks (for any peak surge voltage), and so there would be no surge voltage remaining *across* Cport (see right side of Fig. 11.7)

DC Disconnect calculations (using Mathcad plots) (bulk cap of PSU is available to absorb surge energy)

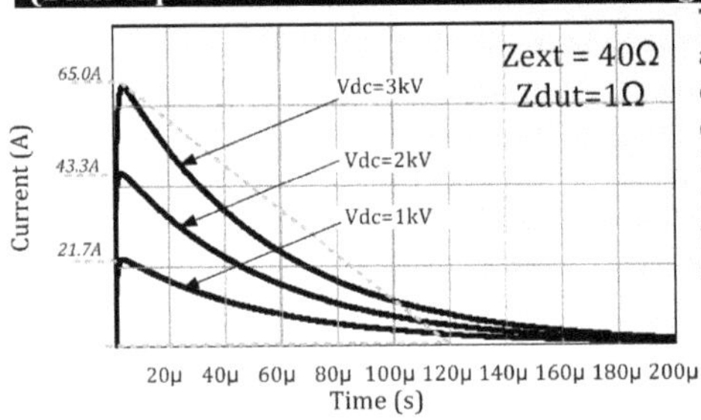

To determine energy delivered, use the approximate current shown here. For example, we can say that for Vdc = 3kV, the current is an average of 65A/2 = 32.5A, for 120μs. So energy is V x I x t = 51 V x 32.5 A x 120 μs = 0.2 Joules. If this is to be absorbed in the bulk cap, and that is allowed to go from 51V to 58V, we get

$$\frac{1}{2} \times Cbulk \times (58^2 - 51^2) = 0.2, \text{ So}$$

$$Cbulk = \frac{2 \times 0.2}{(58^2 - 51^2)} = 524\ \mu F$$

Minm bulk cap for 3kV surge capability

Similarly, for 2kV surge capability, we get 348 μF. So if we set Cbulk to be greater than 348 μF, it will charge up to a max of 58V by the time the applied surge waveform peaks (for 2 kV peak surge voltage), and so there would be no voltage >58V *across* Cport (see left side of Fig. 11.7)

FIGURE 11.10 Plotting the results based on the CWG model, with 40-Ω load connected to it.

The solved component values of the CWG are

$$C = 6.038\ \mu F \quad L = 10.37\ \mu H \quad R1 = 25.105\ \Omega$$

$$R2 = 19.80\ \Omega \quad Rm = 0.941\ \Omega$$

In Fig. 11.10 we plot the waveforms for a peak (Vdc) of 1, 2, and 3 kV. The analysis of that is divided into AC-disconnect and DC-disconnect configurations.

Recommendations for AC Disconnect

We have realized from Figs. 11.5 and 11.7 that the problem with AC disconnect is that because of the AC-disconnect diode blocking in a reverse direction, access to the bulk cap of the PSU (and input 58 V

TVS, not shown) is not available. The surge waveform can therefore play havoc with the front end, especially the 0.1-μF port caps and from there to the port pins or FETs.

The easiest way to handle this is to have a *small* Y-capacitance. The idea behind that is that the applied surge waveform has a certain dV/dt, and if the Y-capacitance charges up at the same rate or a little faster, we will be in a "comfort zone," where on an instantaneous basis, the applied voltage will almost equal to the voltage across the Y-capacitance (same rate of rise), so there will be in effect, no "left-over" voltage which adds to the existing port capacitor voltage. That is the underlying principle here.

In Fig. 11.10 we see that all the waveforms peak at 3.2 μs, but their initial rise extrapolates to the 1-μs mark. We therefore demand that the Y-cap charge up fully to the max applied voltage in 1 μs.

$$Cy = \frac{I \times \Delta t}{\Delta V} = \frac{I \times 1\mu s}{Vdc}$$

and $I \approx V/R = Vdc/Zext$, where $Zext = 4\ \Omega$ (including CWG impedance of 2 Ω into this). So

$$Cy = \frac{\cancel{Vdc} \times 1\ \mu s}{42 \times \cancel{Vdc}} = 24\ nF$$

We thus need to restrict the net lumped capacitance to 24 nF for any surge voltage.

In Chap. 9 we had shown how to estimate and measure the net lumped Y-capacitance.

Note that with the Y-capacitance sized in this manner, surge survivability no longer depends on whether the pass-FET is ON or OFF. So path D in the top half of Fig. 11.5 need not be present. That is good because under a surge event, such a huge current passing through the FET may cause its protection circuitry to switch it OFF which we know is otherwise not good.

Recommendations for Common-Mode Filter Position

In DC disconnect, access to the bulk capacitor of the PSU is available to absorb the huge surge energy. See Figs. 11.5, 11.7, and 11.10. This is fortuitous, but it also means we should not place an unnecessary impedance or barrier between the PSE chip (that we wish to protect) and the bulk cap (its protection). For this reason, it is not advisable to place any exotic inrush-limiting circuitry, or even common-mode filters, *between the PSE chip and the bulk cap of the PSU*. A common-mode filter in the port has pros and cons. We now discuss some key points regarding a common-mode filter in order to make the best decision.

1. It is usually added to just make up for poor filtering in the PSU, so it is better to get the PSU designed correctly.
2. It does help ballasting out imbalances as discussed in Chap. 9.
3. It impedes access to the reservoir of protection (the output bulk cap of the PSU). So, if deemed really necessary, it should preferably be *between the RJ-45 and the PSE chip/FETs.*

Recommendations for DC Disconnect

Assuming we have easy access to the bulk capacitor of the PSU, we can calculate how much energy is being delivered by the surge, and how much bulk capacitance is available to absorb this. A sample calculation is presented in Fig. 11.10. We have basically assumed a triangular current waveshape of height Imax (numerical values provided in the figure) and 120 μs (extrapolated average duration as shown). Then, based on an average current of Imax/2, lasting for 120 μs, we can equate the energy delivered by the surge event, to the change in stored energy based on ½ × CV^2. We thus get 524 μF for 3 kV. The closed-form equation we can also use is

$$C_{\text{bulk}} = \frac{V_{\text{pse}} \times I_{\text{max}} \times 120\ \mu}{V_f^2 - V_{\text{pse}}^2}$$

Here V_{pse} is the normal operating PSE voltage (say 51 V) and V_f is the max voltage we want to see on the cap, say 58 V—keeping headroom for a few additional volts coming from the drop across the ESR of the bulk cap, and finally basing it on 74-V Abs Max rating (process limit).

For example, for achieving 2-kV surge capability with DC disconnect, the output cap of the PSU must be at least

$$C_{\text{bulk}} = \frac{51 \times 43.3 \times 120\ \mu}{58^2 - 51^2} = 347\ \mu F$$

We can pick a standard 330-μF (nominal) value as it is in the ballpark, within the calculation's inherent errors/tolerance.

Are there limits on the Y-capacitance in this DC disconnect case? Not much, since we have calculated that *all* the surge energy gets absorbed in the bulk cap, so we are not relying on the Y-capacitors to charge up and terminate the surge current early. Nevertheless, we should keep to reasonable limits on Y-capacitances since they also need to bleed in a reasonable time. On that basis, it was empirically established that a good target is to keep to less than 0.2 μF of net Y-capacitance, distributed evenly on the port lines, on any side of the pass-FET—a maximum 0.1 μF on each line to Earth ground. We should not forget the voltage rating of these caps have to be commensurate with their expected surge voltage capability. Large-capacitance values with large-voltage ratings are expensive (film caps) and hard to find, in general. They should ideally also have high dV/dt capability and self-healing properties. The best option is to redesign the "48V" PSU for low EMI, as mentioned.

Note that in DC disconnect, the bulk cap can be accessed by the surge from the positive rail (paths A, B, and C in the top half of Fig. 11.5). So surge survivability no longer depends on whether the pass-FET is ON or OFF, irrespective of the value of Y-capacitance. In fact, we could have done the surge test with no PD load connected. The results would have been the same (pass thresholds unaltered).

Surviving the 10/700-μs Surge Test

As mentioned previously, the most recent EN 55024 standard now seems to ask for testing immunity of telecommunication lines using a CWG with an open-circuit voltage waveform that is 10/700 μs (the corresponding short-circuit current waveform is 5/320 μs). This is an alternative to the usual 1.2/50-μs open circuit voltage (8/20-μs short-circuit current) test. EN 55024 also states that the 10/700-μs test is "*applicable only to ports which according to manufacturer's specification may connect directly to outdoor cables*" (see Table 2 of the EN 55024 standard). It also provides another loophole of sorts in the footnotes of Table 2, where it states that "where the coupling network for the 10/700-μs waveform affects the functioning of high-speed data ports, the test shall be carried out using a 1.2/50 (8/20)-μs waveform and appropriate coupling network." Keep in mind that Ethernet and PoE are in effect built around indoor cables. So the applicability of the 10/700-μs waveform to PoE is not very clear. Nevertheless here are the pros and cons and ways to meet the requirement, if so desired.

1. The first thing to look for is the specified impedance of the surge generator. The EN 55024 standard specifies this to be 40 Ω as previously. So the short-circuit currents are the same as before: $I = V/R = 1000\ \text{V}/40\ \Omega = 25\ \text{A}$.
2. However, we also know that if our total Y-cap is small enough (< 25 nF), we manage to charge up the Y-cap in about 1 μs with that level of short-circuit current $dt = (C/I)\ dV = (25\ \text{n}/25\ \text{A}) \times 1\ \text{kV} = 1\ \mu\text{s}$. So in 1 μs the surge current flow would stop entirely because the Y-cap was *fully charged*. Looking at it dynamically, it meant that with a 1.2/50 μs waveform, the Y-cap charged up at almost the same rate as the rising voltage waveform. So there was no significant voltage accumulating differentially across the PoE lines and PoE circuitry. Now, with an even slower rise time (between 5 and 10 μs as with the new 10/700 μs profile), as proposed by the recent EN 55024 standard, we can essentially increase the Y-cap *by at least 5 times* (to 125 nF), and still be well-protected by the simultaneous

charging up of the Y-capacitance. In other words, the softer "attack time" of the new profile actually allows us much higher Y-capacitances, for the same PoE voltage stresses. And that is true for both AC and DC disconnect. So for that reason alone we can say "in the case of limited Y-capacitance, the 10/700-μs profile is actually an easier test to meet than the 1.2/50-μs test."

3. But if the Y-cap is not controlled, we have to imagine the surge current continues as long as the surge waveform is applied. So, in this case, there is no Y-cap charging up and causing the surge current to stop earlier. This case would occur if there was a dead short between PoE ground and chassis ground. It is worst-case, but impracticably so. However, assuming this worst-case for now, if we have to store the entire surge energy in the bulk cap (at the output of the AC-DC power supply), we can calculate that we need at least 347 μF bulk capacitance—for passing 2 kV of (1.2/50-μs) surge, *with no limit on the Y-capacitance* when using DC disconnect (so that the bulk cap can be accessed by the surge). The equation to use is

$$C_{\text{bulk}} = \frac{V_{\text{pse}} \times I_{\text{max}} \times 120\ \mu}{V_f^{\,2} - V_{\text{pse}}^{\,2}}$$

The 120 μs is based on the extrapolated decay curve of the 1.2/50-μs waveform. Keep in mind that this equation assumes that the short-circuit current is $I_{\text{max}}/2$ on an average for the entire duration of the surge. We can now redo the same calculation using only 25 A (for 1 kV as per EN 55024), but this time using an extended time of 1000 μs (based on the 10/700 μs and the short-circuit current value of 320 μs). We get

$$C_{\text{bulk}} = \frac{51 \times 25 \times 1000\ \mu}{58^2 - 51^2} = 1670\ \mu F$$

So with the much wider surge waveform now proposed by EN 55024, if we assume unlimited Y-caps, we get very large bulk cap requirements (and we also must use DC disconnect).

A more practical alternative is to reduce the Y-cap (to ~ 100 nF) too; this would once again significantly limit the energy required to be stored in the bulk cap. So we would then be able to achieve 1-kV surge protection for the 10/700-μs waveform, even with a small bulk cap (~10 μF), *with both AC or DC disconnect*. So *the best way is to control the Y-capacitance as we have constantly recommended, especially for AC disconnect, but now also for DC disconnect, so as to avoid using impractically large bulk-capacitance values.*

If we do happen to have DC disconnect, we could in principle, increase *both* the Y-cap and the bulk capacitance *together*, starting from 100 nF and 10 μF, respectively, but this time very judiciously. That would lead to a practical solution to meet the 1 kV -10/700 μs requirement (if applicable).

Protecting the PD from Surges

In a very similar manner, we can test the PD for surges too. The logic is the same actually, because looking in from the cable, the PD looks (coincidentally) very similar to the the PSE. We "see" a bunch of Y-caps, a port cap of 0.1 μF, a TVS, and also a bulk cap—the latter being the input cap of the DC-DC converter that follows. See Fig. 11.11. Just as in the case of a PSE wired in DC disconnect.

This leads to an equivalent circuit akin to a PSE with DC disconnect as in Fig. 11.7. So it should be handled the same way too. In particular, to handle a 2-kV surge, the electrolytic bulk cap at the input of the DC-DC converter must be 330 μF. Just as for a PSE.

But the bulk cap can be much less too! Because *in a PD, we may be able to reduce the Y-capacitance significantly*. In very rare cases, a low-power PD may in fact have *no* connection to Earth ground. It may even have a two-prong AC plug. However, since it has a DC-DC switching converter inside it, eliminating of Earth ground may pose serious problems meeting EMI-radiated and conducted emission limits. The PD may be encased in plastic and have no user-accessible metal surfaces, allowing for relaxing the isolation requirements, yet for conforming to radiated-EMI emission limits. In particular, there will likely be a metallic foil or metallic spray coating inside the plastic, and this would need to be connected to the Earth ground (through the middle prong of the AC plug). So though the Y-capacitance in a PD may be significantly less than a multiport switch (in which all the Y-caps of the magjacks aggregate together as explained in Fig. 9.12), the Y-capacitance of a PD cannot be eliminated altogether. That said, it is much easier for a PD to survive surge testing than a PSE, and this is the reason: If the Y-cap charges up quickly, and we have calculated that 24 nF will charge up fully within 1 μs, the surge current flow will stop, and we will be able to reduce the bulk cap from 330 μF to much less than 180 μF. From Chap. 5, we know that less than 180 μF is considered desirable (worst case 400 μF for meeting inrush), particularly when we consider inrush and power-up events. But it is also now clear, that to reduce bulk capacitance to that desirable and cost-effective level, we need to significantly reduce the Y-capacitance too, for ensuring surge reliability.

How high should the PDs voltage rating be? It seems to have become fashionable for PD-chip vendors to almost brag about their 100-V process and the excellent field reliability of their PDs as a

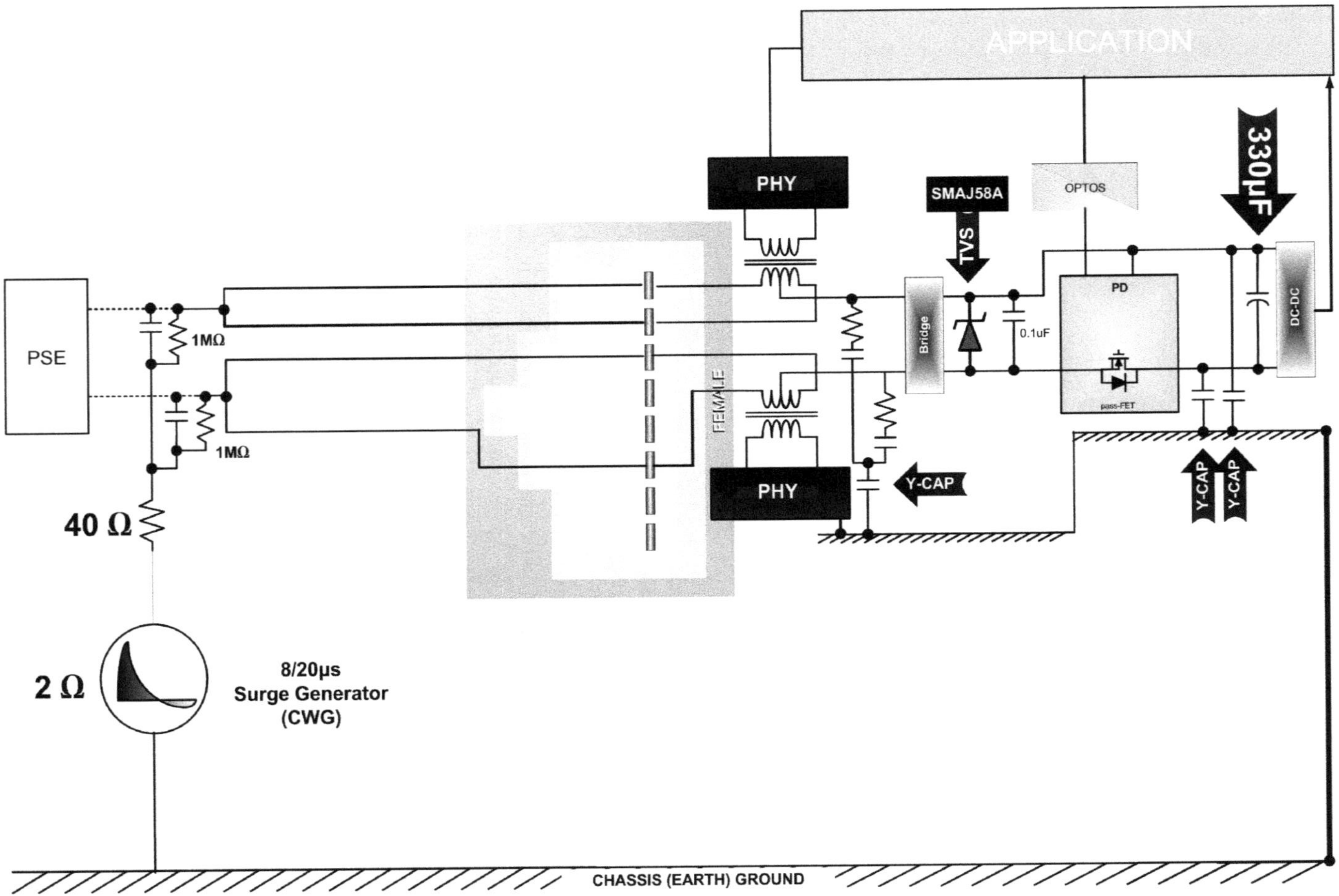

Figure 11.11 Surge test on a PD.

result thereof. However, as mentioned, the PD survives the surge test much more easily than the PSE usually. When a surge strikes, it is on a cable that has a PSE on one end and a PD on the other. They are effectively in parallel from the viewpoint of the surge. Further, the weakest link is usually the PSE because of its complex architecture. If that can survive (with some design skill and almost no added cost) using just a typical 74-V fabrication process, why do we need a 100-V process for a PD, which already has a much reduced Y-capacitance to start with? Keep in mind that when a surge event occurs, the 330 μF recommended bulk capacitance inside the PSE (assuming it uses DC disconnect), will likely absorb a good part of the surge energy. So eventually, the PD has a much better chance of surviving with just a small bulk capacitance of its own. In other words, the PSE, being the weakest link, effectively protects the PD (much like a fuse in an output overvoltage crowbar latch). Therefore, far more attention should be paid to ensuring higher surge reliability for the PSE to start with. That is a much bigger challenge. And so, just using a 100 V rated PD chip does nothing to improve the overall *system reliability*. It is just marketing.

Semiconductors for Protection and Some PCB Recommendations

So far, we have deliberately avoiding mentioning the 58-V transient voltage suppressor (TVS) habitually placed in the PD and PSE. The question is how much good does that really do?

Despite common belief, this has rather limited use. *In fact, it cannot even work on its own.* The reason is that this diode has a max peak-current rating of only 4.3 A, whereas the measured surge currents are closer to 20 to 40 A at 2-kV, depending on amount of Y-capacitance used (it can theoretically be as high as $2000\ \text{V}/42\ \Omega \approx 50\ \text{A}$ with higher Y-capacitances). Further, with only 4.3 A max passing through it, this TVS clamps not at 58 V as commonly assumed *but at 93.6 V*. So not only is its current rating inadequate, so is its voltage rating (and its energy rating too). Basically, the TVS only works to *supplement* the bulk cap's action as discussed previously. Since the bulk cap is usually physically far away, with long traces or wires intervening, and there may even be an ill-advised common-mode filter en route as discussed, *the TVS serves to clamp the voltage to a safe value until the bulk cap starts to act.*

So the TVS only serves to absorb high-frequency spikes, or just the incoming edge of the lightning-surge waveform. But is not so quick-acting either. Because in a typical PSE there are several controller chips, all sharing the same TVS (at their common inputs). Further, the TVS needs to be physically close to all the chips for maximum effectiveness (low-intervening trace inductances). But that is impossible—the TVS cannot be close to *all* PSE chips at the same time. We will typically place it roughly at the the center of several PSE chips, but it cannot service all of them equally well. Therefore, we also need *local decoupling* at the chip level—untill the TVS itself starts to act. And that takes the form of 0.1-μF ceramic caps (X5R or better) placed on each chip. See Fig. 11.12.

Note that vendors of protection devices have a bunch of products they claim solve all problems in surge protection. There are, for example, diode + TVS (*bidirectional clamping*) arrays with ultralow capacitance too, which will not affect data and therefore can be placed directly across the two wires of a twisted pair just as the pairs comes in into the RJ-45. Then there are diode + Sidactor arrays too. A Sidactor is similar to a thyristor or a gas-discharge tube. On being triggered, it crowbars to a low voltage, pulling in a lot of current (usually intending to blow out a series fuse and thereby rendering the equipment nonoperational, but "safe"). We also have MOVs (metal-oxide varistors) in SMD packages nowadays. And so on. But these are still habits of the past. Such arrays/devices were used to protect ISDN/DSL/telephony lines for years. In applying them to PoE, we need to keep in mind the following points.

1. A voltage differential applied across a single twisted pair will not affect PoE since PoE is at the center tap of the transformer (symmetrical). This voltage differential can however certainly get transmitted across the data transformer's isolation barrier and damage the PHY. But for that we can actually put the protection arrays on the PHY-side, closer to the chip they are protecting. It is not a PoE issue in any case. We can also wonder how the voltage appeared differentially across a twisted pair to start with. In fact, that can happen in a cable ESD (CESD) setup as we will soon see.
2. We can put bidirectional diode protection array from one of the data lines, or one of the PoE lines, to Earth ground, in an effort to shunt away the surge energy. But very likely we will fail the mandatory hi-pot test, unless we too use some questionable "loophole" that some Asian vendors claim they are using: of disconnecting all such line-to-ground TVS arrays before a hi-pot test, then reconnecting them after passing the test! This seems to be common practice at some major ODMs.
3. Almost out of force of habit, some ODMs are also known to have placed an expensive bidirectional TVS across the PoE port (from Port_P to Port_N). However, they did not realize that there is a reverse-polarity protection diode in parallel, and that will conduct in one direction anyway, so there is just no use of bidirectional protection in parallel to it. See Path A in Fig. 11.12.
4. Some PSE vendors replace the reverse polarity protection diode with a 58-V TVS, same as the TVS at the input of the PSE.

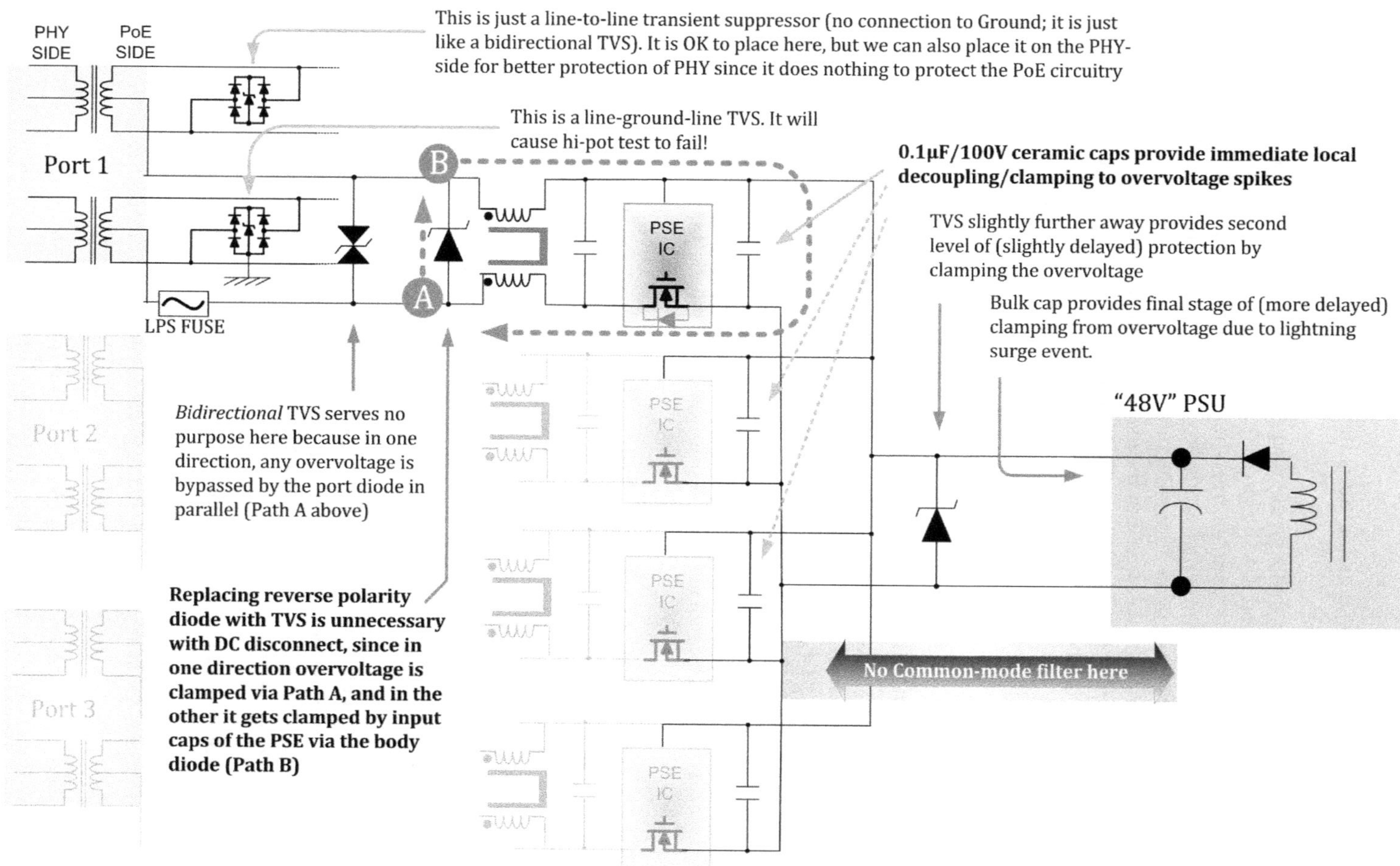

Figure 11.12 Overview of recommended suggestions and layout for surge protection.

This has questionable advantages, but some major OEMs believe it helps. However take a look at Path B in Fig. 11.12.

5. Keep in mind that a weak front-end "protection" can do more harm than good if it fails prematurely at lower levels of energy. And *any* component failure, wherever it comes from, will bring down the entire switch/hub. So, we must make sure that in the most hazardous locations, such as those closest to the RJ-45, we place not the weakest, but the *toughest* protection devices. *Most semiconductors do not have the ability to sink a lot of energy and will develop internal hot-spots and melt. So in this case we recommend it is better to let the energy in, and then let it be fully absorbed by the PSU's output bulk cap.*

PoE Is an Intrabuilding Standard

PoE is essentially an intrabuilding standard. There are some confusing references in IEEE 802.3at to Environment A and Environment B:

In Section 33.4.1.1, titled "Electrical isolation environments," and in Section 33.4.1.1.2:

> There are two electrical power distribution environments to be considered that require different electrical isolation properties. They are as follows:
>
> - Environment A: When a LAN or LAN segment, with all its associated interconnected equipment, is entirely contained within a single low-voltage power distribution system and within a single building.
> - Environment B: When a LAN crosses the boundary between separate power distribution systems or the boundaries of a single building....
>
> .....
>
> Environment B requirements: The attachment of network segments that cross Environment A boundaries requires electrical isolation between each segment and all other attached segments as well as to the protective ground of the NID.

Just above that, in Section 33.4.1 it says:

> Conductive link segments that have differing isolation and grounding requirements shall have those requirements provided by the port-to-port isolation of network interface devices (NID).

The truth is no one seems to know how to do port-to-port isolation effectively or economically. Do we need separate 48-V supplies for each port? In effect, the PoE standard has remained Environment A (intrabuilding).

Keep in mind that even if a tiny section of the Ethernet cable goes out to say, an IP camera mounted on the outside wall of the building, the environment is no longer Environment A. And technically speaking, we are then likely to end up with equipment designed for Environment A operating in an environment it wasn't designed for. It is also now more intensely exposed to atmospheric discharges. Warranties should not apply in this case, should they?

GR-1089 (Telcordia) Requirements

GR-1089-Core (1999) standard relates to testing of lightning and AC- power fault surges in a telecommunications central office environment. It requires "primary protection" from surges and lightning strikes at the service entrance. But there is some "let-through," which is handled by "secondary protection."

Primary protection is the first line of defense. It is required at a facility's (service) entrance. Several means are employed to implement it. Carbon blocks are the oldest type of overvoltage protective device, originally used to protect against overvoltage in telephone installations. They work by forming a spark gap with two pieces of carbon in close (3 to 6 mils) proximity. The gap flashes over at around 600 V. One side is tied to earth and the other to the circuit being protected. Carbon blocks unfortunately degrade with each use, and the only indication that they are not working anymore is *equipment damage.* So gas-discharge tubes are often used instead. These are sealed and rely on electrodes in a mixture of noble gases (argon, neon, and so on). The voltage across them collapses when they break down, allowing them to carry huge current with limited self-dissipation—because of the lowered voltage (dissipation is $V \times I$). These are widely used for primary protection. Solid-state crowbar (thyristor-based) devices (similar to gas-discharge tubes) are also used to clamp transient voltages. They have a fast response time, low capacitance, and high reliability. They are an excellent choice in protecting telecommunication lines. The three devices mentioned above take care of overvoltage conditions. Overcurrent conditions are sometimes handled by fuse links. But fuse links are not intended to provide a current limiting function for network equipment. That is the job of secondary protection.

After being clamped by primary protection, some energy gets through anyway. So we need secondary protection. Secondary protection involves the use of overvoltage and overcurrent devices. Examples are (smaller) solid-state crowbar devices, gas-discharge tubes, and metal-oxide varistors (MOVs). Overcurrent devices are used to interrupt harmful currents, or to provide a high impedance to the protected circuits. Examples are fuses, PTCs (positive-temperature-coefficient polymeric devices), power/line feed resistors, or flameproof resistors.

We must be clear that without primary and secondary protection, there will (soon) be no telecommunications network. Bellcore (now Telcordia) therefore developed a series of tests that go beyond the upper voltage and current limits equipment will normally see. The underlying philosophy is the same as the one we had declared in going well beyond EN 61000-4-5 and CISPR 24 requirements. It adds additional headroom to the basic test, creating a much more robust solution.The idea is that, if a piece of network equipment can survive the Telcordia tests, it will survive in the field for many years.

Let us look at what all this specifically means in terms of protecting PoE equipment (inside the building) from surges. It turns out that GR-1089 is actually not much different from EN 61000-4-5/CISPR 24. In Section 4.5.9 under "Intrabuilding Lightning Surge Tests" it specifically mentions that a 8/20-μs generator can be used as per the IEEE C62.41, which is actually the underlying standard for EN 61000-4-5 too.

The differences between EN 61000-4-5/CISPR 24 and GR-1089 are:

1. The GR-1089 standard asks for only ± 800 V (1 zap) with a 6-Ω resistor, and the next level is ± 1500 V (1 zap) with a 12-Ω limiting resistor. The surge is applied in "metallic" fashion, i.e., differential-mode, which in effect is almost the same as the one-sided shorted Y-cap case we have already been discussing.
2. In contrast, EN6100-4-5/CISPR24 asks for only 1 kV minimum with 40 Ω, and 5 zaps of positive polarity followed by 5 zaps of negative polarity.

Running our CWG's Mathcad file with *Zdut* = negligible (i.e., all Y-caps bypassed by thick copper as worst case), *Zext* = 6 Ω and *Vdc* = 800 V, we get peak current as 87.6 A. Then changing to the next level with *Zext* = 12 Ω and *Vdc* = 1500 V, we get 133.3 A peak. We can calculate the corresponding bulk capacitor requirement (in going from 51 V to a max of 58 V), as 703 μF and 1069 μF, respectively. But if we can somehow permit maximum cap voltage to reach 65 V instead of just 58 V, say by using a heftier TVS with a higher nominal clamping voltage level, but one with a much sharper *V-I* knee, so as to ensure that the 74 V Abs Max rating of a typical PSE chip is still not exceeded, then the calculated bulk capacitances can be reduced to 330 μF and 502 μF, respectively. This is a huge advantage. But to make this happen, besides using a hefty TVS, we may also need to ensure that the ESR of the bulk cap is low enough, because this "ESR bump" actually adds to the voltage seen by the PSE chip. And finally, if nothing else works, simply raising the Abs Max of the PSE chip from 74 to 80 V is useful. And in all cases, a controlled amount of Y-capacitance will greatly help, because though the GR-1089 standard may in effect be asking us to bypass Y-capacitances for the test, in an actual systems board, that is really not a realistic state of affairs. In other words, the EN standard certainly seems a little too mild and optimistic, but the GR-1089 seems overly stringent and pessimistic. At the end of the day, we must not forsake engineering judgment. It is not an unthinking rush to comply with a given standard that may not even be mandatory or realistic. Eventually, all equipment manufacturers basically just want to achieve higher field reliability at lowest added initial and recurring costs.

ESD Protection of ICs

Just as a steady excess of voltage or current constitutes overstress, rate of change of stress is also a possible overstress (for example, dV/dt-induced stress). The most common example of this is electrostatic discharge (ESD). ESD can cause many types of failures. For example, it can induce latchup. When we walk over a carpet, we can pick up enough electric charge to kill a semiconductor by actual physical contact (contact discharge) or near-contact (air discharge). Therefore, ESD handling has become a major concern in modern manufacturing and test environments.

All modern ICs are designed with rather complex ESD-protection circuitry built around their pins. The idea is to divert or dissipate electrostatic energy safely. Today, all ICs also have published ESD ratings. For example, a typical datasheet will declare that an IC withstands 2 kV ESD as per HBM (human body model), and 200 V as per MM (machine model). The HBM tries to simulate ESD from humans, and actually has two versions. As per the more benevolent and more widely used (military) standard, MIL-STD-883, HBM is a 100-pF cap discharging into the device through a 1.5-kΩ series resistor. The rise time of the resulting current pulse is less than 10 ns and reaches a peak of 1.33 A. However, the international ESD specification, IEC 61000-4-2 (in Europe that becomes the European norm EN 61000-4-2), calls for a 330-Ω resistor and a 150-pF capacitor, which gives a peak current of 7.5 A with a rise time of less than 1ns. This is actually much harsher than the MIL-STD-883 HBM profile. Note that the IEC standard was originally called IEC 801-2 and was also originally intended only as an acceptance condition for end equipment (the system), but it now also does double-duty as an ESD test for ICs.

To dispel a popular myth, CMOS/BiCMOS chips are not the only components that are susceptible to permanent ESD damage. Bipolar and linear chips can also be damaged. PN junctions can be subjected to a hard-failure mechanism called thermal secondary breakdown, in which a current spike (which can also come from ESD) causes microscopic localized spots of overheating, resulting in near-melting temperatures. Low-power TTL, as well as conventional opamps, can be destroyed in this manner.

The Machine Model tries to simulate ESD from production equipment, and therefore uses a 200-pF cap with a 500-nH inductor in series (instead of a resistor). Finally, data and telecom equipment also need to pass system-level (not component-level) cable ESD (CESD) testing, also called cable discharge event (CDE) testing. Unlike ESD, there is no industry standard for CESD/CDE testing yet. The intent of the standard is, however, clear—to protect the equipment under test from the following type of event: an operator pulls an unconnected coaxial cable across a carpet, and the cable develops electrostatic charge relative to Earth ground. When the cable is plugged into the equipment, the stored charge gets dumped into the equipment. Modern equipment needs to typically survive up to 2-kV CESD on the output ports. Note that here there is very low-limiting resistance (cable resistance), but significant line inductance/impedance to limit the peak current and its rise time. There is also a lot of ringing caused by transmission-line effects as the energy goes back and forth the cable in waves. So the overall stress profile is less severe, but relatively more sustained than regular ESD.

ESD does not necessarily cause immediate failure. It is known that a latent failure in a CD4041 IC (a CMOS quad-buffer), tucked-deep inside a satellite system assembled in 1979, surfaced five years later in 1984 just as it was being readied for launch. Therefore it is quite possible that we often mistake similar latent failures as "poor quality" or "bad components."

Cable ESD (CDE)

Cable-sourced electrostatic discharge (CESD or cable ESD), also called a cable discharge event (CDE), becomes a possibility when an Ethernet cable becomes electrostatically charged. The cable could be charged primarily because of tribocharging (for example, by dragging it on a carpet or tiled floor), or even by induction (for example, from an already-charged person touching or holding it). This charged cable is suddenly discharged into a circuit when the cable is plugged into network equipment such as a switch, hub, repeater, and so on. This is most likely to occur in a new installation, or during an upgrade of an existing installation. "Good" cables are ironically the worst offenders as they ceate an excellent dielectric (good insulation) between the two plates of this charged capacitor.

The first plate of this charged capacitor is (all) the wires inside the cable (all are charged to the same potential). The second plate is the Earth ground (the aluminum-metal plate we used in surge testing too). In an actual test, typically 150 to 200 m of cable are taken (to provide some headroom) and all 8 wires are galvanically tied together at both ends to create one long copper plate. The outer jacket of the Ethernet cable is the main dielectric. Clearly, the cable needs to be laid down properly on a long metal conduit to create a good capacitor to Earth ground.

A formal standard to define a cable-discharge event with an industry-wide test method has yet to be established, but most manufacturers use internal CDE test setups to evaluate their designs; though a few cut corners and just test to IEC 61000-4-2 Level-4 specifications.

The assumption that equipment is capable of withstanding CDE if it passes IEC 61000-4-2 Level-4 discharges, is not really accurate. That is because the charged capacitances in the two tests are very different. In the IEC ESD spec we have 150 pF, whereas in CDE we have a much larger capacitance depending on the length of the cable involved and also the cable elevation over the Earth ground. There is also a *transmission line effect* in CDE (multiple reflections), caused by distributed capacitance (separated by distributed inductance), as opposed to a lumped capacitance in regular ESD. Therefore, CDE typically dumps much more energy than IEC Level-4 discharges into the EUT, though this energy comes in waves because of transmission-line reflections, quite unlike a lightning surge event.

The other major differences between a CDE test and a surge test are

1. There is no PD connected to the RJ-45 of the PSE (since a charged cable is to be plugged into it).
2. When a cable is plugged in, we can't be sure which pin of the male RJ-45 plug will be in contact first with the female. In other words, unlike a surge event, we can't rely on twisted pair properties to predict only common-mode disturbances. So in fact, CDE is differential mode—rather it has a differential mode component—it is applied between *each pin* of the RJ-45 and the aluminum metal sheet that serves as the Earth ground.

The CESD test can be done in many ways. But its basic intent at least, is clear from the following rather generic CESD test procedure applied to PSEs. An Ethernet cable, with all its eight wires connected together at (only) one end, is electrostatically charged up with respect to earth ground (the aluminum plate it rests on), using a hi-pot voltage generator (Megger). The Megger is then suddenly disconnected by means of a relay in series with it, present on its side of the cable. Almost immediately afterward, a relay on the other side of the cable is activated by an electronic circuit, and in doing so connects one of the separated eight wires (not two of a pair as we do in surge testing), to its corresponding pin within the RJ-45 of the PSE under test. Note that at this specific moment when the voltage is suddenly applied to the PSE, the PSE would still be in the middle of some sort of predetection/initialization mode, since there is no PD connected to it (unlike the surge test in which the PSE's pass-FET is assumed fully conducting). This test is repeated on all eight pins/wires of the RJ-45. After each zapped pin, we should check that the switch/hub/PSE is working (data and power).

This test, when correctly done, is capable of inflicting severe damage to the PHY (and PSE, of course) because the current does not flow in the transformer symmetrically as in a surge test. See Fig. 11.13. This current is through one *half* of the transformer so the other half does not contribute in creating any flux-cancelation. The drive transformer gets severely saturated during this event but usually recovers. However, because there is no flux-cancelation, there is transformer action, and so a huge voltage can be transferred across to the winding on the PHY-side, causing damage to it.

Note that there are major companies that do a much softer version of this test. For example see National Semiconductor Application Note 151, titled"Cable Discharge Event." What they are doing, in effect, is shown in Fig. 11.14. This applies the same voltage to two pins of a twisted pair at the same time. This is common-mode, just like a surge test. There is no voltage reflected on the host/PHY side. Clearly, this method is not realistic, since we cannot really guarantee that in reality, any two wires of a cable will both contact their corresponding pins of the RJ-45 simultaneously. This test does not conform to the basic intent of a CESD test as described above. It is a test that is very easy to pass and cannot correlate with any field reliability numbers either.

It is pointed out that though clearance and creepage requirements have to be increased, commensurate with the typically high 4-kV cable ESD rating being demanded by some major OEMs, because of the lower associated energy, other changes are not required. A system that has passed a 1.5-to 2-kV surge test, as described previously, can easily comply with a 4-kV ESD test.

Lastly, keep in mind that as the test is usually done, there is a fixed polarity: the copper of the cable is always charged to a positive voltage with respect to Earth ground. We can argue about the validity of that assumption, but as mentioned, there is no industry-wide standard for CESD testing. It is still evolving.

Port Protection Diode in PoE: Any TVS Required?

We finally take up one last lingering question that is being asked nowadays, and is getting increasingly debated: Does it make sense to replace the usual output port diode of a PSE with a more expensive TVS, as is practiced by a few, but very large, OEMs today?

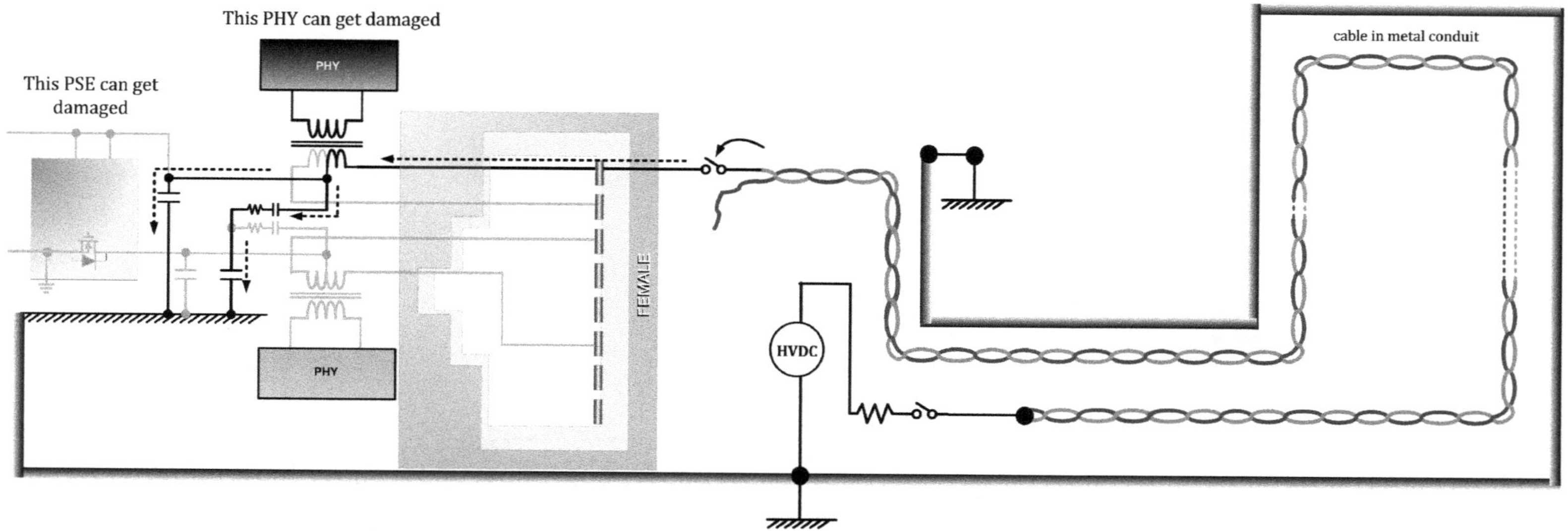

FIGURE 11.13 Recommended cable ESD test setup.

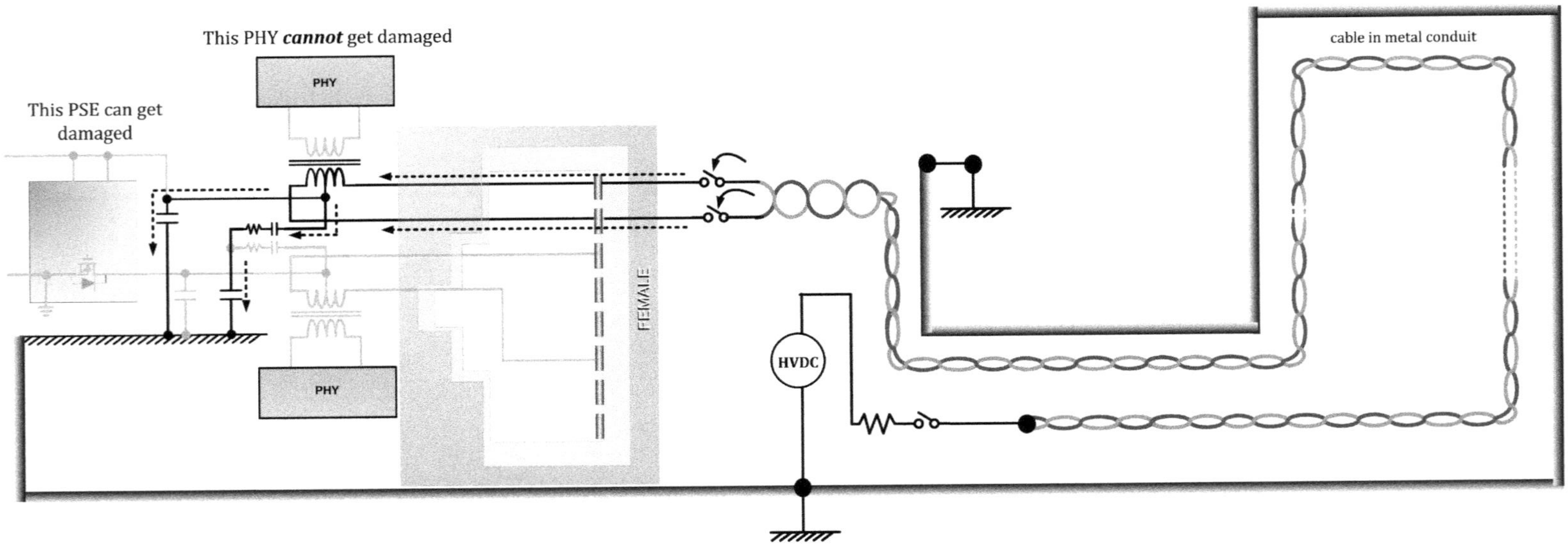

FIGURE 11.14 Not recommended cable ESD test setup.

The Purpose of the Output Port Diode of the PSE

On page 38 of IEEE 802.3af-2003, the schematics shown in Fig. 11.15 appear. Of key interest here is the diode D1. One of the widespread assumptions about this diode is it is for some sort of reverse polarity protection. Yes, that is true, but only *indirectly* so. Note that on the same page, the standard actually makes it clear that the primary function of D1 is to present a nonstandard signature for any "reversed voltage PSE-to-PSE connection." In other words, the purpose of the diode is to altogether prevent the other PSE from even turning ON and applying potentially damaging reverse-voltage on the port. This is a PSE-to-PSE case, but with reverse polarity.

It is also true that if the "other PSE" does turn on by mistake (despite getting a wrong signature), the diode D1 would certainly help prevent a catastrophic situation. It would protect the PSE by literally bypassing it (through the diode D1, rather than, say, through a dangerous sneak path through an ESD structure in the PSE chip). Also, provided the LPS (limited power source) fuse is present on all PoE ports and is placed *between* the diode and the RJ-45, the fuse may blow, thus opening the path through which the abnormal current was flowing. The reverse-voltage source would then get disconnected, but if the fuse blows, the PSE would need to go in for repairs.

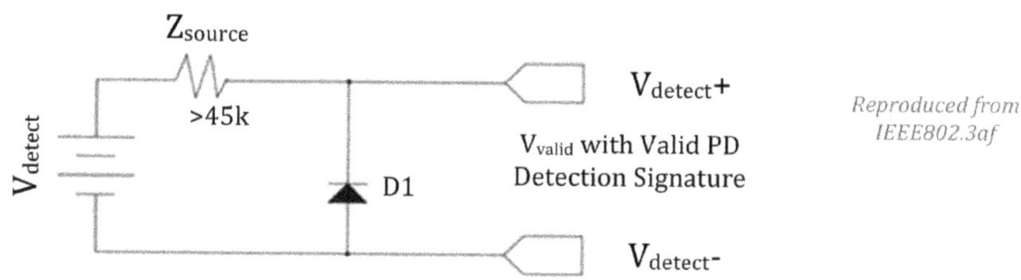

Figure 33-8 – PSE detection source

A functional equivalent of the detection circuit that has no source impedance limitation, but restricts the PSE detection to the first quadrant, is shown in Figure 33-9

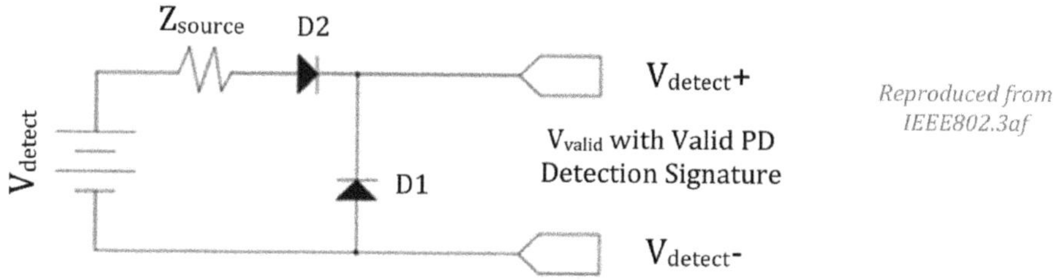

Figure 33-9 – Alternative PSE detection source

FIGURE 11.15 Extract from the AF standard showing the "reverse polarity diode."

Nothing in the standard indicates the most appropriate position for the LPS fuse, and equipment manufacturers do tend to put it at very different places. But what is certain is that diode D1 is universally present on PoE ports and is always strapped between the positive (Port_P) and negative (Port_N) lines of the PSE port.

Should D1 Be a TVS?

Historically, some OEMs went a step further and started suggesting that this low-cost standard slow-port diode (typically 1A/100 V) be replaced by a transient-voltage suppressor (TVS) diode (a TVS is just a rugged zener diode). Note that this new suggested port TVS diode is in addition to the single TVS routinely placed at the input side of the PSE board, that is, between the "48V" power supply and the PSE chips (the usual SMAJ58A). The thought is that using a TVS will somehow provide some protection from an undefined "spike" that may appear on the output (cable side).

This new port TVS will certainly add to the cost since it is 3 to 5 cents per port. It is yet to be seen what real advantage it offers, and under what specific conditions. In the next sections, we try to analyze that.

Spike Protection

In general, the kind of spikes that can be applied on the ports can be either voltage spikes or current spikes. Further, these can be applied either differentially (symmetric mode, i.e., between the port lines) or in common-mode manner (asymmetric mode, i.e., between one or both the port lines tied together and the system/chassis ground). We have seen that there are mandatory thresholds (as for surge currents in accordance with EN 61000-4-5) and also some customer- or market-dictated test regimens (like ESD). The important point to note here is that whatever we do, the PSE chip itself can get damaged only by *excessive voltage* on any of its pins with respect to its IC ground pin. In other words, damage due to voltage only occurs by a differential voltage across the pins of the IC, not with respect to Earth ground. So the question remains, why should we use a more expensive TVS in the position of D1 in Fig. 11.15? Does it really help reduce (or clamp) any voltages or spikes appearing across the pins of the IC?

One of the theories is based on the fact that the *cable has some inductance*. So if we are passing full-load current (say 600 mA) and we suddenly unplug the cable at the PSE end, it would cause an *inductive kickback* (voltage spike, differential in nature) that could damage the Port_P and Port_N pins of the IC. The first question is, does the cable really have so much inductance? We know that each pair is twisted to reduce the inductance for data transmissions.

But here we are talking not about the inductance of the wires of a given pair, but the inductance as seen by the PoE source, which is a current loop formed by two adjacent pairs, with some insulation between them. That is not negligible. What exactly are the cable characteristics from the viewpoint of PoE (not data). How do we measure them?

Cable Characteristics

In Fig. 11.16, we present the first step: characterizing the cable.

On the basis of several measurements over different cable lengths, we can safely say that every *100 m of cable presents about 90 μH inductance and also not an insignificant 4.5 nF of capacitance.* The good news is that along with the (distributed) inductance there is also a (distributed) capacitance. And we know that capacitances tend to absorb voltage spikes. So on the face of it we already expect that the "inductive kickback" we feared may not be as serious as first thought. Subsequent data proves that.

The Spike as We Unplug the PSE

In Fig. 11.17, we simplify the equivalent diagram and show what it looks like in steady state (before unplugging), and then just after we unplug the cable. We are passing the full 802.3at Type 2 current prior to unplugging, just to maximize any spikes caused by inductive kickback. Note that the inductance will try to force the current to continue briefly, and a high voltage is created across the "spark gap." This is the usual historic principle of the camera flash and the automotive spark plug. Note also that the duration of this arc is very small (Δt), and the current will ramp down to zero very steeply during that interval—all the way from 600 to 0 mA (ΔI = 600 mA). The voltage across the spark gap will follow the simple equation $V = L \times \Delta I / \Delta t$.

So there will certainly be a voltage spike as the connection breaks, maybe not as severe as we thought (because of the distributed capacitance). The key question is *where* does this spike appear? It can cause damage to the PSE chip only if it appears on the left side (PSE side) of the broken connection. Not on the cable (right) side!

The analysis in the figure shows that the output port capacitor of 100 nF will provide (most of) the current during the short-time Δt, and its value is large enough to keep the voltage across the port constant. So the spike will appear on the cable side, not on the PSE side.

Note that the current does not flow through the diode as we may intuitively conclude. Because the anode of the diode is at 0 V and its cathode is at 50 V (thanks to the 100-nF capacitor). So it is, in effect, reverse-biased. Yes, if the 100-nF cap would not have been present, nature would have forced current continuity in the cable inductance

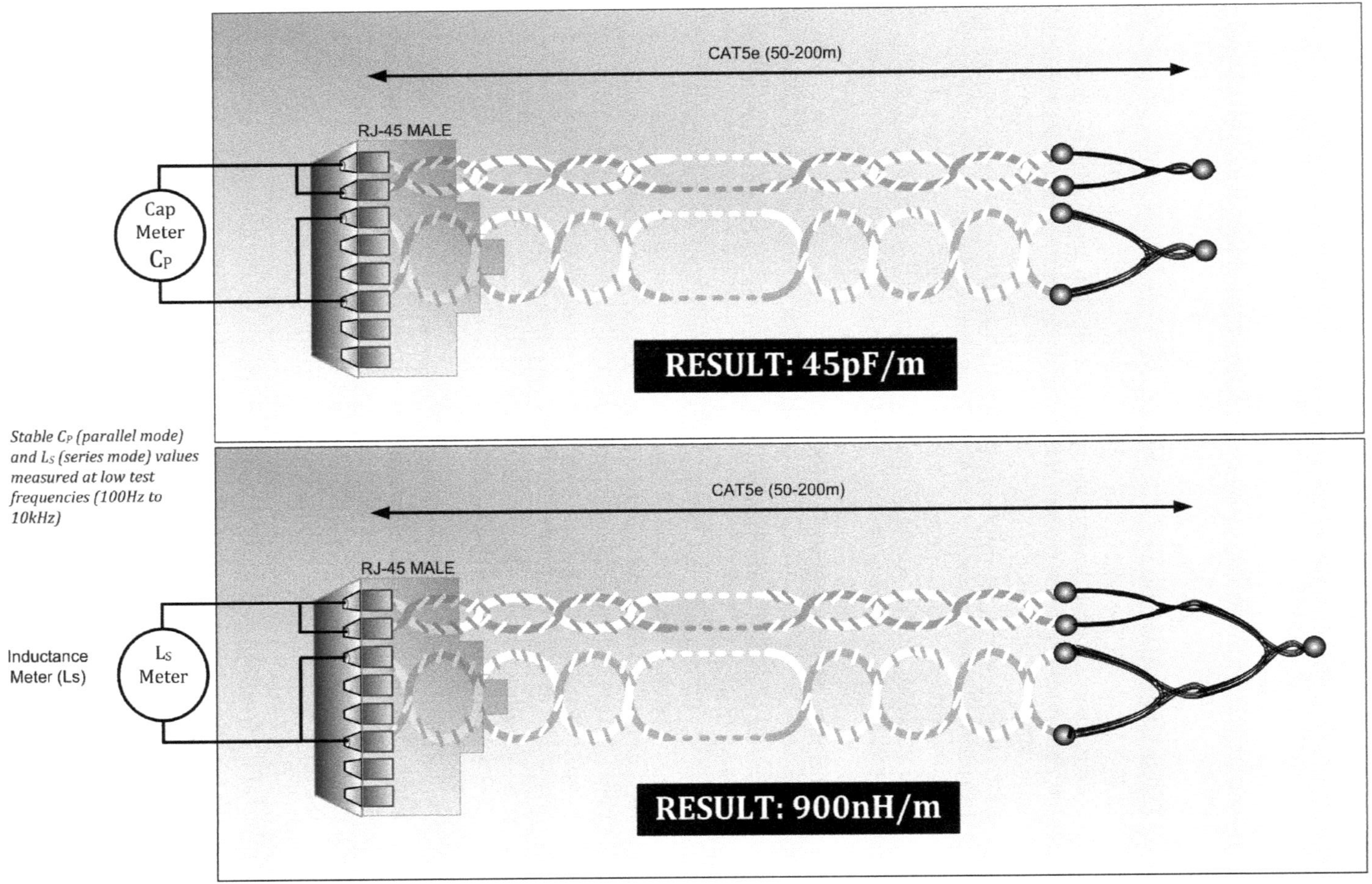

FIGURE 11.16 Characterizing the distributed capacitance and inductance of a CAT5e cable.

by forcing D1 (the reverse-polarity port diode) to conduct. At which point Port_P (its cathode) would have been dragged to about-0.6 V with respect to Port_N (its anode). This indeed could have caused substrate currents in the PSE chip resulting in malfunction and possible damage. All that doesn't happen, because of the 100-nF port capacitance! That is the critical component whose role we need to understand.

Measuring the Spike

For worst-case analysis, we take a PSE in AC-disconnect mode. We use 100 m CAT5e cable with 600 mA through it using high-wattage resistors at the other end. The oscilloscope is floated (battery operation) to avoid any stray interactions through the Earth ground wiring. On unplugging the cable we capture Fig. 11.18. Notice a negative spike on the cable side of the broken connection, just as the analysis in Fig. 11.17 predicted. We see a very small bump in the voltage on the PSE side of the RJ-45. On the other hand, the cable side gets the inductive kickback. But it is not very severe—the swing is only about 56.8 V + 50 V = 106.8 V measured from the instant just prior

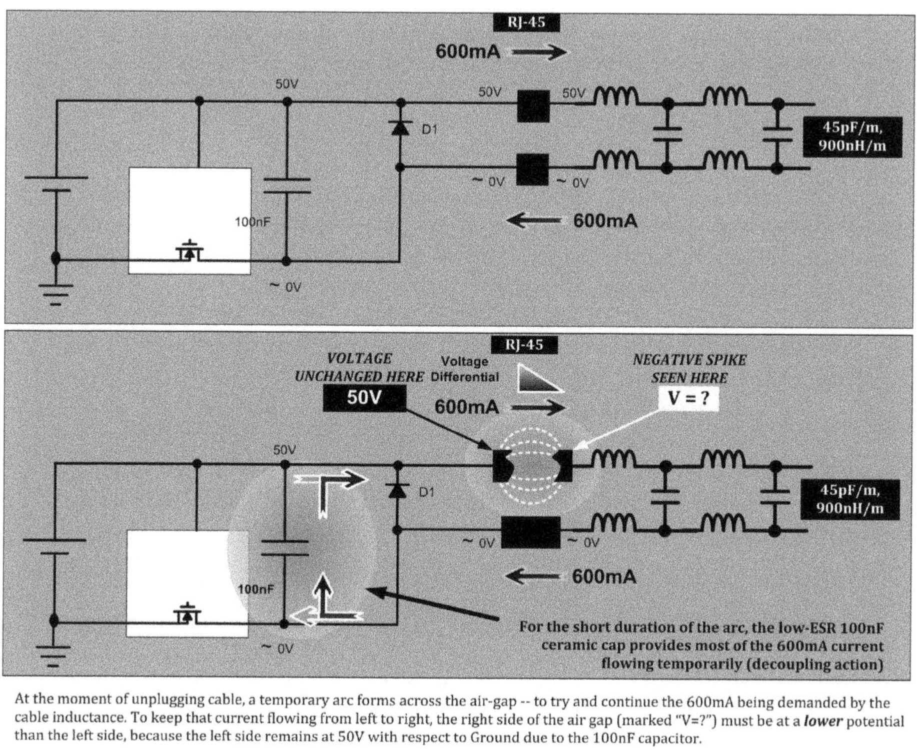

FIGURE 11.17 What happens as we unplug the cable: spike on the cable, not PSE.

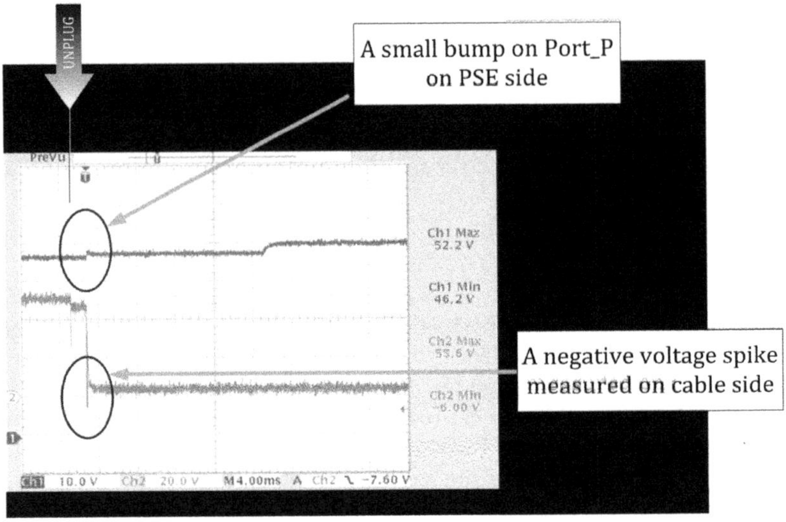

FIGURE 11.18 Waveforms on either side of the broken connection as we unplug the cable.

to unplugging. Clearly this is because of the distributed capacitance on the cable that "snubs" the spike significantly.

Unplugging the PD

If we unplug the cable on the PD side, for the same reason, the PD itself sees no spike. But there is a spike on the cable. Does this spike reach the PSE? Similarly, if we unplug the cable on the PSE side, the spike we see on the cable side in Fig. 11.18: does that reach the PD? The PD is relatively well protected because it has a TVS at its input, which clamps spikes. So we need not worry about that. On the other hand, when we unplug the PD, we do get a spike on the PSE side via the cable. Is that cause for concern?

At experiments done at a major OEM, it was seen that the spike was severely attenuated for long cable lengths because of the intervening cable inductance and capacitance. If the cable was too small, it had too little stored energy to worry about. But at about 20 m of cable length, the spike was maximum.

Their conclusions were that under this particular scenario it does help somewhat to replace the port diode D1 with a TVS. However, keep in mind that as we saw for surge tests and CESD tests, if we use DC-disconnect, we gain access to the huge bulk cap at the output of the silver-box AC-DC power supply, and that absorbs the energy. See path B in Fig. 11.12. So in that case, output TVS diodes are certainly not required.

Appendix: Modeling and Analysis of the Combination-Wave Generator Used for Surge Testing (EN 61000-4-5)

The CWG Circuit

The CWG circuit was shown in Fig. 11.9. That is the one to refer here.
From Fig. 11.9 we see that in the s-plane

$$Za = Zext + Zdut$$

$$Zb = \frac{R2 \times (Zext + Zdut)}{R2 + Zext + Zdut}$$

$$Zc = Rm + sL$$

$$Zd = (Rm + sL) + \frac{R2 \times (Zext + Zdut)}{R2 + Zext + Zdut}$$

Mathematical Analysis

At time $t = 0$, the capacitor C is fully charged (to Vdc). The initial condition of the cap is described as a step function:

$$Vc(0) = Vdc \times \xi(t) = \frac{Vdc}{s}$$

At the very moment the switch is thrown, we also assume that the applied-voltage source, Vdc, is simultaneously removed. This is equivalent to assuming a large enough value for Rc, so we can ignore it in the following analysis. $I1$ therefore gets fully sourced from the capacitor. And the voltage across the cap, in frequency domain (s-plane), is then described as

$$Vc = \frac{Vdc}{s} - \frac{I1}{sC}$$

Note that this expresses Vc, but in terms of another unknown—the current $I1$. To eliminate that we need to solve for $I1$

$$I1 = C \times (Vdc - sVc)$$

We note that we also have the following equations describing the currents at the first node

$$I2 = \frac{Vc}{R1}$$

$$I3 = I1 - I2$$

Solving, we get for *I3*

$$I3 = C \times Vdc - sC \times Vc - \frac{Vc}{R1}$$

or

$$I3 = [C \times Vdc] - \left[\left(sC + \frac{1}{R1}\right) \times Vc\right]$$

Now, we know we have another equation for *I3*

$$I3 = \frac{Vc}{Zd}$$

So equating the two equations for *I3*, we can eliminate *I3*, and thus express *Vc* in terms of *Vdc* and the impedances only (no dependence on currents).

$$C \times Vdc - sC \times Vc - \frac{Vc}{R1} = \frac{Vc}{Zd}$$

$$Vc = \frac{C \times Vdc}{\frac{1}{Zd} + \frac{1}{R1} + sC}$$

We are interested in ultimately calculating the voltage across *Zdut*, which we call *Vo* here. That is simply

$$Vo = I5 \times Zdut$$

In turn, the sum of the currents *I5* and *I3* is equal to *I3*. We can also say that the current *I3* splits up in inverse proportion to the impedance in the two branches. It can be shown from basic principles that the solutions for the current splits in any such case are as follows:

$$I5 = I3 \times \frac{R2}{R2 + Za} \text{ and } I4 = I3 \times \frac{Za}{R2 + Za}$$

So, using the equation above for *I5*, we get

$$Vo = I5 \times Zdut$$

$$Vo = I3 \times \frac{R2}{R2 + Za} \times Zdut = \left[\frac{Vc}{Zd}\right] \times \frac{R2 \times Zdut}{R2 + Za}$$

$$Vo = \left[\frac{\dfrac{C \times Vdc}{\dfrac{1}{Zd} + \dfrac{1}{R1} + sC}}{Zd}\right] \times \left[\frac{R2 \times Zdut}{R2 + Za}\right] = \left[\frac{1}{Zd} \times \frac{C \times Vdc}{\dfrac{1}{Zd} + \dfrac{1}{R1} + sC}\right] \times \left[\frac{R2 \times Zdut}{R2 + Za}\right]$$

$$Vo = \frac{C \times Vdc \times R2 \times Zdut}{\left(1 + \dfrac{Zd}{R1} + sC \bullet Zd\right) \times (R2 + Za)} = \frac{C \times Vdc \times R2 \times Zdut}{\left[1 + Zd \bullet \left(sC + \dfrac{1}{R1}\right)\right] \times (R2 + Za)}$$

$$Vo = \frac{C \times Vdc \times R2 \times Zdut}{\left[1 + \left\{\left(Rm + sL + \dfrac{R2 \times (Zext + Zdut)}{R2 + Zext + Zdut}\right) \bullet \left(sC + \dfrac{1}{R1}\right)\right\}\right] \times (R2 + Za)}$$

Note that if the branch impedance, Zb, does not involve anything other than resistors, this is preferably kept simpler for now as follows:

$$Vo = \frac{C \times Vdc \times R2 \times Zdut}{\left[1 + \left\{(Rm + sL + Zb) \bullet \left(sC + \dfrac{1}{R1}\right)\right\}\right] \times (R2 + Za)}$$

Solving and simplifying:

$$Vo = \frac{C \times Vdc \times R2 \times Zdut}{\left[\left(1 + \dfrac{Rm}{R1} + \dfrac{Zb}{R1}\right) + s\left(Rm \bullet C + Zb \bullet C + \dfrac{L}{R1}\right) + s^2 LC\right] \times (R2 + Za)}$$

$$Vo = \frac{C \times Vdc \times R2 \times Zdut}{\left[\left(\dfrac{1}{LC} + \dfrac{Rm}{R1 \bullet LC} + \dfrac{Zb}{R1 \bullet LC}\right) + s\left(\dfrac{Rm}{L} + \dfrac{Zb}{L} + \dfrac{1}{R1 \bullet C}\right) + s^2\right] \times (R2 + Za) \times LC}$$

$$Vo = \frac{\dfrac{C \times Vdc \times R2 \times Zdut}{(R2 + Za) \times LC}}{\left[\left(\dfrac{1}{LC} + \dfrac{Rm}{R1 \bullet LC} + \dfrac{Zb}{R1 \bullet LC}\right) + s\left(\dfrac{Rm}{L} + \dfrac{Zb}{L} + \dfrac{1}{R1 \bullet C}\right) + s^2\right]}$$

$$Vo = \frac{\dfrac{R2 \times Zdut}{(R2 + Za) \times L}}{\left[\left(\dfrac{1}{LC} + \dfrac{Rm}{R1 \bullet LC} + \dfrac{Zb}{R1 \bullet LC}\right) + s\left(\dfrac{Rm}{L} + \dfrac{Zb}{L} + \dfrac{1}{R1 \bullet C}\right) + s^2\right]} \times Vdc$$

where

$$Za = Zdut + Zext$$

and

$$Zb = \frac{R2 \times (Zext + Zdut)}{R2 + Zext + Zdut}$$

To map this into the time domain, we need to reformat the above result into the following standard form:

$$Vo = \frac{\beta\gamma}{(s+\alpha)^2 + \beta^2}$$

Comparing, we get:
2α = coefficient of s in the denominator. So

$$\alpha = \frac{1}{2} \times \left(\frac{Rm}{L} + \frac{Zb}{L} + \frac{1}{R1 \bullet C} \right)$$

Also, β^2 = (the constant term in the denominator) – α^2, so

$$\beta = \sqrt{\frac{1}{LC} + \frac{Rm}{R1 \bullet LC} + \frac{Zb}{R1 \bullet LC} - \alpha^2}$$

or

$$\beta = \sqrt{\frac{1}{LC} + \frac{Rm}{R1 \bullet LC} + \frac{Zb}{R1 \bullet LC} - \frac{1}{4} \bullet \left(\frac{Rm}{L} + \frac{Zb}{L} + \frac{1}{R1 \bullet C} \right)^2}$$

And γ = the numerator divided by β, so

$$\gamma = \frac{R2 \times Zdut \times Vdc}{\beta \times (R2 + Za) \times L}$$

or

$$\gamma = \frac{\dfrac{R2 \times Zdut \times Vdc}{(R2 + Za) \times L}}{\sqrt{\dfrac{1}{LC} + \dfrac{Rm}{R1 \bullet LC} + \dfrac{Zb}{R1 \bullet LC} - \dfrac{1}{4} \bullet \left(\dfrac{Rm}{L} + \dfrac{Zb}{L} + \dfrac{1}{R1 \bullet C} \right)^2}}$$

Then using the standard form of inverse Laplace transform, the solution in time domain is finally (with due credits for the Laplace Transform portion of the above derivation to Mr. Nanoo Staal)

$$Vo(t) = \gamma \times e^{-\alpha t} \times \sin(\beta t)$$

Plotting the Results

We can plot the results in Mathcad, using the following iterated values derived in "Introduction to Voltage Surge Immunity Testing" by Bryce Hesterman and Douglas Powell, available at http://www.denverpels.org/Downloads/Denver_PELS_20070918_Hesterman_Voltage_Surge_Immunity.pdf.

$$C = 6.038\ \mu F$$

$$L = 10.37\ \mu H$$

$$R1 = 25.105\ \Omega$$

$$R2 = 19.80\ \Omega$$

$$Rm = 0.941\ \Omega$$

Case 1: Open-Circuit Case, $Vdc = 500$ V, $Zdut = 10^6\ \Omega$, $Zext = 0\ \Omega$

Using Mathcad, we plot the results for Vo versus time in μs in Fig. 11.19.

We see that

1. The 10 to 90 percent rise time is very close to 1.2 μs.
2. The time to fall to 50 percent of the max, is close to 50 μs.

This is as required and defined for the 1.2/50 μs open-circuit voltage waveform in EN 61000-4-5.

Case 2: Short-Circuit Case, $Vdc = 500$ V, $Zdut = 0.1\ \Omega$, $Zext = 0\ \Omega$

In this case, $Zdut$ is a very small resistor of 0.1Ω placed at the output. Therefore the voltage across it is numerically equal to the short-circuit current divided by 10. We get Fig. 11.20.

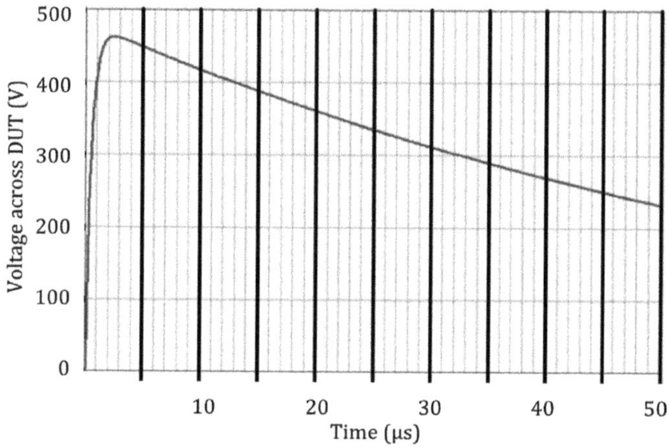

Figure 11.19 Open-circuit voltage using the standard 1.2/50-8/20 μs CWG.

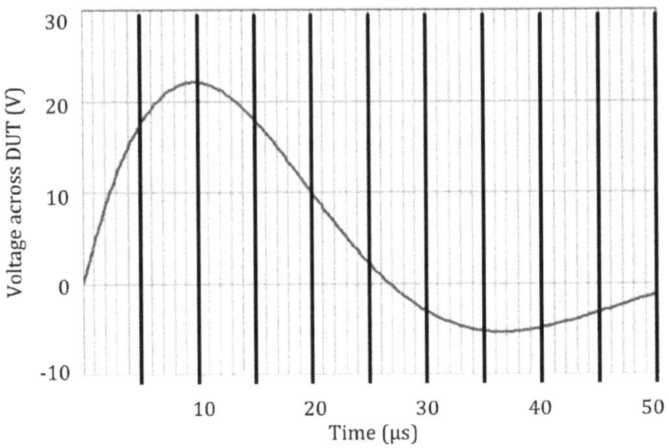

FIGURE 11.20 Short-circuit current (voltage across 1 Ω) using the standard 1.2/50-8/20 μs CWG.

We can see that

1. The current overshoots on the way down by about 30 percent, as described in the standard.
2. The 10 to 90 percent rise time is close to 8 μs, and the time to fall to 50 percent of the max value is close to 20 μs, as required and defined for the 8/20 μs short-circuit current waveform in EN 61000-4-5.
3. The max voltage of *Vo* is 22.1 *V* i.e., the current is 221 A. Therefore, the source impedance is 500/221 = 2.26 Ω, which is very close to the 2 Ω implied by the standard.

A practical case involving surges on data lines (e.g., Ethernet) is discussed next. As per the standards, the effective (total) source impedance is then required to be 42 Ω. This number assumes 2 Ω source impedance coming from the CWG itself, to which is added an external 40 Ω resistor (Zext). We can see that the small error in the source impedance of the CWG, in other words 2.26 Ω vs. 2 Ω, is really quite insignificant in a practical case.

Case 3: Practical Case, *Vdc* = 500 V, *Zdut* = 1 Ω, *Zext* = 40 Ω

It is of great interest to see the actual rise and fall times with an external resistor set to 40 Ω, with the DUT presenting a very small resistance (as it would if it were avalanching). We get Fig. 11.21.

We can eyeball the curve to see that the rise time is slightly more than the open-circuit case—about 2.5 μs as compared to 1.2 μs. And the fall time is smaller—about 38 μs as opposed to 50 μs. This

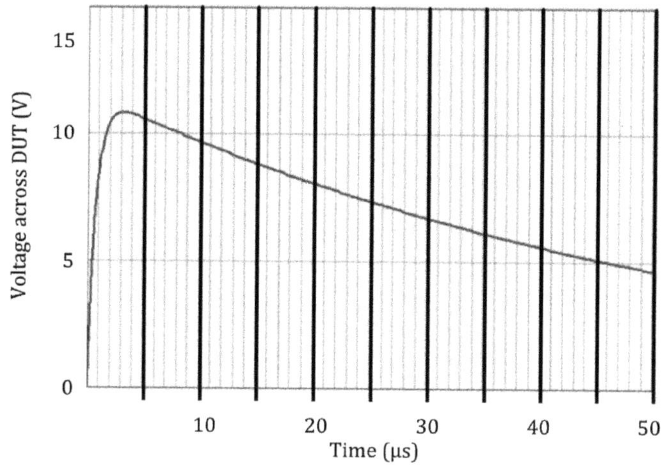

FIGURE 11.21 Voltage across 1 Ω, with 40-Ω resistor in series, using the standard 1.2/50-8/20 μs CWG.

is as expected, because the case of 40-42 Ω is somewhere in between the two extremes of open-circuit and short-circuit. We can see that in going from the former to the latter, the rise time increases from 1.2 to 8 μs, whereas the fall time decreases from 50 to 20 μs.

The current peaks at about 10.8 A in this case, which is roughly commensurate with a crude estimate based on Ohm's law for resistors: $V/R = 500\ \text{V}/42\ \Omega = 11.11\ \text{A}$.

CHAPTER 12

Lab Skills, Thermal Management, and Decoupling

Using Oscilloscopes Wisely (in PoE)

Let us illustrate this with an entirely plausible, yet completely hypothetical story on the perils of using oscilloscopes and probes in the lab without adequate thought or expertise.

In a large multibillion networking company supposedly dedicated to broadband communications via the power of the sinc waveform, a principal hardware engineer, when asked to take a measurement of the load transient overshoot, or the noise and ripple (or even just the ripple) at the output of a switching power supply, would without exception set his scope on 50 Ω input impedance. When questioned, he would reveal that he did this because he was "matching the impedance"—presumably of the oscilloscope to the impedance of the coaxial cable connecting the probe. This senior engineer, with the weight of the organization behind him, went on to have the honor of becoming the company's oldest hardware engineer left supporting PoE. Incidentally, he achieved these remarkable results using just a regular low-cost passive 10X scope probe, (which as you will read below is actually *intended* to be connected to a *high-input impedance*, not 50 Ω). To his organization, well-known for its armory of networking patents, it probably seemed like familiar, old-fashioned, solid engineering.

For the others out there who feel they still need to be reminded, here are the dos and don'ts regarding scope probes in general, available at http://www.williamson-labs.com/scope-probes.htm:

1. Always use 10X probes: they load the DUT (device-under-test) ~ 10 M Ω @ ~10 pf. A 1X probe offers 1 M Ω @ ~50 pf. The designation 10X refers to the attenuation of the signal by the

probe (not gain). In order to attain such light loading by the scope, while maintaining bandwidth, this trade-off is required.

2. Make sure the probes are compensated (adjust trimmer at connector housing) if attaching them to a different scope. This ensures maximum fidelity and bandwidth of the signals being eyeballed.
3. Use the shortest ground lead or clip-lead possible: the shorter the better! Excessive ground lead length introduces unnecessary inductance and can alter the displayed signal, as well as reducing the scope's effective bandwidth (acts like a lowpass filter).
4. When measuring very high frequencies—especially in tight spaces, consider using an RF probe. Also, there are so-called FET or active probes, which are nonloading (almost) wideband probes with built-in amplifiers.
5. When buying probes for your oscilloscope, make sure the probe is of sufficient bandwidth for your particular scope: the probe is the first-order bandwidth determinant of any scope.

   Some scopes have such a wide bandwidth, that no passive probe is able to do it justice, and the only way to use the maximum bandwidth of this type of scope is to drive the scope from a 50 Ω source through a 50-Ω coax, terminated into 50-Ω at the scope's input. In fact, some high performance scopes have a 1 M Ω/50-Ω termination switch for just such occasions.

If we open a regular 10X probe, we will see a 9-M resistor in series with the tip (right next to the tip itself). If we set the input impedance of the scope to 1 M, the signal "sees" a divider of ratio 1 M/(1 M + 9 M) = 1 M/10 M. A modern scope will "know" that a 10X probe is in use and will in effect, amplify the signal 10 times to get back its original amplitude.

But why do this attenuation followed by amplification anyway? Because otherwise, the small natural input capacitances of the scope and cable will provide a low-impedance path for high frequencies, and *the bandwidth will be lower*. Now, by placing a 9-M resistor in series, even the input capacitance is divided down by a factor of 10. So, the bandwidth of the scope goes up significantly when we use a 10X passive probe connected to the 1-M input impedance of scope. In other words, everything works just fine if you use a high-impedance probe with a high-impedance input setting. All bets are off however, if you use a 10X probe with a 50-Ω input impedance setting. On the other hand, if you use a 1X probe (with no 9-M resistor present at its tip), with 50-Ω input impedance setting, the least you will do is to load the signal source (and output of the switching converter being tested). You may even damage the terminations of the scope. But luckily, most scopes now have a large DC-blocking cap in series with the 50-Ω input resistor termination, just like PoE terminations evolved too.

One suitable occasion to use the 50-Ω input impedance setting on the scope is to use a 50-Ω active probe. This has such a high-input impedance (FET-based) that it has negligible "loading" effect on the signal being measured. It contains electronics powered by a low voltage supply, containing an amplifier with low-output impedance that amplifies the sensed signal, and then drives the 50-Ω coaxial cable. Now, to avoid reflections, the scope must be set on a 50-Ω impedance setting. The scope will "know" this probe is in use, and will amplify the signal by a factor of 2—because it knows there is a divider action taking place between the coaxial cable's (high-frequency) impedance of 50 Ω, and the 50-Ω input impedance of the scope—a divide by two ratio.

Then there is something called the 50-Ω *passive* probe. This is just a coaxial wire with either a 450-Ω resistor in series with the tip, or 950 Ω. For either of these, the scope must be set on 50-Ω impedance. Now by divider action, we have either 50 Ω/(450 Ω + 50 Ω) = 1 Ω/10 Ω, or 50 Ω/ (950 Ω + 50 Ω) = 1 Ω/20 Ω. So the scope will amplify the signal 10 or 20 times depending on the probe (which it senses by means of the contacts where the cable plugs into the body of the scope). In the application note titled "Choosing the Best Passive and Active Oscilloscope Probes for Your Tasks" available at http://cp.literature.agilent.com/litweb/pdf/5990-8576EN.pdf, Agilent warns: "The key benefits of this probe include low capacitive loading and very high bandwidth—in the range of a couple of GHz, which helps to make high-accuracy timing measurements. In addition, this is a low-cost probe compared to an active probe in a similar bandwidth range. You would use this probe in applications such as probing electronic circuit logic (ECL) circuits, microwave devices or 50 Ω transmission lines. The one critical trade-off is that this probe has relatively heavy resistive loading (500 Ω for example–author), which can affect the measured amplitude of the signal." "The measurement is a bit intrusive."

When trying to take measurements in an oven, engineers realize that most probes get damaged at high temperatures, so they use a coaxial cable with a BNC connector on the other end. The principal hardware engineer mentioned previously used that clever arrangement too, with a 50-Ω scope impedance setting. Basically, the source of the AC signal was now almost fully loaded by a 50-Ω resistor across it. Further, since a simple BNC connector was now being used, the scope had no way of "knowing" what probe was connected to it, so it would simply display what was coming in with no amplification. The engineer was soon using this arrangement to check everything, including the step load response of a switching converter between "no (50 Ω in this case) load" to max load. He would see no overshoots and report the switching controller was perfectly designed. He also

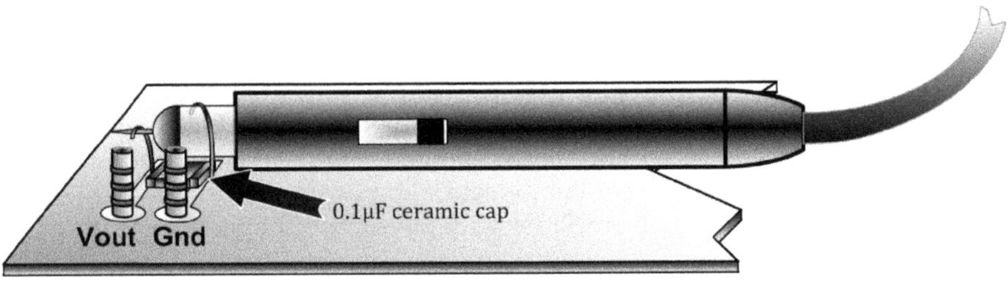

FIGURE 12.1 Correct way to measure noise and ripple.

initiated device verification testing (DVT) on the new PSE flagship chip of the company with every intent of publishing the scope captures in the company datasheet, with support of the entire company (except one). Note that here the engineer could try this: add a 9-MΩ resistor in series with the tip, and use the 1-M impedance setting on the scope to try and make it amplify the signal 10 times.

In PoE, customers will often complain of *detections gone awry* and so on, and we will be trying to track down the source of the "noise" affecting detections. That usually takes us straight to the doorstep of some switching regulator—either the "48V" AC-DC PSU, or some on-board converter, say, for the 3.3-V rail. Therefore it is important to know how to measure the noise and ripple correctly. But even with a 10X probe and a 1-M impedance setting, we are not assured of an accurate reading—because the ground lead of the scope probe in particular, can pick up a lot of high-frequency noise through the air, acting like an antenna. Therefore, a recommended way to do this measurement is shown in Fig. 12.1.

Measuring PSE Port Voltage

The IC ground is the Source terminal of the PSE's low-side N-FET (or the other end of the sense resistor connected to it, if any). That is rightfully the PSE-system ground plane too (not to be confused with the overall system ground which is on the host side). Some engineers however prefer to call the upper rail ("48V") as the PSE's system ground, in accordance with the old telecom convention of positive ground. But as explained in Chap. 1, that has no relevance to PoE, so it makes more sense to have the control IC's ground as the PSE system ground—why have multiple ground planes?

But if we place the oscilloscope ground clip on this PSE system ground (Source terminal), we do not see what we really want to monitor: the port voltage. If we place the ground clip on the Drain terminal of the N-FET, we will get very odd results because the Drain is a swinging voltage. In contrast, the positive port rail (equal to "48V" rail if we have DC disconnect) is "quiet" and can serve as the reference

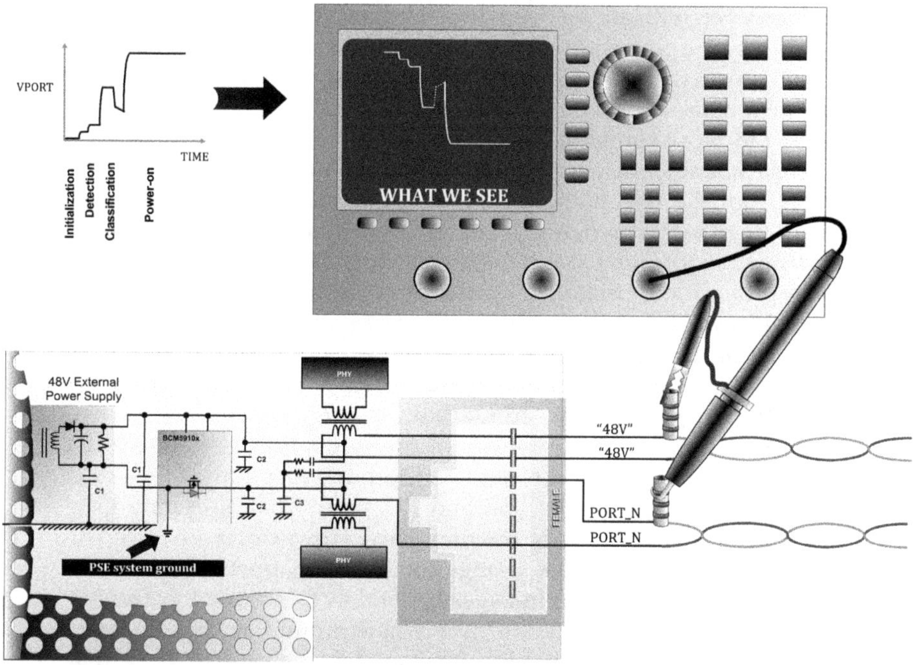

FIGURE 12.2 Using a regular scope to monitor port voltage.

plane for the scope. So if we connect the ground clip of the probe on this and the other end to the Drain, as shown in Fig. 12.2, we get an *inverted* waveform compared to what we presented as the port voltage throughout this book.

But there is a caveat to this particular measurement technique as discussed below.

Earth Ground Loop Issue and Isolating the Oscilloscope

We remember from Chap. 1 that one of the advantages of using data transformers was to physically break up ground loops. In other words, the signal would choose only the paired wires of the cable, because that was the only route available to it, and would then travel down the pair in a strictly differential-mode fashion. On the other hand, there would be no differential-mode *noise* pickup along the cable since (a) the cable was "twisted" and therefore *both* wires of each pair would pick up noise equally, and (b) there could be no significant common-mode noise currents either, because though there are natural leakages to Earth ground all along the length of the cable, we have limited that almost completely by using good insulation, and so on.

However, with all the preceding thoughts, there are several ways we can create a path through the Earth ground without even realizing it. This is usually through the AC wiring of the complex, in particular the Earth wire.

In the most common form, such a path will manifest itself as *detection failures:* for example, we may put a 14-k signature resistor on the PD but the PSE will turn on nevertheless. In overload testing, where there was a switched load, this phenomenon has also caused *mysterious chip failures* combined with inductive spikes. In fact this Earth ground loop issue has caused innumerable oddities over the years. *Whenever we see something odd happening, functionally, we must first rule out Earth ground loops.*

In Chap. 11, we took extraordinary measures to stop sneak-path currents through the Earth ground during surge testing, because we needed to maintain the integrity of the setup, and also, in reality, we always have very *high-frequency* AC ground loops anyway. This is through the Y-caps on the PSE and PD sides. That is exactly how lightning surges affect us in the first place. However, if we have DC ground loops in particular, strange things will happen. These sneak paths are hard to describe and sketch, but they exist. And the proof of that is when steps to eliminate all possible ground leakage currents are carried out, proper PSE-PD functionality is restored. Note that this also tells us that we should not use very large Y-capacitances, not just for surge survivability, but for ensuring proper *functionality* too in PoE-based networking systems.

In an oscilloscope, the ground clip is actually directly connected to the enclosure, which in turn is firmly bolted to the Earth prong of the AC cord. This is a primary source of Earth currents, especially when we have something other than a pure resistive load card (albeit with discrete classification/detection front-end circuitry), on the other end of the cable. So people often remove the middle prong of the AC cord or use a 3-to-2 pin-adapter plug (though that is not recommended for safety reasons). This is called "floating the scope" in lab parlance. It is necessary in the absence of other techniques to the same effect. That is how Fig. 12.2 came about.

Some engineers try to float the scope using an isolation transformer. This may work but we have to be careful because all isolation transformers are not equal, as discussed in Chap. 11. Some have connections to Earth ground on both sides, or internal shields. A good way out is to use a battery-powered scope, or run it off an inverter. But we must make sure the charging cord to the inverter is not left connected to the AC outlet during a critical measurement. Because that can complete a ground loop too, through the mains wiring, and in an actual case this was definitely identified as not only causing functional problems, but also lowered surge survival thresholds.

Another acceptable way is to not float the scope, but use *differential probes.* These have a high-input impedance on both pins, so there is no leakage via the ground clip. Further, the waveforms we will now see are not "inverted" as in Fig. 12.2.

Lab power supplies are another major source of ground leakage currents. For "safety and/or reliability reasons" it seems, lab power supplies from reputed vendors when dissected, revealed not only that the output VDC return terminal was hardwired to the enclosure and from there to the ground prong of the AC plug, but that disc-shaped MOVs were internally connected between all three prongs of the AC inlet. MOVs have significant leakage (~100 μA) even at low voltages unfortunately. This can play havoc with PoE test setups. Keep in mind that removing the AC prong of the AC cord no longer helps here because the leakage goes through the MOV to the neutral wire, and that is connected to Earth ground far away somewhere by the utility. So standard lab power supplies can rarely be used for providing the "48V" rail in PoE test setups. We are better off using a regular "48V" silver-box AC-DC switching power supply.

Thermal Management

Historically, the basic equation for the heat transfer process was

$$\text{Thermal Resistance} = \frac{\text{Temperature Differential}}{\text{Heat}}$$

The equations are analogous to the well-known electrical equation (Ohm's law):

$$\text{Resistance} = \frac{\text{Voltage Differential}}{\text{Current}}$$

Therefore the following electrical to thermal analogy was proposed:

$$\text{Resistance} \longleftrightarrow \text{Thermal Resistance}$$

$$\text{Voltage} \longleftrightarrow \text{Temperature}$$

$$\text{Current} \longleftrightarrow \text{Heat Dissipation}$$

On the basis of these, a resistor network shown in the lower half of Fig. 12.3 describes the heat-transfer process described in the upper half of the same figure. This is the process by which the junction temperature stabilizes with respect to its environment (ambient).

Note *We are calling thermal resistance "Rth." In literature it is often called* θ.

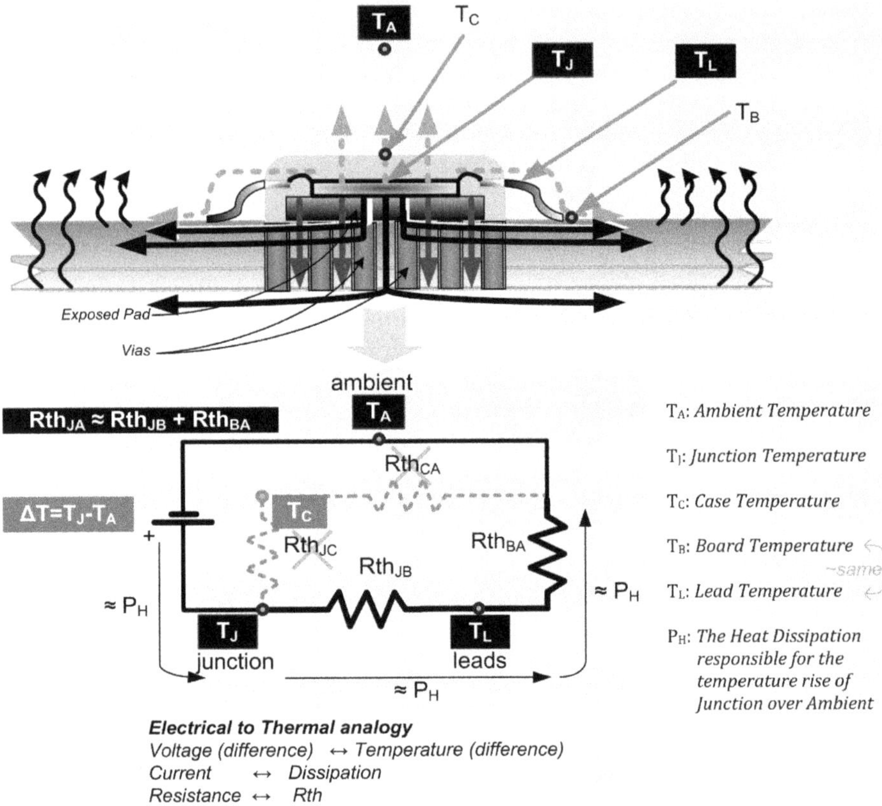

FIGURE 12.3 Example of heat transfer from a leaded package.

Looking carefully at Fig. 12.3, we show the primary path of heat transfer is marked by solid lines, whereas the secondary path is shown by dotted lines. We can usually ignore the heat transfer through the case and write the following basic (historically accurate) equations:

$$Rth_{JB} = \frac{T_J - T_B}{P_H} \ ^\circ\mathrm{C/W}$$

Note that in Fig. 12.3, we see that two thermal-resistance terms have been crossed out (in the secondary heat transfer path). The reason we cannot correctly define these terms Rth_{JC} or Rth_{CA} is that the following equations are *not* true

$$Rth_{JC} = \frac{T_J - T_C}{P_H} \ ^\circ\mathrm{C/W} \qquad \text{Not true}$$

$$Rth_{CA} = \frac{T_C - T_A}{P_H} \ ^\circ\mathrm{C/W} \qquad \text{Not true}$$

The reason they are invalid is we really do not know what is the exact amount of heat going down this path. It certainly is not P_H, which by definition, is the *entire* heat dissipation in the package. Nor is it a significant portion of P_H, since in Fig. 12.3, we are talking about a package with an exposed pad soldered directly on to the PC. We can easily visualize that most of the heat is going through the exposed pad on to the PCB, not through the case. Yes, if we were talking, for example, of a TO-3 power transistor (like the well-known 2N3055 power transistor), things would be completely reversed. In that case, we would conceivably have the metal case connected to a big heatsink. So we can imagine that most of the heat would now go through the case, not through the leads and board. And in that case it would make sense to talk about Rth_{JC}. Similarly it would then not make sense to talk of Rth_{JB}. So it all depends on what exact situation we are talking about.

The key groups involved in the standardization effort for measuring thermal resistance of semiconductors are Semiconductor Equipment and Materials International (SEMI) and Electronic Industries Alliance (EIA). The better-known term JEDEC (Joint Electron Device Engineering Council) is actually just the standardization body of the EIA.

To make things clearer, JEDEC introduced a new thermal resistance term: ψ. Note that this term does not replace Rth, but complements it. As a quick practical guide, the correct term to use is Rth when we are sure most of the heat is going directly down that particular path. However, if the path in consideration is not the primary transfer path, then the term we need to use is ψ. If both paths happen to be almost sharing in the heat transfer process, we need to use both terms ψ and Rth, and we cannot ignore either.

In the case shown in Fig. 12.3, we had realized that Rth_{JC} or Rth_{CA} were not appropriate (as defined historically). However, with the new thermal-resistance term ψ, defined as it is by JEDEC, we can now correctly write

$$\Psi_{JC} = \frac{T_J - T_C}{P_H} \ ^\circ\mathrm{C/W}$$

$$\Psi_{CA} = \frac{T_C - T_A}{P_H} \ ^\circ\mathrm{C/W}$$

However, in general, to predict the junction temperature, we need to focus only on the primary heat transfer path. In our case this is the path junction to board to ambient. In other words, we can completely ignore ψ in our further discussions here, and simply focus on Rth_{JA}, which is a sum of Rth_{JB} and Rth_{BA} as indicated in Fig. 12.3.

Note that we will also be assuming the board temperature and lead temperature are almost identical in our case.

For an exposed pad package, we write

$$Rth_{JA} = Rth_{JB} + Rth_{BA}$$

The JEDEC Standards (JESD)

When trying to estimate temperature from published thermal resistance values in a chip datasheet, or when comparing vendors, it is important to know the following.

Historically, the SEMI organization drafted the initial standards for ceramic packages, integrated circuit packages and semiconductor packages.

Some key SEMI standards to note are

1. SEMI G30-88 "Test Method for Junction-to-Case Thermal Resistance Measurements of Ceramic Packages."
2. SEMI G38-87 "Test Method for Still- and Forced-Air Junction-to-Ambient Thermal Resistance Measurements of Integrated Circuit Packages."
3. SEMI G68-0996 "Test Method for Junction-to-Case Thermal Resistance Measurements in an Air Environment for Semiconductor Packages."

The SEMI standards used theta (θ) for thermal resistance (we have called it Rth). Later, recognizing certain limitations of these early standards, EIA/JEDEC came along in the early 90s, and defined their own standards. In particular, they broke up thermal resistance into the familiar theta (Rth in our case), and a new term psi (ψ) as discussed previously. The main difference is that we use Rth_{AB} for the thermal resistance between two points A and B when we know that the most of the heat flows between A and B, whereas we use ψ_{AB} when there are several, possibly unknown/unquantifiable, paths of heat flow between points A and B.

The key JEDEC standards for thermal resistance are

1. JESD51 Methodology for the Thermal Measurement of Component Packages (Single Semiconductor Devices) (This is an overview document.)
2. JESD51-1 Integrated Circuit Thermal Measurement Method—Electrical Test MethodJESD51-2A Integrated Circuit Thermal Test Method Environmental Conditions—Natural Convection
3. JESD51-3 Low Effective Thermal Conductivity Test Board for Leaded Surface Mount Packages
4. JESD51-4 Thermal Test Chip Guideline

5. JESD51-5 Extension of Thermal Test Board Standards for Packages with Direct Thermal Attachment Mechanisms
6. JESD51-6 Integrated Circuit Thermal Test Method Environmental Conditions—Forced Convection
7. JESD51-7 High Effective Thermal Conductivity Test Board for Leaded Surface Mount Packages
8. JESD51-8 Integrated Circuit Thermal Test Method Environmental Conditions—Junction to Board
9. JESD51-9 Test Boards for Area Array Surface Mount Package Thermal Measurements (Note: example for BGA: Ball Grid Array, and LGA: Land Grid Array devices)
10. JESD51-10 Test Boards for Through-Hole Perimeter Leaded Package Thermal Measurements (Note: example for DIP: Dual In-line packages and SIP: Single In-line packages)
11. JESD51-11 Test Boards for Through-Hole Area Array Leaded Package Thermal Measurements (Note Example for PGA: Pin Grid Array devices)
12. JESD51-12 Guidelines for Reporting and Using Electronic Package Thermal Information

These standards can be downloaded free from www.jedec.org. A quick guide to which JEDEC standard to consult for measurements concerning surface-mount packages like the quad flat pack (QFP or QFN) is as follows:

1. Identifying a steady-state condition is discussed in JESD51-1. Also basic testing methods, in particular dynamic and static modes of testing.
2. The design of the enclosure used for natural convection measurements can be found in JESD51-2A as per Table 1. Also the difference between the thermal-resistance terms θ (Rth) and ψ and how to measure case temperature.
3. The design of the wind tunnel used for forced convection measurements can be found in JESD51-6 as per Table 1.
4. Design of a 1s test board is found in JESD51-3. 1s stands for a single signal-layer board.
5. Design of a 2s2p test board is found in JESD51-7. 2s2p stands for double-signal layer, double-buried power plane.
6. If the package has an exposed pad soldered on to the board (like the BCM5910X), we need to consult JESD51-5 in conjunction with the relevant JESD51 references.
7. The method to measure board temperature is described in JESD51-8.

Types of Test Boards

Two test board cross sections are defined in the JESD51 standards.

1s boards (as per JESD51-3): The first cross section is referred to as the low effective thermal conductivity or 1s board. The 1s refers to the one signal layer on the component side of the board, so the board is sometimes referred to as a single layer board. Limited signals are permitted on the opposite side actually, making it a 2s or two-layer board. The key point is that this board does not have power planes (0p). The signal-layer traces (obviously the exposed sides for such a board) are 70 μm (2 oz/ft^2) finished copper thickness. Sometimes, this board is referred to as a 2s0p board too.

2s2p boards (as per JESD51-7): The second cross section is referred to as the high effective thermal conductivity or 2s2p board. It has significantly more copper than the previous one. The 2s refers to the signal layers on both outer (exposed) surfaces of the board and the 2p refers to two power planes in the board (voltage and ground, both inner layers). The board is sometimes referred to as a four-layer board. The (outer) signal-layer traces are 70 μm (2 oz/ft^2) finished copper thickness and the (inner) power planes are both 35 μm (1 oz/ft^2) finished copper thickness. For packages with ball pitch ≤ 0.5 mm the traces are reduced to 50 μm (1.5 oz/ft^2) finished copper thickness (outer) for both boards (see JEDEC51-10).

The required test boards for different types of packages and PCBs are presented in Table 12.1 along with the JEDEC standard to refer for more information.

We see from Table 12.1 that the two test boards in most common use today, as applicable to the QFN packages too, are described in JESD51-3 and JESD51-7. Of these, the most commonly used board is the four-layer "2s2p" board as described in JESD51-7.

The defined stackup of a 2s2p test board is shown in the top half of Fig. 12.4. Note that the signal layers are on the outside, whereas the power planes are on the inside. This is not necessarily the best arrangement for thermal performance, but is certainly the JEDEC standard by which to *compare package thermal data from different vendors*. Otherwise we could end up comparing apples to oranges. However, in carrying out a practical PoE PCB design it is very common to have a four-layer stackup as shown in the lower half of Fig. 12.4.

Note that we do not recommend a Vcc, Vdd, 3.3 V or 12 V, 48 V power plane—for EMI reasons. That used to be the practice historically. However, with the advent of switching power supplies, the truth is the positive rail is now "noisy," with a lot of high-frequency ripple content. Large noisy copper areas become good radiators of

Package	Spec JESD51-[x]	Board Type
Leaded Surface Mount, Peripheral Leads	[3]	1s
	[7]	2s2p
Leaded Surface Mount Peripheral Leads with direct thermal attach (e.g., QFP with exposed pad)	[3, 5]	1s
	[7, 5]	2s2p
Leadframe-based perimeter array with direct thermal attach (e.g., QFN)	[3, 5]	1s
	[7, 5]	2s2p
Array Surface Mount (e.g., BGA or LGA)	[9]	1s
		2s2p
Through-hole Perimeter Array (e.g., DIP)	[10]	1s
		2s2p
Through-hole Array (e.g., PGA)	[11]	1s
		2s2p

TABLE 12.1 Test Boards as per JESD51-[x]

electric fields. They also couple capacitively to the adjacent ground plane and rather than being "quieted" themselves, have the ability to inject noise into the ground plane, affecting its efficacy too. So make thick traces if need be, but avoid power planes other than ground.

Also, copper filling is good for improving thermal resistance, but be careful not to leave *floating* copper—all such copper fill must be connected to the main ground plane using multiple vias spread evenly around ("stitching").

Improving PCB Thermal Resistance for Exposed Pad Packages

As can be seen from Table 12.1, when talking about packages with exposed pads, we need to refer to JESD51-5 to understand how the standardized 2s2p board is constructed. This would apply typically to PSE chips with integrated (on-chip) pass-FETs in QFN/DFN or even TSSOP packages. The JEDEC standard also specifies that the 2s2p test board size be 76.20 mm × 114.30 mm ± 0.25 mm for packages less than 27 mm on a side. That is 3 in. × 4.5 in., a fairly large board with just one chip being characterized! The standard also recommends the pattern of traces and that a 1-foot square box be used for natural convection measurements as per JESD51-2A.

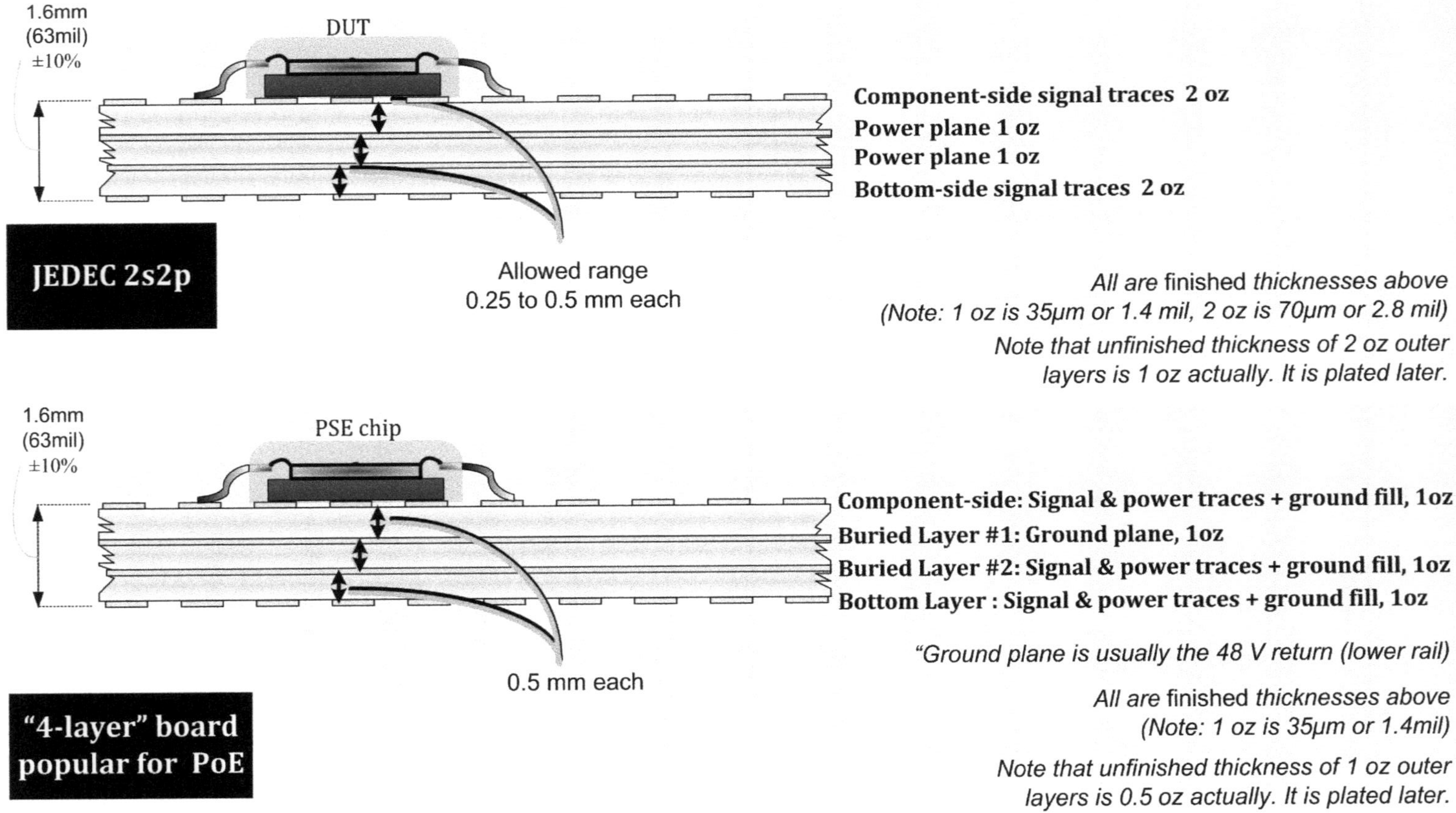

FIGURE 12.4 JEDEC 2s2p stackup compared with PoE-recommended four-layer stackup.

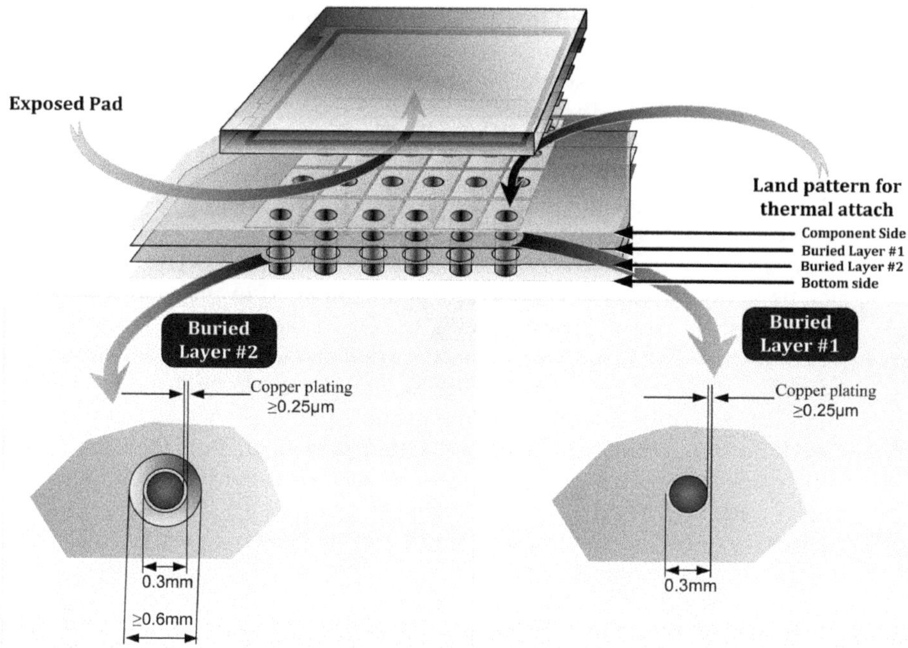

FIGURE 12.5 A typical thermal characterization PCB design for exposed pad package in accordance with JESD51-5.

As per the JEDEC 2s2p board recommendations

- The allocated area for the exposed pad (thermal attach) must be equal to or slightly *larger* than the actual exposed pad of the package. However, it cannot be more than 1 mm larger.
- As indicated in Fig. 12.5, the allocated area is broken up into 1-mm squares separated by 0.2 mm. That means the pitch (from center to center) is 1.2 mm.
- At the center of each such square, there is a via of 0.3-mm barrel diameter. The copper plating through the barrel must be more than 25 μm thick.
- All the thermal vias connect to the first buried layer (just below the component side).
- Connection of the vias to the signal layer on the bottom side is optional.
- Green-masking is optional (though it must *not* be present on the thermal attach area of course).

A notable and discussion-worthy feature of the PCB suggested by the JEDEC standard is the design of thermal vias under the exposed pad solder joint as shown in Fig. 12.5. Keep in mind that the JEDEC board is for standardization, and all its recommendations do not necessarily lead to the best thermal performance. For example, in an actual PoE board, we do not break up the area under the exposed pad into copper squares as in Fig. 12.5, unless deemed really unavoidable. Some may create such a pattern if they want to implement "thermal relief," but that is usually not necessary in PoE boards.

Note that as per Fig. 12.4, in an actual PoE board, the first buried layer is usually the "48V" return (lower rail of supply). This could also be the IC ground in some chip architectures, as in the case of no sense resistor present (instead, using the R_{DS} of the pass-FET for current sensing). The IC ground is usually the exposed pad of the QFN package. So as per Fig. 12.5, the JEDEC standard asks that the thermal vias be firmly connected to the ground plane. And that recommendation is actually good for an actual (practical) PoE PCB too. However, in the 2s2p board there is no electrical contact with the power plane further below (buried layer number 2). And also, no copper is connected to the thermal vias on the bottom layer. So, two improvements are possible here.

1. First, keep in mind that in an actual PoE board we do not recommend making buried layer number 2 into a + "48V" plane, for EMI reasons. Rather we use it for signal/power trace routing and then do a copper fill. But all the filled copper islands must be stitched rather well (using several vias not far apart), to the the upper ground plane.
2. Another major improvement over Fig. 12.5 occurs when we sink the thermal vias to the bottom layer and connect all of them to a square copper island there, one that literally shadows the thermal attach area on the component side under the IC. We also carry out copper filling on the bottom layer wherever possible, stitching the fills to the buried ground plane (not to the copper fills on the plane directly above).

Practical Thermal Resistances

As an experiment, the standard 2s2p board was modified to test the impact of the two key additional PCB recommendations above—no grid under exposed pad, and generous copper filling, particularly on bottom layer for maximizing conduction and convection aided heat loss. However, other things were unchanged. The same-sized board was used in the same recommended JEDEC box, with the same position of thermocouple, and so on. A chip with a published 2s2p thermal resistance (junction to air) of 24°C/W, when placed on this 2s2p "look-alike board" showed a much improved thermal resistance of only 10.5°C/W.

The JEDEC board and the "look-alike board" created above were both large. What happens on a typical compact daughtercard, with multiple PSE chips? We are back to square-one! In other words, with all the helpful PCB recommendations above, which more than halved the thermal resistance of a 2s2p standard board, if we now make the actual PoE board much smaller, and just large enough to accommodate all the components for a typical 24-port solution, the thermal resistance goes back up to about 25°C/W. No gain, no loss. That is with a four-layer board. *With a low cost two-layer board (no ground plane possible, but generous copper filling on top and bottom layers), the measured thermal resistance was 40° to 45°C/W*. That is almost double that of a four-layer board.

Sizing Copper Traces

There are complicated curves available for copper versus temperature rise of PCB traces in the now-obsolete standard MIL-STD-275E. These curves have also found their way into more recent standards like IPC-2221 and IPC-2222. Engineers often try to create elaborate curve fit equations to match these curves. But the truth is the earlier curves can be easily approximated by simple linear rules as follows.

The required cross-sectional area of an external trace is approximately

1. 37 mils^2 per Amp of current for 10°C rise in temperature (recommended).
2. 25 mils^2 per Amp of current for 20°C rise in temperature (aggressive, but OK).
3. 18 mils^2 per Amp of current for 30°C rise in temperature (usually not recommended).

For the *traces in inner layers, multiply the calculated width of an external trace by 2.6 to get the required width.*

To calculate width of a trace from the cross-sectional area, keep in mind that 1 oz copper is 1.4 mils thick and 2 oz copper is 2.8 mils thick.

It is very important to size the traces correctly, as this can contribute significantly to the overall dissipation and temperature rise.

Calculating Junction Temperature

We use the excellent thermal resistance number of the JEDEC "look-alike board" mentioned previously. We know that its thermal resistance was around

$$Rth_{BA} = 10.5°\text{C}/\text{W}$$

The thermal resistance of the chip from junction to lead (or board) is published as 0.5°C/W. We must add typically 15°C to the room ambient to get the "local (chip) ambient" T_A (see Fig. 12.3). Further, if AC disconnect is being used we need to add another 10°C to the local ambient, to account for the heating up of the PCB by the AC disconnect diodes.

For example, suppose the chip's dissipation is 1.6 W and its datasheet declares: $Rth_{JB} = 0.5$°C/W and $Rth_{JA} = 25.5$°C/W (we ignore any published Rth_{JC} for reasons discussed previously). We also keep in mind that these Rth numbers were measured/computed on a JEDEC 2s2p board. When we go from a JEDEC approved board to a JEDEC look-alike board, made in accordance with preferred PCB practices, we can safely assume that the term Rth_{JB} is constant. So the only term that changes for us is $Rth_{BA} = 10.5$°C/W. At a room ambient of 55°C we thus estimate the junction temperature as follows:

$$\begin{aligned} T_J &= P_H \times (\theta_{JB} + \theta_{BA}) + T_A \\ &= 1.6 \times (0.5 + 10.5) + 55 + 15 + 10 \\ &= 97.6°\text{C} \end{aligned}$$

Note that to achieve this figure, the recommended PCB guidelines must be followed and the four-layer board itself must be 3 in. × 4.5. in. (same as the standard JEDEC 2s2p board). If not, we will get closer to

$$\begin{aligned} T_J &= P_H \times (\theta_{JB} + \theta_{BA}) + T_A \\ &= 1.6 \times (0.5 + 25) + 55 + 15 + 10 \\ &= 120.8°\text{C} \end{aligned}$$

where we have used $Rth_{BA} = 25$°C/W based on experiments on a small four-layer daughtercard with all our PCB guidelines implemented.

Different Ways of Specifying Maximum Operating Temperature

When comparing datasheets, we need to know there are different ways to specify the maximum "operating temperature" of a device. Some vendors take "operating temperature" as the *ambient temperature*. But some consider this as *junction temperature*. Customers often get confused by this and tend to think the higher numbers in the latter case implies a better temperature range. Whereas, that may not be so on closer examination. Let us do some estimates to understand this better.

For example, suppose Datasheet A (from Vendor A) states that its recommended range for Chip A is –40° to 85°C—using a 2s2p board as per JESD51-7 guidelines. Whereas Datasheet B for Chip B provides electrical characteristics tables guaranteed over the junction temperature range of –40° to 125°C. So prima facie, it looks like Datasheet A is inferior to Datasheet B. But let us do some math. The junction to ambient thermal resistance Rth_{JA} of Chip A is stated to be 25.5°C/W using a 2s2p board. This amounts to a temperature differential of $1.6 \times 25.5 = 40.8$°C. Note that this assumes Chip A has a dissipation of 1.6 W. Its junction temperature at the maximum rated ambient of 85°C is 85°C + 40.8°C = 127.2°C. This is very close to the maximum junction temperature of 125°C specified for Chip B. So the parts do have the same junction temperature range. Neither is better or worse in terms of operating temperature range. However, if Chip B has a much smaller chip dissipation compared to the delared 1.6 W of Chip A, the truth is Chip B can be operated to much higher ambient temperatures than the upper limit of 85°C stated for Chip A. So there are two yardsticks actually: the maximum operating junction temperature and the maximum operating ambient temperature. We need to do some calculations to decide which vendor is really better, and in what regard, if they have different ways of specifying thermal performance.

Fan Speed

We have to be cautious of claims of high fan speed, because of fan backpressure effects. The engineers may think the fan is very fast, say 400 linear feet per minute (LFM), whereas in reality, it could be only half of that, say 200 LFM. Or the engineer thinks it is 200 LFM, whereas it really is only 100 LFM. And so on.

Generally, we can assume that with an *actual* 100 LFM, the junction temperature falls by 5°C compared to the no forced-air case. With an *actual* 200 LFM, we can assume 8°C advantage. Admittedly, these are conservative estimates. Temperatures may actually be lower if there is air turbulence. But since we can't depend on that always, the above guidelines are appropriate for first (worst-case) estimates.

Proper Chip Decoupling

We will keep this simple: every modern chip has a lot of things going on "under the hood." Gates and circuit blocks may suddenly turn on or off in nanoseconds, either demanding a sudden surge of current, or turning that demand off. If local decoupling is not provided right next to the supply pins of the IC, there will be sudden dips and overshoots of the supply voltage on its pins, because the supply rails comes from a regulator usually quite far away, with intervening trace inductances, and the regulator therefore cannot cater to the sudden demands of the

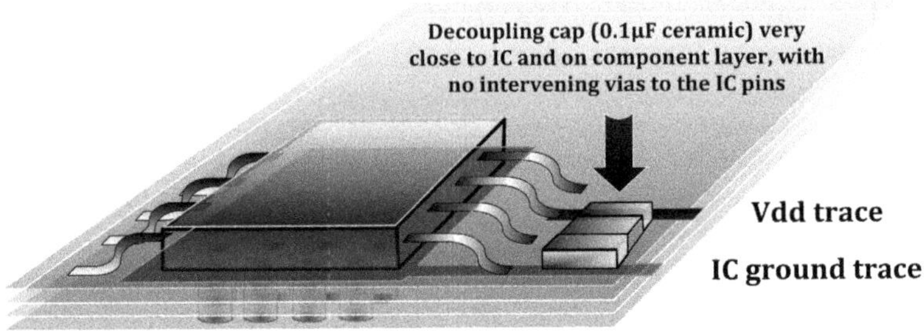

FIGURE 12.6 A preferred decoupling scheme.

chip right away. These supply variations (spikes) can damage the IC (overshoot) or send it into power-on reset (undervoltage). So a low ESR, low-ESL ceramic capacitor, preferably 0.1 μF is required very close to the positive supply pin as shown in Fig. 12.6. Note that it is on the same side of the PCB as the chip. That is a preferred arrangement, as compared to putting the decoupling cap on the bottom layer, from where it connects to the IC through vias. Vias have non-insignificant inductances and have been known to have cause a certain lack of decoupling that manifested itself as a rather small but undeniable field failure rate. So the best way is for the supply lines to first come to the decoupling cap, from there on to the pins of the IC, and from there the thermal under the exposed pad connecting to the ground plane. This is shown in Fig. 12.6.

If we have analog and digital rails for example, we need to provide local decoupling for each, in the manner describe above.

We should also place a 10 μF/63 V or 10/100-V electrolytic bulk capacitor on the "48V" rail—somewhere centered close to all the PSE chips on the PoE board. This provides bulk decoupling, until the bigger bulk caps on the output of the AC-DC power supply (which are much further away) can take over the task, and also replenish all the downstream decoupling caps.

CHAPTER 13

N-Pair Power Delivery Systems

Overview

In Chap. 1 we learned that local area networking (LAN) started years ago, took different forms, then slowly evolved into its present incarnation which we call Ethernet. The idea of sending power over data cables too started much the same way, quite disparately, before it settled down to what we know today as PoE. To a systems and IC designer, that is sometimes a blessing and a disguise: He or she knows very well what to do, unfortunately that is *all* that can be done. Functioning within the constraints of a carefully defined standard may seem to stem creativity at times. In the interest of ensuring interoperability, a certain amount of excitement and innovation may have gone missing, lost forever somewhere in ether.

A few years ago, the author was working at a certain networking company where the team was raring to go and create the next PSE chip. They would carry their brief proposal for approval to the VP/GM who would listen patiently, then lean back and ask quite a valid question: "What's the wow factor?" That was typically greeted by silence. What more can you do other than design to IEEE 802.3at? Doesn't *that* tell you everything that you need to do, and *can* do (but no more)? Yes, we could make the silicon a little smaller, maybe go from a quad to an octal, maybe even get it to cool itself with thermoelectric technology, but the questions would then be: Who would buy that? Would it sell? Would there be a "second source" available to an OEM who went with our solution? No OEM likes to be at the mercy of one chip vendor. A wow factor serves exactly what *purpose* here?

Innovation does occur, however slowly, even under the seeming stranglehold of an open standard. Fortuitously, open standards usually come *after the fact*. Quite like words getting legitimized in dictionaries long after they had already made it to slang. Similarly, LAN technologies were already around when the 802.3 committees were set up. Power over Ethernet was already around (mainly from PowerDsine, now Microsemi, and Cisco) before the AF committee got to work. In a similar vein, Cisco recently came up with UPOE (Universal Power over Ethernet), and yet another consortium or alliance, is chasing HDBaseT.

Since all these emerging technologies are using the same cabling infrastructure, it is important to maintain *backward compatibility*. We want to ensure a new X-PoE-X PSE recognizes another X-PoE-X PD. Also hopefully, that an X-PoE-X PD behaves as an IEEE 802.3at PD when connected to an IEEE 802.3at PSE. And also that an X-PoE-X PSE behaves as an IEEE 802.3at PSE when connected to an IEEE 802.3at PD. In other words, a proprietary PSE and a proprietary PD based on some new standard should be able to *mutually identify* each other, and then activate perhaps higher power or some new features. Since we can't ever control what gets plugged into the opposite end of the cable, this proprietary device (PSE or PD) must be able to operate in an IEEE-standards-compliant mode in all other cases. That is much like a Japanese couple who relish sushi when they are together, but can equally well enjoy pasta with an Italian colleague or bratwurst with a German friend. A kind of multiculturalism, at an electronics, systems level.

Proprietary solutions will try to come up with proprietary detection and classification schemes too, because it is preferable to complete mutual identification at the physical layer (Layer 1) itself. But if that is not possible, then just as AT devices are allowed to do, they can usually start up in AF mode, and later mutually identify each other via LLDP (Layer 2), and then activate higher power or new features as desired. Unless it is for example, a computer booting up—it cannot even afford to initiate the Power-up sequence unless it is sure there is adequate power at the end of the road to see the entire boot-up process through. So identification at the physical layer becomes very attractive, even though there are huge backward compatibility issues to iron out.

While we try to innovate in PoE, there are constraints we must operate within, rules we must abide by. Unbridled innovation must be reined in somewhat, to stay firmly within already-established safe-operation boundaries. For example, we can't play with the 1500 VRMS isolation requirement applicable to TNV-1. Nor can we exceed the SELV level of 60 V. Further, TIA/IEEE collected a lot of data on temperature rise during the framing of the AT standard, and that poses a limit on maximum current, *but not maximum power* (as we will soon learn). There is a subtle dichotomy often overlooked: Higher power is not necessarily commensurate with higher current. We don't necessarily push more power into the PD by pushing more current into the cable. If we believe that, nothing may distinguish us from Dr. Edward Wildman Whitehouse, who, as we described in Chap. 1, had a simple theory of electrical propagation: The further that electricity has to travel, the larger the kick it needs to send it on its way. We remember that this "simple theory" seems to have crippled the first transatlantic cable, in the 19th century.

One of the boundaries of PoE that we *can* try to defy without compromising safety is the 100-m limit on cable length. Keep in mind that is an artificial limit, not created or limited by the basic concept of power over data, but by Ethernet itself. So, if we can extend Ethernet beyond 100 m (long-reach Ethernet), or use DSL techniques instead, we will be left with the task of delivering power over longer distances (whether we then call it PoE or not). Things will certainly change. It is very likely that classification as we know it today will need to be tweaked, or even abandoned. But the overriding interest here is that there is a lot of telephony cable out there that is much longer than 100 m, and people want to use that somehow. That is the reason why PSE chip vendors often receive requests to send power (and data) over much longer cables. Being part of older infrastructure, these cables are usually of inferior grade: AWG 26, corresponding to CAT3 cable resistance. There are also cases where the available cable is only single-pair. There are also cases where the available infrastructure is not twisted-pair cabling but *coaxial* cable (RG-6). For example, there can be coaxial wiring network originally used for connecting TV monitoring cameras, and now there may be a desire to use the very same wiring to operate IP cameras instead. So, in general, we need to understand how to transmit not only data, but power over *long* stretches of cable, single pair or otherwise. We will realize there is in fact far more to transmitting power over long distances than just $V = IR$. We will also see in this chapter that there are several rather nonintuitive aspects of what we thought was a "simple" power delivery problem.

Another artificial PoE boundary we can try to tear down is the number of pairs. We have already seen that single-pair power delivery is often desired. But what if we just want to reduce losses? We can send 0.6 A (for Type 2 applications) down all four pairs of the Ethernet cable. The cable losses will be much lesser for the same 30 W PSE power, and we will also get some more power over to the PD-side. But we could also just try to push more power (60 W), by sending 1.2 A down all four pairs. The AT standard has in fact left a loophole for that. But in any case, TIA temperature data supports the fact that 1.2 A (60 W from the PSE) will lead to a roughly 10°C rise in temperature of the cable bundle, so 60 W is also acceptable. That was the basis for UPOE from Cisco. Here two pairs share the 1.2-A forward current, and two pairs return the 1.2 A.

In that spirit, in this penultimate chapter we will take a somewhat unfettered view and simply try to perceive what lies ahead for PoE.

Starting with Resistance

One thing is sure: The medium we have available is copper within a cable. The only other thing certain is we have to share it with data (phantom power techniques).

Let us write the equations for predicting the resistance of a certain length of wire of a certain wire gauge (AWG). The equation which relates diameter in mils to AWG is

$$d_{\text{mils}}(\text{AWG}) = 5 \times 92^{\frac{36-\text{AWG}}{39}} \quad \text{(accurate)}$$

The following equation looks like the correct equation, but is actually an approximation (metric is alien to AWG and mils):

$$d_{\text{mils}}(\text{AWG}) \approx \frac{1000}{\pi} \times 10^{\frac{-\text{AWG}}{20}} \quad \text{(approximate)}$$

To go from diameter in mils to mm we use

$$d_{\text{mm}}(\text{AWG}) = \frac{d_{\text{mils}}(\text{AWG})}{39.37}$$

To calculate resistance we use

$$R_o(\text{AWG}) = \frac{\rho_o}{\text{Area}_{\text{sq.meter}}(\text{AWG})} \Omega/\text{m}$$

As a function of temperature T

$$R(\text{T}, \text{AWG}) = [1 + \alpha(\text{T} - 25)] \times R_o(\text{AWG})\ \Omega/\text{m}$$

ρ_o is resistivity of copper $= 1.72 \times 10^{-8}\ \Omega/\text{m}$;
α is the temperature coefficient of the resistivity of Cu $= 3.9 \times 10^{-3}\ {}^{\circ}\text{C}^{-1}$ (i.e., $\approx 4\%$ every 10°C)

The results are as follows (for a single strand of 100-m long wire):

1. AWG 26/typ. CAT3: Resistance of 15.18 Ω→ approximated to 20 Ω
2. AWG 24/typ. CAT5e: Resistance of 9.548 Ω→ approximated to 12.5 Ω

(The approximation is as per 802.3af/at and includes patch cables, contact resistances, tolerances, temperature effects, and so on)

NOTE *To avoid confusion, note that in literature, we often find a table for telecom cable resistance reproduced in Table 13.1. As mentioned, it is the loop resistance of a single pair cable of 1 ft. For example, from the table, AWG 24 (CAT5e) is 19.1 ft/Ω. 100 m is about 330 ft. So, the (loop) resistance of a 100 m (single-pair) cable is 330 ft/19.1 ft = 17.3 Ω. Further, this is at 20°C. We know that resistance of copper goes up 4 percent every 10°C. And since $(1.04)^4 = 1.17$, that means resistance of Cu goes up 17 percent from 20°C to 60°C. Therefore 100-m single-pair cable loop resistance is 1.17 Ω × 17.3 Ω = 20.2 Ω at elevated temperatures. Since this is for single-pair application, for two-pair PoE applications the actual loop resistance is 20.2/2 = 10.1 Ω. As mentioned, to account for tolerances, contact resistances, and so on, we take 12.5 Ω for PoE using 100 m of AWG 24 (CAT5e).*

Loop Resistances for N-Pair Power Delivery

In Fig. 13.1, we first calculate the loop resistances, assuming simply that 100 m of a single strand of CAT3/AWG26 is 20 Ω, and 12.5 Ω for CAT5e/AWG24. We see that as a coincidence, in the case of two-pair power delivery, the loop resistance is the same numerically. For one pair it is double that, and for four pair it is half of that. We use these numbers in the calculations that follows.

Power Estimates for N-Pair Power Delivery

In Figs. 13.2 and 13.3, we do some simple calculations for voltage drop and power reaching the PD, and also the cable losses. Note that we have also done this for "efficient PoE" where we do not double the current for four-pair operation as in UPOE, but keep IEEE-compliant currents (Type 1 = 0.35 A; Type 2 = 0.6 A), but use four pairs to reduce losses. The key results of the figures are summarized in Table 13.2. In the table we have only quoted the results for max

AWG	feet/Ω (at 20°C)
22	30.3
24	19.1
26	12.0
28	7.55

This is for a single-pair cable; resistance (in feet/Ω), is the loop resistance. The length stated above (feet) is however the length of the cable (the loop length is actually twice that).

TABLE 13.1 Resistance versus AWG Table Found in Literature

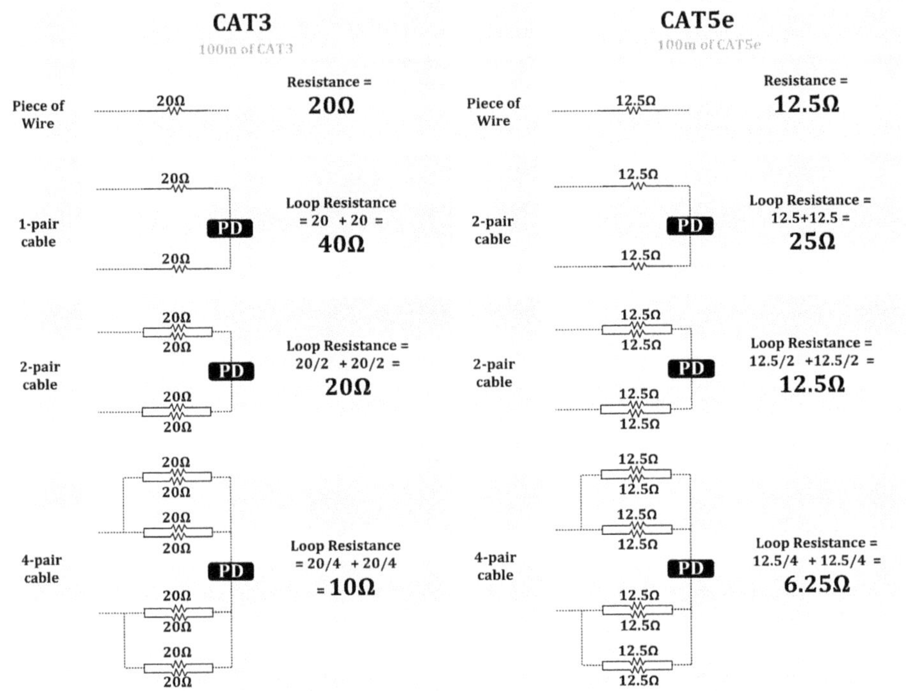

FIGURE 13.1 Loop resistances for 100-m cable calculated.

	AWG 26, CAT3 0.175 A/strand, Vport = 44 V			AWG 24, CAT5e 0.3 A/strand, Vport = 50 V		
	V_{CABLE}	P_{CABLE}	$P_{PSE} \rightarrow P_{PD}$	V_{CABLE}	P_{CABLE}	$P_{PSE} \rightarrow P_{PD}$
1-pair	7 V	1.225 W	7.7 W→6.475 W	7.5 V	2.25 W	15 W→12.75 W
2-pair	7 V	2.45 W	15.4 W→12.95 W	7.5 V	4.5 W	30 W→25.5 W
4-pair	7 V	4.9 W	30.8 W→25.9 W	7.5 V	9 W	60 W→51 W

TABLE 13.2 Summary of Results for Maximum Safe Current Pushed through N-Pairs

allowed current per strand, which is 0.175 A/strand for AWG 26 (CAT3) and 0.3 A/strand for AWG 24 (CAT5e). In other words, since the AT standard says we can pass 0.6 A through two pairs, we get 0.3 A per strand—there are two strands for the forward current (0.3 A + 0.3 A = 0.6 A) and two strands for the return current (0.3 A + 0.3 A = 0.6 A). So for UPOE, we use the same current/strand and we get a total port current of $2 \times I$, which is 0.7 and 1.2 A for CAT3 and CAT5e, respectively. That is how we get 50 V × 1.2 A = 60 W at the PSE, as in UPOE.

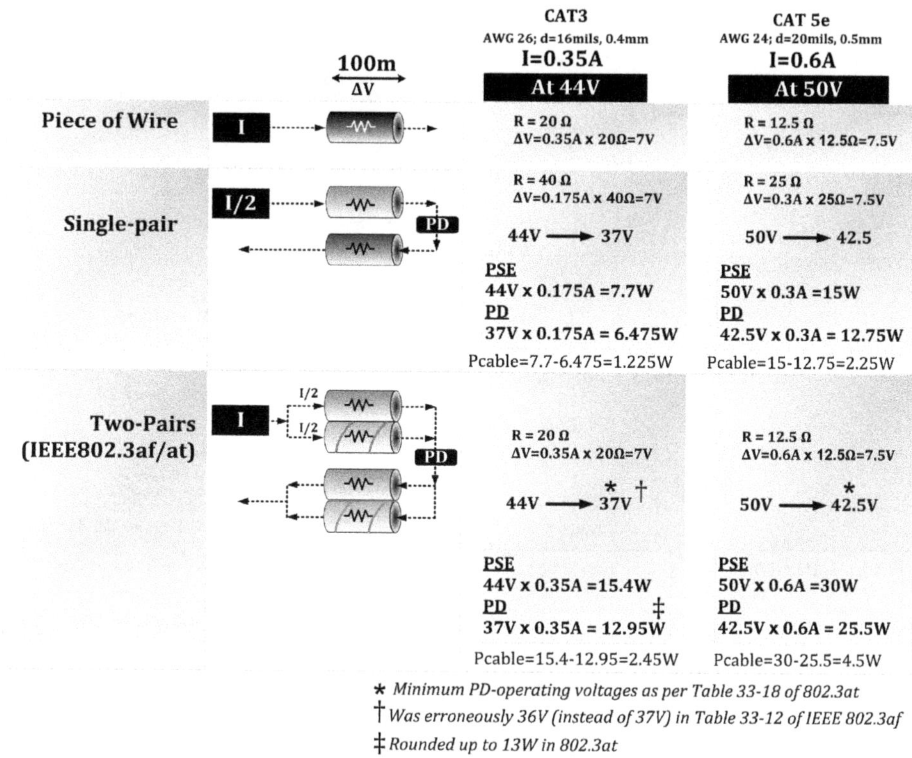

FIGURE 13.2 Power delivery calculations for one-pair and two-pair operations.

Maximum Power Delivery over Long Distances Using Available PSEs

Here we are restricting ourselves to *available* PSEs. So we have two types available basically: AF PSEs that can support 0.35 A, and AT PSEs that can support 0.6 A. We will drive this current either through single pair or two pairs. Note that by fixing the port current, the current per strand is perhaps no longer completely "safe" for the single-pair case, because ideally we would halve the port current for that as we did in Fig. 13.2. But we keep in mind that since single-pair cables are not necessarily found in bundles similar to Ethernet cabling, this current level may very well be acceptable in terms of its temperature rise. Let us *assume* for now, that it is acceptable.

Having fixed the port current, we now extend the cable from 100 to 200 m and then to 300 m, doing calculations at each stage to see how much power is available at the PD end. We imagine this is an easy problem. See Fig. 13.4. As an example, let us take the ungrayed cell. This is the equivalent of an AF PSE connected in normal two-pair

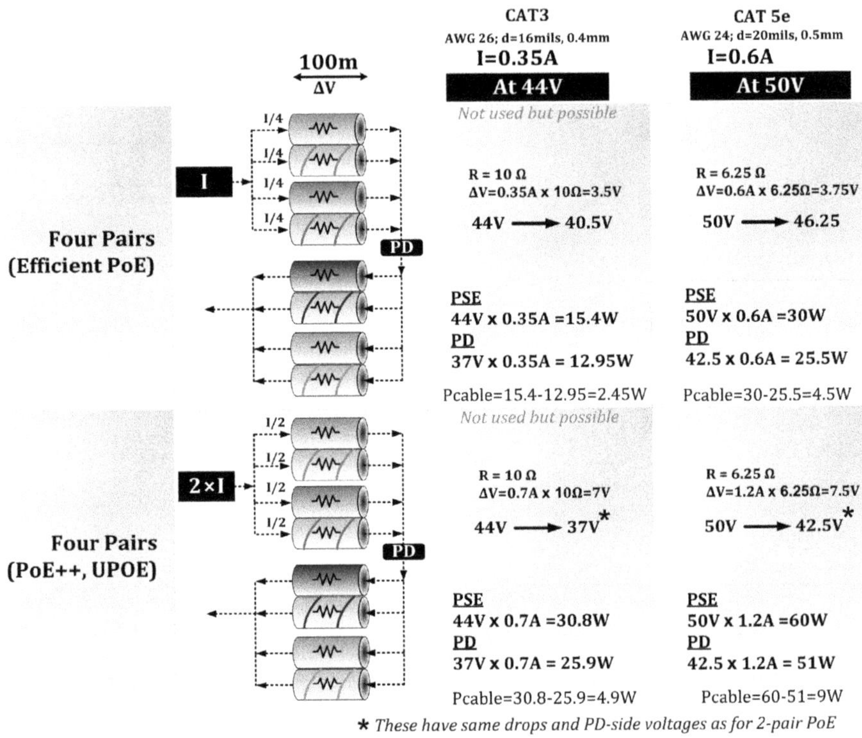

FIGURE 13.3 Power delivery calculations for four-pair operation.

fashion to 300 m of CAT5e. The port voltage is 50 V. The (loop) resistance of 100 m is 12.5 Ω, so 300 m is 37.5 Ω. Passing 0.35 A, the cable drop is $37.5\ \Omega \times 0.35\ A = 13.125$ V. That leaves a voltage of $50 - 13.125 = 36.875$ V at the PD. So the PD power is $36.875\ V \times 0.35\ A = 12.9$ W, as indicated. There are several observations based on this table within Fig. 13.4.

1. It is really interesting that going to 300 m using two-pair power delivery over CAT3, we actually get more power into the PD if we push in only 0.35 A instead of 0.6 A. Using one-pair, we do not even have a solution at 300 m using 0.6 A.
2. Returning to the ungrayed cell, we can confirm that in fact there are two possible solutions for the same PD power (12.9 W), except that this second solution gets enabled only at port currents of 0.983 A. Also the PD-side voltage is then 13.125 V. This corresponds to a huge cable loss of 36.3 W. This is obviously an unsafe condition. However, putting a PD-side UVLO of 30 V (lowest possible value as per AT standard) and current limiting to below 0.983 A will not allow this second solution to exist.

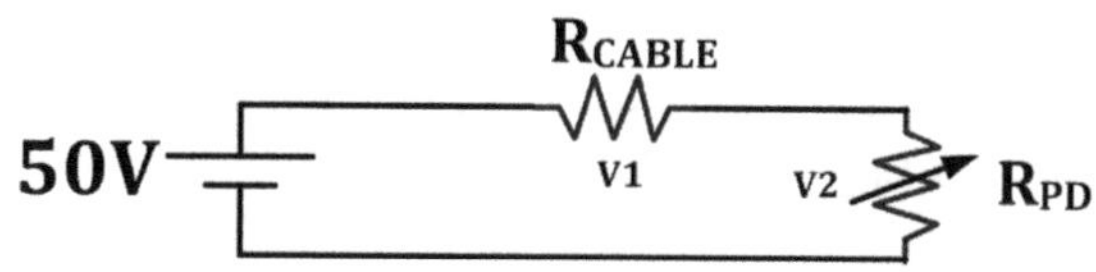

		2-pair 24AWG Cat5e 12.5Ω/100m	2-pair 26AWG Cat3 20Ω/100m	1-pair 26AWG POTs 40Ω/100m
100m	Iport=0.6A	V1 = 0.6×12.5 = 7.5V PD = 42.5×0.6 = 25.5W	V1 = 0.6×20 = 12V PD = 38×0.6 = 22.8W	V1 = 0.6×40 = 24V PD = 26×0.6 = 15.6W
	Iport=0.35A	V1 = 0.35×12.5 = 4.38V PD = 45.62×0.35 = 16W	V1 = 0.35×20 = 7V PD = 43×0.35 = 15W	V1 = 0.35×40 = 14V PD = 36×0.35 = 12.6W
300m	Iport=0.6A	V1 = 0.6×37.5 = 22.5V PD = 27.5×0.6 = 16.5W	V1 = 0.6×60 = 36V PD = 14×0.6 = 8.4W	V1 = 0.6×120 = 72V PD = No Solution
	Iport=0.35A	V1 = 0.35×37.5 = 13.125V PD = 36.87×0.35 = 12.9W	V1 = 0.35×60 = 21V PD = 29×0.35 = 10.15W	V1 = 0.35×120 = 42V PD = 8×0.35 = 2.8W

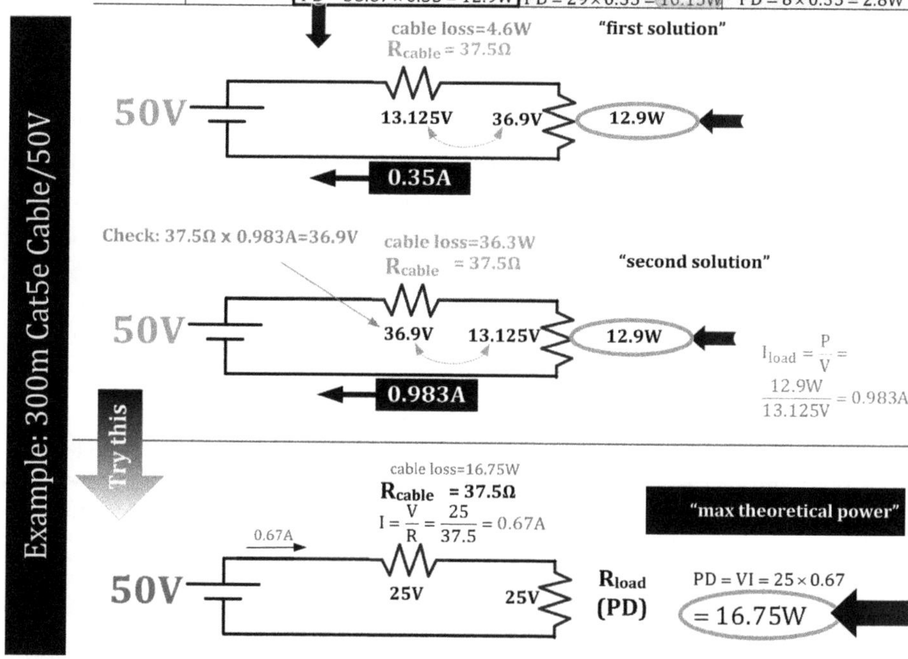

The cable and the PD (load) have the same resistance ⇒ max power delivery

Figure 13.4 Extending the cable length using standard PSEs.

We see that both accurate current limiting and PD-side UVLO are good things to have. But are these set optimally in the standard to make it future-proof? We will discover that is not necessarily so. Not in the case of the UVLO.

One thing is emerging quite clearly: there is much about power delivery we still do not understand. Luckily the second solution will not occur. However, returning to the problem of maximizing power delivery, in the same figure, we throw in another possibility for this 300-m case, one that increases the PD power from 12.9 to 16.75 W. This is a substantial jump. And it can be done with only 0.67 A (still within the capability of a Type-2 PSE), provided we allow the PD to operate down to 25 V. Why is that?

We can do all the calculations we want, but we will finally realize that: *Maximum power delivery occurs when the load impedance (in this case* R_{PD}*) is exactly equal to the source impedance (in this case* R_{CABLE}*).*

In our case, since the source voltage was 50 V, by setting $R_{LOAD} = R_{PD,}$ we got exactly 25 V across each. This is a fundamental property of DC power transfer, very similar to the concept of "impedance matching" in high frequency (AC) circuits to pass the maximum signal across interfaces with no reflections. The same concept applies here too.

Impedance Matching for Maximum Power Delivery

With reference to Fig. 13.5:

- If R is infinite, current is zero, therefore no power is transferred out of the battery and none into R either (since the current through R is zero). *No power into load.*
- If R is 0 Ω, and we somehow pull out 2 A out of the battery, we will get 50 V × 2 A = 100 W out of the battery (max value). Yet *none* of it gets delivered into R, since R is zero and so the voltage across it is also zero. *No power into load.*

This is actually the principle behind a switching converter: Wattage is $V \times I$, so if either is zero, the wattage is zero. We switch the MOSFET (load in our case) between fully nonconducting ($I = 0$) and fully conducting ($V = 0$) states, and in either case there is (ideally) no dissipation.

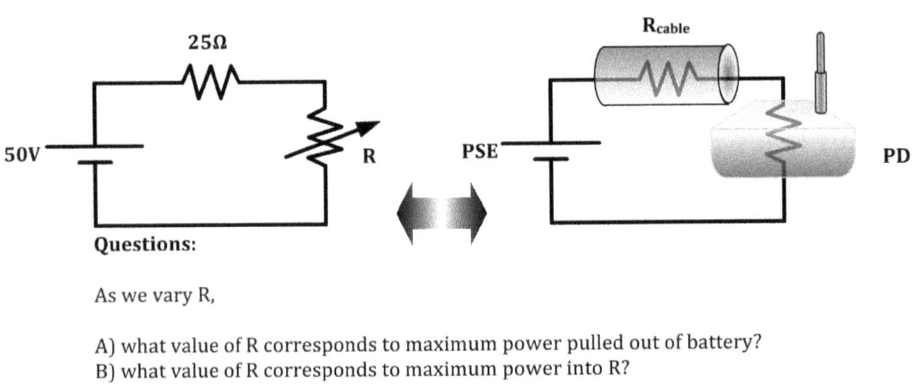

FIGURE 13.5 Maximizing power delivery.

To transfer maximum power we have to maximize not just V or I, but the product $V \times I$.

Suppose R equals 25 Ω, we will get a current of 1 A. The power transferred out of battery is 50 V × 1A = 50 W. Power into R is 25 V × 1 A = 25 W. This is the max power possible into R.

This is the DC (power) version of the usual AC impedance matching principle that we use in high-frequency (AC) transmission line analysis to maximize signals.

This is also the principle that determines the maximum power that can be delivered to a PD (resistance, R) over a long length of cable (100 m of CAT3 is 20 Ω). Assuming the cable (loop) resistance is 20 Ω, max power delivery occurs from a 50-V source occurs when its resistance is also 20 Ω and the voltage across it is also 25 V (equal resistance divider). Therefore, max power that can be delivered is $V^2/R = 25^2/20 = 31.25$ W. The corresponding current is $V/R = 50/40 =$ 1.25 A. Unfortunately the dissipation in the cable at this point is also 31.25 W. It is way too high.

The safe value of current is only 0.35 A as per IEEE 802.3af/at, at which we get a cable loss of $I^2R = 0.35^2 \times 20 = 2.45$ W. For CAT5e, resistance is 12.5 Ω for 100 m, and max allowed current is 0.6 A, which gives a cable dissipation of $I^2R = 0.6^2 \times 12.5 = 4.5$ W. Data collected at Linear Technology suggests that the thermal resistance of a cable bundle in a wall is about 1.8°C/W, so 4.5 W will give a temperature rise of 1.8°C/W × 4.5 W = 8.1°C. This conforms to the typical observed temperature rise as per various experiments conducted by TIA and ISO to support ratification of IEEE 802.3at, and their maximum suggested temperature rise of 10°C to avoid long-term degradation of cabling infrastructure (besides adverse effects on return loss, etc.).

So the theoretical limits on max power delivery are ultimately capped at lower levels by concerns on dissipation in the cables and the recommendation to stick to a maximum temperature rise of 10°C.

The Power Delivery Problem

We are now realizing that for every solution of PD power, we can have another possible solution, which may not get activated for various reasons but is nevertheless present and we therefore need to make sure that we stay out of that, since this second solution is usually very lossy.

In Fig. 13.6, we show that there is a very easy way to "discover" this solution. It is present even with 100-m cable. Let us take the normal IEEE-compliant case of 100 m of CAT5e. We know that we get 42.5 V at the PD side. That leaves 7.5 V across the cable. This is the cased stated in the middle row of Table 13.2 too.

Now flip the voltages so that we now have 7.5 V across the PD and 42.5 V across the cable. To get that voltage in the cable, the current must be $I = V/R = 42.5$ V/12.5 Ω = 3.4 A. This 3.4 A passing through the PD, delivers $P_{PD} = V \times I = 7.5$ V × 3.4 A = 25.5 W. This is the same as

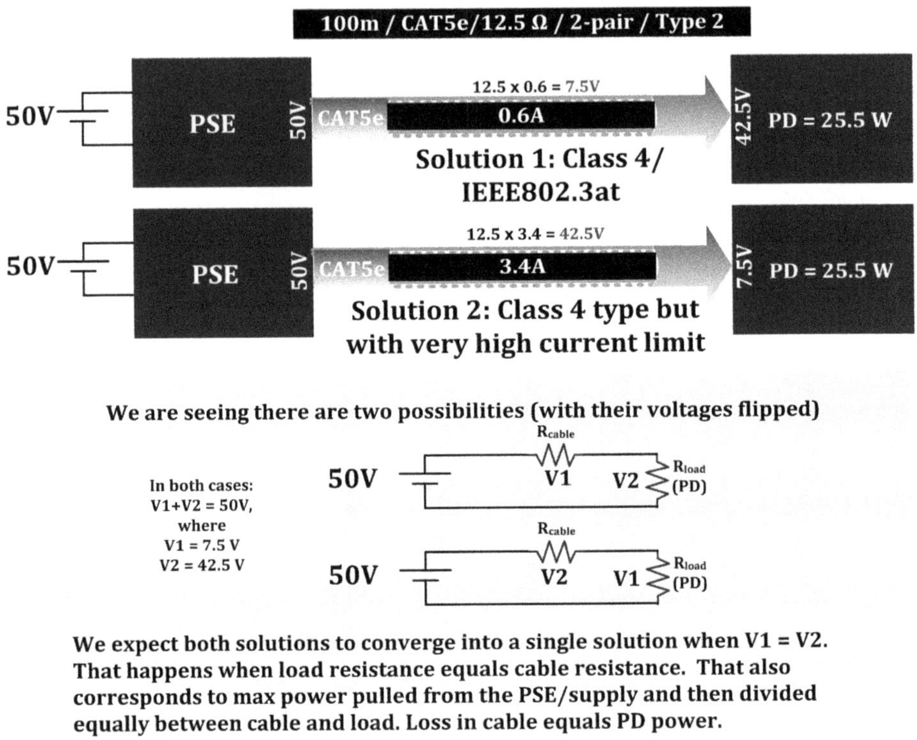

FIGURE 13.6 The symmetrical solution to the power delivery problem.

before, except that we know the PSE will not allow 3.4 A and the PD will not function at 7.5 V. So all the dots are connected now.

Why are we laboring over this? The reasons are

1. Up to 100 m, our regular solution to the power delivery problem is not "close" to the second solution, so we never saw it or realized its presence.
2. However as we increase the cable length and try to maximize the power too, the second solution creeps up very close to the first solution, and we do not want to get inadvertently locked into another solution that we did not even know existed. So we need to *not* let the first solution get too close to the second solution, otherwise the system may also exhibit something similar to oscillations as it *flips between the two solutions*.
3. The maximum power delivery occurs when the two solutions coincide. They become one at that point. After that point no solution exists at all—and that was the reason for the "no solution" annotation in the extreme right-hand column of the table in Fig. 13.4.

Mathematical Solution

We have the input voltage (PSE voltage) dividing between cable and load.

$$V_{IN} = I \times R_{cable} + \frac{P_{PD}}{I}$$

We get a quadratic equation from this

$$I^2 - \frac{V_{IN}}{R_{cable}} I + \frac{P_{PD}}{R_{cable}} \Rightarrow aI^2 + bI + c$$

where a = 1, $b = -V_{IN}/R_{cable}$, $c = P_{PD}/R_{cable}$

$$I = \frac{-b - \sqrt{b^2 - 4ac}}{2a}$$ first solution (lower current, higher PD voltage)

$$I = \frac{-b + \sqrt{b^2 - 4ac}}{2a}$$ second solution (higher current, lower PD voltage)

We will have no solution if b^2 is less than $4ac$. And the two solutions converge when b^2 equals $4ac$. That is when

$$\left(\frac{-V_{IN}}{R_{cable}}\right)^2 = 4 \times 1 \times \frac{P_{PD}}{R_{cable}} \Rightarrow P_{PD} = \frac{\left(V_{IN}/2\right)^2}{R_{cable}}$$

And since $P_{PD} = V_{PD}^2/R_{PD}$, we get $V_{PD} = V_{IN}/2$, and $R_{cable} = R_{PD}$.

In Fig. 13.7 we have plotted this out for 100 m of CAT5e, both for standard two-pair PoE, but also four-pair PoE. We see that if we restrict the port current to less than 1.75 A (the lower threshold of current limit for Type 2 as per the standard, see Fig. 6.2), we only see a part of the entire solution curve.

In all cases, we can confirm that the point of intersection of the two solutions (maximum PD power) is where the PSE voltage is evenly distributed between cable and PD (impedance matching).

On the left side of Fig. 13.7, we have made some conclusions:

1. If we fix the port current to 0.6 A even for four-pair solution (the "Efficient PoE" case in Fig. 13.3), we deliver only 27.5 W – 25.5 W = 2 W more to the PD. But that is certainly 2 W less dissipated in the cable. So based on the 1.8°C/W

measured at Linear Technology (as mentioned previously), this will reduce the temperature of the cable bundle by about 3.6°C. That is not insignificant.

2. However, the main advantage of using four-pairs is that the two-pair cable cannot even achieve over 50 W into the PD (disregarding all safety/temperature aspects for the moment). That is because it is past the point of intersection of the two solutions (in its imaginary solution region). Whereas for four-pairs, we are still well within the 100 W limit of the four-pair case as we can see from its curve on the right-hand plot of Fig. 13.7. So four-pair UPOE (the 1.2-A case) can (safely) achieve just over 50 W into the PD using 100 m of CAT5e. Here *51 W is not even possible using two-pair PoE.* This last sentence is usually not understood very clearly by many engineers.

Lowering the PD Undervoltage Lockout

The AT standard does say that the PD must turn OFF before the port voltage falls to 30 V. Though there is a valid concern if this is an unintended error in the standard as discussed in Chapter 5. That aside, it also does not specify that the PD must continue operating down to at least, say 32 V or even 36 V. And that is the problem, because as we go to longer cable lengths we maximize power delivery *provided we allow the PD to continue to function down to close to half the PSE-side port voltage* (25 V for V_{PSE} = 50 V or 30 V for VPSE ≈ 60 V). In that sense, the IEEE committees seem to have missed an opportunity to future-proof the standard for longer than 100-m cable lengths. The logic should have been based on "impedance matching"—at a DC level too. This is something very basic that is often overlooked.

However, we must keep in mind that getting close to the max power delivery point also has some risks: the "second solution" comes very close to the "first solution," and we must avoid oscillating between two states.

So we may be able to activate a lowered UVLO threshold in a proprietary PD. We may need to do that, in fact, if we want to maximize PoE reach.

Right now, if we assume the PD works only down to 36 V, the solution becomes trivial—by the need to keep the remote end of the cable above 36 V to avoid turning OFF the PD. So , in effect, we have to derate the port currents. Here is a sample calculation.

Suppose we have 300 m of CAT5e cable. The PSE port voltage is 52 V and the PD's UVLO is 36 V. The maximum voltage drop we can tolerate is 52 V − 36 V = 16 V. Since the loop resistance is 3 × 12.5 Ω − 37.5 Ω, we get the current as $I = V/R = 16\text{ V}/37.5\ \Omega = 0.43$ A. This can be supported by a Type 2 PSE. The power into the PD is therefore $V \times I$ = 36 V × 0.43 A = 15.5 W. The PSE-side power is 52 V × 0.43 A = 22.4 W.

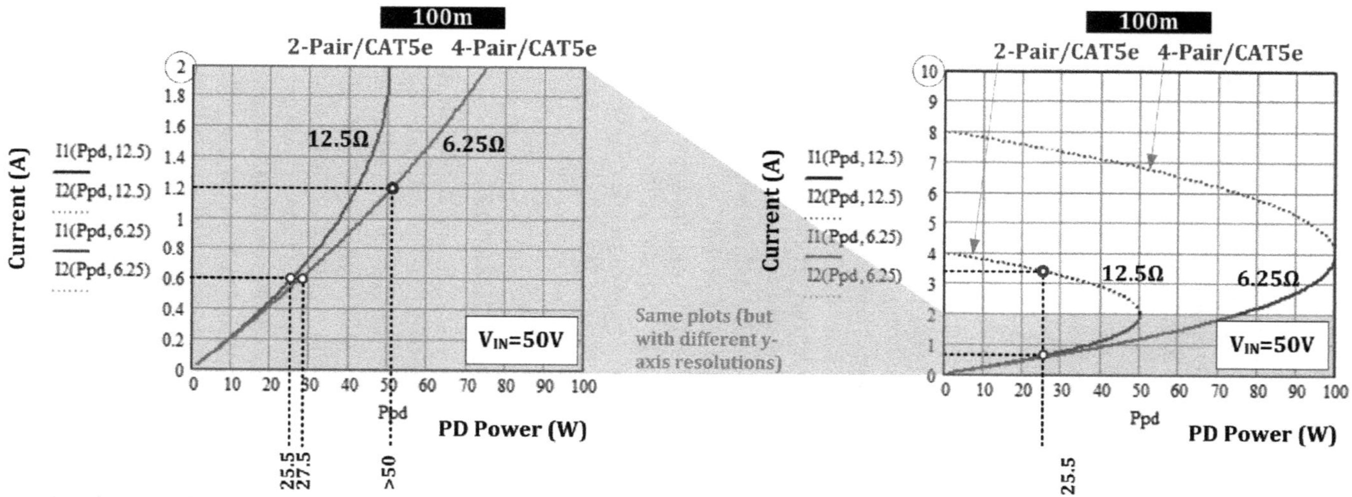

Conclusions: Not much advantage for 25.5W or even 30W PDs in going from 2-pair to 4-pair, and the additional complexity is not worth it. But 2-pair solution cannot even support PDs greater than 50W, whereas 4-pairs can --- at 1.2A, still assumed safe for cable. In fact 4-pairs can go up to 100W, if cable temperature is not the limiting issue.

Conclusions: Even for 25.5W using 2-pairs (Type 2) , there are actually two possible (real) solutions for a given PD wattage! One corresponds to 0.6A, another to 3.4A. The latter corresponds to a cable drop of 3.4 x 12.5 = 42.5V. So the voltage at the PD end is 50 – 42.5 = 7.5V. In general, it is important to have UVLO on the PD side to avoid entering such dissipative modes. But a low-enough current limit on the PSE side will also prevent such spurious high-current modes. The IEEE802.3af normal (protective) current limiting, does double duty here (not recognized as such by the standard).

FIGURE 13.7 Plotting the solutions for the complete power delivery problem.

If we had lowered the UVLO to allow it to operate down to 52 V/ 2 V = 26 V, we can redo the steps above. The maximum voltage drop we can tolerate is 52 V – 26 V = 26 V. Since the loop resistance is 3 × 12.5 Ω – 37.5 Ω, we get the current as $I = V/R = 26\ \text{V}/37.5\ \Omega = 0.69$ A. This is a little too high for a Type 2 PSE. So let us limit the current to 30 W/52 V = 0.58 A now. With 0.58 A, the cable drop is 37.5 Ω × 0.58 A = 21.75 V. So the PD-side voltage is now 52 V– 21.75 V = 30.25 V. This may require a PD with lowered UVLO, but it can certainly be supported by a Type 2 PSE. The power into the PD is therefore $V \times I = 30.25\ \text{V} \times 0.58\ \text{A} = 17.5$ W. The PSE-side power is 52 V × 0.58 A = 30 W.

In other words, we have increased the power into the PD from 15.5 to 17.5 W. This 2 W extra may be able to support more applications or features at the end of 300 m. All we have to do is stay flexible about the UVLO threshold of the PD.

For quick reference, we have provided power-delivery curves in Fig. 13.8, using existing AF and AT PSE's and a PD set at the bare-minimum UVLO of 32 V. Note that RG-6 coaxial cable is actually a generic type of cable, and comes in many different variations, mainly related to the AWG and type of shielding. We have assumed the worst case for power delivery estimates, by using the thinnest wire gauge and the most sparse shielding found across several RG-6 products surveyed in the marketplace. Keep in mind that coaxial cable, if used, is applicable only to single-pair power and data delivery.

In Fig. 13.8, the curved portions in the lower half, where the curves coincide are based on the current derating requirement mentioned above—to keep the voltage at the PD end above the UVLO level. So *at that point it really does not matter anymore whether the PSE being used is AT (capable of 600 mA) or AF (capable of 350 mA).*

Plotting Power Delivery Curves over Long Distances

Finally, we plot out both solutions of power delivery for various cases. In Fig. 13.9, we have the solutions for two-pair operating using CAT3. In Fig. 13.10, we have the solutions for two-pair operating using CAT5e. In Fig. 13.11, we have the solutions for four-pair operating using CAT5e. In Fig. 13.12, we have the solutions for one-pair operating using CAT3. Note that in each case, on the lower side we have a dark-gray highlighted region. This corresponds to existing equipment (with appropriate "safe" current limiting and a average UVLO of 36 V). Right above it is a region in light gray that shows how the power-delivery region expands just by allowing the UVLO to go to 25 V. There is no compromise in safety because the current limits are intact. Outside both these regions, we have the remaining theoretical curves with no restrictions on current or UVLO. The upper triangle has the "second solution" curves and is basically just a "keep-off" region.

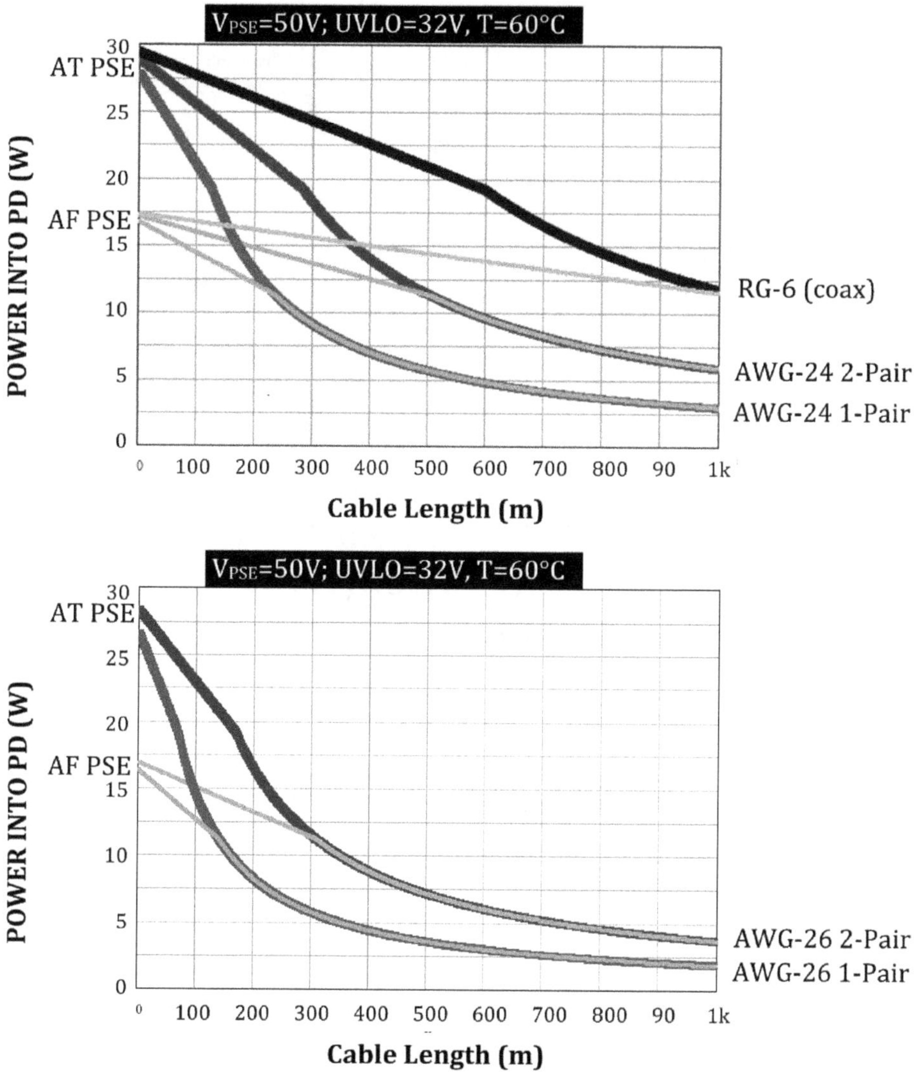

FIGURE 13.8 Plotting the power-delivery curves using existing (IEEE-compliant) PSE-PD.

Sample Numerical Calculations for N-Power Delivery

Let us do some simple calculations to illustrate the principles we have learned. Let us stick to *single-pair* here.

Case 1: No PSE-side current limit, no PD-side UVLO

This would be the maximum theoretical power delivery. Suppose we have *300 m of single-pair CAT3-type cable* (AWG 26). We can take the

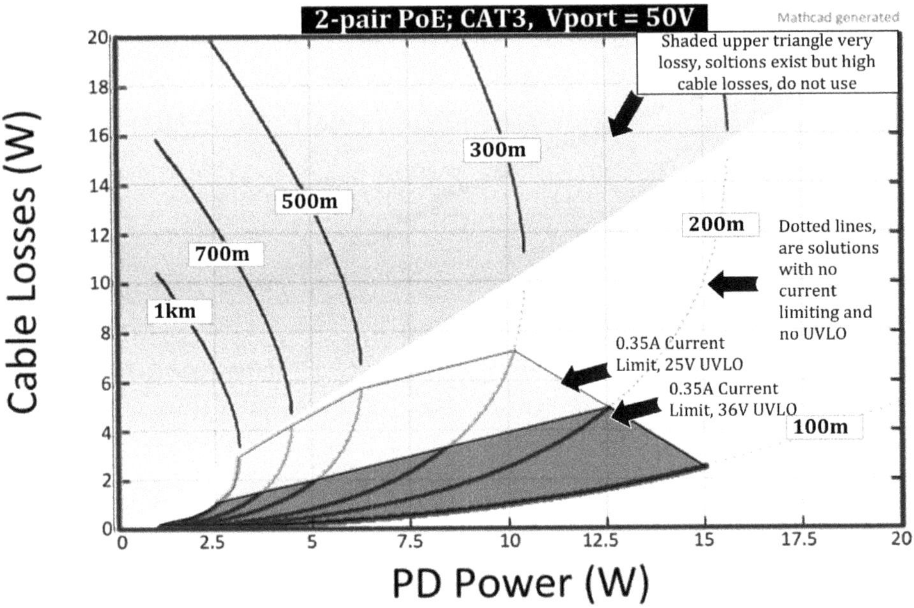

FIGURE 13.9 Two-pair PoE solutions using CAT3 cable.

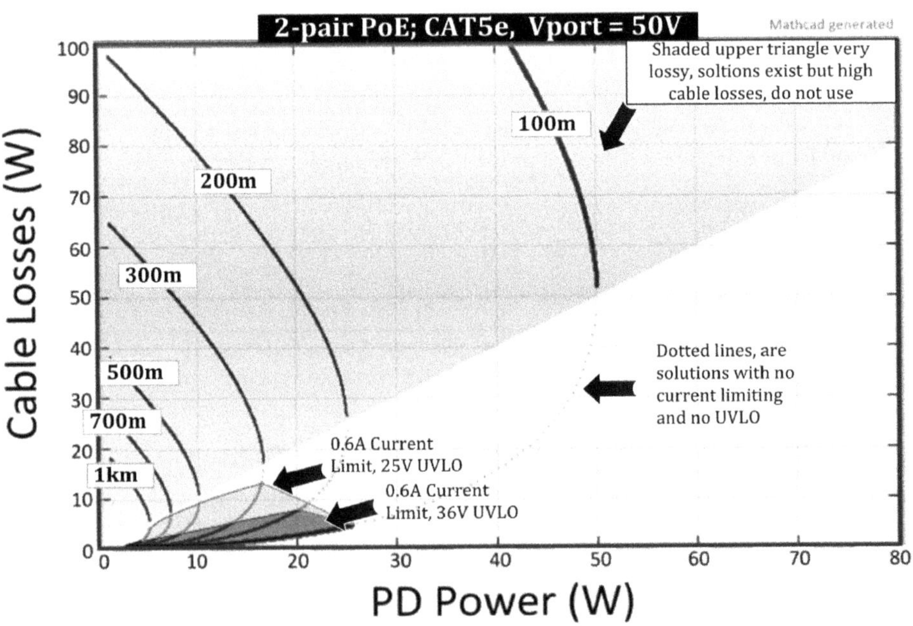

FIGURE 13.10 Two-pair PoE solutions using CAT5e cable.

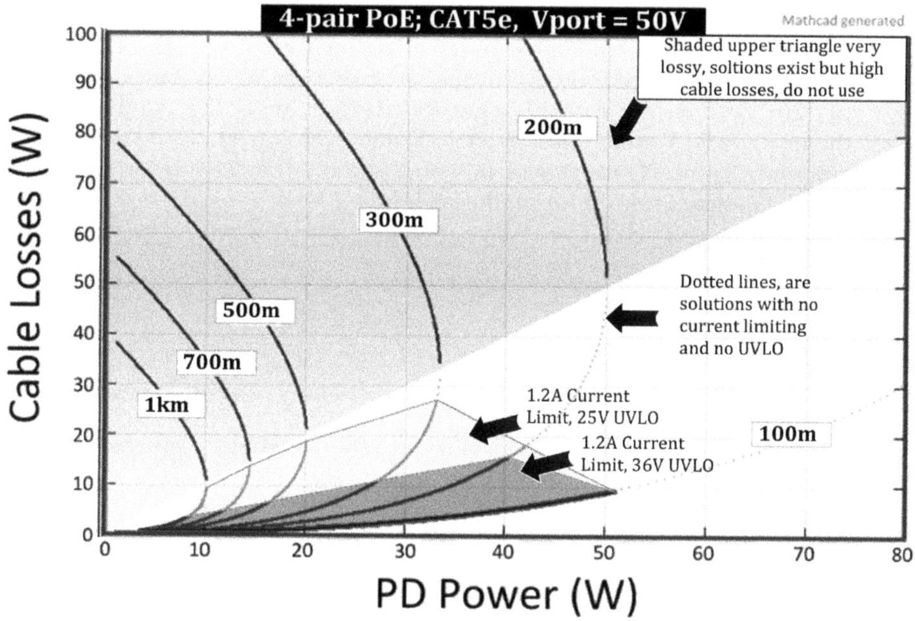

FIGURE 13.11 Four-pair PoE solutions using CAT5e cable.

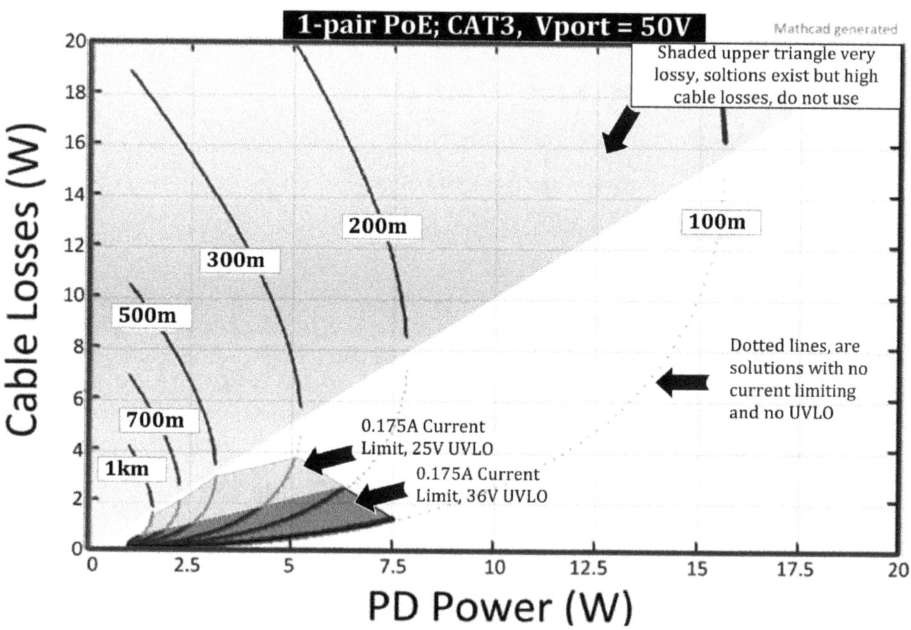

FIGURE 13.12 One-pair PoE solutions using CAT3 cable.

two-pair resistance either as 20 Ω/100 m, which actually includes patch cables and several contact resistances. Or we can take 15 Ω/100 m for AWG 26. Let us take the former number just to give us some margin. So, for single-pair, we get 40 Ω/100 m, or 120 Ω/300 m. If the input is 50 V, maximum power transfer occurs when cable resistance equals the PD resistance. In that condition, the PD-side voltage is 25 V. So the current through the cable is

$$I = \frac{V}{R} = \frac{25}{120} = 0.21 \text{ A}$$

Therefore the cables/PD losses are

$$P_{PD} = \frac{V^2}{120} = \frac{25^2}{120} = 5.2 \text{ W}$$

This coincides with the point where the dotted and solid lines meet (border of gray triangular area), in Fig. 13.12. That is the maximum theoretical power.

Now, the IEEE standard asks us to limit the current to 175 mA per pair for CAT3. The 5.2 W number in the equation above was achieved by forcing 210 mA into it. Let us limit the current next.

Case 2: PSE-side safe current limit, no PD-side UVLO

Now we first calculate the voltage drop across the cable, using the safe current limit of 175 mA.

$$V = I \times R = 0.175 \times 120 = 21 \text{ V}$$

So the voltage at the PD end is

$$50 - 21 = 29 \text{ V}$$

The PD-side dissipation, therefore, is

$$P_{PD} = V \times I = 29 \times 0.175 = 5.075 \text{ W}$$

This coincides with the point where the light-gray pentagonal area ends (on the 300-m line), in Fig. 13.12. But the PD-side voltage is less than 36 V (IEEE-type PD). So we now limit the PD-side voltage too.

Case 3: PSE-side safe current limit and PD-side UVLO

The voltage drop across the cable must be 50 V − 36 V = 14 V. This determines the current

$$I = \frac{V}{R} = \frac{14}{120} = 0.117 \text{ A}$$

Therefore PD-side dissipation is

$$P_{\text{PD}} = V \times I = 36 \times 0.117 = 4.2 \text{ W}$$

This coincides with the point where the dark gray-pentagonal area ends (on the 300-m line), in Fig. 13.12.

How Far Will a Given PD Operate?

We can ask the reverse question: we have a certain PD with a certain requirement in terms of watts. How much distance can we operate it at using a one-pair configuration, with AWG 24 or AWG 26? We provide these "constant power" design curves in Figs. 13.13 and 13.14. Note that we have taken the basic resistance values calculated in the first sections of this chapter, rather than buttressing those with connector and patch cord resistances. Also, *these are purely theoretical curves which still do not tell us what the required current is (whether it is "safe" or not) and what the voltage at the PD end is.* Once we fix a point, however, we can easily work all that out as described in this chapter, and ascertain its validity based on available PSEs and cabling.

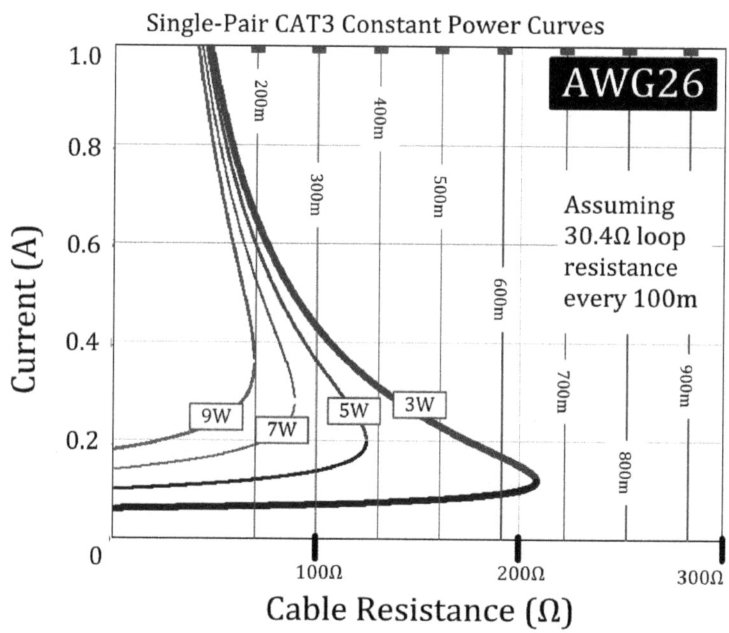

FIGURE 13.13 How much distance can be theoretically attained for a given PD wattage using one-pair configuration (AWG 26).

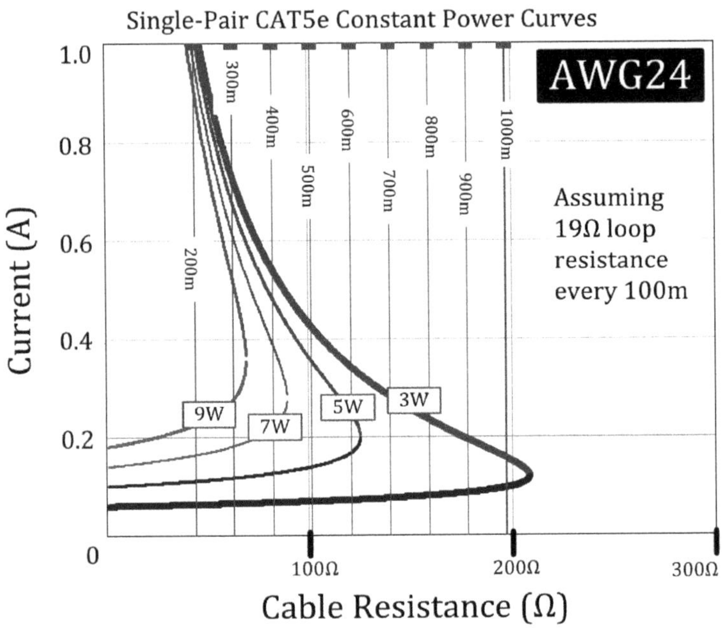

FIGURE 13.14 How much distance can be theoretically attained for a given PD wattage using one-pair configuration (AWG 24).

Learning from Telephony

There are so many lessons we have, and can still learn, from techniques used since a century ago. One question as we struggle to exceed 100 m (330 ft) in Ethernet is: in POTs (plain old telephone system), how did we ever achieve up to 5.5 km runs using low-grade (AWG 26), single-pair wire?

First, the current consumption was very low. A typical old nonelectronic (carbon microphone based) phone could work down to at least about 20 mA of current (with no connection to the AC mains of course). In fact, Bell Telephone specifications stated that *three* telephones should work in parallel on a 20-mA loop. Note that European telephone companies often state that phones working in parallel is "technically impossible" and is therefore discouraged. But most of their telephones do actually work in parallel.

Let us do a simple calculation here: 100 m of AWG 26 (CAT3) Ethernet cable has a loop resistance of 20 Ω. So single-pair 100 m is 40 Ω. Therefore, 5 km of single-pair AWG 26 would have a loop resistance of 50 × 40 = 2000 Ω. Passing 20 mA of current produces a cable drop of 20 mA × 2000 Ω = 40 V. That still leaves a voltage of about 48 V – 40 V = 8 V at the subscriber end. Old phones could easily work down to at least 3 V. And in fact, carbon mic (transmitter) telephones continue to work down to a fraction of a volt. In effect, they had no UVLO. That is something we have been focusing on in the previous sections.

With the advent of electronics, the situation changed, at times in the *wrong* direction. For example, electronic phones (those with no connection to the mains) could barely work below a certain threshold of around 3 V. Even where they did work, they exhibited the "cliff effect," whereby they abruptly stopped working when the line voltage fell below the critical level. In particular, this meant that one telephone on a party line may tend to hog all the line current, cutting others off. With older carbon microphones, all receivers on the same line would still operate, just with reduced output.

Another great reason for the old successes was that old telephones had no bridge rectifiers (with wasted voltage drops) at their input—they worked with either polarity. Keep in mind that IEEE 802.3at provided for a 2-V diode bridge offset.

In Fig. 13.15, we present one of the oldest schematics to explain telephony as it developed. In looking at it carefully, we see how signal and supply were coupled on a single twisted pair. And that is the inspiration behind using a single-pair Ethernet cable for power. We no longer need center-tapped transformers (at least not two). With a single transformer, with no center-tapping, we can couple the "48V" supply on to the single-pair data line, using two inductors and two blocking caps (to establish symmetry). See Fig. 13.16 for an implementation of one-pair Ethernet, based on a century-old technique in telephony. Since center-tapping cannot be used on one-pair systems, we need two capacitors to prevent the DC from flowing into the transformer and burning it out. There are also two inductors to prevent the PSE from shorting out the data lines from the signal viewpoint. Note that (all) the paired inductors can be wound on a single magnetic core as indicated in the blurb. But note the polarity dots: this is not a common-mode filter. It is actually two differential-mode filters on a single core. And there is no flux cancelation from the forward and return currents of PoE source. In fact, the flux reinforces so the core volume must be large enough to not saturate the core. Yes, we can also add a common-mode filter to this configuration. That can be small. This is all basic magnetics at work.

In the same figure we show an alternative to center-tapping on two pairs! However now the current from the PSE to the PD flows in two strands that are *not* part of the same twisted pair. In general, the imbalance in resistance between twisted pairs is much worse (~8 percent) than the imbalance between two strands of the same twisted pair (~5 percent). So the currents will not divide as well as in the traditional method of PoE center-tapped injection. Note that we have one bridge rectifier for *each* pair now, whereas in the traditional method there was one for the data pair and one for the spare pair.

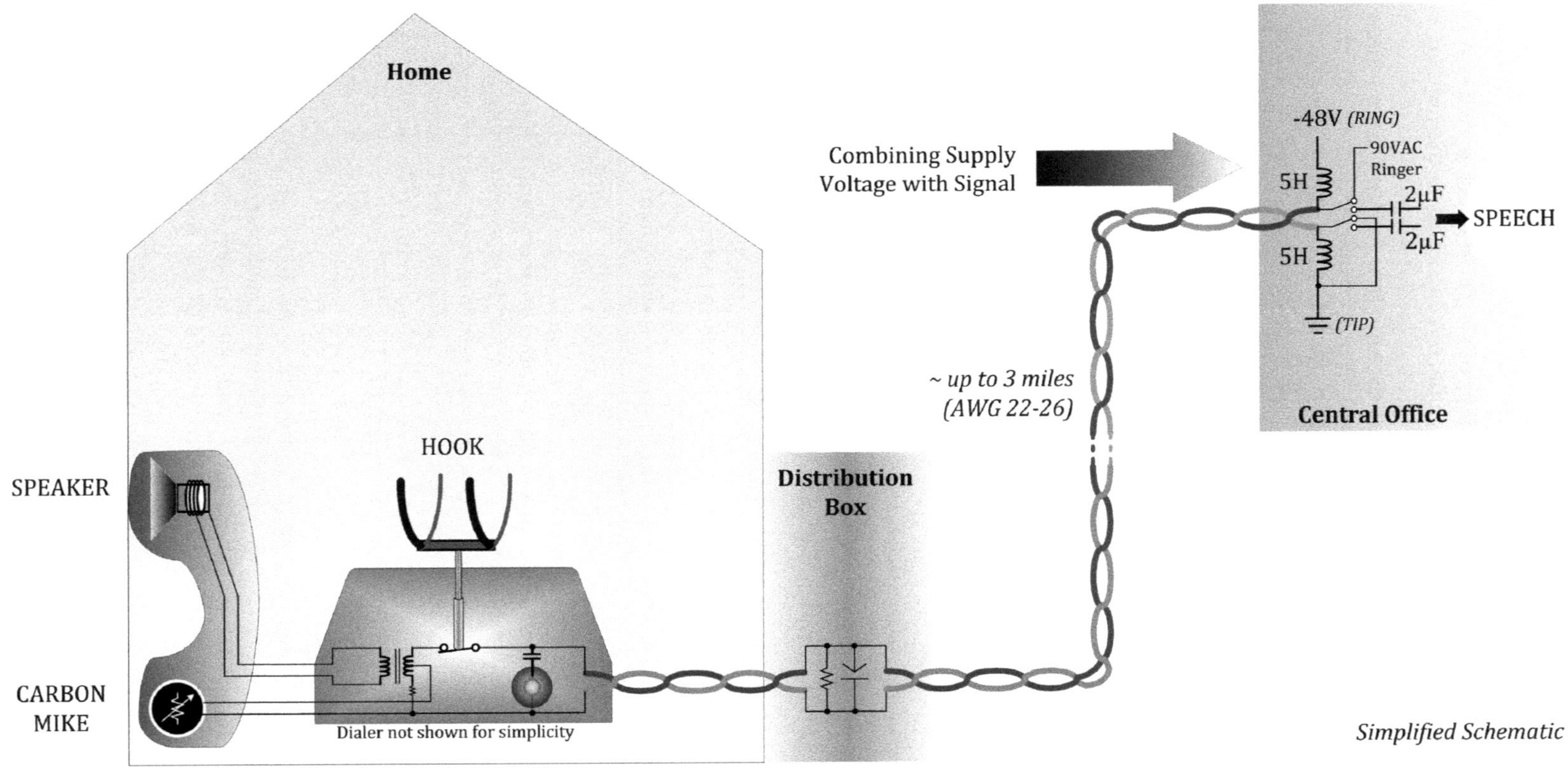

FIGURE 13.15 An old telephony circuit showing how to combine DC and signal on a single pair of twisted cable, just as we can do in one-pair Ethernet.

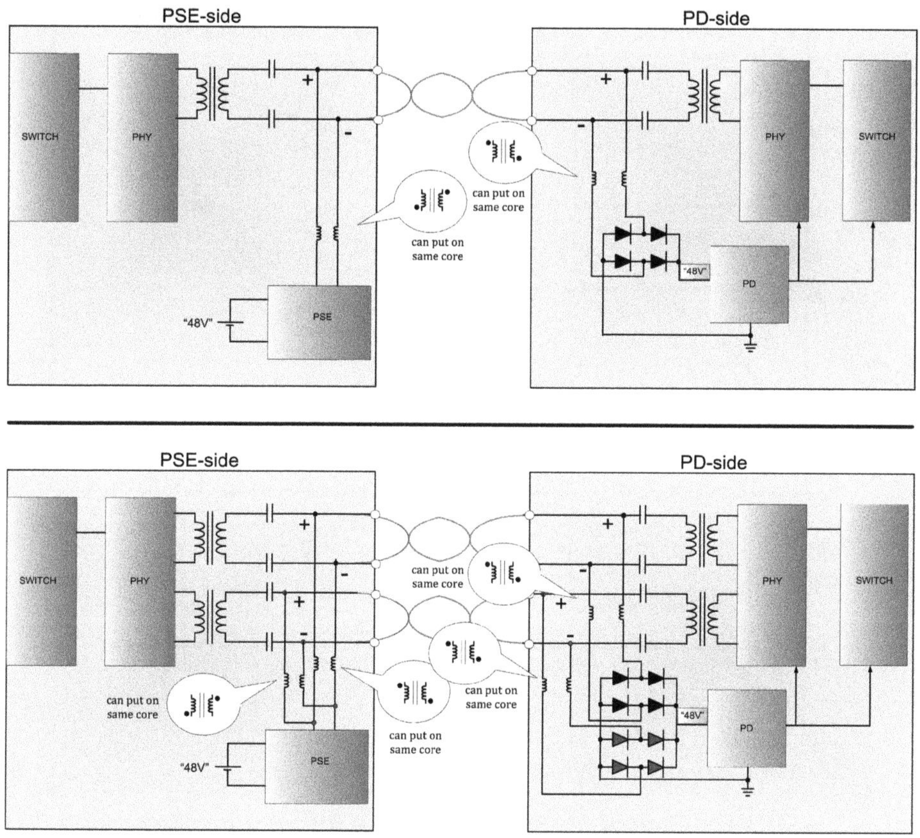

FIGURE 13.16 Implementing PoE on one-pair and an alternative to center-tapping in two-pair Ethernet (not good sharing though).

Four-Pair Implementations

In the AT standard, Section 33.1.4.1 titled "Type 2 cabling requirement" says:

> ...Under worst-case conditions, Type 2 operation requires a 10°C reduction in the maximum ambient operating temperature of the cable when all cable pairs are energized at I_{Cable} (see Table 33-1), or a 5°C reduction in the maximum ambient operating temperature of the cable when half of the cable pairs are energized at I_{Cable}.

So though the AT standard left a "hook" for four-pair operation, that is all it did. Market forces are however already at work and we have UPOE, HDBaseT among others, driving a lot of power into the cables. UPOE for example is basing its abilities on the temperature

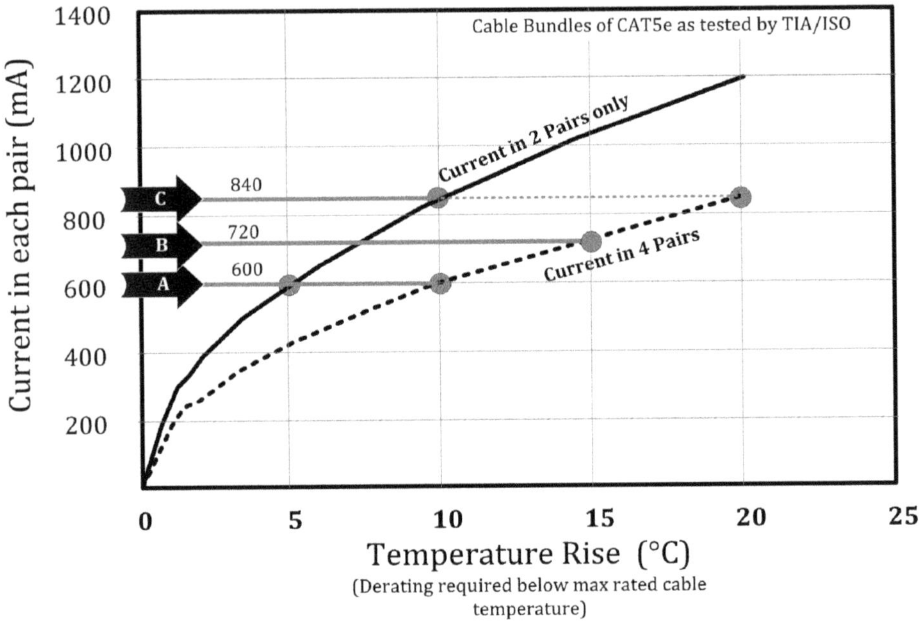

FIGURE 13.17 Temperature data collected on CAT5e cable bundles (TIA/ISO).

data collected by ISO/TIA as described in Fig. 13.17. We have placed three marked arrows that we now discuss

1. The guiding rule now is to accept a 10°C rise in temperature in the interest of maintaining life of the cabling infrastructure. So 600 mA was selected. However, in reality that was 600 mA passing through all four pairs (in PoE that would amount to 1200 mA forward current through two pairs and 1200 mA return current through the remaining two pairs). That is what leads to 10°C rise. Currently we use 600 mA through only two pairs and that actually causes only 5°C rise (data from others like Siemon seems to indicate 7°C to 8°C rise for two pairs and 10°C for four pairs). So UPOE increased the current to 1.2 A, which at 50 V is 60 W at the PSE end.
2. During the deliberations leading up to the AT standard, there was talk about accepting 15°C rise. That is produced by 720 mA (though through four pairs). So the draft AT standard had temporarily fixed 720 mA for high-power PoE, which was 50 V × 0.72 A = 36 W, instead of 30 W. Later, it was felt that 15°C was too much, and they went back to 10°C.
3. At some stage there was talk about high-power PoE being 840 mA. That was based strictly on data in which only two of the four pairs had 840 mA current through them (one pair

forward current, one pair return). Yes and that caused only 10°C rise, not 15°C. But it effectively closed the door on using existing PSEs for four-pair operation because if someday someone tried to pass 840 mA through each of the two pairs of the four pairs, the temperature rise would be something over 20°C. So it was not future-proof and was dropped.

Another high-power market drive standard is PoH, standing for Power over HDBaseT (also called HDBT). The spec was released in September 2011 by a consortium including Sony, LG, Samsung, and Valens Semiconductor. Other alliance partners include PoE pioneers Microsemi (formerly PowerDsine). PoH seeks to avoid worst-case estimates of temperature rise as per IEEE/TIA, but does use their data for a *smaller number of cables/bundle*. But it is still restricted to 100 m. In Fig. 13.18 we have summarized the losses of the present and evolving standards. We have not included single pair here as that is not meant for 100 m (usually), and its wattage depends so much on distance (length of cable).

All of the evolving high-power standards have an Achilles heel: *imbalance in currents* carried between pairs. As mentioned previously,

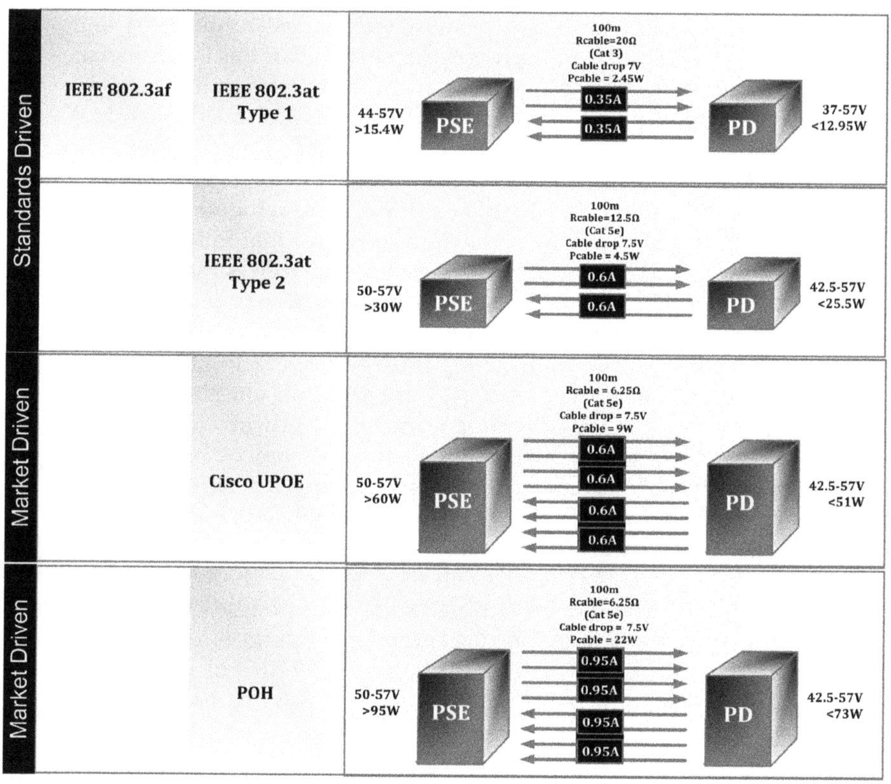

FIGURE 13.18 Summary of present and evolving standards (over 100 m).

the imbalance in resistance *between twisted pairs* is much worse (~ 8 percent) than the imbalance *between two strands of the same twisted pair* (< 5 percent, typically 3.5 percent). We should also consider what may happen if one of the paralleled pairs develops a fault and all the current gets diverted into the unbroken pair. We need to be able to detect severe imbalances including faults too.

In Fig. 13.19 we have a generic four-pair possibility. It is obviously not as per the standard, but is meant to conform to the overall "spirit" behind the standard. Note that a given PSE (say PSE_1) will turn on power, which will then charge up the bulk cap (input of the DC-DC converter). The charge from this can however flow backward through the body diode of the pass-FET of the unpowered PD (dotted arrow) (and also through its ESD structures through other pins, if connected together from the two PDs), charging up the 0.1-μF port cap at its input (near the RJ-45), and finally reverse-biasing its bridge rectifier BR_2. When that happens, PSE_2 will be unable to detect the 25-k resistor of PD_2 and will not power up. We will be left with two-pair operation. To avoid this phenomenon, OR-ing diodes will need to be added as shown.

Admittedly, this figure hides more than it reveals. There are several proprietary detection (and classification) techniques around to try and inform the PSE and PD at the physical layer itself that this is a four-pair configuration. Some are using "three-event" classification for example.

Note that to try and reduce imbalances and also to aid in mutual identification, it is normal to *not* connect GND_1 and GND_2 in Fig. 13.19. After passing through the PD interfaces, however, the power grounds (PGND_1 and PGND_2) are connected and together form the lower rail (ground plane) of the DC-DC converter that follows.

To ensure active current balancing, there are others that take two DC-DC converters, and combine their outputs using load-share ICs based on the historic industry workhorse: UC3907 from TI.

Sometimes, people try to create four-pair operation by sending power over the data pairs (as in a normal PoE-enabled switch), and *also* trying to make a Midspan unit inject power via the spare pairs. However, this greatly exaggerates imbalances because the length of the cables connecting the switch and the Midspan may be very different. Also, their "48V" rails may be very different. If we do the math carefully we will see that even a couple of volts difference in their "48V" rails can lead to huge changes in their current distribution. So the simple "rule" that people often use of doubling the power just because number of pairs is doubled, is clearly wrong or overly optimistic. *Only if 100 percent current balancing is assured*, and that *cannot* happen merely by passive techniques (it requires active current sharing), can we say that we will get 2 × 25.5 W = 51 W at the PD end after 100 m of cable. Looked at it in a reverse manner, if we cannot promise 51 W into the PD, surely the PSE can't be supplying 60 W either. So, without a lot of additional circuitry to stand behind it, the "51 W to PD" or "60 W from PSE"

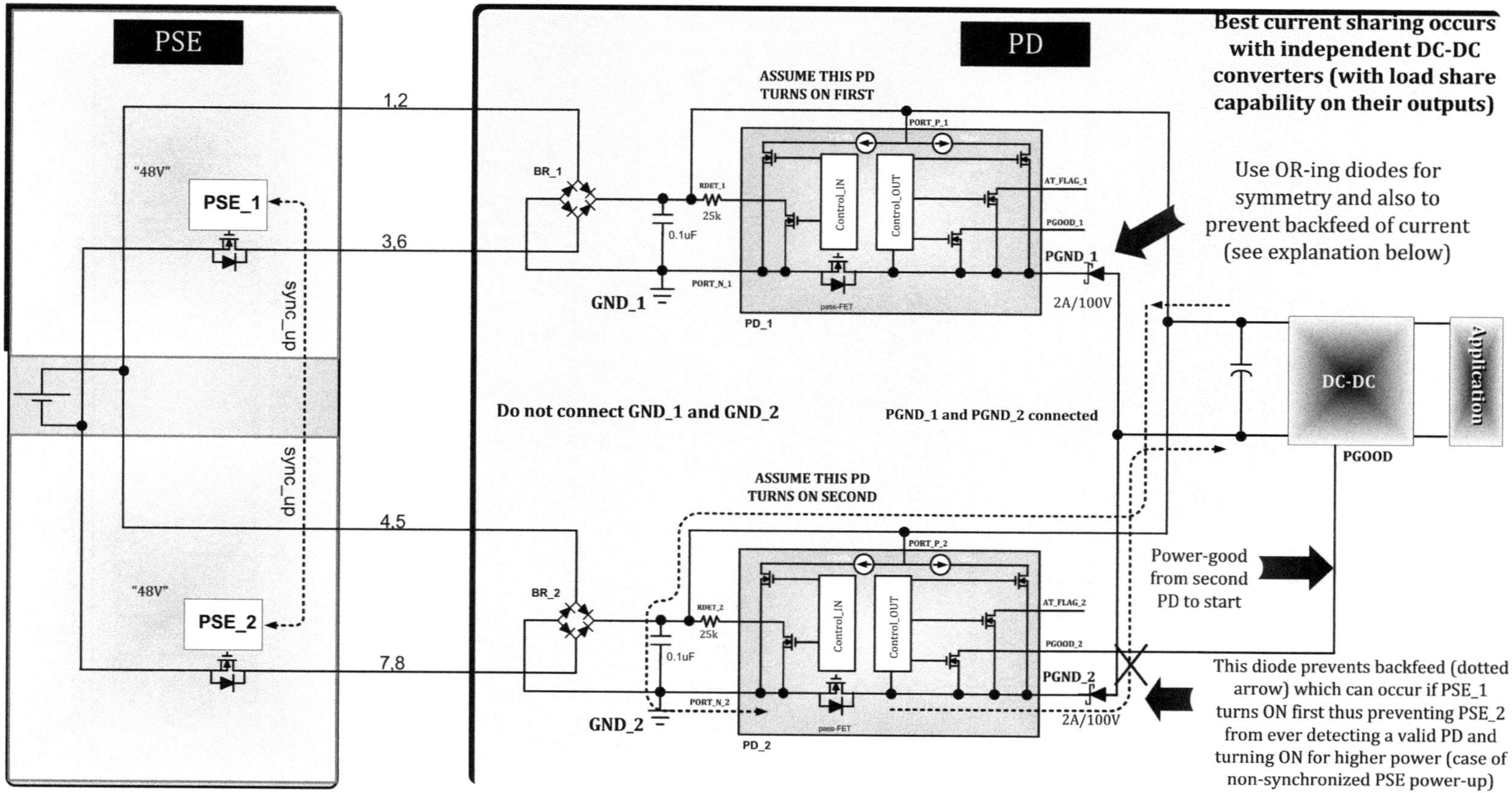

Figure 13.19 Four-Pair PSE-PD configuration.

claims are just marketing. More power, yes. How much more? That is not fully answered yet.

One of the things we must continue to keep in mind is that in PoE, PD power is the power entering its RJ-45. How much useful power this eventually gets converted to depends on the losses in the PD, and that not only includes the losses in the pass-FET of the PD and in its switching converter, but also the input bridge. In two-pair operation, though we still had two bridge rectifiers, only one actually conducted current. In four-pair operation, both the diode bridges conduct. Though the energy loss is still proportional to the total power, the dissipation in two bridges adjacent to each other on the PCB can be very high, and from a thermal standpoint, that is not very good. So a lot of effort is going into minimizing the bridge rectifier losses. One option is to use 100-V Schottky bridges. Another is to use P-FETs in what is called an "active bridge." It can be easily done with discrete components (see AN4006 from ST Microelectronics), but IC-based (monolithic) active bridges are also expected to appear soon. Keep in mind however, that most PD chips are designed with detection and classification thresholds based on assuming a 2-V drop from the diode bridge. So their thresholds are literally translated from the IEEE values. However, when we use an active bridge, the diode drop in now negligible and therefore serious interoperability issues can arise from that.

Future Innovation

Just as we thought we had figured it all out, we come across Fig. 13.20, a development based on the phantom circuit principle we introduced in Chap. 1. It is a DSL-based idea from 2-Wire Inc. (US Patent Number

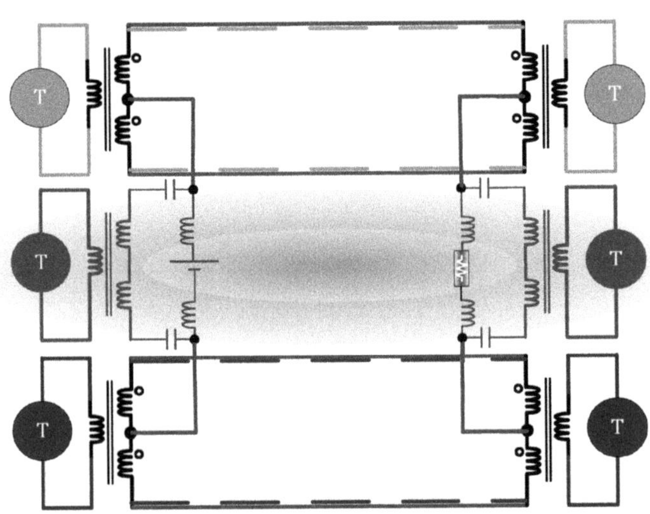

FIGURE 13.20 A phantom inside a phantom (and with PoE).

7190716, inventors Andrew Norrell and others). Note that an additional phantom-data channel has been created on two twisted pairs. PoE can be injected with a combination of inductor-feeding (as in one-pair PoE) and the usual center-tapping method. This can also be extended to four PoE, providing six (not four) data channels.

Future innovation will draw on techniques and tricks used in the past century. For example in Fig. 13.21, we have used the well-known "hybrid transformer" or auto-multiplexer, used in traditional

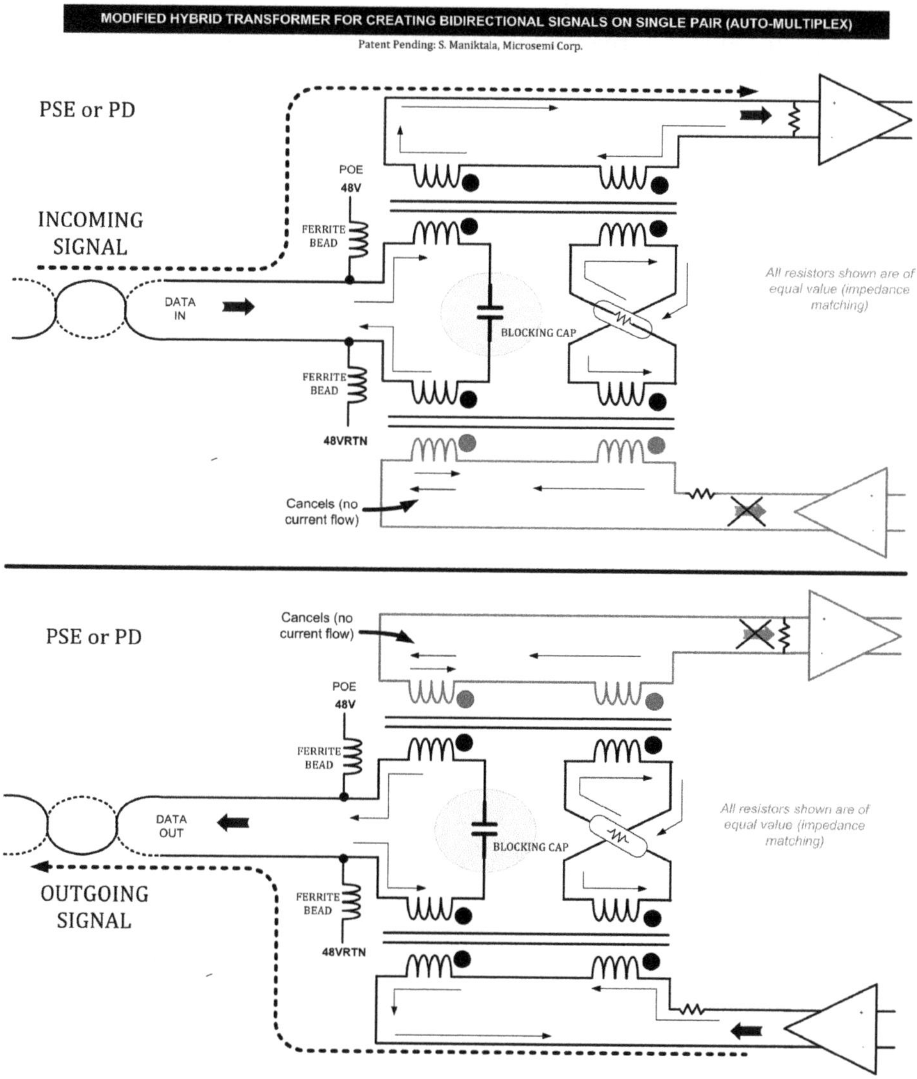

FIGURE 13.21 A "hybrid transformer" modified for injecting PoE.

telephony decades ago. Modern "hybrid" circuits that connect a single twisted pair to a PHY transceiver are basically electronic circuits that simulate the original behavior of this hybrid transformer. We can see from the figure that an outgoing signal does not reach the adjacent receiver. On the other end of the line however, it becomes an incoming signal, and gets automatically directed to the receiver on the other side. Similarly, when the transmitter on the other side sends a signal, it does not go the receiver on that side, but to the receiver on the opposite end of the cable. In effect, we have bidirectional communication over a single pair (multiplexing). The author added one blocking cap to prevent the DC from the PoE from burning out the transformers, and also used ferrite beads for injecting PoE. Though the symbol for a bead is the same as an inductor, beads actually use high-frequency *resistance*, not inductive impedance. They achieve *better* results in this application (no spurious oscillations of LC-tank circuit) as compared to inductor-injection of PoE commonly used in one-pair.

CHAPTER 14

Auxiliary Power and Flyback Design

Overview

At the end of this concluding chapter, we will carry out a sample design of a flyback regulator. But this book will not teach switching-power conversion from scratch. Because that is a complex area in itself, deserving of an entire book. We recommend that the relatively inexperienced reader refer to *Switching Power Supplies A-Z*.

In the initial part of this chapter we will discuss another very important PD systems issue: that of auxiliary power sources (e.g., wall-adapters) in PDs.

Many PoE applications employ auxiliary power sources, typically an AC "wall wart" (or solar cell) connected to the Powered Device (PD). Integrating auxiliary power can be a very challenging design task and the PoE designer must understand the various methods and inherent tradeoffs that exist with each method of implementation.

Three configurations are commonly used to add auxiliary power to PoE systems. See Fig. 14.1 for broad, introductory schematics of each. Note also how we can always go from a configuration with OR-ing diodes placed on the upper (positive) rail, to a configuration with diodes attached to the lower (return) rail. Both are equivalent methods, and the underlying logic presented below, remains the same for either configuration.

Option A: at the PD's front end (before the pass-FET). Often called the Front Aux technique, or FAUX Pin method. There are actually several suboptions here as we will discuss.

Option B: at the input of the PD's switching converter (behind the front end). Often called the Rear Aux technique, or RAUX pin method. Note that in this case, there is, in particular, an additional suboption that includes an additional OR-ing diode (marked "optional" in Fig. 14.1). The resulting behavior gets somewhat changed as discussed later.

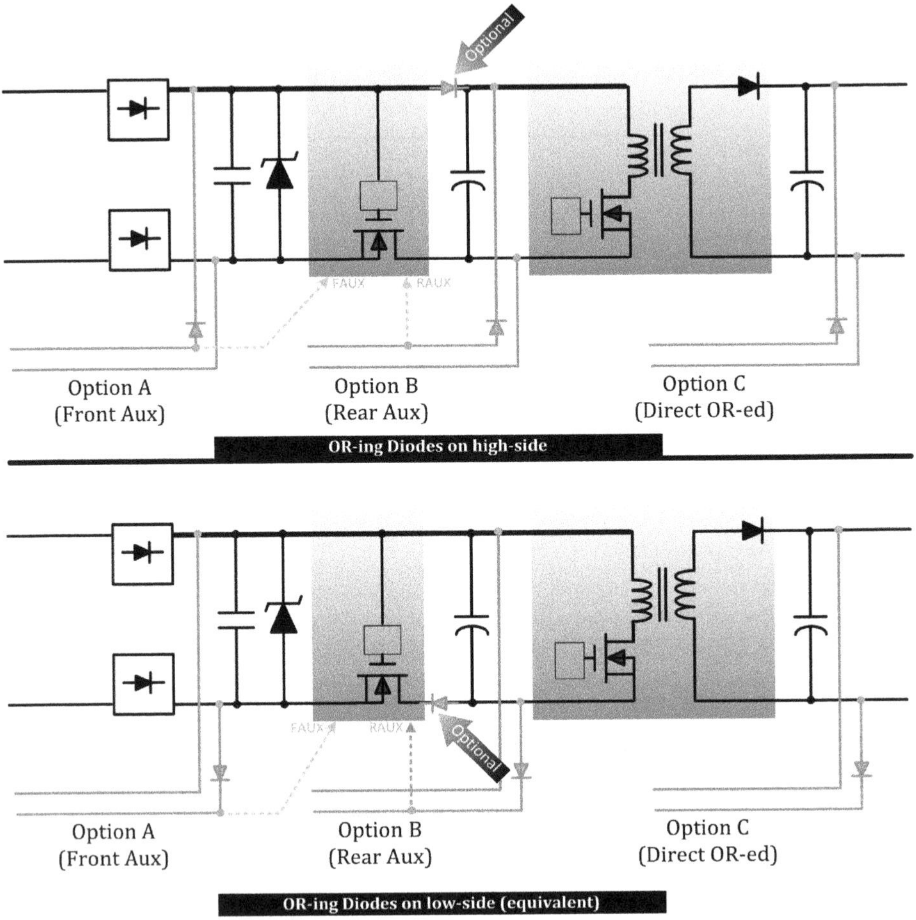

FIGURE 14.1 Adding auxiliary power.

Note that both Option A and Option B (the latter without the extra diode suboption) *share the upper rail*. So the schematic difference is really only where the return wire of the adapter is connected: on the Drain of the pass-FET (Option B) or the Source of the pass-FET (Option A).

Option C: at the PD's switching converter's output voltage. We can call this direct OR-ed or output OR-ed method.

Each of the above methods has its pros and cons, discussed as follows. As mentioned, the OR-ing diodes can be placed either on the high side or low side as indicated, it changes nothing about the logic presented below.

Auxiliary Power Option A (Front Aux or FAUX Pin Method)

The behavior and the requirements depend on the following cases.

V_{AUX} within PoE Range

In the following discussion, pay close attention to Fig. 14.2. In general, this method is said to support only V_{AUX} between 42 and 57 V. But there are ways around that restriction as discussed here.

It is clear that if V_{PSE} is present (i.e., between 44 and 57 V), V_{AUX} needs to be greater than V_{PSE} for the OR-ing diode D1 in Fig. 14.2 to conduct. Therefore wall-adapter priority, something that is usually preferred in applications, is *not* assured. V_{AUX} will dominate only if it is the *larger* of the two. And when that happens, it will cause D1 to conduct. But if V_{AUX} isn't larger than V_{PSE}, D1 will not conduct, and the application will continue to be powered from the PSE rail.

However, if V_{PSE} is not present initially, or it momentarily drops out for whatever reason, V_{AUX} will then be able to cause D1 to conduct. However, since this power is injected at the input of the front end, for the *application* to actually receive power from the AUX rail, the pass-FET Q1 must conduct. We know that most PD ICs turn ON the pass-FET somewhere between 36 to 42 V, since the PoE standard requires that any PD must activate on a rising voltage waveform before the voltage reaches 42 V. In other words, if the AUX rail is higher than 42 V, the pass-FET will certainly conduct and the application will then receive power from the AUX rail.

If V_{AUX} is providing power, now if the PSE tries to turn ON, it will *not* detect the 25-k signature resistor anymore, since V_{AUX} would have charged up the port capacitance in parallel to the 25 k. So now it is a case of "first come, first served." If the PSE is providing power, the AUX rail can take over only if it is greater than PSE rail, so the PSE will continue to provide power indefinitely. If the AUX rail is providing power, the PSE rail can never come up. And only if the AUX rail drops out, will the PSE rail be able to take over.

NOTE *If the AUX rail is high enough, it is actually possible that when D1 conducts, the bridge rectifier may not get immediately or fully reverse-biased. So the PSE may not turn OFF right away. Depending on the cable resistance, both the wall-adapter and the PSE may continue to deliver power—in some ratio. On closer examination, we will realize that D1 starts to conduct when V_{AUX} exceeds the port voltage on the PD-side, but for the PSE to stop delivering power completely (bridge rectifier reverse-biased), V_{AUX} must eventually equal or exceed the port voltage at the PSE* end. *In fact, as per the IEEE standard, the PSE will stop delivering power only when the current it is pushing through, drops below 5 mA. And once that happens, it will be unable to come up again unless of course, the adapter is powered down or unplugged.*

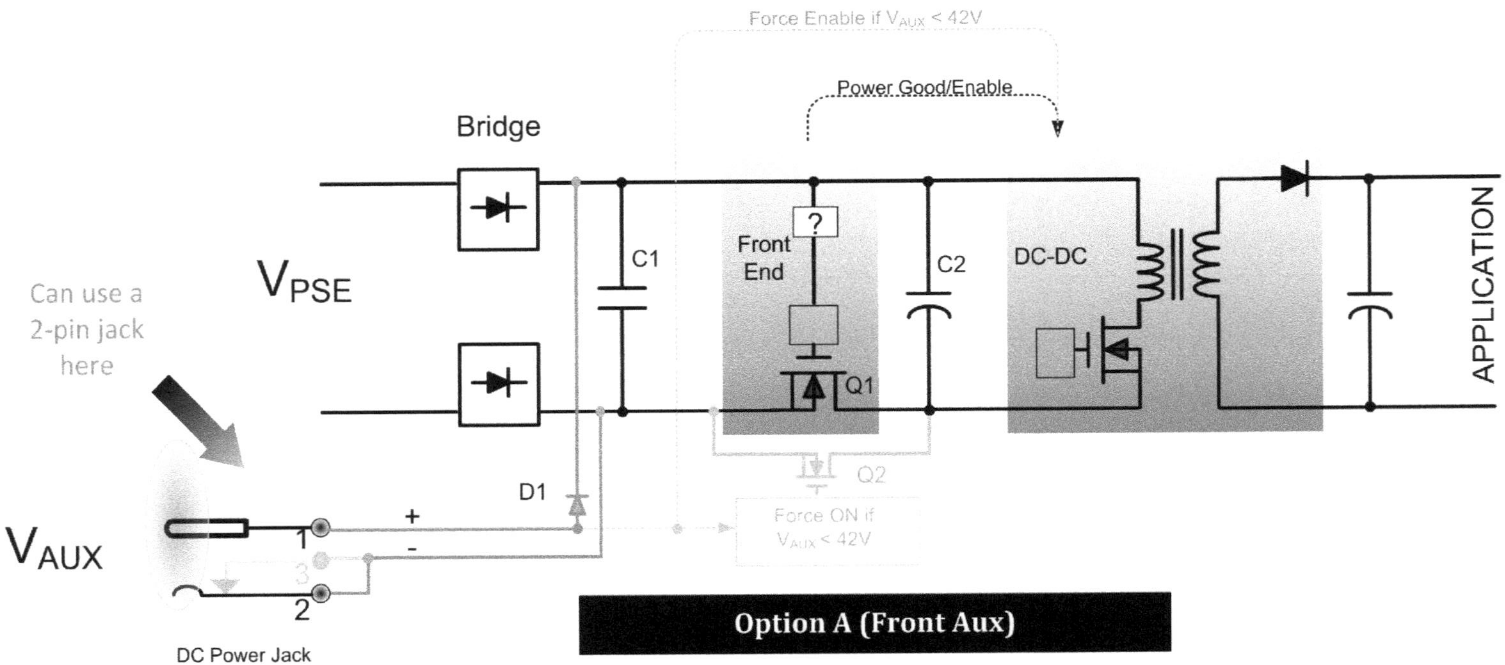

Figure 14.2 Front Aux method in more detail.

Summarizing the cases so far (ignoring diode forward-drops for simplicity):

1. Large V_{AUX}: 44 V < V_{PSE} < V_{AUX} < 57 V (e.g., V_{AUX} = 54 V, V_{PSE} = 50 V). Wall-adapter priority (Aux dominance).
   Seamless transfer? PD application will *not* get reset if adapter is plugged in. But the PD application *will* get reset if adapter is unplugged.
2. Smaller V_{AUX} (but in PoE range): 42 V < V_{AUX} < V_{PSE} < 57 V (e.g., V_{AUX} = 45 V, V_{PSE} = 52 V). First come, first served.
   Seamless transfer? PD application will *not* get reset if PSE is turned OFF (provided adapter is already plugged in). But the PD application *will* get reset if the adapter is unplugged.

Controlling Inrush Current

One problem we should be getting aware of already is that of inrush currents during "hot-swap" (changeover from PSE to AUX and vice versa). In general, we have two voltage sources of *unequal* voltages. So if one suddenly takes over (under any conducive condition), it may find a large partially discharged capacitor (C2, at the input of DC-DC stage) that needs charging up almost immediately. So the inrush current into C2 can be very high. Now, luckily, OR-ing diodes rarely get damaged when thus operated, because diodes typically have very high one-shot (nonrepetitive) surge current ratings. However, any FET in the path of this inrush can easily get damaged.

We remember from Chap. 5 that the IEEE PoE standard actually does *not* demand that Q1 have any active current limiting whatsoever (if C1 + C2 + C_{PSE} is less than 180 μF). But we also showed that for ensuring a smooth Power-up, it is actually necessary to include a dominant (lower) current limit in the PD as compared to the PSE's current limit, during the critical Power-up phase. Otherwise Q1 will conduct fully, possibly dragging the port voltage below 30 V and typically causing the PD front end to turn off (UVLO). This will lead to unacceptably jerky Power-up behavior. So we concluded that, in fact, the PD *must* have current limiting during inrush and Power-up (for any C2, greater than 180 μF or less).

But keep in mind that the PD-side current limiting feature seems to be no longer necessary *after* Power-up has been achieved. So does that mean we can dispense with Q1's current-limit function completely after Power-up is achieved? Perhaps we can. But it is not advisable. So most good commercial PD chips retain some higher-protective current limiting even after Power-up, usually just for abundant caution. For example, this current limit may therefore be set rather high, at say 1.8 A typically, and may also be mentioned in the datasheet with no guaranteed/declared min-max values. So it is not taken very seriously unfortunately. In fact there are some "good" commercial PD chips that just rely on thermal protection (under faults) once Power-up is achieved. That is even less desirable.

We are now beginning to realize that to support hot-swap (between PoE rail and AUX rail), especially in the case of Front Aux support in which we can see that the inrush current will necessarily pass through Q1, we *must* include effective and defined current limiting in Q1 even after Power-up. Under normal PoE operation, this default current limit should be high enough to remain "transparent" under normal operation, but under hot-swap conditions, it will enter the picture to keep the inrush in control and thus protect Q1.

We therefore conclude that to support the Front Aux option, the pass-FET of the front end must have a well-defined current limit always. For example, for a PD chip that supports only AF loads, we can set the inrush current limit of its pass-FET (Q1) to ~ 200 to 300 mA (during the regular PoE inrush phase), then let the current limit rise to ~ 500 mA after Power-up is achieved. However, if a wall adapter is detected, we can raise the current limit to say 1 A. The last step will protect the FET under hot-swaps but also allow more power to the PD application as discussed in (2) in the following section.

Similarly, for a PD chip that supports PoE+ loads, we can set the inrush current limit of its pass-FET (Q1) to ~ 200 to 300 mA (during the regular PoE inrush phase), then let the current limit rise to ~ 1000 mA after Power-up is achieved. However, if a wall adapter is detected, we can raise the current limit to say 1.8 A. That will protect the FET under hot-swaps, but also allow more power to the PD application as discussed in (2) in the next section.

V_{AUX} Outside PoE Range

So far we have discussed the case of V_{AUX} within the normal PoE voltage range. But what if V_{AUX} is *less* than 42 V? There are *two* problems associated with that:

1. As mentioned, the PD front end may not turn ON (i.e., Q1 will not conduct). However, this can be overcome by "fooling" the front end by introducing a small Boost-converter placed in the block marked "?" in Fig. 14.2. This will wake up the front end, which will then hopefully make Q1 conduct.

   Another way is to design the front end with a FAUX pin as indicated with a dashed line in Fig. 14.1. This pin can be used to detect a wall-adapter voltage present (by sensing voltage on the remote side of the OR-ing diode), and then step in to defeat/negate the IEEE-compliant > 30 V UVLO of the front end. The FAUX pin can also be used to simultaneously increase the default current limit of Q1, to say 1.8 A, for PoE+ loads as discussed above. So using FAUX pin support, it is possible to allow AUX voltages typically down to ~ 13 V using the Front Aux method.

But note that there is some danger in defeating the input UVLO in particular, especially if we are not 100 percent sure that D1 is actually conducting (passing current), and the bridge is completely reverse-biased. Because it could happen that though the wall-adapter voltage is "present" (on one side of the OR-ing diode), the PoE rail was there first, and is therefore perhaps still delivering power (partially or fully). So defeating the UVLO would then amount to IEEE noncompliance. The design of the FAUX feature is certainly not so trivial.

2. We mentioned above that we could use a small boost converter to "fool" the IC to turn Q1 ON. That would work, but usually with a limitation on output power. The reason is that in that case, the default current limit may not be as high as we need it to be, or as high as we can set it to be with formal FAUX pin support present in the chip. *Why is the current limit such a problem anyway*? Because for the same output power, if the voltage is less, the current needs to be much higher. Otherwise full power cannot be guaranteed. So, despite using a boost converter, the max current that Q1 can handle may not be commensurate with a lower (AUX) voltage. However, if the IC supports the FAUX pin feature, and if we set the default current limit to 1.8 A as suggested above, and provided Q1 is designed with a low-enough R_{DS} so as *not* to enter thermal shutdown with higher currents, then for V_{AUX} of 13 V, we can support AUX power up to 13 V × 1.8 A = 23.4 W. But note in this case, 1.8 A must be the *guaranteed min* of the default current limit. With just a "typical" current limit value declared or tested, the user cannot really guarantee any AUX power for AUX rails lower than the normal PoE range. One alternative, to have *no* current limit in the front end at all after Power-up is achieved, is not advisable as discussed previously.

NOTE *Depending on the design of the front end, the DC-DC stage may not get a Power Good signal under the above "solutions" for low V_{AUX}. Therefore additional external circuitry may be required to force the DC-DC stage ON via the force-enable path shown in Fig. 14.2.*

3. There is in fact another option to support low AUX voltages. We can completely bypass the pass-FET Q1 by an additional external FET ("Q2"). See "Q2" in Fig. 14.2. We will still need to force-enable the DC-DC stage (i.e., defeat its IEEE-compliant UVLO if any).

But note that we should take great care to never turn ON the bypass FET Q2 while still operating (partially or fully) under PoE power because that would defeat the recommended (desirable) current limiting function present inside Q1. We should wait untill we are completely sure that *only* D1 is conducting (zero current through the bridge rectifier), before we turn Q2 ON.

Also, to control inrush currents, since Q2 typically has no current-limiting, to protect it better from the high-inrush currents, it is advisable to insert a small current-limiting resistor in series with the OR-ing diode D1, or better still, we can use an NTC (negative temperature coefficient) device in series with the OR-ing diode.

Brute-Force Wall-Adapter Priority

There is a brute-force method to ensure wall-adapter priority (AUX dominance) under *any* condition: by simply disconnecting the PSE power whenever the wall adapter is plugged in. And that is best accomplished by using a three-terminal adapter plug. This is actually better known as a *two-conductor, interrupting DC-power jack,* and it is available from several vendors, such as Switchcraft. Its chosen current rating should be around 5 A, to avoid the contacts from wearing out prematurely. Such a mechanical switch is illustrated in Fig. 14.3. Note that on inserting the male, the contact between pins 2 and 3 gets broken. In a variation of this, the switch is guaranteed to be "break before make." So, if that condition is met, there is no need for any OR-ing diodes either, since PSE power and AUX power will never coexist on the same copper wire. In other words, with break before make, before the wall adapter is plugged in, the power comes from the PSE, but on inserting the adapter plug, the PSE power gets interrupted, *then* the AUX power is applied.

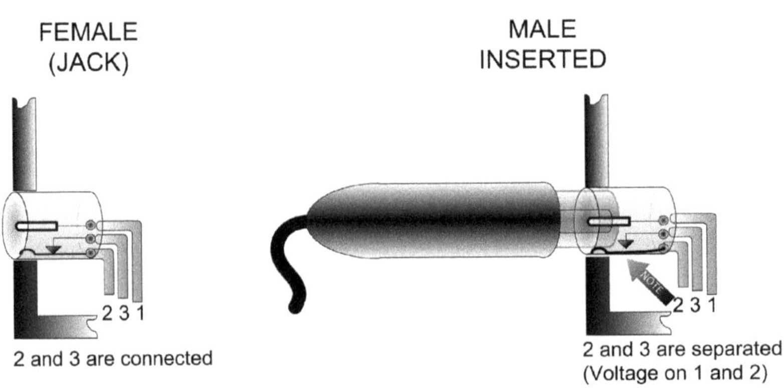

FIGURE 14.3 Two-conductor interrupting DC power jack.

The complete scheme with this option/enhancement is shown in Fig. 14.4. Note that all the other accompanying techniques discussed above are still required along with this new switch. The switch only adds brute-force *adapter priority*.

There is a limitation however. Now the PSE rail cannot even come up till the adapter is not just powered down (i.e., unplugged from the AC outlet), but actually unplugged from the PD too (the male DC plug removed from the jack). Also, nothing about the hot-swap process is seamless anymore, and the PD application will very likely get reset whenever the adapter is plugged into the PD, or when it is unplugged.

There is a way out of this limitation too, and that is by means of a relay as shown in Fig. 14.5. Now we can use a two-terminal jack, and the male plug does not have to be manually plugged in or unplugged. Only when there is voltage present on the AUX rail will the relay activate and create adapter priority. But once again, if V_{AUX} is outside the normal PoE operating range, we will need to (a) force Q1 to conduct (or bypass it with Q2), (b) force-enable the DC-DC converter. But despite that if we cannot allow more current through Q1 or Q2, the total power will be significantly limited if the AUX rail voltage is too low.

Despite all the solutions, mainly because of *power limitation*, the Front Aux method is usually *not* preferred. Instead Rear Aux is the most popular choice.

Auxiliary Power Option B (Rear Aux or RAUX Pin Method)

In the following discussion, pay close attention to Fig. 14.6. Application of Rear Aux can be very useful, but also confusing.

Rear Aux has the following advantage from a *heuristic* viewpoint: it introduces the ability to cause the PSE to disconnect power if desired, by just turning the pass-FET of the front end ("Q1") OFF. If the PSE current then falls below 5 mA, the PSE will typically disconnect, though it will likely keep attempting detection.

Rear Aux method has another major advantage that when the AUX rail is delivering power that current does not pass through the PD's pass-FET (Q1), so the current cannot get limited. Therefore, so long as the DC-DC stage can function down to very low voltages (~ 10 V) (if necessary by deactivating any DC-DC stage UVLO), then full power to the application can be assured and maintained even for very low AUX rails.

As in the Front Aux method, we still have two *natural* cases when the AUX rail is within the normal PoE range (ignoring diode forward drops for simplicity, and assuming optional diode D2 is not present so far):

1. Large V_{AUX}: 44 V < V_{PSE} < V_{AUX}< 57 V (e.g., V_{AUX} = 54 V, V_{PSE} = 50 V). Wall-adapter priority (AUX dominance).

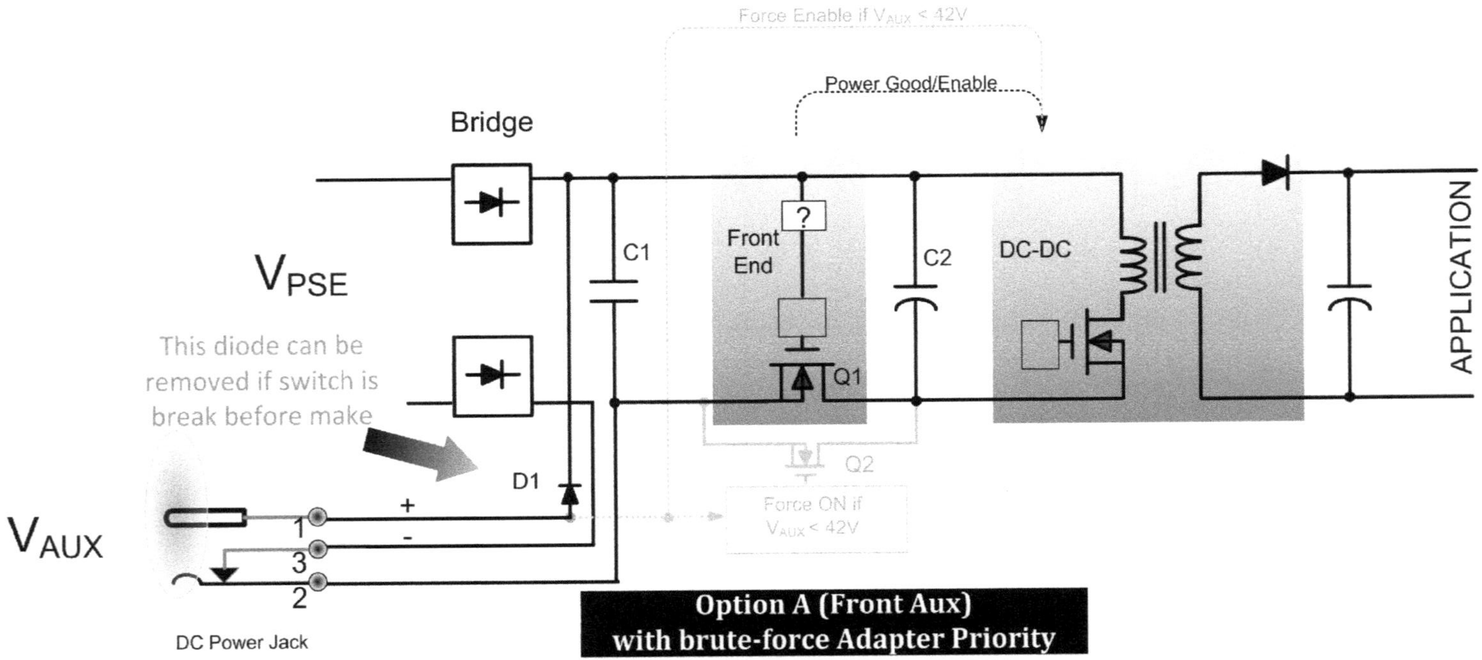

FIGURE 14.4 Front Aux method in more detail (with brute-force adapter priority).

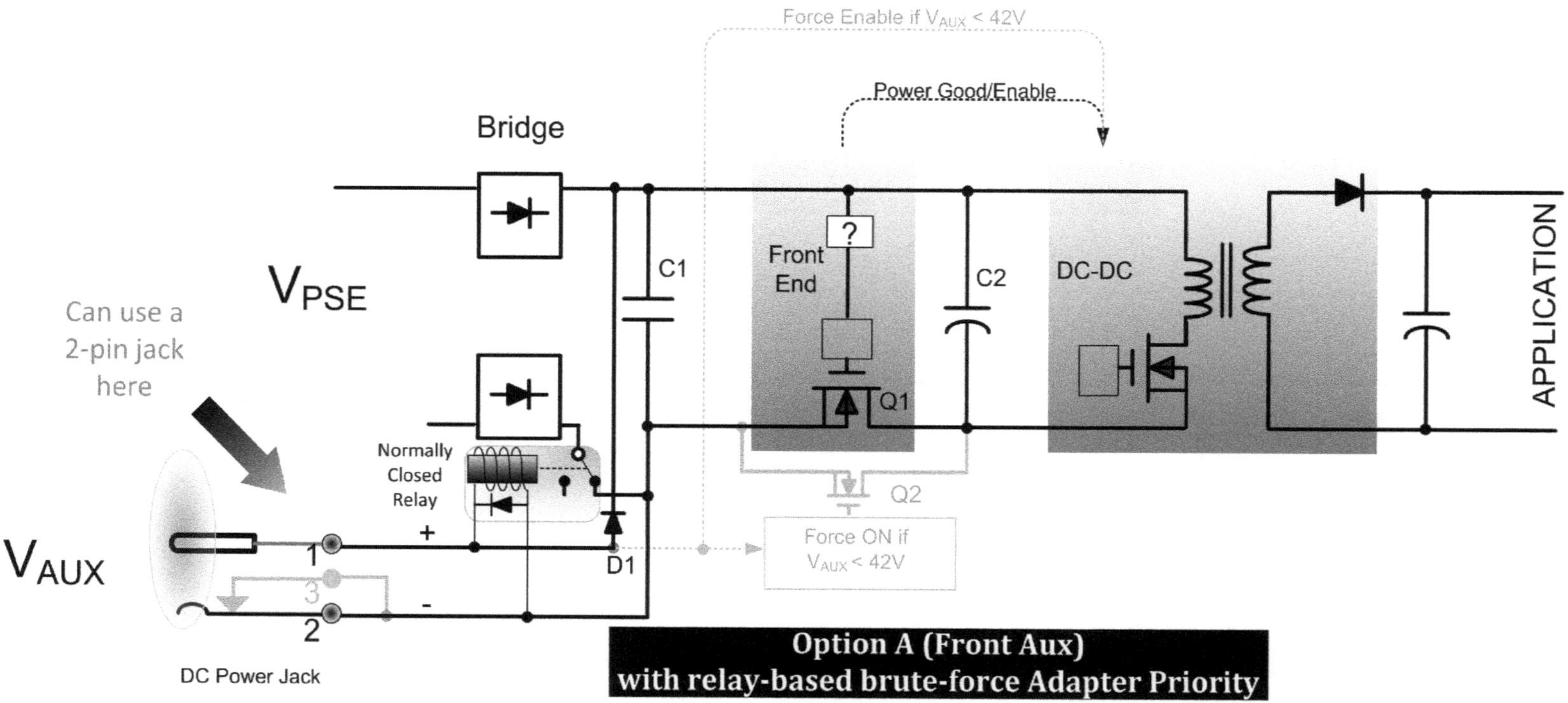

FIGURE 14.5 Front Aux method in more detail (with relay-based brute-force adapter priority).

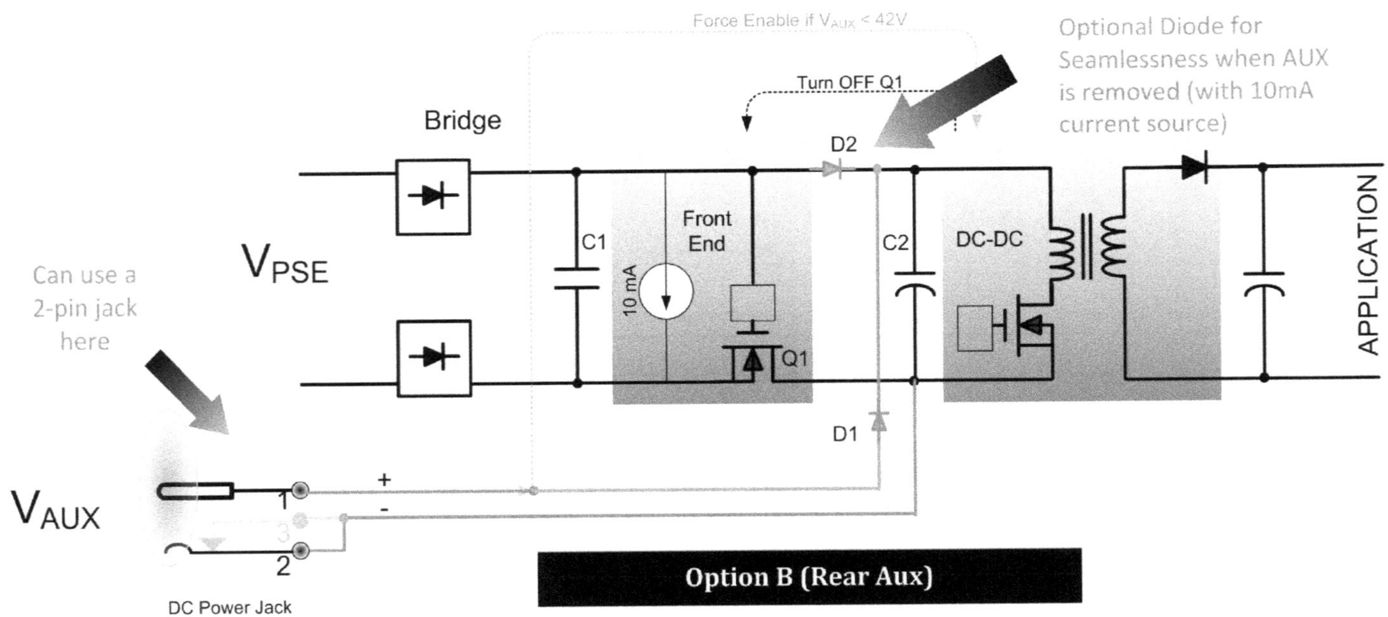

FIGURE 14.6 Rear Aux method in more detail.

Seamless transfer? PD application will *not* get reset if adapter is plugged in. But the PD application *will* get reset if adapter is unplugged.

2. Smaller V_{AUX} (but in PoE range): 42 V < V_{AUX} < V_{PSE} < 57 V (e.g., V_{AUX} = 45 V, V_{PSE} = 52 V). First come, first served.

Are there any changes/improvements over the Front Aux method? Yes, because in case (2) above, with Rear Aux method, we do not need to force the pass-FET of the front end (Q1) ON anymore to deliver power. So the AUX rail can be lowered even down to 10 V typically. The only condition is that the DC-DC converter can function down to that voltage (though as mentioned, we may need to add some external circuitry to override its UVLO and force-enable it). So in fact, case (2) above can be restated with a wider-voltage range for the Rear Aux method.

2. Smaller V_{AUX} (even outside PoE range): ~ 10 V < V_{AUX}< V_{PSE} < 57 V (example: V_{AUX} = 12 V, V_{PSE} = 52 V). First come, first served.

   Another question: Is (2) really still stuck on "first come, first served"? Not necessarily so. Because with Rear Aux method we have an additional degree of freedom (and also control): we can forcibly turn the pass-FET of the front end OFF, thereby disconnecting the PSE rail at will. Of course we need a pin on the front end to be able to do that. If so, in effect, we can enforce adapter priority always. Yes, the PD chip needs to be explicitly designed to support this feature, in a manner very similar to the FAUX pin support. In other words, we need to provide a RAUX pin on the front end, which will have the following behavior:

   By sensing the voltage on the remote side of the OR-ing diode D1 as per the dashed line marked RAUX in Fig. 14.1, we can detect the presence of the AUX rail. So, even if V_{AUX} is less than V_{PSE}, we can force the AUX rail to dominate for any condition, and at all times, by simply turning OFF the pass-FET of the front end (disconnecting the PSE from the application).

   When we turn OFF the pass-FET Q1, there are actually two possibilities going forward—the two "suboptions" within this Rear Aux option as mentioned previously:

   1. **Diode D2 not present**: The PSE will get disconnected and stay disconnected. Because the AUX voltage will flow back into the port capacitance, in parallel to the 25-k signature resistor, and that will prevent the PSE from ever detecting a valid PD and turning ON the PoE rail (keeping it in standby, waiting for AUX rail to be removed). This conserves system power, optimizes power delivery, and allows port power to be allocated by the host elsewhere where required (other

ports). But that "green" feature can also be a disadvantage on resumption of PoE power. Because PoE power may not be available immediately if it was committed elsewhere. So though asserting adapter priority will be seamless, causing no PD-side reset, when the adapter is either powered down or unplugged, there *will* be a PD-side reset on resumption of PoE power.

2. **Diode D2 present**: With this addition, we can choose to keep alive the PoE rail as a sort-of UPS standby in case the wall adapter is not delivering power. The diode D2 will prevent the negation of the 25-k signature by the AUX rail, so the PSE will be able to power up. However, to then keep it up and alive, we need to design the front end such that when the wall adapter is sensed via the RAUX pin, we force the front end to draw > 10 mA from the PSE, preventing MPS disconnect. See Figs. 14.6, 14.7, and 14.8. Now we have the advantage that if the AUX rail is removed, the PoE rail can be smoothly (seamlessly) take over power delivery (perhaps with the help of some architectural soft-transitioning features). This will ensure no interruption in power and consequent PD reset.

Summarizing, for Rear Aux option:

1. Potentially seamless transitioning from PoE power to AUX power and back is possible, provided the front end is designed to support this feature with a dedicated sense pin on the front end. In addition, we need to provide the diode D2 shown in Figs. 14.6, 14.7, and 14.8. Otherwise, the AUX voltage will flow back into the port capacitance, in parallel to the 25-k signature resistor, and that will prevent the PSE from ever detecting a valid PD.
2. Wall Adapter priority is *naturally* possible if $V_{AUX} > V_{PSE}$. But with Rear Aux method we can easily ensure unconditional adapter priority with a dedicated RAUX pin that turns OFF the pass-FET of the front end. As with Front Aux method, we can also brute-force adapter priority by using a three terminal switch as shown in Fig. 14.7. Or we can use a relay as shown in Fig. 14.8. Unfortunately, all the limitations of the mechanical switch and relay that we discussed in the Front Aux option, are applicable here too. For one, we are likely going to lose seamlessness under hot-swap conditions.
3. Inrush current protection is not available with Rear Aux method. The adapter must protect itself well under these conditions. This can be problematic when dealing with FET-derived diodes in particular, since they have lower surge ratings than simple PN diodes.

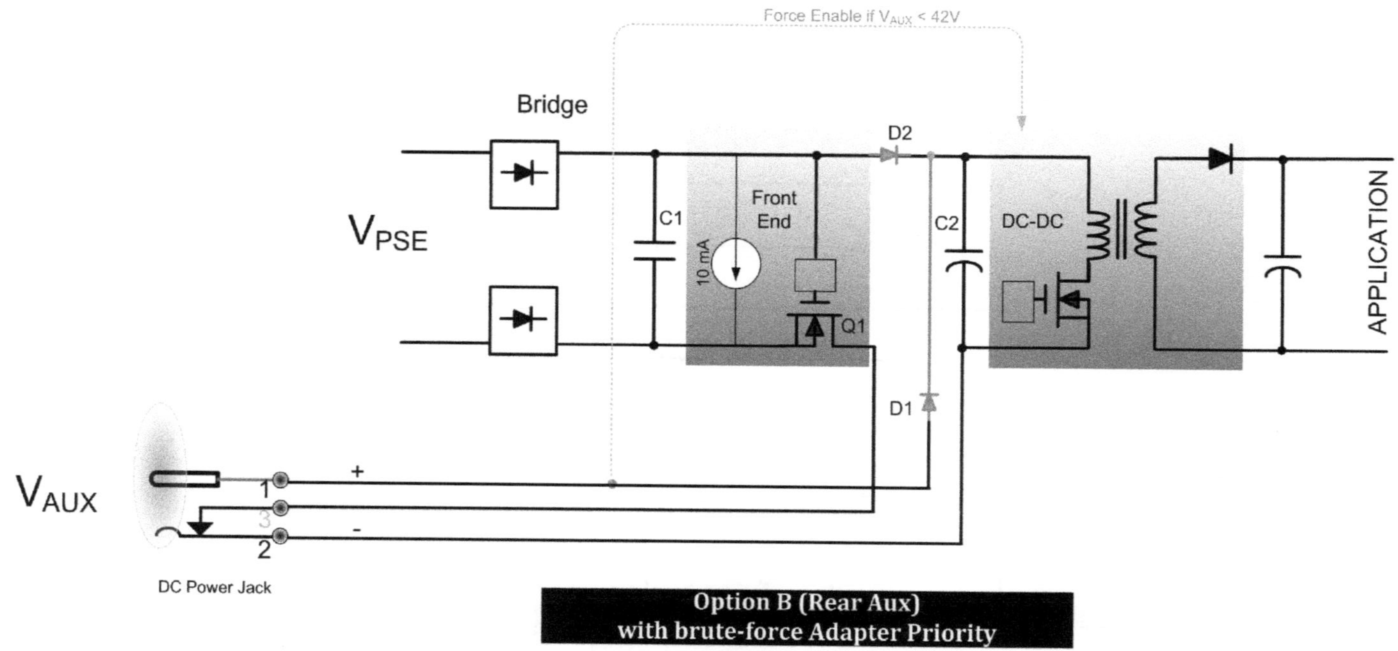

Figure 14.7 Rear Aux method in more detail (with brute-force adapter priority).

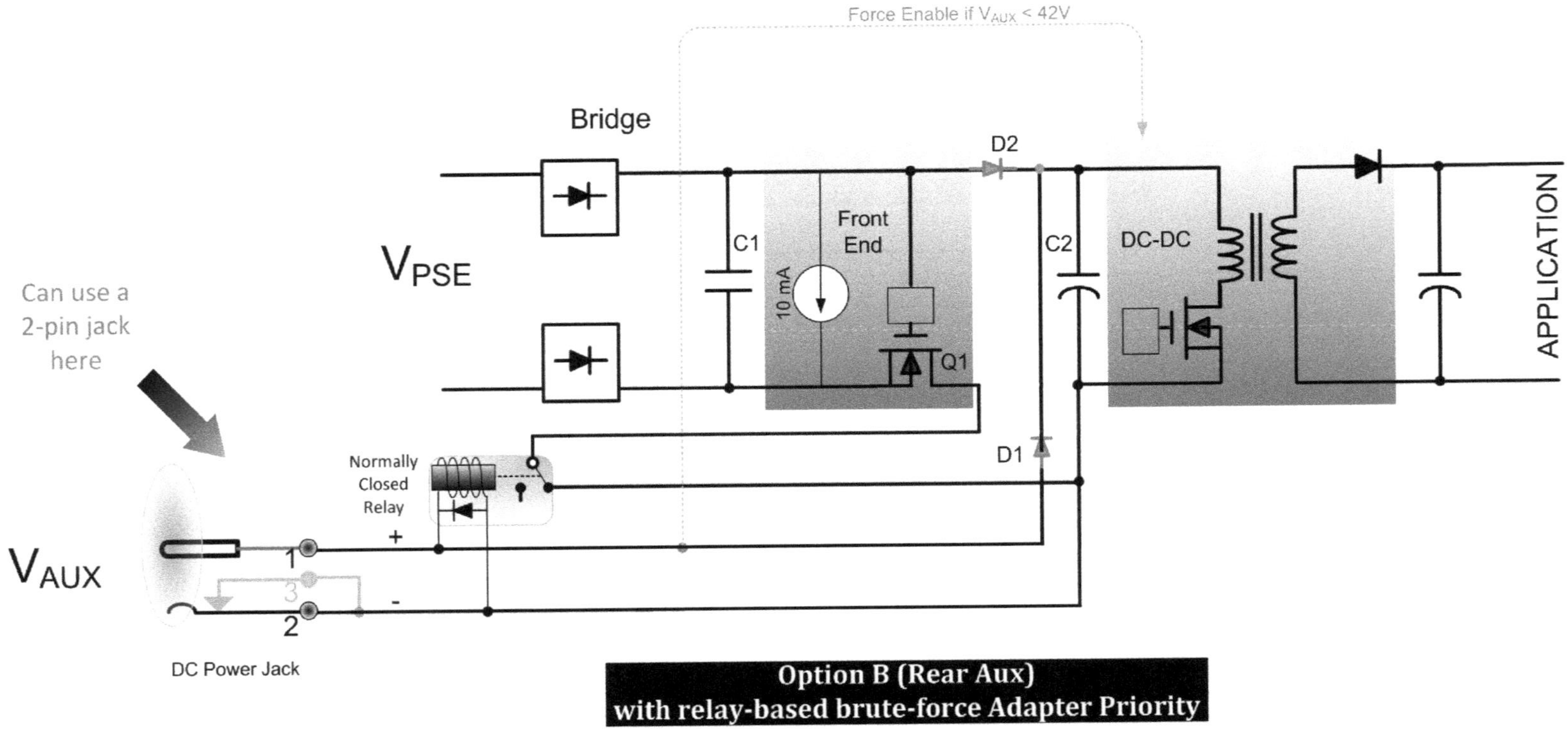

Figure 14.8 Rear Aux method in more detail (with brute-force relay-based adapter priority).

Auxiliary Power Option C (Direct OR-ed)

The final option, where auxiliary power is OR-ed directly to the output of the PD power supply, is a simple solution but offers few advantages. It requires an AUX power supply that is designed to deliver the current and *regulated* voltage rails required by the PD load. So it has to have a built-in DC-DC converter. In addition, this configuration usually requires additional components and some duplication of functions. It adds cost to the adapter.

Summarizing: Option C is not preferred, mainly because it requires the AUX rail to be fully regulated. In all previous options, the AUX rail was unregulated and applied to the input of the PWM/DC-DC stage, which would carry out the required regulation. We also may need an OR-ing diode in series with the DC-DC converter of the PD, to ensure proper operation of this converter. So this will lead to a constant loss term even under regular PoE operation.

Flyback Design Procedure

As a brief introduction to this topology, see the waveforms in Fig. 14.9 as we go along.

A switch (usually a FET) toggles ON and OFF repetitively. When the switch is ON (fully conducting), current ramps up in the primary winding (marked P). By the polarity dots on the windings shown in the figure, we see that both the dotted ends of the two windings go low with respect to the nondotted ends, when the switch is ON. Therefore the anode of the catch diode, also called output diode—the one connected to the secondary winding, goes low with respect to its cathode. The diode is therefore reverse-biased, or OFF (fully nonconducting) whenever the FET is ON, and vice versa.

In terms of energy, this means that energy drawn from the input V_{IN} is building up in the transformer during the converter's ON-time (the ON-time of the FET), and during this duration no energy gets transferred over to the output. This is a key property of the flyback (transformer-based) topology, and also of the buck-boost (nonisolated, inductor-based) topology on which the flyback is based.

When the FET turns OFF, the polarity of the voltage across the primary winding suddenly reverses in an attempt to drive the current in the same direction it was flowing in just prior to the switch transition. Earlier, the inductor was, in effect, a passive element with input voltage just dropping across it. Now it becomes a source of potential itself, much like a battery, as it tries to drive inductor current in the same direction. Therefore the voltage across its ends flips. This is explained in the upper section of Fig. 14.10, using the simpler buck-boost (inductor-based) topology, on which the flyback (with isolating transformer) is based.

We have just described the most fundamental property of inductors: We cannot change the direction and magnitude of the *current* in

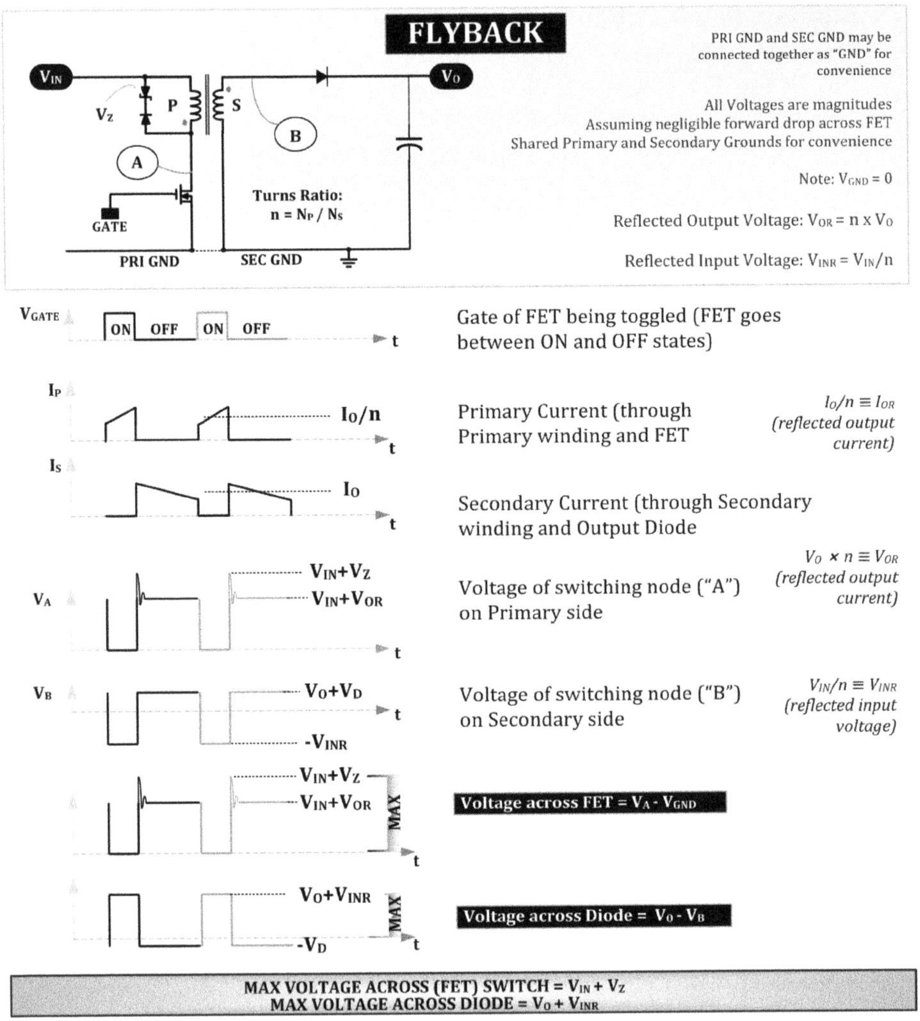

FIGURE 14.9 Waveforms of a flyback.

an inductor instantaneously. Because that current is related to stored energy inside the inductor, equal to $\frac{1}{2} \times L \times I^2$, and cannot be instantaneously wished away or changed (conservation of energy law). The inductor current is certainly allowed to slew up or slew down, but if we try to force it to do anything else, or prevent it from trying to remain "smooth," it will generate a huge voltage spike that can create flashovers and destroy semiconductors. That fact is in fact usefully exploited too, and is the underlying principle behind the camera flash. But clearly, in our case, we do not want to see the FET go up in a white flash. So we want to maintain the inductor current smooth and not jerky. And that is the basic reason we always provide a catch

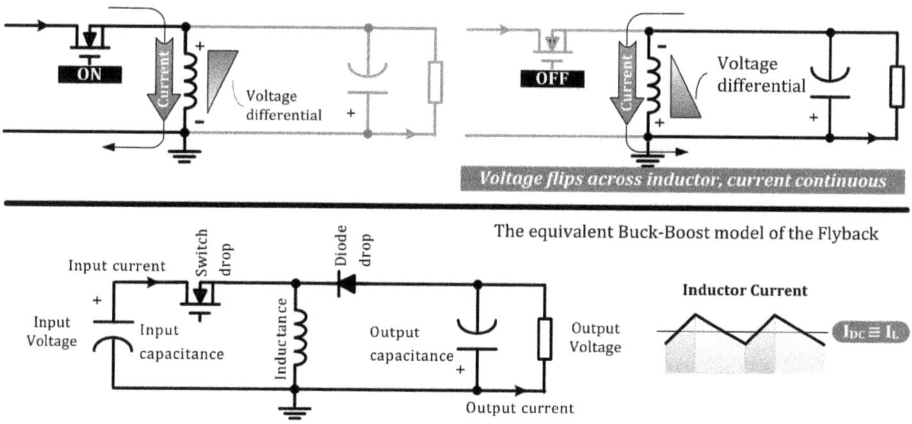

	Primary-side equivalent Buck-Boost	Secondary-side equivalent Buck-Boost
Input Voltage	V_{IN}	$V_{INR}=V_{IN}/n$
Input current	I_{IN}	$I_{INR}=I_{IN} \times n$
Input capacitance	C_{IN}	$n^2 \times C_{IN}$
Inductance	L_P	$L_S=L_P/n^2$
Switch drop	V_{SW}	V_{SW}/n
Output Voltage	$V_{OR}=V_O \times n$	V_O
Output current	$I_{OR}=I_O/n$	I_O
Inductor Current (I_{DC})	$I_{OR}/(1-D)=I_O/[n \times (1-D)]$	$I_O/(1-D)$
Output capacitance	C_O/n^2	C_O
Diode drop	$V_D \times n$	V_D
Duty Cycle	D	D
Current Ripple Ratio	r	r

FIGURE 14.10 The equivalent nonisolated (inductor-based) buck-boost models of the flyback.

diode (the output diode of the flyback) in any switched-inductor topology.

Note that when we take the inductor of a buck-boost topology and replace it with a transformer, we get the flyback topology. The transformer follows the same principles of an inductor, with one important difference: It does not "care" which of its particular windings we are drawing current from. We can immediately stop current in one winding, provided we continue it in another winding. But since windings can have different number of turns, we need a sort-of *weighted* current term to be preserved at switch transitions—and that is the product of the amperes and the number of turns (ampere-turns) through which that current flows through. In general we say:

1. The total *amperes* flowing in any inductor cannot be discontinuous (jerky) during switch transitions.
2. The *ampere-turns* flowing through a transformer cannot be discontinuous (jerky) during switch transitions.

So, as the FET turns OFF, the voltage polarity flips in both windings of the transformer in Fig. 14.9, in an effort to keep the current (or rather ampere-turns in this case) continuous. On the secondary side, the anode of the catch diode therefore jumps up in voltage and the diode gets forward-biased. Now all the energy stored during the preceding ON-time interval finds a way out—it flows into the output cap of the flyback.

By varying the ratio of the ON-time to the OFF-time we can vary the output voltage. The reason is, if we apply a certain voltage V_{ON} across the inductor during the ON-time, and we then fix the OFF time to a certain T_{OFF}, thereby establishing the ratio T_{ON}/T_{OFF}, the system, when steady, will create a reverse inductor voltage V_{OFF}, based exactly on the following simple rule, called the "volt-seconds law"

$$V_{ON} \times T_{ON} = V_{OFF} \times T_{OFF}, \text{ so } \quad V_{OFF} = V_{ON} \times \frac{T_{ON}}{T_{OFF}}$$

Where does this rule come from? The basic inductor equation $V=VdI/dt$ leads to

$$V = L\frac{\Delta I}{\Delta t} \text{ during the ON-time and OFF-time}$$

So

$$V_{ON} = L\frac{\Delta I_{ON}}{T_{ON}}, \quad \Delta I_{ON} = \frac{V_{ON} \times T_{ON}}{L}$$

$$V_{OFF} = L\frac{\Delta I_{OFF}}{T_{OFF}}, \quad \Delta I_{OFF} = \frac{V_{OFF} \times T_{OFF}}{L}$$

And if the volt-seconds law is true, we have $V_{ON} \times T_{ON} = V_{OFF} \times T_{OFF}$, and so $\Delta I_{ON} = \Delta I_{OFF}$. So talking in terms of magnitudes as we always do, the increase in current during the ON-time, that is ΔI_{ON}, must equal the decrease in current during the OFF-time, that is ΔI_{OFF}. And that is the reason we get a steady-state to start with, because if there is net current increment or decrement at the end of every switching cycle, we are, by definition, not in steady-state. In other words, the volt-seconds law is just another way of expressing a steady-state, and the volt-seconds law in turn, predicts the voltage reversal, and also the voltage across the inductor during the OFF-time, which is what transfers to the output and is seen by us as the output rail. We also realize that if we alter the ratio of the ON-time to the OFF-time, we indirectly vary V_{OFF}, and thereby vary V_O, the output.

In general, instead of talking about the ratio of the ON-time to the OFF-time, in power conversion we prefer to talk in terms of the ratio of the ON-time to the *total* time, and we call that the "duty cycle" of the converter. So, we write

$$D = \frac{T_{ON}}{T_{ON} + T_{OFF}} = \frac{T_{ON}}{T} = T_{ON} \times f$$

where T is the time period and f is the switching frequency. This relation is true for *all* topologies. For a flyback, specifically, based on the fact that $V_{ON} = V_{IN}$ (voltage across inductor during on-time equals the input rail), and $V_{OFF} = V_O$ (voltage across inductor during the off-time equals the output rail), using the volt-seconds law, the exact relationship (called the DC transfer function of the converter) becomes

$$D \approx \frac{V_O}{V_{IN} + V_O} \quad \text{(buck-boost)}$$

$$D \approx \frac{nV_O}{V_{IN} + nV_O} = \frac{V_{OR}}{V_{IN} + V_{OR}} \quad \text{(flyback)}$$

where n is the turns ratio of the transformer ($n = N_P/N_S$), and $V_{OR} = n \times V_O$ is called the *reflected output voltage*—that is, the effective output voltage as "seen" by the switch from the primary side of the transformer. The approximate sign is inserted above, because these are "ideal" equations—corresponding to the case of *no* losses at all (100 percent efficiency). In reality, the exact relationship is

$$D = \frac{\eta V_O}{V_{IN} + \eta V_O} \quad \text{(buck-boost)}$$

$$D = \frac{\eta V_{OR}}{V_{IN} + \eta V_{OR}} \quad \text{(flyback)}$$

where η is the actual (measured) efficiency of the converter. By definition

$$\eta = \frac{P_O}{P_{IN}}$$

where P_O is the output power ($V_O \times I_O$) and P_{IN} is the input power ($V_{IN} \times I_{IN}$). This can also be expressed as a percentage by multiplying it by 100.

The flyback is a popular topology for low-cost applications. In a PD, in general, we need to have isolation (1500 VRMS) between the MDI/PI side and the user-accessible application/host side. This was discussed in Chap. 10. The flyback is well-suited for a low-cost isolated topology. However, there is another option for an isolated topology, though it is more commonly used for slightly higher powers: the Forward topology. But note that (in its simplest implementation), the Forward topology is not allowed to exceed a "duty cycle" of 50 percent. The reason is we need to give enough OFF-time for the primary winding to get demagnetized (reset) during the OFF-time, otherwise we will see a flux-staircasing condition because of violation of the volt-seconds law as applied to the transformer, causing core-saturation of the transformer after just a few switching cycles.

With that brief background on the principles of power conversion, we start looking at the best way to initiate a design. The entry point of any switching converter design is the r, or current ripple ratio. In a nonisolated DC-DC (inductor-based) topology this is defined very simply as

$$r = \frac{\Delta I}{I_L}$$

where ΔI is the total swing in inductor current (see Fig. 14.11), and I_L is the center of the inductor current ramp. It is recommended that r be set around 0.4 to 0.5—that value represents an optimum for the entire converter in terms of the size of the inductor and associated power components like the input and output capacitors.

When we come to the flyback, we do not have an inductor, but a transformer. However, as shown in Fig. 14.11, from either side of the transformer we can mentally think of the waveform as part of an equivalent inductor current waveform and define an r in much the same way. The r thus defined, is the same for the primary side as for the secondary side, because the current waveform on the secondary side is simply scaled by a factor n from the primary, whereas by definition, r is just a geometrical ratio (shape factor), so scaling the entire waveform does not change its r value. Having defined r, we can set it to 0.4 to 0.5 typically, and proceed with the design.

If the input voltage has a wide range, as is rather typical, the basic question is at what input voltage should we set r to 0.4 to 0.5 as suggested previously? The answer is *at the lowest-input voltage* because in a flyback, the lowest-input voltage point causes the highest currents, and we want to also ensure that the transformer core is

I_L is the same as I_{DC} (it is the average inductor current, i.e. the center of ramp)

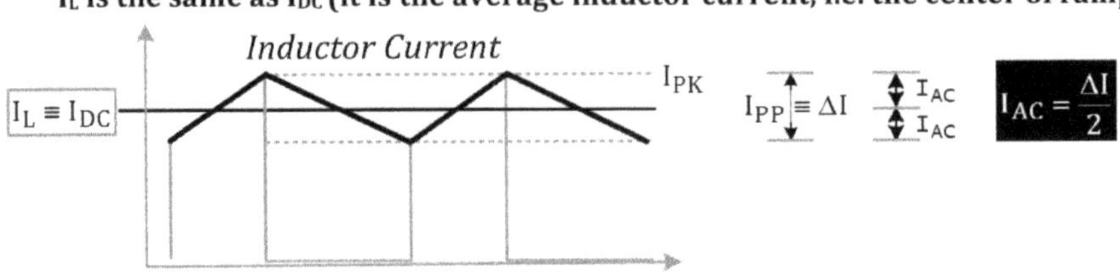

All Topologies:

$$r = \frac{\Delta I}{I_L} \equiv \frac{2 \times I_{AC}}{I_{DC}}$$

Current Ripple Ratio

All Topologies:

I_{PK} ⇒	$I_{DC} + I_{AC}$	$I_L \times \left(1 + \frac{r}{2}\right)$
I_{PP} ⇒	$\Delta I \equiv 2 \times I_{AC}$	$I_L \times r$

Figure 14.11 Definition of r for nonisolated (inductor-based) DC-DC converters.

selected at the lowest input so as to be sure it does not saturate at that worst-case voltage end. So very simply put: Any flyback design is always started at the lowest input, by setting *r* to about 0.4 to 0.5 at that point.

We can start a practical design now. We have selected the LX7309 from Microsemi. It is a versatile controller IC, which can be used either as a flyback or as a Forward because its duty cycle is limited to a max of 50 percent.

Suppose we want to design a PD for a four-pair PoE application. *The requirement is to get an isolated 12 V at 4 A. That is a 48 W output.* The input to the DC-DC converter (flyback) is allowed to vary between 32 to 57 V. The target efficiency is 0.85 (85 percent) at high line (57 V) and 0.8 (80 percent) at low line (32 V). We can always use more advanced techniques to improve efficiency, like synchronous rectification (putting a FET in parallel to the output diode), and so on, but the basic design procedure we describe below is unchanged as a result of all that, so we will ignore other possibilities here.

The transformer-based flyback is a little daunting to some. So it is important to know that the flyback can be reduced to a simpler, inductor-based, equivalent buck-boost. In fact, *two* equivalent buck-boost models can be created, one for working out the currents and stresses in the primary side of the flyback by reflecting all the secondary-side components (and voltages and current too) over to the primary side. Another model for working out the currents and stresses in the secondary side of the flyback is by reflecting all the primary-side components over to the secondary side. This interesting technique, with the mapped equivalent components and voltages/currents is shown in Fig. 14.10. The only thing this model ignores is the leakage inductance of the transformer. That has to be tackled separately. More importantly, this leads to a huge voltage spike that can "kill" the FET, so we always need to clamp it to safe values by means of either a zener clamp (shown in Fig. 14.9) or an RCD clamp-consisting of a resistor, capacitor, and diode (not discussed any further here).

We start by picking a 150-V FET. The reason for that is we need to have sufficient headroom above V_{INMAX}, to be able to handle the reflected output voltage V_{OR}, and the leakage inductance spike that appears on the Drain of the FET, as shown in the row marked V_A in Fig. 14.9. Higher and higher FET voltage ratings not only cost more in terms of money, but they have a negative-performance impact after a point, because of sluggish response and higher R_{DS} (increased conduction and switching losses). However, high-FET ratings can help too in one specific way, by significantly reducing the dissipation in the leakage-inductance clamp.

In general, we need a careful compromise in terms of cost and performance, so we may need to tweak the voltage rating of the FET later.

Note that the leakage-inductance clamp, whatever form it takes, is connected to the Drain of the FET as indicated in Fig. 14.9 and also in Fig. 14.12. It is used for literally burning up the energy in the leakage inductance, because unfortunately, primary-side leakage is not coupled to the secondary side *by definition*, so it cannot transfer over its energy to the secondary side and then use the catch diode from there to avoid any voltage spike. Therefore, when the FET turns OFF, we will see a huge voltage spike on the Drain of the FET. It can easily destroy the FET if we do not have a well-designed clamp (zener or RCD type). Yes, there are exotic techniques to try and recover this leakage energy and push it back into the input bulk cap of the converter (active clamps), but that is out of scope here too.

Having selected a 150-V FET, we need some derating. We fix that at 0.8. So, in effect, to us the available voltage range is only 150 V × 0.8 = 120 V. How much headroom do we have? At V_{INMAX} of 57 V, that gives us a headroom of 120 V – 57 V = 63 V. Here we can pick a zener that clamps at around 50 V, keeping in mind that its clamp level will be higher than 50 V based on zener tolerances and also the instantaneous current through it, and so we need to leave some margin for that too. But if we do that we will realize that the duty cycle at low-line will exceed 50 percent, which the LX7309 does not allow. So we back off on the zener rating and pick a zener of 28 V. So V_Z = 28 V. We will confirm later that this is appropriate.

Having fixed the clamping level, we need to fix the reflected output voltage V_{OR}. That basically is just the output voltage (12 V in our case) multiplied by the transformer turns ratio n. From the duty-cycle equations presented previously, we realize that the duty cycle depends on the turns ratio. So, in effect, by fixing V_{OR}, we fix the turns ratio and the duty-cycle. We realize that V_{OR} is one of the most important parameters to fix for a flyback, and must be done with great care. Significant tweaking of that may be necessary.

For optimizing the efficiency, we need to reduce the clamp dissipation. For that, a good thumb-rule is to ensure that the clamp (V_Z) is about 40 percent higher than the V_{OR}. The relative magnitudes of the two are shown by the gray curly brackets on the left side of Fig. 14.12.

We therefore get $V_{OR} = V_Z/1.4 = 28\text{ V}/1.4 = 20$ V. So turns ratio is $n = V_{OR}/V_O = 20/12 = 1.67$.

The reflected output current is $I_{OR} = I_O/n = 4/1.67 = 2.4$ A.

Based on this we can solve for duty cycle at low line and high line.

$$D_{VINMIN} = \frac{n \times V_O}{(\eta_{VINMIN} \times V_{INMIN}) + (n \times V_O)}$$

$$= \frac{1.67 \times 12}{(0.8 \times 32) + (1.67 \times 12)} = 0.439$$

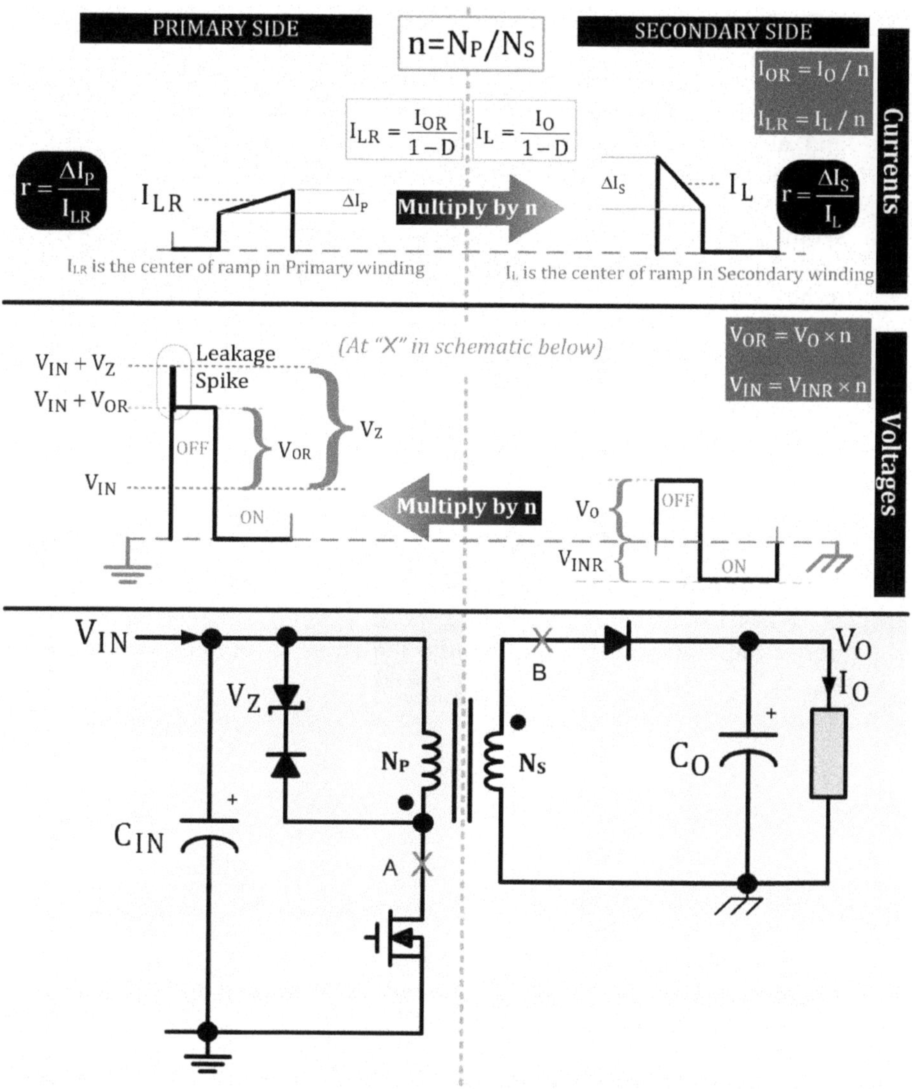

Figure 14.12 Close-up of the key waveforms of a flyback, and definition of r for flyback.

$$D_{\text{VINMAX}} = \frac{n \times V_O}{(\eta_{\text{VINMAX}} \times V_{\text{INMAX}}) + (n \times V_O)}$$

$$= \frac{1.67 \times 12}{(0.85 \times 57) + (1.67 \times 12)} = 0.292$$

We can see that our max duty cycle is about 0.44. This is correct because the LX7309 states that the max duty cycle actually has a spread between 0.44 to 0.5. So to guarantee max power, we need to be below 0.44. That is why we picked $V_Z = 28$ V. If we had picked a higher value we could be in trouble with the duty-cycle limit. We may still need to tweak this further, based on actual efficiency figures and also zener characteristics and tolerance. A flyback design is always very tricky and essentially iterative.

Let us set the switching frequency to $f = 200$ kHz. With this we can set the primary-side inductance based on the design target $r = 0.5$ as explained previously. Since we are setting this at minimum input, let us be clear and write it more explicitly as $r_{\text{VINMIN}} = 0.5$ here. From this point onward, we are relying on the equations available in the book *Switching Power Supplies A-Z*.

$$L_P = \frac{V_{\text{OR}}}{I_{\text{OR}} \times r_{\text{VINMIN}} \times f} \times (1 - D_{\text{VINMIN}})^2$$

$$= \frac{20}{2.4 \times 0.5 \times 200k} \times (1 - 0.439)^2 = 26.3\ \mu H$$

Now we can pick the core size-assuming ferrite with $\mu = 2000$, $B_{\text{sat}} = 3000$ Gauss, and a gap factor $z = l_e/(l_e + \mu lg) = 1$. The ferrite core, under the worst-case, has to handle the entire incoming power of $P_{\text{IN\_}}V_{\text{INMIN}} = Po/(\eta \times V_{\text{INMIN}}) = 48/0.8 = 60$ W.

$$V_{e\_\text{cm}^3} = \frac{31.4 \times P_{\text{IN\_VINMIN}} \times \mu}{z \times f_{\text{MHz}} \times B_{\text{sat\_GAUSS}}^2} \times \left[r_{\text{VINMIN}} \times \left(\frac{2}{r_{\text{VINMIN}}} + 1 \right)^2 \right]$$

$$V_{e\_\text{cm}^3} = \frac{31.4 \times 60 \times 2000}{10 \times 0.2 \times 3000^2} \times \left[0.5 \times \left(\frac{2}{0.5} + 1 \right)^2 \right] = 2.617\ \text{cm}^3$$

A possible choice is EFD25/13/9 with a *Ve* of 3.3 cm^3. It has an effective area of *Ae* = 0.58 cm^2. number of primary turns is (using 3000 Gauss = 0.3 T)

$$N_P = \left(1 + \frac{2}{r_{\text{VINMIN}}}\right) \times \frac{\text{V}_{\text{INMIN}} \times D_{\text{VINMIN}}}{200 \times B_{\text{sat\_TESLA}} \times f_{\text{MHz}} \times A_{e\_\text{cm}^2}}$$

$$N_P = \left(1 + \frac{2}{0.5}\right) \times \frac{32 \times 0.439}{200 \times 0.3 \times 0.2 \times 0.58} = 6.73$$

We round this up to 7 turns. Since turns ratio is 1.67, the number of secondary turns is $7/1.67 = 4.19$. We round this up to 4 turns. The air gap can be found from

$$l_g = (z - 1) \times \frac{l_e}{\mu}$$

where we have set $z = 10$ and effective length l_e for the selected core is $Ve/Ae = 3300/58 = 56.9$ mm. So

$$l_g = (10-1) \times \frac{56.9}{2000} = 0.256 \text{ mm}$$

This means we can use standard gapless (unground) core halves and insert spacers of thickness $0.256/2 = 0.13$ mm on the outer limbs.

Now we will evaluate its current stresses. But first, we will also evaluate the current ripple ratio at high line, based on the selected inductance. We have

$$r_{\mathrm{VINMAX}} = \frac{V_{\mathrm{OR}}}{I_{\mathrm{OR}} \times L_P \times f} \times (1 - D_{\mathrm{VINMAX}})^2$$

$$r_{\mathrm{VINMAX}} = \frac{20}{2.4 \times 2.63 \times 10^{-5} \times 2 \times 10^5} \times (1 - 0.292)^2 = 0.795$$

Since this is less that 2 (critical conduction boundary), it is still operating in continuous conduction mode (CCM) at high line, so we can continue to use the same CCM equations as for V_{INMIN}.

RMS Current in Switch and Primary Winding

The general equation is

$$I_{\mathrm{SW\_RMS}} = \frac{I_{\mathrm{OR}}}{1-D} \times \sqrt{D \times \left(1 + \frac{r^2}{12}\right)}$$

At low line and high line, respectively, we get

$$I_{\mathrm{SW\_RMS\_VINMIN}} = \frac{2.4}{1-0.439} \times \sqrt{0.439 \times \left(1 + \frac{0.5^2}{12}\right)} = 2.861 \text{ A}$$

$$I_{\mathrm{SW\_RMS\_VINMAX}} = \frac{2.4}{1-0.292} \times \sqrt{0.292 \times \left(1 + \frac{0.795^2}{12}\right)} = 1.88 \text{ A}$$

RMS Current in Input Cap

The general equation is

$$I_{\mathrm{CIN\_RMS}} = \frac{I_{\mathrm{OR}}}{1-D} \times \sqrt{D \times \left(1 - D + \frac{r^2}{12}\right)}$$

At low line and high line, respectively, we get

$$I_{\text{CIN\_RMS\_VINMIN}} = \frac{2.4}{1-0.439} \times \sqrt{0.439 \times \left(1-0.439+\frac{0.5^2}{12}\right)} = 2.16 \text{ A}$$

$$I_{\text{SW\_RMS\_VINMAX}} = \frac{2.4}{1-0.292} \times \sqrt{0.292 \times \left(1-0.292+\frac{0.795^2}{12}\right)} = 1.598 \text{ A}$$

Ensure we parallel enough caps to handle this RMS current. For example, if each cap is rated for 1.2 A RMS, we need $2.16/1.2 \approx 2$ (two) such caps in parallel.

RMS Current in Output Cap

The general equation is

$$I_{\text{COUT\_RMS}} = I_{\text{O}} \times \sqrt{\frac{D+\frac{r^2}{12}}{1-D}}$$

At low line and high line, respectively, we get

$$I_{\text{COUT\_RMS\_VINMIN}} = 4 \times \sqrt{\frac{0.439+\frac{0.5^2}{12}}{1-0.439}} = 3.619 \text{ A}$$

$$I_{\text{COUT\_RMS\_VINMAX}} = 4 \times \sqrt{\frac{0.292+\frac{0.795^2}{12}}{1-0.292}} = 2.792 \text{ A}$$

Ensure we parallel enough caps to handle this RMS current. For example, if each cap is rated for 0.75 A RMS, we need $3.619/0.75 \approx 5$ such caps in parallel.

RMS Current in Output Diode and Secondary Winding

The general equation is

$$I_{\text{D\_RMS}} = I_{O} \times \sqrt{\frac{1+\frac{r^2}{12}}{1-D}}$$

At low line and high line, respectively, we get

$$I_{\text{D\_RMS\_VINMIN}} = 4 \times \sqrt{\frac{1+\frac{0.5^2}{12}}{1-0.439}} = 5.394 \text{ A}$$

$$I_{\text{D\_RMS\_VINMAX}} = 4 \times \sqrt{\frac{1+\frac{0.795^2}{12}}{1-0.292}} = 4.878 \text{ A}$$

Selection of Wire Gauge for Primary Winding

We refer to Fig. 14.13 which contains a useful nomogram. For this we need to know the center of ramp of the primary-side current. That is equal to $I_{OR}/(1-D)$. We call it I_P:

$$I_p = \frac{I_{OR}}{1-D} = \frac{2.4}{1-0.44} = 4.3 \text{ A}$$

where we have used the numbers for V_{INMIN} because we know that the worst-case occurs at low line. From the nomogram

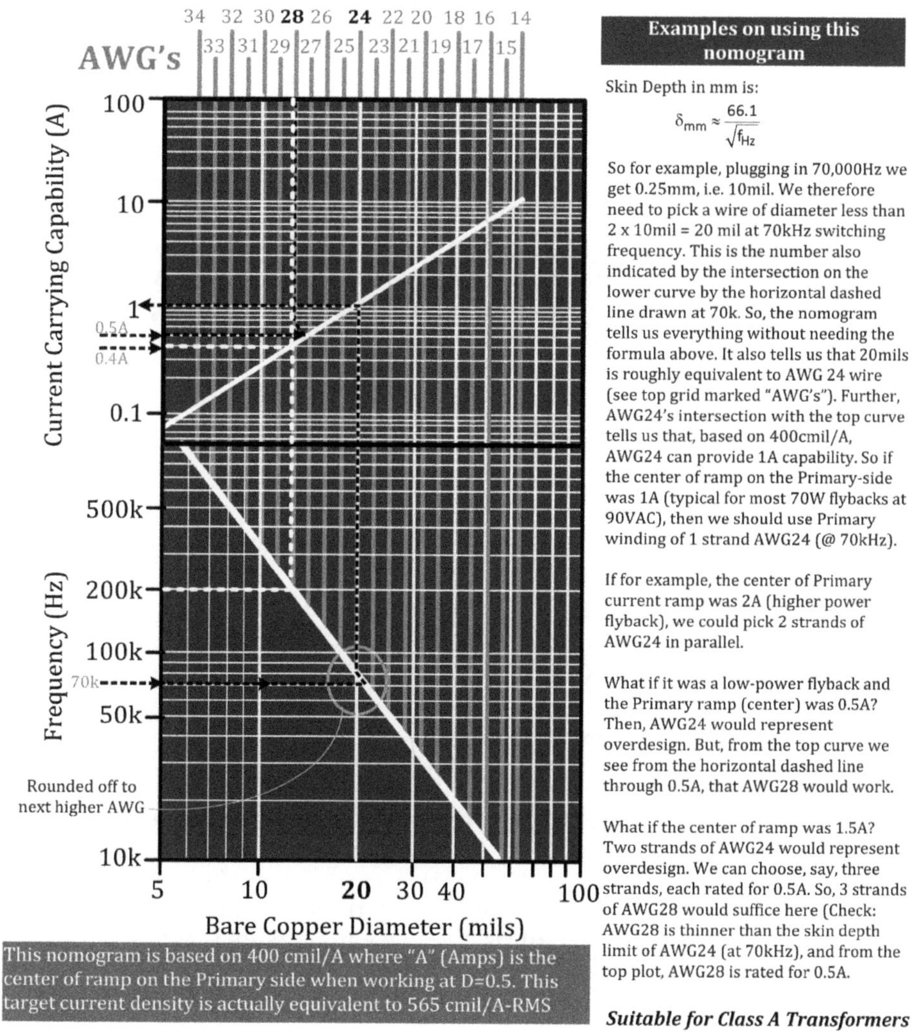

FIGURE 14.13 Wire-gauge selection nomogram for flyback.

(bold white-dashed line), we see that at 200 kHz, because of skin-depth considerations, we need to pick AWG 28. Each strand of that, can carry 0.4 A, based on a target current density of 400 cmil/A. Therefore, to handle a center-of-ramp of 4.3 A, we need 4.3/0.4 ≈ 10 strands of AWG 28. We can break this up into more manageable bundles of five wires, and use two bundles in parallel. We know that we need seven turns for the primary winding.

Selection of Wire Gauge for Secondary Winding

We need to know the center of ramp of the secondary-side current. That is equal to $I_O/(1–D)$. We call it I_S:

$$I_S = \frac{I_O}{(1-D)} = \frac{4}{1-0.44} = 7.14 \text{ A}$$

where we have used the numbers for V_{INMIN} because we know that the worst-case occurs at low line. From the nomogram (bold white dashed line), we see that at 200 kHz, because of skin depth considerations, we need to pick AWG 28. Each strand of that, can carry 0.4 A, based on a target current density of 400 cmil/A. Therefore, to handle a center-of-ramp of 7.14 A, we need 7.14/0.4 ≈ 18 strands of AWG 28. We can break this up into more manageable bundles of five wires, and use four bundles in parallel (20 strands). We know that we need four turns for the secondary winding.

Alternatively, we can consider foil windings, as discussed in *Switching Power Supplies A-Z*.

Zener Dissipation

This is maximum at low line. Theoretically, the equation for zener dissipation is

$$P_{\text{clamp}} = \frac{1}{2} \times L \times I_{\text{PEAK}}{}^2 \times f \times \frac{V_Z}{V_Z - V_{OR}}$$

The peak current (at low line) is

$$I_{\text{PEAK\_VINMIN}} = I_{\text{P\_VINMIN}} \times \left(1 + \frac{r_{\text{VINMIN}}}{2}\right) = 4.3 \times \left(1 + \frac{0.5}{2}\right) = 5.375 \text{ A}$$

In practice, if we measure the peak current into the zener clamp at the moment the switch turns OFF, we will measure that it is typically only 0.6 to 0.8 times the peak switch current, because some of the peak current is diverted into the parasitic capacitances inside the transformer. Applying this "fudge factor" (0.7) into the theoretical zener equation, we get

$$P_{\text{clamp}} = \frac{1}{2} \times L_{\text{LK}} \times (0.7 \times I_{\text{PEAK}})^2 \times f \times \frac{V_Z}{V_Z - V_{\text{OR}}}$$

$$\approx \frac{1}{4} \times L_{\text{LK}} \times I_{\text{PEAK}}{}^2 \times f \times \frac{V_Z}{V_Z - V_{\text{OR}}}$$

Assuming that the leakage inductance L_{LK} is typically 1 percent of the primary inductance (L_p = 26.3 μH), we get the final estimate of clamp dissipation to be

$$P_{\text{clamp}} \approx \frac{1}{4} \times \frac{26.3\ \mu}{100} \times 5.375^2 \times 200\ \text{k} \times \frac{28}{28 - 20} = 1.33\ \text{W}$$

Design of the RCD clamp can also be found in *Switching Power Supplies A-Z.*

Index

Note: Page numbers followed by *f* denote figures; page numbers followed by *t* denote tables; page numbers followed by *n* denote notes.

A

R

S

U

Lightning Source UK Ltd.
Milton Keynes UK
UKHW020105121221
395508UK00002B/39

9 780071 798259